喀斯特地区玉米高产高效生产理论

吕巨智 周勋波 主编

中国农业科学技术出版社

图书在版编目（CIP）数据

喀斯特地区玉米高产高效生产理论 / 吕巨智，周勋波主编. --北京：中国农业科学技术出版社，2024.6.
ISBN 978-7-5116-6879-0

Ⅰ.S513

中国国家版本馆 CIP 数据核字第 20249G5A49 号

责任编辑　张国锋
责任校对　李向荣
责任印制　姜义伟　王思文

出 版 者	中国农业科学技术出版社
	北京市中关村南大街 12 号　　邮编：100081
电　　话	（010）82109705（编辑室）　　（010）82106624（发行部）
	（010）82109709（读者服务部）
网　　址	https://castp.caas.cn
经 销 者	各地新华书店
印 刷 者	北京建宏印刷有限公司
开　　本	185 mm×260 mm　1/16
印　　张	21.25
字　　数	440 千字
版　　次	2024 年 6 月第 1 版　2024 年 6 月第 1 次印刷
定　　价	128.00 元

━━◀ 版权所有・翻印必究 ▶━━

《喀斯特地区玉米高产高效生产理论》
编写人员名单

主　编：吕巨智　周勋波
副主编：杨　丽　谭贤杰　唐国荣　李发桥
编　委：钟昌松　范继征　程伟东　迟宇新
　　　　石达金　贺囡囡

前　言

玉米集饲、粮、工业用于一身，是广西壮族自治区（以下简称广西）仅次于水稻的第二大粮食作物，广西玉米人均消费是中国最多的地区之一。玉米是广西规模化养殖和农村家庭养殖的主要原料，随着广西畜牧业持续稳定增长，饲料需求量不断增加，饲用玉米消费将呈刚性增长，目前无法自给自足，每年需从广西外调进 300 万 t 玉米作为饲料原料，供需缺口大。据报道，2022 年全国玉米单产为 429 kg/亩（1 亩 ≈ 667 m^2），而广西玉米单产仅为 303 kg/亩，不到全国平均水平的 3/4。目前广西玉米的种植面积趋于饱和，只有快速提高玉米单产才能有效解决广西玉米原料短缺问题。

广西大学玉米栽培生理团队和广西农业科学院玉米研究所依托国家现代农业产业技术体系南宁综合试验站团队、国家玉米改良中心广西分中心、农业农村部粮食作物种质资源创新与利用重点实验室、广西创新团队玉米栽培兼病虫害防治岗位功能团队、广西农业环境与农产品安全重点实验室、广西高校作物栽培与生理重点实验室培育基地等平台，经过团队成员的共同努力，紧密围绕玉米高产栽培、农田水分利用及耕地地力提升方面开展了比较系统的长期研究，特别是在玉米群体分布与作物生产力、群体内竞争互补、光能与水资源利用等方面进行了一系列创新性研究，先后在国内外重要学术期刊上发表了数十篇学术论文。为了阶段性总结经验和促进学术交流，纪念历届团队成员付出的努力，也为进一步深入研究提供借鉴，特编辑出版此书。在编辑分类过程中，主要按照玉米密植高产栽培生理机理和土壤碳氮循环两个研究方面分为上下篇，上篇着重于密植高产群体的生理特征、高产群体质量指标、高产高效协同机制等；下篇着重于耕层固碳培肥机理、土壤微生物多样性和温室气体排放等。以上这些研究为科学地建立玉米高产高效栽培奠定了一定的理论基础。

在本书编写过程中，得到本领域专家和团队成员的大力支持，在此感谢他们的辛勤劳动。由于作者水平所限，不足之处在所难免，敬请批评指正。

编　者
2024 年 3 月

目 录

上 篇

不同密度与种植方式对玉米主要农艺性状和产量的影响
 ………………………………………范继征,闫飞燕,石达金,钟昌松,吕巨智 (3)
栽培模式对玉米叶绿素含量、叶面积指数及产量的影响
 ………………………………………范继征,闫飞燕,石达金,钟昌松,吕巨智 (10)
10个玉米品种耐密性分析及其对主要性状的影响
 …………范继征,程伟东,闫飞燕,石达金,吕巨智,张玉,钟昌松,刘永红 (17)
应用灰色关联度分析法分析西南区玉米新组合
 ………………………………………范继征,石达金,吕巨智,唐国荣,程伟东 (23)
水分和施氮量对亚热带一年两作玉米灌浆及产量的影响
 ………………………………………胡雨欣,王桂阳,梁修仁,毛祥敏,吴海燕,周勋波 (32)

下 篇

Straw return and nitrogen fertilization regulate soil greenhouse gas emissions and global warming potential in dual maize cropping system ……………L. Yang, I. Muhammad, Y. X. Chi, Y. X. Liu, G. Y. Wang, Y. Wang, X. B. Zhou (45)
Straw return and nitrogen fertilization to maize regulate soil properties, microbial community, and enzyme activities under dual cropping system
 ………………………L. Yang, I. Muhammad, Y. X. Chi, D. Wang and X. B. Zhou (76)
Effect of water conditions and nitrogen application on maize growth, carbon accumulation and metabolism of maize plant in subtropical regions
 ……………………………Y. X. Chi, F. Gao, I. Muhammad, J. H. Huang, X. B. Zhou (101)
Effects of nitrogen and water stress on the rehydration, endogenous hormonal regulation and yield of maize ………………………………………Y. X. Chi, S. Ahmad, K. J. Yang, J. Fu, L. Yang, X. B. Zhou, H. D. Zhu (118)
Low irrigation water minimizes the nitrate nitrogen losses without compromising the soil fertility, enzymatic activities and maize growth …………I. Muhammad, J. Z. Lv, L. Yang, S. Ahmad, S. Farooq, A. Khan, M. Zeeshan, X. B. Zhou (140)
Nitrogen Fertilizer Modulates Plant Growth, Chlorophyll Pigments and Enzymatic Activities under Different Irrigation Regimes …………I. Muhammad, L. Yang, S. Ahmad, S. Farooq, A. A. Al-Ghamdi, A. Khan, M. Zeeshan, M. Elshikh, A. M. Abbasi, X. B. Zhou (159)

Effect of Previous Crop Nitrogen Application on Yield of Following Maize Under Different
　　Planting Patterns ……………………………………………………Y. X. Zhao, X. M. Mao,
　　　　　　　　　　　　　　　　J. H. Huang, D. H. Jiang and X. B. Zhou（185）
Impact of the mixture verses solo residue management and climatic conditions on soil
　　microbial biomass carbon to nitrogen ratio: A systematic review ……I. Muhammad, J. Wang,
　　　　　　　　　　A. Khan, S. Ahmad, L. Yang, I. Ali, M. Zeeshan, S. Ullah, S. Fahad,
　　　　　　　　　　　　　　　　　　　　　　　　　　　　S. Ali and X. B. Zhou（195）
Effects of Spatial Distribution on Photosynthesis and Yield of Summer Maize
　　………………………………X. M. Mao, P. J. Shen, Y. X. Zhao and X. B. Zhou（223）
Photosynthetic Characteristics of Summer Maize under Different Planting Patterns and the
　　Responses to Nitrogen Application of Previous Crop
　　………………………X. B. Zhou, P. J. Shen, Y. X. Zhao, D. H. Jiang, J. H. Huang（231）
Planting Pattern Effects on Soil Water and Yield of Summer Maize
　　………………………………………………X. Y. Wang, X. B. Zhou, Y. H. Chen（241）
Regulation of soil microbial community structure and biomass to mitigate soil greenhouse
　　gasses emission ………………………………………………I. Muhammad, J. Z. Lv, J. Wang,
　　　　　　　　　　　　　　　　S. Ahmad, F. Saqib, S. Ali, and X. B. Zhou（251）
Row Spacing Effects on Radiation Distribution, Leaf Water Status and Yield of Summer
　　Maize ……………………………………………J. Q. Liu, M. D. Li and X. B. Zhou（295）
不同耕作方式对玉米田土壤物理性质及产量的影响
　　…………范继征，闫飞燕，石达金，吕巨智，张玉，钟昌松，程伟东，刘永红（307）
不同耕作方式对土壤水分及玉米生长发育的影响
　　………………吕巨智，钟昌松，范继征，石达金，程伟东，刘永红，闫飞燕（315）
水氮条件对南亚热带玉米产量及农田土壤有机碳氮组分的影响
　　…………………………………………………………刘涌鑫，毛祥敏，周勋波（324）

上 篇

不同密度与种植方式对玉米主要农艺性状和产量的影响

范继征,闫飞燕,石达金,钟昌松,吕巨智

(广西农业科学院玉米研究所,广西 南宁 530007)

摘要:[目的]针对目前广西玉米生产密度低、种植方式混杂,高产品种增产潜力发挥不足等问题,系统开展不同种植模式(行距、密度)下主栽品种的主要农艺性状、叶面积指数、叶绿素相对含量和产量等变化规律的研究。[方法]采用随机区组设计,研究2种密度和3种种植方式对玉米品种"迪卡008"的农艺性状和产量的影响。[结果]随着密度的增加,株高、茎粗、穗长、秃尖、行粒数和百粒重呈下降趋势;穗位高、穗粗、穗行数和产量呈上升趋势。随着种植方式的变化,株高、穗长和行粒数呈上升趋势;穗位高、穗行数和百粒重呈先升后降的趋势;茎粗、秃尖长和穗粗呈先降后升的趋势。在不同的密度与种植方式下,叶绿素相对含量在四个不同生育时期均表现出先升高后降低的变化趋势,叶面积指数在三个生育时期均表现出逐步升高的趋势。本研究中最高产量的种植模式是57 000株/hm² 密度下0.65 m 等行距种植,产量为9 677.35 kg/hm²,与传统种植习惯比较,增产了13.54%。[结论]结合品种自身特性,合理选择种植模式能够促进玉米产量的提高。

关键词:玉米;种植密度;种植方式;农艺性状;产量

Effects of Maize different Planting Density and Cultivation Mode on Agronomic Characteristics and Yield

Fan Jizheng, Yan Feiyan, Shi Dajin, Zhong Changsong, Lü Juzhi

(Maize Research Institute, Guangxi Academy of Agricultural Sciences, Nanning 530007, Guangxi, China)

Abstract:[Objective] At present, there were in some problems on maize production, such as low planting density, miscellaneous cultivation mode and insufficient productive potentials etc. In different cultivation mode, the change law about agronomic characteristics, leaf area index, SPAD and yield were studied on main cultivated varieties. [Method] The effects of maize 'Dika 008' different planting density and cultivation mode on agronomic characteristics and yield was tested in field experiments using randomized blocks design. [Result] The results showed that, when increasing planting density, plant height, stem diameter, ear length, barren tip, kernels per row

---

基金项目:广西科学技术厅基本科研业务专项项目[200802(基)];广西农业科学院基本科研业务专项项目[201104(基)]。

作者简介:范继征(1982—),女,助理研究员,主要从事玉米栽培与耕作研究工作,E-mail: fiona-fiona-happy@163.com。

and 100-kernel weight reduced gradually; ear height, ear diameter, rows per ear and yield increased gradually. shelling percentage and yield increased first and then decreased. With cultivation mode changing, plant height, ear length and kernels per row increased gradually; ear height, rows per ear and 100-kernel weight increased first and then decreased; stem diameter, barren tip and ear diameter decreased first and then increased. On different planting density and cultivation mode, SPAD increased first and then decreased in four different plant growth periods; leaf area index gradually increased in three different plant growth periods. In this experiment, the cultivation of high yield mode was 57,000 plants/ha density with 0.65 m equal planting spacing, compared with traditional cultivation method, the yield was 9 677.35 kg/ha and the increase rate was 23.27%. [Conclusion] Combined with varieties of their own features, it could promote the increase of corn yield by selecting rational planting mode.

Keywords：Maize; Planting density; Cultivation mode; Agronomic characteristics; Yield

【研究意义】目前,玉米是广西仅次于水稻的第二大粮食作物,常年播种面积约55万 hm^2,占广西粮食总播种面积的14%~16%,常年产量约210万 t[1]。2010年全区玉米种植面积约为53.86万 hm^2,平均单产258 kg,依然在低水平徘徊。而如何提高玉米单产继而提高全区玉米产量仍然是一个亟待解决的问题。除选育高产品种之外,如何从栽培角度提高现有品种的产量水平,对于当前玉米生产具有重要意义。【前人研究进展】在玉米生产中,种植方式是协同高密度条件下,个体通风透光条件、营养状况并最终影响产量的因素之一。国内学者关于种植密度和种植方式对产量的影响已经开展了大量的研究,杜国田(2010)研究表明,玉米品种郑单958以行距72.0 cm和60.0 cm,密度67 740株/hm^2和60 480株/hm^2进行组合,产量最优[2];胡萌等(2010)认为随密度增加叶面积指数(LAI)下降,产量构成因素表现为单位面积有效穗数先上升后下降,穗粒数下降,百粒重变化因品种而异[3];范秀玲等(2010)研究表明偏垄宽窄行种植方式单位面积产量显著高于常规种植方式[4];温日宇等(2011)认为合理密植有利于提高玉米产量,在一定范围内,随着种植密度的增加,玉米产量随之增加[5];丰光等(2011)认为选择适宜品种并结合品种自身特性,合理提高种植密度才能促进玉米产量的提高[6];李洪等(2011)研究表明,普通株型玉米品种可以通过株行距合理配置的方式,提高种植密度,从而达到增产效果[7]。【本研究切入点】当前,增加种植密度和改变株行距种植方式作为玉米生产上的增产途径,越来越受到人们的关注。而增加密度主要是通过耐密植品种的应用来达到这一目的。广西大部分玉米种植区的土壤干旱瘠薄,长期以来农民在生产上以稀植大穗型品种为主,种植密度在48 000株/hm^2左右。在当地生产上还没有耐密植品种,普通株型玉米品种有没有增加种植密度的空间,能否从栽培学方面通过改变种植方式来提高普通株型玉米的种植密度,进一步提高玉米产量。

【拟解决的关键问题】针对广西玉米种植的特点,本试验设计了48 000株/hm^2和57 000株/hm^2两种密度以及0.65 m等行距、0.75 m+0.55 m宽窄行和0.80 m+0.40 m宽窄行三种种植方式,采用随机区组设计,探索试验品种农艺性状及产量的变化规律,从而为普通株型玉米品种提高产量寻求最佳种植方式提供依据。

1 材料与方法

1.1 试验材料

供试品种为"迪卡008"。

1.2 试验地基本情况

试验于2011年上半年在广西壮族自治区农业科学院玉米研究所试验地进行。试验地地势平坦，肥力中等，全N 0.096 5%、全P 0.085 4%、全K 0.119 1%、碱解N 80.75 mg/kg、速效P 17.98 mg/kg、速效K 158.75 mg/kg、有机质 18.74 g/kg、pH 6.72。

1.3 试验方法

试验采用随机区组设计，A因素为两种种植密度，B因素为不同株行距的3种种植方式，共6个处理（表1），3次重复。采用直播方式，每小区种植6行，行长8 m。生育期内记录倒伏情况，测量乳熟期第3节茎粗、株高和穗位高；收获后测量穗长、秃尖长、穗粗、行数、行粒数、百粒重、折算14%含水量后的单位面积产量；分别测量三叶期、大喇叭口期、开花吐丝期、灌浆期、成熟期籽粒的叶绿素相对含量（SPAD）、鲜重和干重；分别测量三叶期、大喇叭口期、开花吐丝期的叶面积指数（LAI）。

表1 不同密度与种植方式的配置

编号	密度（株/hm²）	种植方式	产量（kg/hm²）
1	A1：48 000*	B1：0.65 m+0.65 m 等行距*	8 523.50
2	A1：48 000*	B2：0.75 m+0.55 m 宽窄行	8 730.09
3	A1：48 000*	B3：0.80 m+0.40 m 宽窄行	8 258.83
4	A2：57 000	B1：0.65 m+0.65 m 等行距	9 677.35
5	A2：57 000	B2：0.75 m+0.55 m 宽窄行	8 252.43
6	A2：57 000	B3：0.80 m+0.40 m 宽窄行	9 304.88

*为当地习惯种植密度和种植方式。

1.4 数据分析

采用Excel和DPS软件对试验数据进行统计与分析。

2 结果与分析

2.1 不同种植密度对农艺性状和产量的影响

随着密度的增加，该品种的农艺性状呈现不同的变化趋势（图1）。株高、茎粗、穗长、秃尖、行粒数和百粒重随着密度的增加呈下降趋势；穗位高、穗粗、穗行数和产量随着密度的增加呈上升趋势。随着密度的增加，使得群体中每个植株对于光照、温度、水分、养分的竞争加剧，进而影响到了个体植株的营养生长和生殖生长，具体表现为农艺性状的表型出现了差异。株高、茎粗、穗长和百粒重呈现下降趋势，原因是密度增加后，单个植株营养生长受到一定程度的限制，表现出高度降低，茎粗变细，穗长缩短，行粒数减少，籽粒饱满度降低。随着密度增加，植株之间彼此遮阴，导致穗位高度增加。此外，果穗变粗，相应穗行数增加。虽然植株果穗变短、行粒数减少、百粒重降低，但由于植株总数的增加，果穗总量的增加，产量依然表现为增产。

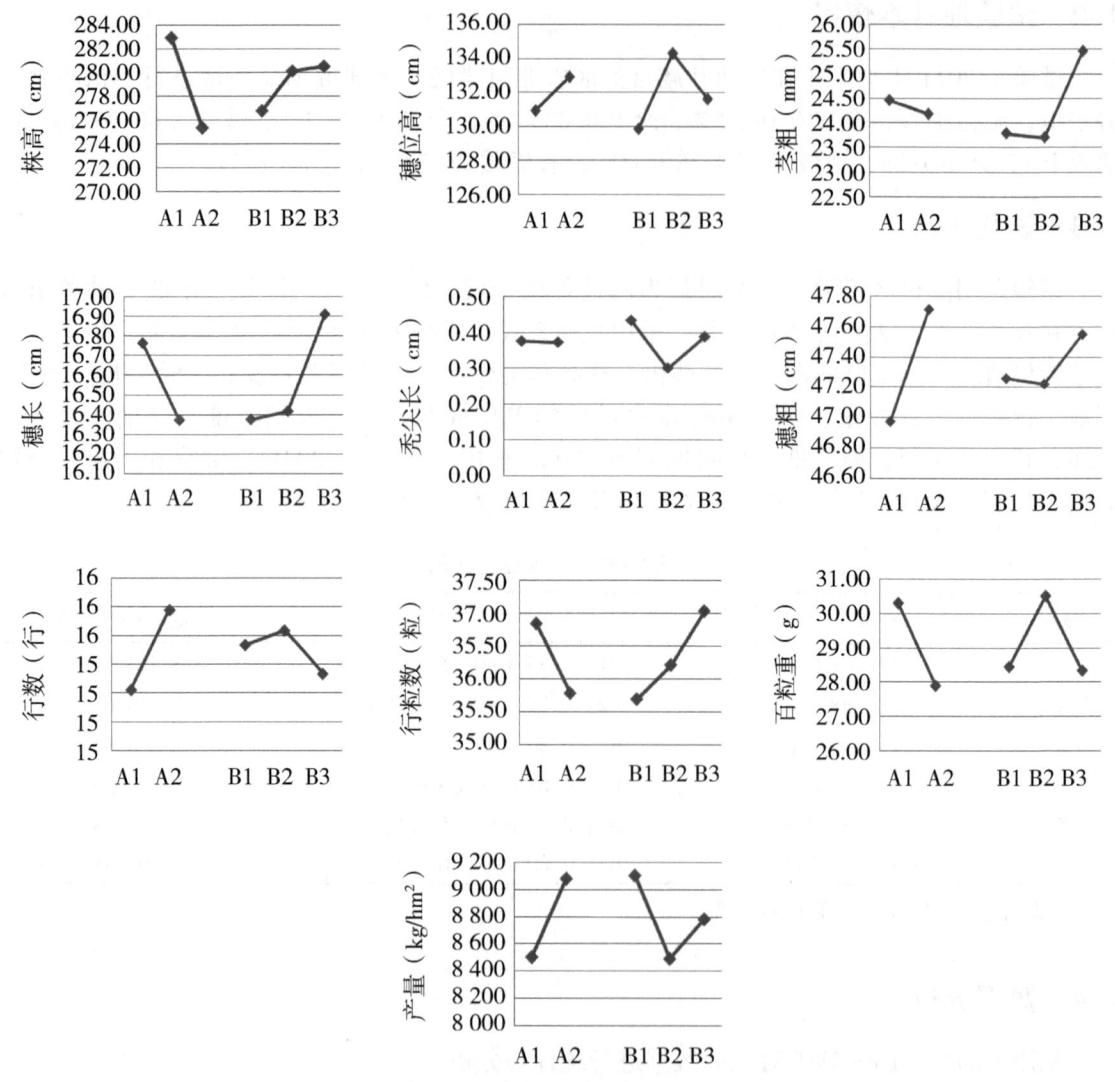

图 1　品种农艺性状和产量随密度增加和种植方式不同的变化趋势

2.2　不同种植方式对农艺性状和产量的影响

随着种植方式的不同，该品种的农艺性状有不同的变化趋势（图1）。株高、穗长和行粒数随着种植方式的不同呈上升趋势；穗位高、穗行数和百粒重随着种植方式的不同呈先升后降的趋势；茎粗、秃尖长和穗粗随着种植方式的不同呈先降后升的趋势。产量随着种植方式的不同呈现先降后升的趋势。随着种植方式即行距的变化，植株之间的空间距离发生改变,由等行距逐渐改变为宽行逐渐变大,窄行逐渐变小,对于各种营养条件的竞争也随之改变。株高、穗长、行粒数随着行距变异表现为不断增加,原因是窄行之间的竞争起到了主要作用,植株彼此之间的遮阴竞争导致植株高度增加,宽行之间的边际效应致使果穗变长,行粒数增加。穗位高、穗行数和百粒重随着种植方式的不同呈先升后降的趋势,原因是随着行距的进一步变宽和变窄,由宽行效应起主要作用逐渐变为窄行效应起主要作用。茎粗和穗粗随着种植方式的不同表现为先降后升,原因是当窄行效应起主要作用时,表现为下降趋势,当宽行效应起主要作用时,表现为上升趋势。秃尖长的变化原因则与之相反。产量表现为先降后升,原因是在同一密度下,宽窄行之间的效应对产量有影响,窄行效应致使产量降低,随着行距的进一步改变,宽行效应又占据主导地位,表现为产量提高。

2.3 不同种植密度对各个时期 SPAD、LAI 的影响

在两种不同的种植密度下,SPAD 和 LAI 呈现出相同的变化趋势(图 2)。SPAD 在 4 个不同生育时期表现出先升高后降低的趋势。在前 3 个时期,低密度下的 SPAD 均高于同期高密度下的 SPAD 值;在成熟期,则表现出高密度下的 SPAD 高于低密度下的 SPAD。LAI 在 3 个生育时期表现出逐步升高的趋势。在 3 个不同的生育时期,高密度下的 LAI 均高于低密度下的 LAI,且两者之间差距有逐步扩大的趋势。

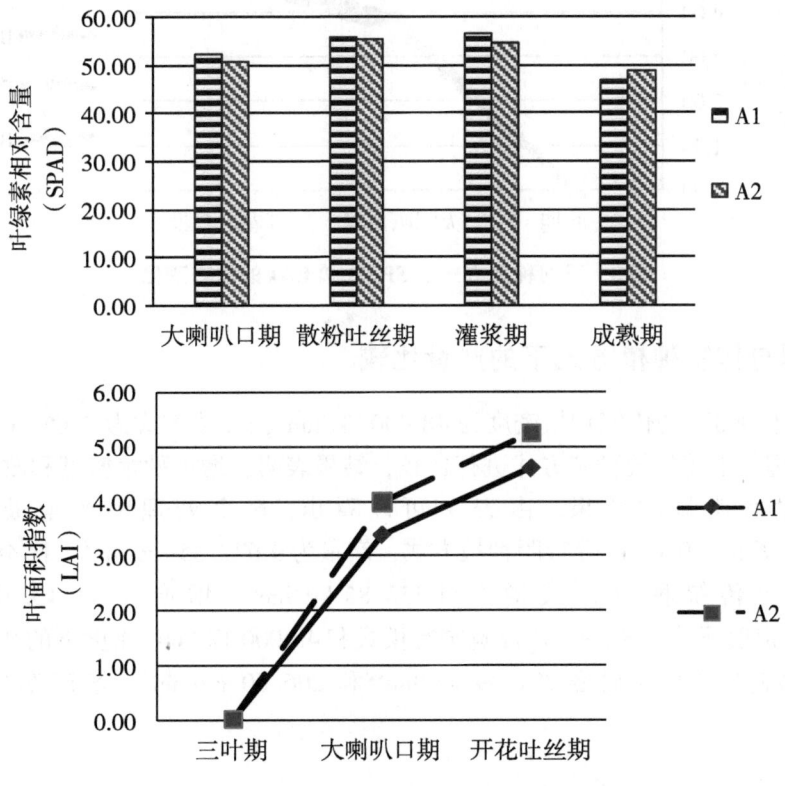

图 2　不同种植密度下 SPAD 和 LAI 的动态变化

2.4 不同种植方式对各个时期 SPAD、LAI 的影响

在 3 种不同种植方式下,SPAD 和 LAI 呈现相同的变化趋势(图 3)。SPAD 在 4 个不同生育期表现出先升高后下降的趋势。在大喇叭口期,SPAD 在 3 种模式下表现为 B3>B1>B2;

在散粉吐丝期,则表现为 B1>B3>B2;在灌浆期,三者之间几乎没有差别;在成熟期,则表现为 B1>B2>B3。

LAI 在 3 个生育期表现出逐步升高的趋势。在三叶期,3 种模式下的叶面积指数之间没有明显的差异,在大喇叭口期和开花吐丝期均表现为 B3>B1>B2,且彼此之间的差距有逐步扩大的趋势。

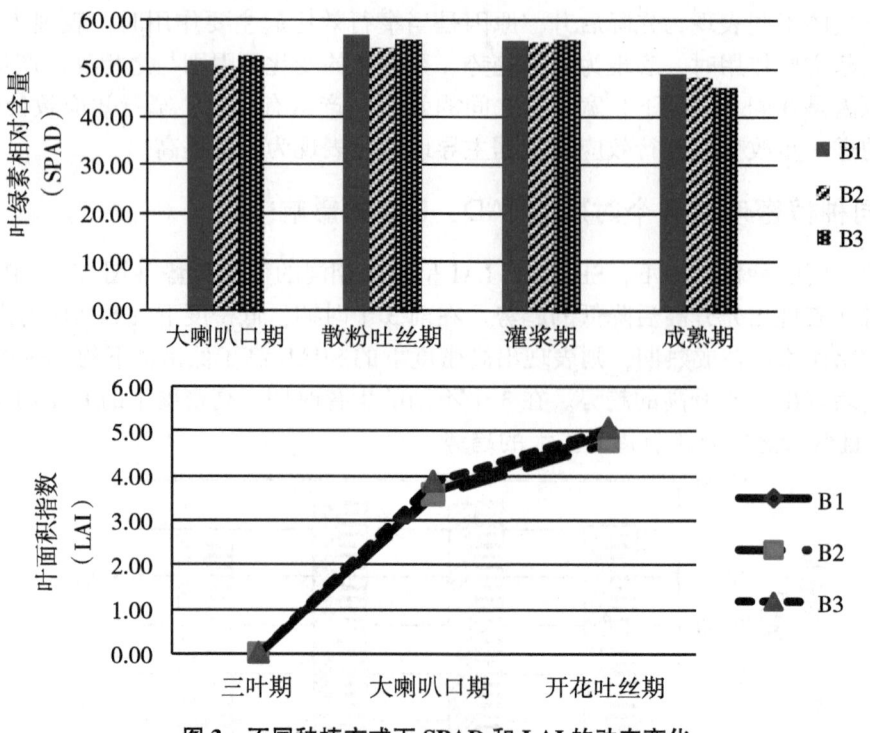

图 3 不同种植方式下 SPAD 和 LAI 的动态变化

2.5 各处理与传统种植方式下的产量比较

广西"迪卡 008"的传统种植密度为 48 000 株/hm², 种植方式为 0.65 m 等行距,笔者用该试验各处理与上述传统种植方式进行比较,结果表明,增加种植密度和改变种植方式对提高玉米产量有明显的效果。由表 1 可以看出,6 个处理中产量最高的处理是 57 000 株/hm² 密度下 0.65 m 等行距种植方式,产量为 9 677.35 kg/hm²,比 48 000 株/hm²、0.65 m 等行距传统种植方式增产 1 153.85 kg/hm²,增产率为 13.54%;其次是 57 000 株/hm² 密度下的 0.8 m+0.4 m 宽窄行模式和 48 000 株/hm² 密度下的 0.75 m+0.55 m 宽窄行模式,增产幅度分别是 781.39 kg/hm² 和 206.60 kg/hm²,增产率分别为 9.17% 和 2.42%。

3 讨论

玉米与其他植物一样,群体间存在避阴现象,当群体密度变大时,植株个体间为减少竞争而争先向上生长,以此来获得更多的光照和热量,结果使得植株自身株高和穗位高较正常情况下有所增加[8]。本文中随着密度的增加,株高降低,穗位高增加,两者之间没有表现为一致变化,原因可能是与品种的特性有一定的关系。不同的种植方式对植株个体有不同的效应,前人研究认为宽窄行种植可以克服密植带来的植株个体间彼此遮光不透气的不良状况,从而减轻倒伏并获得增产[9-11]。在本试验中,宽窄行处理的增产效果不是很明显,原因可能是供试品种"迪卡008"本身是一个边行优势较小的高产稳产型品种。还需要进一步进行多品种和多年份的试验。此外,本研究选取试验品种较少,而且不同品种有不同的适宜区域,在单一地区进行试验一定程度上会影响结果的准确性。为此,应进行多品种、多点、多年份的深入研究,使结果更为准确。

4 结论

试验结果表明,与其他种植模式相比,玉米品种"迪卡008"在密度为57 000株/hm^2,行距为0.65 m+0.65 m的种植模式下产量最高,可达9 677.35 kg/hm^2。因此,结合品种自身特性,合理选择种植模式能够促进玉米产量的提高。

参考文献

[1] 孔祥林,闫飞燕. 广西玉米生产现状及发展思考 [J]. 中国种业, 2008, 1:12-14.

[2] 杜国田. 玉米不同密度和行距对比试验研究 [J]. 现代农业科技, 2010, 6:62.

[3] 胡萌,魏湜,杨猛,等. 密度对不同株型玉米光合特性及产量的影响 [J]. 玉米科学, 2010, 18 (1): 103-107.

[4] 范秀玲,李凤海,史振声,等. 玉米偏垄宽窄行种植方式的增产作用和生理特性研究 [J]. 玉米科学, 2010, 18 (1): 108-111.

[5] 温日宇,郭耀东,刘建霞,等. 不同密度和种植方式对玉米产量的影响 [J]. 山西农业科学, 2011, 39 (8): 814-815.

[6] 丰光,李妍妍,景希强,等. 玉米不同种植密度对主要农艺性状和产量的影响 [J]. 玉米科学, 2011, 19 (1): 109-111.

[7] 李洪,王斌,李爱军,等. 玉米株行距配置的密植增产效果研究 [J]. 中国农学通报, 2011, 27 (9): 309-313.

[8] Ioannis S, Tokatlidis. Variation within maize lines and hybrids in the absence of competition and relation between hybrid potential yield per plant with line traits [J]. The Journal of Agricultural Science, 2000, 134: 391-398.

[9] 郭国亮,李洪,栗红生. 不同株型玉米品种的结实性及其受光态势的研究 [J]. 山西农业科学, 1998, 26 (1): 19-23.

[10] 张永科,何仲阳,马永平,等. 玉米密植和营养改良之研究——Ⅱ. 行距对玉米产量和营养的效应 [J]. 玉米科学, 2006, 14 (2): 108-111.

[11] 张永科,何仲阳,马永平,等. 玉米密植和营养改良之研究——Ⅲ. 玉米营养和产量的相关分析 [J]. 玉米科学, 2006, 14 (3): 129-132.

栽培模式对玉米叶绿素含量、叶面积指数及产量的影响

范继征[1]，闫飞燕[2]，石达金[1]，钟昌松[1]，吕巨智[1]

(1. 广西农业科学院玉米研究所，广西 南宁 530007；
2. 广西农业科学院农产品质量安全与检测技术研究所，广西 南宁 530007)

摘要：以广西主推玉米品种正大619为研究材料，采用裂区设计，2种施肥方式为主处理，6种耕作方式为副处理，比较研究玉米叶绿素含量、叶面积指数和产量的变化规律，为广西玉米生产提供科学依据。结果表明：施肥和耕作方式对叶绿素含量和叶面积指数的影响较小，两者对产量的影响较大。常规三次施肥条件下，深松25 cm+免翻耕处理的产量最高，达到了8 273.62 kg/hm^2，比对照旋耕25 cm处理增产16.11%。从施肥角度来看，常规三次施肥处理的玉米产量高于一次性施肥处理的玉米产量。从耕作角度来看，深松处理可以显著提高玉米产量。

关键词：玉米；叶绿素含量；叶面积指数；产量

Effect of different cultivation mode on the chlorophyll content, leaf area index and yield of maize

Fan Jizheng[1], Yan Feiyan[2], Shi Dajin[1], Zhong Changsong[1], Lü Juzhi[1]

(^{1}Maize Research Institute, Guangxi Academy of Agricultural Sciences Nanning 530007, Guangxi, China; ^{2}The Quality and Safety of Agricultural Products and Testing Technology Research Institute, Guangxi Academy of Agricultural Sciences Nanning 530007, Guangxi, China)

Abstract: The main extending varieties Zhengda 619 in Guangxi was used as the research material. Split block design was applied in this experiment. The chlorophyll content, leaf area index and yield of corn were research on the two fertilizer and six tillage methods, which would provide a scientific basis for maize production in Guangxi. The results showed that the fertilizer and tillage methods had less effect on the chlorophyll content and leaf area index, both had bigger influence on the production. On the conventional fertilization conditions, deep loosening 25 cm + free ploughing treatment had the highest yield reached 8 273.62 kg/ha. Comparing with rotary tillage 25 cm, the yield of that increased 16.11%. From a fertilization standpoint, the yield of conventional fertilization

---

基金项目：粮食安全关键技术研究与应用示范项目（桂科攻1123001-1J）；广西农业科学院基本科研业务专项（桂农科2014YZ20）。

作者简介：范继征，女，助理研究员，主要从事玉米栽培与生理研究；闫飞燕为通信作者，副研究员，主要从事玉米栽培与生理研究。

treatment was higher than one-off fertilization treatment production. From a tillage standpoint, deep loosening treatment could significantly improve corn production.

Keywords：Maize；Cultivation mode；Chlorophyll content；Leaf area index；Yield

近些年来，随着农村劳动力转移和城镇化的发展，一方面，提高了农村劳动者的收入，缓解了人地矛盾关系，促进了城市的繁荣发展；另一方面，农作劳动力过度流失也影响到农业生产[1]。由于劳动力缺乏及生产成本增加，广西许多地区群众自发采用了玉米免耕技术和一次性施肥的生产方式。此外，受机械动力及传统观念因素的影响，土壤耕层深度一般在15~25 cm，特别是近几年采用免耕的方式，土壤耕层深度更浅，年复一年耕层变浅、犁底层变坚硬，从而限制了作物的生长发育，不但土壤蓄水保水能力下降，而且作物易倒伏、早衰。目前，国内已有大量关于施肥方式和耕作方式的研究报道，主要内容有耕作方式对土壤特性的影响[2]、一次性施肥技术[3]、深耕对土壤和产量的影响[4]、深耕技术的优势[5]、平作与垄作的产量表现[6]、耕作方式对光合特性的影响[7]、常规耕作、免耕、深翻、深松对土壤和玉米产量的影响[8]等方面。针对广西玉米施肥与耕作技术应用研究相对滞后的现状，开展施肥方式与耕作方式对玉米产量的影响试验，可为广西玉米生产提供科学参考。

1　材料与方法

1.1　试验材料

试验玉米品种为正大619。试验所用化肥为：恩泰克®缓效肥（$N：P_2O_5：K_2O = 21：7：11$）；"鲁西牌"复合肥（$N：P_2O_5：K_2O = 15：15：15$），"群山牌"尿素，总氮含量≥46.4%。

1.2　试验地基本情况

试验于2012年下半年在广西壮族自治区农业科学院明阳基地进行。试验地前作为玉米，地势平坦，肥力中等（表1）。

表1　供试土壤理化性质

全N（%）	全P（%）	全K（%）	碱解N（mg/kg）	速效P（mg/kg）	速效K（mg/kg）	有机质（g/kg）	pH
0.079	0.085	0.751	57.5	34	181.8	18.8	6.66

1.3　试验设计

采用裂区设计，以施肥方式为主处理，耕作方式为副处理。（1）主处理。A1：一次清施肥（播种时一次性施入缓效肥恩泰克，50 kg/亩）；A2：常规施肥（分3次施肥，种肥10 kg/亩复合肥鲁西、苗肥5 kg/亩鲁西+5 kg/亩尿素、追肥15 kg/亩鲁西+15 kg/亩尿素）。（2）副处理。B1：深松25 cm+旋耕25 cm；B2：深松35 cm+旋耕25 cm；B3：旋耕25 cm；B4：深松25 cm+免翻耕；B5：深松35 cm+免翻耕；B6：免翻耕。种植密度3 800株/亩，每小区种植10行，行距0.7 m，行长33 m，小区行间不留走道。试验田四周设保护行。田间管理同当地大田生产。

1.4 测定项目与方法

同一重复内的有关测定在同一天进行。（1）叶绿素含量SPAD值：用日本产SPAD-502型叶绿素计测定主茎倒三叶叶绿素SPAD值；（2）叶面积指数测定方法：叶面积指数=绿叶总面积÷占地面积；（3）产量：实收中间4行计产，并折算成标准含水量（14%）的产量。

1.5 数据处理

试验中所获得的数据采用Microsoft Excel 2007、DPS 7.05等统计分析软件进行数据处理与统计分析，以及采用Origin8.0软件进行作图，显著性检验采用LSD法[9]。

2 结果分析

2.1 试验期降水与温度特征

由图1可知，2012年秋玉米全生育期间（8—12月）降水量为595.9 mm。玉米种植后期又一次大的降水，降水量达到了104 mm。秋玉米生育期间的平均温度为22.1℃。玉米出苗后期的9月9—19日经历了一次明显的降温过程，对秋玉米的生长造成了一定影响。

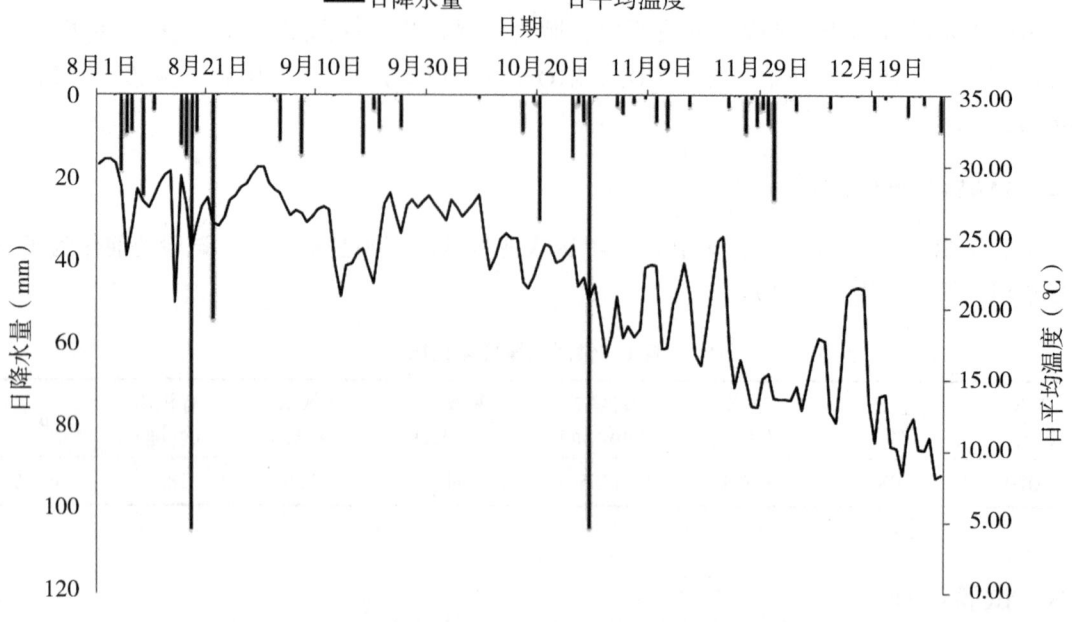

图1 2012年秋季玉米生育期间降水量与平均温度

2.2 不同耕作方式对玉米生长发育不同时期叶绿素含量的影响

从图2可以看出，在玉米生长发育的不同时期，叶绿素含量在一次清施肥和常规三次施肥两种条件下均表现出了先升高后降低的趋势；在两种不同施肥方式下，6种不同耕作方式处理的叶绿素含量变化趋势基本一致；在同一时期，一次清施肥的各个处理叶绿素含量均高于常规三次施肥的处理。在一次清施肥条件下，6 种不同耕作处理的叶绿素含量在苗期、大喇叭口期和灌浆期的差异达到了显著水平，在成熟期差异不明显。其中，深松 35 cm+免翻耕处理在苗期的叶绿素含量最高；免耕处理在大喇叭口期的叶绿素含量最高；深松 25 cm+免翻耕在灌浆期和成熟期的叶绿素含量最高。在常规三次施肥条件下，6 种不同耕作处理的叶绿素含量在苗期的差异达到了显著水平，在大喇叭口期、灌浆期和成熟期的差异不显著。其中，深松 25 cm+免翻耕处理在苗期和成熟期的叶绿素含量最高；深松 35 cm+免翻耕处理在大喇叭口期和灌浆期的叶绿素含量最高。

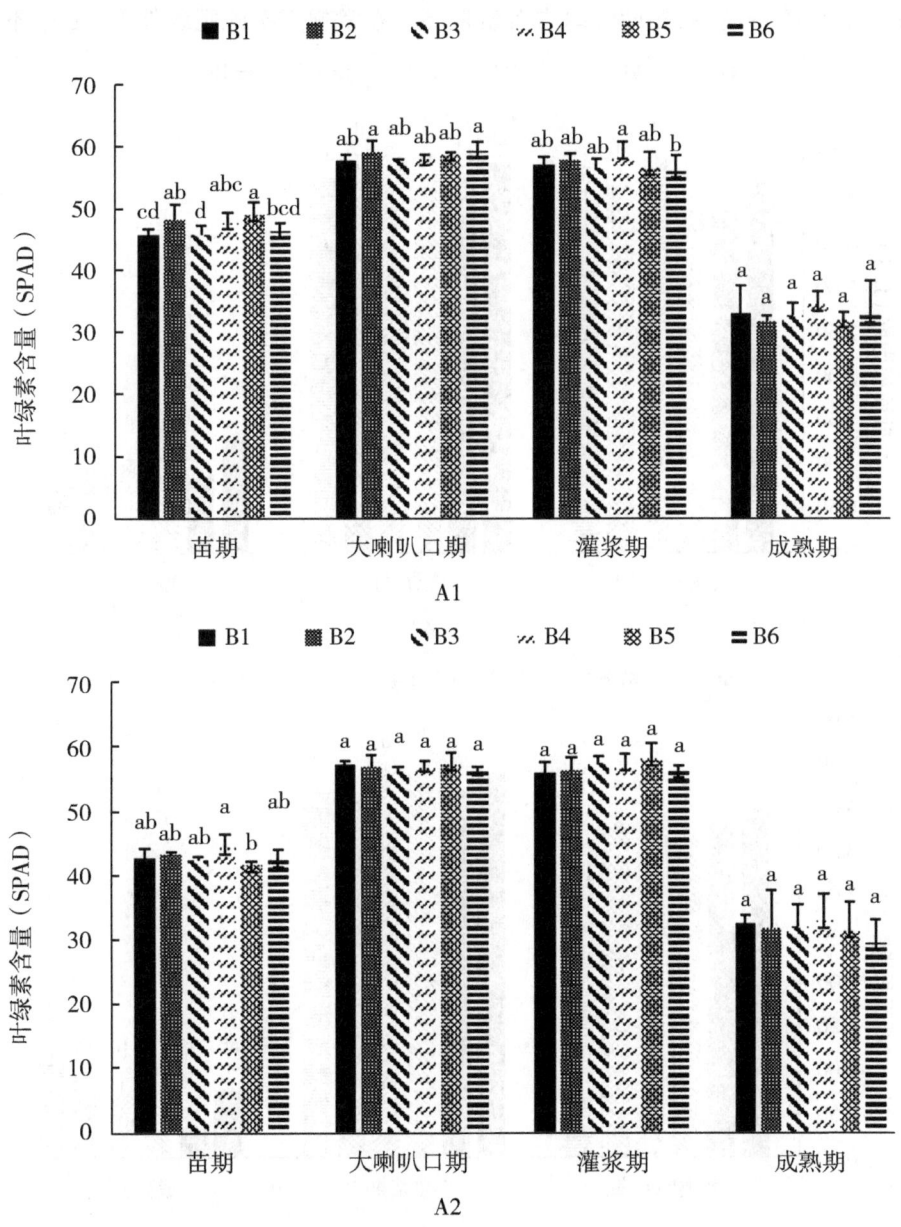

图 2　不同施肥与耕作方式对玉米不同生育期叶绿素含量的影响

2.3 不同耕作方式对玉米生长发育不同时期叶面积指数的影响

叶面积指数一方面可反映作物生长状况，另一方面可反映叶片对光能的利用情况。从图 3 可以看出，在玉米生长发育的大喇叭口期、灌浆期和成熟期，叶面积指数在一次清施肥和常规三次施肥两种条件下均表现出了先升高后降低的相同趋势；在两种不同施肥条件下，6 种不同耕作处理的叶面积指数变化趋势基本一致。在一次清施肥条件下，6 个不同耕作处理的叶面积指数在大喇叭口期和灌浆期的差异达到了显著水平，在成熟期的差异不显著；其中，耕作处理深松 25 cm+免耕在 3 个时期的叶面积指数与其他处理相比均较高；耕作处理深松 25 cm+旋耕 25 cm 在成熟期叶绿素含量最高。在常规三次施肥条件下，6 个不同耕作处

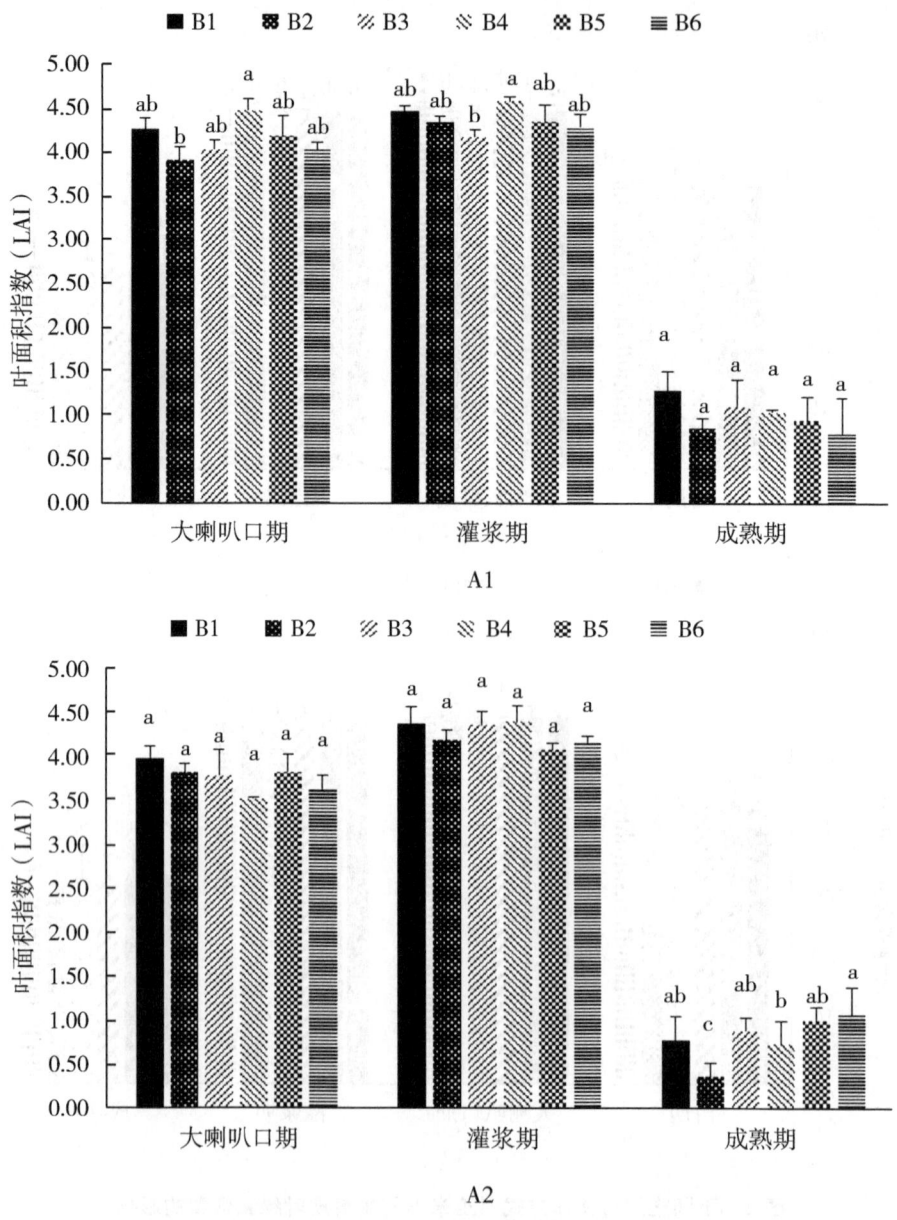

图 3 不同耕作方式对玉米不同生育期叶面积指数的影响

理的叶面积指数在大喇叭口期和灌浆期的差异均不显著，在成熟期的差异达到了显著水平；其中，耕作处理深松 25 cm+旋耕 25 cm 在大喇叭口期的叶面积指数最高，耕作处理深松 25 cm+免翻耕在灌浆期的叶面积指数最高，免翻耕处理在成熟期的叶面积指数高于其他处理。

2.4 不同耕作方式对玉米产量的影响

从表 2 可以看出，不同施肥和耕作方式对玉米产量的影响较大，常规三次施肥的各个处理产量均高于一次性施肥相对应的各个处理；其中，处理旋耕 25 cm 在两种施肥条件下的产量表现差异较小，其他处理的差异较大。在一次清施肥条件下，以广西常规耕作方式旋耕 25 cm 为对照，其他各个耕作处理与对照的产量差异不明显，深松 35 cm+旋耕 25 cm 处理与免耕处理之间的差异达到了显著水平。其中，深松 35 cm+旋耕 25 cm 处理的产量最高，达到了 7 239.62 kg/hm^2，比对照增产 2.10%；免耕处理较对照减产 2.73%。在常规三次施肥条件下，以广西常规耕作方式旋耕 25 cm 为对照，其他各个耕作处理与对照的产量差异达到了极显著水平，各个处理的产量均表现增产。其中，深松 25 cm+免翻耕处理的产量最高，达到了 8 273.62 kg/hm^2，比对照增产 16.11%；免耕处理较对照增产 0.4%。

表 2 不同耕作方式对玉米产量的影响

	处理	产量（kg/hm^2）	比对照增产（±%）
A1：一次清施肥	B1：深松 25 cm+旋耕 25 cm	7 080.60 abA	-0.14
	B2：深松 35 cm+旋耕 25 cm	7 239.62 aA	+2.10
	B3：旋耕 25 cm	7 090.66 abA	—
	B4：深松 25 cm+免翻耕	6 925.55 abA	-2.33
	B5：深松 35 cm+免翻耕	6 942.12 abA	-2.09
	B6：免耕	6 896.96 bA	-2.73
A2：常规三次施肥	B1：深松 25 cm+旋耕 25 cm	7 440.38 bcBC	+4.42
	B2：深松 35 cm+旋耕 25 cm	7 596.38 bB	+6.61
	B3：旋耕 25 cm	7 125.60 dC	—
	B4：深松 25 cm+免翻耕	7 629.03 bB	+7.07
	B5：深松 35 cm+免翻耕	8 273.62 aA	+16.11
	B6：免耕	7 153.80 cdC	+0.40

3 讨论与结论

综合上述结果，不同施肥和耕作方式对玉米叶绿素含量、叶面积指数、产量均有不同程度的影响。其中，施肥和耕作方式对叶绿素含量和叶面积指数的影响较小，两者对产量的影响较大。从施肥角度看，常规三次施肥的玉米产量高于一次性施肥的玉米产量。从耕作角度看，深松处理可以显著提高玉米产量。

由于玉米整个生育期对养分的需求因生长发育阶段的不同表现出较大差异，一次性施肥要满足玉米不同生长发育阶段期的营养需要，对肥料的供肥性能提出了更高的要求。由于一次性施肥时玉米产量还受多重因素的影响，本研究结果与崔涛和喻猛[10]、高强等[3]的研究结论并不一致。试验中采用的生物缓效肥由于后期供肥不足，一次性施用情况下玉米产量较分次合理施肥下降，该缓效肥总养分低于常规分次施肥总养分，是导致一次性施肥减产的重要原因之一,因此土壤水肥条件差的地块不建议采用此类肥料进行一次性施肥,以免造成减产。

深翻和深松分别通过铧式犁和深松铲疏松土壤,虽然都对土壤产生了扰动,但深翻翻转了土壤,深松不翻转土壤。李永平等[8]研究表明,深翻或者深松对改善土壤结构、提高土壤水分和养分含量、增加玉米产量均具有重要作用。关于免耕对作物的增产和减产效应均有较多报道。邱红波等[11]研究认为,免耕栽培玉米较翻耕栽培玉米倒伏率高、有效株数少,从而导致单位面积产量显著低于翻耕栽培玉米。冯延江等[12]研究认为,免耕条件下玉米产量比传统耕作有所增加。本研究结果认为,深松对增加玉米产量具有重要作用,免耕在常规三次施肥条件下较常规耕作有小幅度增产,在一次性施肥条件下有小幅度减产。

参考文献

[1] 鲁奇,杨春悦,张超阳.少数民族地区农村劳动力转移的调查研究——以广西壮族自治区为例[J].山西大学学报:哲学社会经济版,2007,30(4):1-6.

[2] 许迪,Schmid,Mermoud.夏玉米耕作方式对耕层土壤特性时间变异性的影响[J].水土保持学报,2000,14(1):64-70.

[3] 高强,李德忠,汪娟娟,等.春玉米一次性施肥效果研究[J].玉米科学,2007,15(4):125-128.

[4] 闫惊涛,康永亮,田志浩.土壤耕作深度对旱地冬小麦生长和水分利用的影响[J].河南农业科学,2011,40(10):81-83.

[5] 王景琴,朱秀章,刘通.耕地深松深耕技术的优势及完善措施[J].现代农业科技,2011,19:137-139.

[6] 刘玉涛,王宇先,张树权,等.耕作方式对半干旱地区玉米生长和产量的影响[J].黑龙江农业科学,2012(7):19-21.

[7] 刘武仁,郑金玉,罗洋,等.不同耕作方式对玉米叶片冠层光合特性的影响[J].玉米科学,2012,20(6):103-106,111.

[8] 李永平,王孟本,史向远,等.不同耕作方式对土壤理化性状及玉米产量的影响[J].山西农业科学,2012,40(7):723-727.

[9] Steel R G D, Torrie J H. Principles and Procedures of Statistics: A Biometrical Approach [M]. 2nd ed. New York: McGraw-Hill, 1980.

[10] 崔涛,喻猛.玉米一次深施复合肥施肥效应研究[J].辽宁农业科学,2006(5):22-24.

[11] 邱红波,何腾兵,龙友华,等.免耕栽培对玉米根系性状及其产量的影响[J].贵州农业科学,2011,39(9):55-57.

[12] 冯延江.免耕覆盖对玉米生长发育及产量的影响[J].黑龙江农业科学,2008(3):32-33.

10个玉米品种耐密性分析及其对主要性状的影响

范继征[1]，程伟东[1]，闫飞燕[1]，石达金[1]，吕巨智[1]，张玉[1]，钟昌松[1]，刘永红[2]

(1. 广西农业科学院玉米研究所，广西 南宁 530007；
2. 四川省农业科学院作物研究所，四川 成都 610066)

摘要：对10个玉米品种进行耐密性及密度对主要农艺性状的影响进行分析。结果表明：不同品种的株高、穗位高、茎粗、穗长、穗粗均随密度的增加而降低，出籽率随密度增加而升高；不同品种的空秆率、秃尖长、穗行数、行粒数、百粒重和产量随密度增加而呈现不同变化。综合分析来看，品种"中单901"适宜密植且产量高，在密度为69 000株/hm^{2}时，产量最高可达10 138.70 kg/hm^2。

关键词：玉米；耐密性；农艺性状；产量

Density-tolerance Analysis of 10 Maize Varieties and the Effects on Maize Agronomic Traits

Fan Jizheng[1], Cheng Weidong[1], Yan Feiyan[1], Shi Dajin[1],
Lü Juzhi[1], Zhang Yu[1], Zhong Changsong[1], Liu Yonghong[2]

([1]Institute of maize research, Guangxi Academy of Agricultural Sciences, Nanning 530007, Guangxi, China; [2]Crop Research Institute, Sichuan Academy of Agricultural Sciences 530007, Chengdu 610066, Sichuan, China)

Abstract: Density and variety were chosen as testing factors study the influences of different planting densities on yield and main agronomic characters of 10 maize varieties. Results showed that the different varieties of plant height, ear height, stem diameter, ear length, ear diameter were decreased with increasing of the density, seed percentage increased with the increase of density. And rate of empty stem of different vacieties, and bald tip length, rows per ear, kernels per row, hundred grain weight and yield increased or decreased with the increase of density. Comprehensive analysis showed that the variety "Zhongdan 901" was suitable for condensed planting and higher yield, the output ceached up to 10,138.70 kg/ha in the density of 69,000 plants/ha.

Keywords: Maize; Density-tolerance; Agronomic Traits; Yield

基金项目：国家现代玉米产业技术体系南宁综合试验站资助项目（CARS-02-73）；广西农业科学院基本科研业务专项资助项目（桂农科2015YM37）；公益性行业（农业）科研专项经费资助（20150312713）。

作者简介：范继征（1982—），女，助理研究员，主要从事玉米栽培与耕作研究工作，E-mail：fiona-fiona-happy@163.com。

合理密植是提高玉米产量的主要途径，也是当前玉米栽培的发展趋势[1-4]。近年来，针对玉米不同密植条件下对农艺性状、籽粒灌浆、机播产量、产量构成因子、高产稳产性、适应性的影响等相关内容开展的研究报道较多[5-9]。本研究针对当前西南玉米产区大面积生产种植密度较低，销售品种多、乱、杂的问题，选取10个品种为研究材料，开展不同玉米品种耐密性试验，研究种植密度对玉米主要农艺性状和产量的影响，以期为广大农户选择生产上安全低风险的玉米品种提供参考依据。

1 材料和方法

1.1 试验概况

试验于2015年春在广西壮族自治区农业科学院明阳基地（22°36′34″N，108°14′33″E）进行，试验地土壤为黏土，肥力中等，前作为玉米。供试玉米品种10个，分别是会单四号、贵单8号、华优168、青青515、渝单8号、中单901、桂单165、桂单0810、太平洋98、钻卡巴巴。3月2日播种，7月6日收获。施肥量及管理方式同当地大田生产。

1.2 试验设计

试验采用裂区设计，密度为主处理，设置两种密度，处理A为54 000株/hm²（常规密度）和处理B为69 000株/hm²（相比常规密度，每公顷增加15 000株）；品种为副处理，共20个处理。小区行长6 m，行距0.7 m，每小区6行，小区面积为25.2 m²，3次重复。

1.3 调查项目

蜡熟期在田间测定株高、穗位高、茎粗等。收获前记载每个小区的空秆率，收中间2行脱粒后计产。收获后选取有代表性的10穗进行室内考种，测量穗长、穗粗、秃尖长、穗行数、行粒数、百粒重、出籽率等性状。

1.4 数据处理和统计分析

试验数据的处理分别采用Microsoft Excel 2013和DPS 7.05统计软件进行分析。

2 结果与分析

2.1 不同品种植株农艺性状对密度的响应分析

2.1.1 不同玉米品种株高、茎粗、穗位高对密度的响应　与常规低密度相比，在高密度条件下，各个品种的株高、茎粗和穗位高均呈降低的趋势（表1）。方差分析表明，株高、穗位高和茎粗在密度间差异依次表现为差异极显著、差异显著和差异不显著，在品种之间差异均达到极显著水平。从单个品种来看，"会单四号"和"钻卡巴巴"的株高、穗位高、茎粗受密度影响最大，品种"华优168""太平洋98"受密度影响最小。多重比较分析表明，株高、穗位高的变化在品种间表现出较大差异性，说明株高、穗位高主要受遗传因素控制。但在大范围内与品种比较而言，密度对株高、穗位高的影响更大。

表1 不同品种株高、茎粗和穗位高的多重比较

品种	株高（cm）		茎粗（mm）		穗位高（cm）	
	54 000株/hm²	69 000株/hm²	54 000株/hm²	69 000株/hm²	54 000株/hm²	69 000株/hm²
贵单8号	272.13 ABab	245.13 Bb	17.06 Bbc	16.71 BCbc	133.20 Aa	118.00 Aab
桂单0810	285.53 Aa	269.60 Aa	16.16 Bc	15.70 BCbcd	139.93 Aa	126.20 Aa
桂单165	259.07 BCbc	237.00 BCbc	18.11 ABab	16.43 BCcd	131.73 Aa	118.07 Aab
华优168	245.60 Cd	245.33 Bb	17.56 ABbc	17.79 BCbc	117.33 BCbd	116.00 Ab
会单四号	228.13 De	193.20 De	19.05 Aa	17.19 ABab	89.40 Ef	73.20 Cd
青青515	252.40 Ccd	235.33 BCbcd	19.41 Aa	19.00 ABbc	110.07 BCDcde	94.53 Bc
太平洋98	245.20 Cd	243.33 Bb	17.45 Bbc	17.21 Aa	102.40 De	98.53 Bc
渝单8号	249.80 Ccd	225.93 Ccd	17.38 Bbc	16.63 ABbc	112.60 BCDbcd	100.93 Bc
中单901	245.67 Cd	225.27 Cd	16.39 Bc	14.98 BCbcd	107.13 CDde	94.00 Bc
钻卡巴巴	267.80 Bb	245.33 Bb	18.65 Aab	16.63 Cd	118.47 Bb	101.93 Bc

注：同列不同小写字母表示差异显著（$P<0.05$），不同大写字母表示差异极显著（$P<0.01$）。表2同。

2.1.2 不同玉米品种空秆率对密度的响应 由图1-A可以看出，各个品种在高低密度条件下植株空秆率变化差异明显，除品种"青青515"外，其余品种空秆率均随着密度的增加而升高。其中，品种"青青515"的空秆率随着密度的增加而大幅度下降，差异达到了极显著水平；品种"渝单8号"的空秆率随着密度的增加而急剧升高，增幅最大；品种"贵单8号""桂单165"和"中单901"次之，品种"会单四号""桂单0810""太平洋98"和"钻卡巴巴"较低，品种"华优168"的空秆率受密度影响最小。

2.2 不同品种穗部性状对密度的响应分析

2.2.1 不同品种穗长、穗粗和秃尖长对密度的响应 由图1B~D可以看出，不同品种的穗长、穗粗和秃尖长均受到密度的影响，表现出不同的变化趋势。其中，穗长和穗粗均随着密度的增加而减小，不同品种的秃尖长表现各异。从单个品种表现来看，品种"贵单8号"穗长受密度影响最大，表现为穗长明显变短，品种"桂单0810"穗长变化最小，几乎不受密度影响；品种"华优168"穗粗受密度影响最大，穗粗明显变细，品种"桂单0810"和"青青515"穗粗变化最小；品种"会单四号"和"青青515"的秃尖长受密度影响最大，前者表现为秃尖长明显减小，后者则相反，秃尖长明显增大，品种"钻卡巴巴""桂单0810"和"渝单8号"秃尖长受密度影响较小，秃尖长变化不明显。

2.2.2 不同品种穗行数、行粒数、百粒重和出籽率对密度的响应

由图1E~H可以看出，不同品种的穗行数、行粒数、百粒重和出籽率均受到密度的影响，表现出不同的变化趋势。其中，不同品种的穗行数随着密度的增加增减程度不一致，品种"钻卡巴巴"穗行数降低幅度最大，品种"太平洋98"的穗行数无变化，其余品种介于两者之间；大多数品种的行粒数和百粒重随着密度的增加而降低，品种"青青515"的行粒数减少最多，"中单901"次之，而品种"桂单0810"的行粒数有小幅度的增加；品种"华优168"和"钻卡巴巴"的百粒重降幅最大，而品种"中单901"和"桂单0810"均有小幅升高；所有品种的出籽率均随着密度增加而升高，品种"太平洋98"的增幅最大，而"桂单0810"的增幅最小。

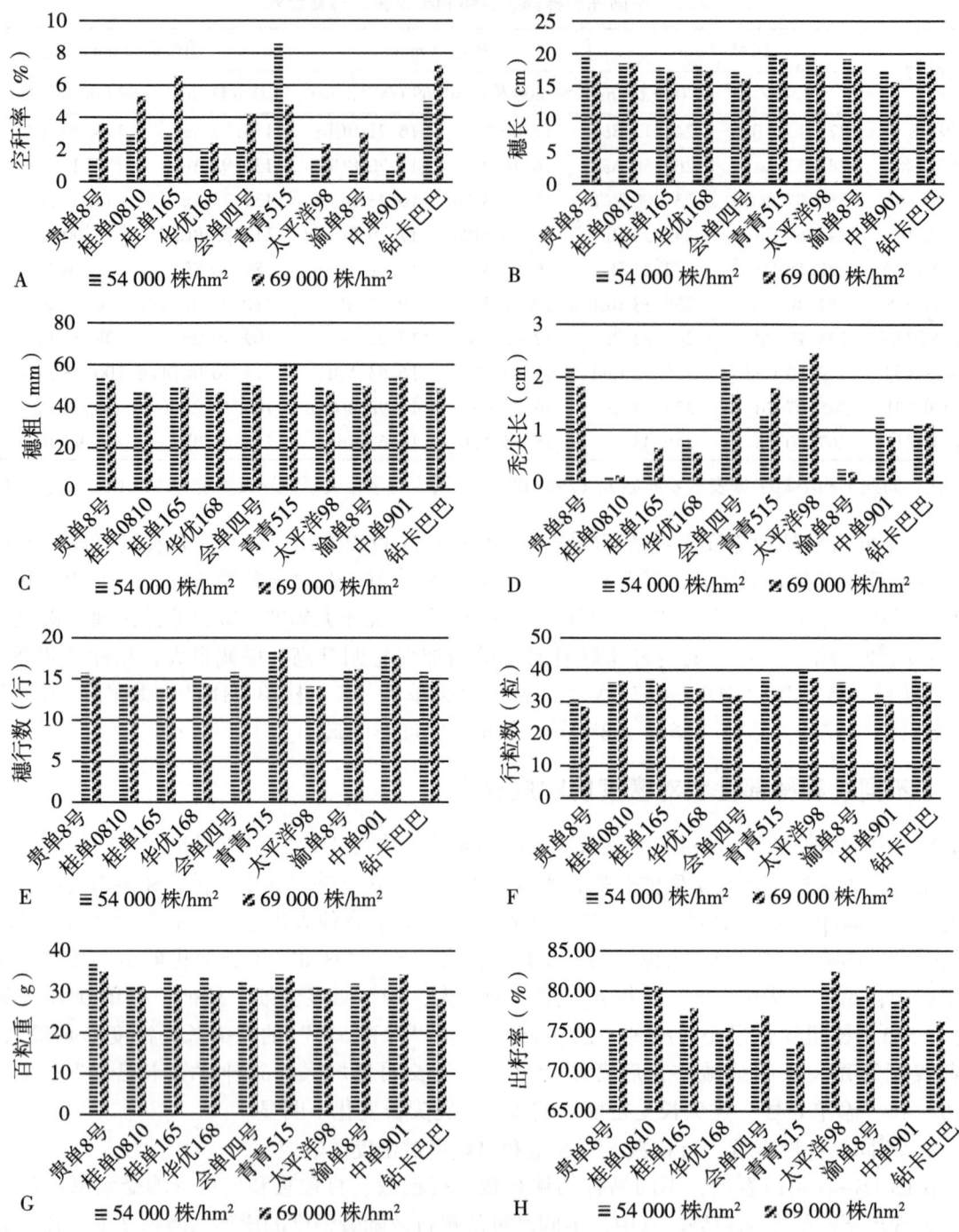

图 1　在高低密度条件下的植株主要农艺性状的变化

2.3　不同品种产量对密度的响应分析

由表 2 可以看出，随着种植密度的增加，大多数品种的产量均表现增产，增产幅度各异。其中，品种"桂单 165"增产幅度最大，差异达到了显著水平，增幅为 9.65%；其次，品种"华优168""桂单0810"和"青青515"的产量也有大幅度提高,增幅分别为8.69%、8.37%和7.42%;品种"渝单8号"增幅最小,仅为1.48%。与之相反,品种"贵单8号"和"钻卡巴巴"两者表现为减产,前者减产幅度最大,达到了8.60%,后者减产幅度较小,仅为1.51%。从单一密度各品种产量表现来看,品种"中单901"的产量均最高,在54 000株/hm²和69 000株/hm²株密度条件下,产量分别达到了9 652.71 kg/hm²和10 138.70 kg/hm²。

表2　不同品种产量的多重比较

品种	产量（kg/hm²） 54 000株/hm²	产量（kg/hm²） 69 000株/hm²	比低密度增产（±%）
贵单8号	8 457.74±491.26 CDEcd	7 730.52±187.08 De	-8.60[ns]
桂单0810	8 320.04±234.56 DEd	9 016.48±395.18 BCcd	8.37[ns]
桂单165	8 426.51±297.40 CDdE	9 239.29±276.10 BCbc	9.65*
华优168	8 009.17±197.23 EFd	8 704.87±425.45 Cd	8.69[ns]
会单四号	7 427.53±124.55 Fe	7 591.44±300.06 De	2.21[ns]
青青515	8 974.90±566.17 BCbc	9 641.09±336.80 ABb	7.42[ns]
太平洋98	8 865.54±489.70 BCDc	9 311.32±207.58 Bbc	5.03[ns]
渝单8号	9 350.00±348.27 ABab	9 488.16±493.14 Bbc	1.48[ns]
中单901	9 652.71±201.18 Aa	10 138.70±302.05 Aa	5.03[ns]
钻卡巴巴	7 987.14±794.01 EFd	7 866.82±576.11 De	-1.51[ns]

3　结论与讨论

从耐密性角度而言,在当地适宜高密度种植的品种有"桂单165""华优168""桂单0810""青青515""太平洋98"和"中单901",适宜低密度种植的有"贵单8号""钻卡巴巴""渝单8号"和"会单四号";从产量表现来看,优先种植的品种依次为"中单901""青青515"和"渝单8号"。综合分析,品种"中单901"不仅适宜高密度栽培,而且产量最高,表现最好。

不同玉米品种的农艺性状受到密度的影响,响应不同。株高、穗位高、茎粗、穗长、穗粗均随密度的增加而降低,出籽率随密度增加而升高。前人研究结果表明,株高和穗位高随着密度的增加而升高[2,5],与本研究结果不一致,可能原因是试验地土壤肥力均等情况下,随着密度增加导致单个植株养分不足所致。空秆率、秃尖长、穗行数、行粒数和百粒重随密度增加表现出不同的增减幅度,其中大部分品种的变化趋势与前人研究结果相同[2,10],部分品种表现不一致,这主要与品种特性以及本研究设置密度有关。

增加密度是当前提高玉米产量的主要途径之一。已有研究结果表明,玉米产量随种植密度的增加而提高,但当密度达到一定程度后,产量又随密度的增加而降低[11]。本研究中部分品种的产量随着密度的增加而增加,也有部分品种的产量随着密度增加而降低。因此,不同的品种应选用适宜的种植密度,才能获得较高产量。

参考文献

[1] 范继征,闫飞燕,石达金,等.不同密度与种植方式对玉米"迪卡008"主要农艺性状和产量的影响[J].南方农业学报,2012,43(6):759-763.

[2] 张丽宏,李新,罗湘宁,等.不同类型玉米耐密性分析及对主要性状的影响[J].西北农业学报,2014,23(3):36-39.

[3] 路海东,薛吉全,郝引川,等.基于群体产量及相关性状的玉米耐密性评价[J].西北农业学报,2014,23(7):46-50.

[4] Duvick D N, Cassman K G. Post-green revolution trends in yield potential of temperate maize in the north-central United States[J]. Crop Science, 1999, 39: 1622-1630.

[5] 丰光,李妍妍,景希强,等.玉米不同种植密度对主要农艺性状和产量的影响[J].玉米科学,2011,19(1):109-111.

[6] 陈尚洪,陈红琳,沈学善,等.密度和施氮量对丘陵区机播夏玉米产量及倒伏影响研究[J].西南农业学报,2012,25(3):805-808.

[7] 王晓燕,张洪生,盖伟玲,等.种植密度对不同玉米品种产量及籽粒灌浆的影响[J].山东农业科学,2011,4:36-38.

[8] 王楷,王克如,王永宏,等.密度对玉米产量(>15 000 kg/hm^2)及其产量构成因子的影响[J].中国农业科学,2012,45(16):3437-3445.

[9] 王磊,高杰,渠建洲,等.两种密度下不同玉米品种的高产稳产及适应性分析[J].玉米科学,2016,24(2):136-141.

[10] 汤彬,李宏志,曹钟洋,等.不同种植密度对13个玉米品种产量及主要农艺性状的影响[J].湖南农业学,2013(1):17-21.

[11] 陈传永,侯玉虹,孙锐,等.密植对不同玉米品种产量性能的影响及其耐密性分析[J].作物学报,2010,36(7):1153-1160.

应用灰色关联度分析法分析西南区玉米新组合

范继征,石达金,吕巨智,唐国荣,程伟东

(广西农业科学院玉米研究所,广西 南宁 530007)

摘要:采用灰色关联度分析方法,基于15个农艺性状数据对2017年国家玉米产业技术体系西南区55个玉米新组合进行分析比较。结果表明,不同玉米新组合农艺性状与产量的关联度大小依次为单株粒重>穗粗>籽粒含水量>百粒重>空秆率>穗行数>出籽率>穗长>倒折率>行粒数>茎粗>株高>穗位高>秃尖长。其中,单株粒重的关联度大于0.9,秃尖长的关联度小于0.7,表明单株粒重是提高玉米产量的主要因素,秃尖长是降低玉米产量的主要因素。以产量最高组合1703为理想品种,综合评价表现较优的组合有成单388、绵1308、川单1751和荣玉1756。

关键词:玉米;西南区;灰色关联度;产量

Grey correlative analysis of new combinations from maize region of southwest China

Fan Jizheng, Shi Dajin, Lü Juzhi, Tang Guorong, Cheng Weidong

(Maize research institute, Academy of Guangxi Agricultural Sciences, Nanning 530007, Guangxi, China)

Abstract: Based on the data of 15 agronomic traits, the method of grey correlative analysis was used to analyze and compare about 55 new corn combinations in maize region of southwest China in 2017. Results showed that the order of correlation degree between agronomic traits and yield from large to small was: grain weight per plant > ear diameter > grain water content > hundred grain weight > barren stalk rate > ear row number > rate of seed > ear length > rupture rate > row grain number > stem diameter > plant height > ear height > bald tip length. The correlation degree of grain weight per plant was greater than 0.9, and bald tip length was less than 0.7, which indicated that the grain weight per plant was the main factor to increase maize yield and bald tip length was the main factor to decrease maize yield. Compared with ideal varieties with the highest yield combination 1703, the combinations with better comprehensive evaluation performance were Chengdan 388, Mian 1308, Chuandan 1751 and Rongyu 1756.

Keywords: Maize; Southwest China; Grey correlative analysis; Yield

---

基金项目:国家玉米产业技术体系南宁综合试验站(CARS-)。

作者简介:范继征,女,助理研究员,主要从事玉米栽培生理研究。E-mail:fiona-fiona-happy@163.com。

西南区玉米常年种植面积在466.67万hm²，主要分布在广西、云南、贵州、四川和重庆等地，是我国第三大玉米产区[1]。该地区以山地和丘陵为主，地形地貌复杂、生态环境多样，用常规方法进行统计学分析，来评价参试玉米品种的优劣，常会造成一定程度的偏差。而灰色关联度分析法具有所需样本少、方法简便、信息量大等优点[2]，已在玉米品种鉴定中取得良好效果[3-7]。本研究对2017年国家玉米产业体系西南区55个玉米新组合进行产量与主要农艺性状之间的灰色关联度分析，明确各个性状与产量之间的关系，综合评价各个新组合的优劣，旨在为西南区玉米新组合评价提供科学依据。

1 材料与方法

1.1 试验材料

供试材料为国家玉米产业体系西南区各个单位提供的55个玉米新组合（表1），以及对照品种桂单162和渝单8号。试验于2017年在广西农业科学院明阳基地进行。

表1 试验材料

序号	组合	序号	组合	序号	组合	序号	组合
T1	荣玉1754	T16	青青711	T31	1702	T46	GD1676
T2	南T931	T17	苏科1618	T32	成单388	T47	HZ1722
T3	绵TX1701	T18	荣玉1755	T33	青青722	T48	1703
T4	渝单701	T19	南T928	T34	苏科1614	T49	成1702
T5	靖2017-1	T20	绵732	T35	荣玉1756	T50	青青733
T6	德3239	T21	渝单841	T36	南T605	T51	苏科玉1604
T7	黔1701	T22	靖2017-2	T37	绵1308	T52	渝单836
T8	川单1751	T23	德10	T38	渝单703	T53	黔1705
T9	恩1077	T24	黔1702	T39	靖2017-3	T54	渝单896
T10	CS17-1	T25	川单1752	T40	恩912	T55	恩1010
T11	云瑞506	T26	恩1124	T41	黔1703	CK1	桂单162
T12	S183	T27	CS17-2	T42	川单1753	CK2	渝单8号
T13	HZ1720	T28	云瑞007	T43	恩932		
T14	1701	T29	桂A115	T44	黔1704		
T15	成单658	T30	HZ1721	T45	云瑞1207		

1.2 试验方法

试验采用双对照间比法排列，不设重复，每5个参试品种设2个对照。种植密度为$5.7×10^4$株/hm²，小区面积为21 m²，5行区，行长6 m，行距0.7 m。试验田四周设大于5行保护行，田间管理同当地水平。

1.3 调查项目

调查株高、穗位高、茎粗、穗长、穗粗、秃尖长、穗行数、行粒数、百粒重、出籽率、籽粒含水量、单株粒重、倒折率、空秆率和产量等 15 个性状。果穗性状按照单穗平均法，收获时取有代表性的 20 个样穗进行考种，取平均值比较。

1.4 数据分析

采用 Excel 2016 和 DPS V14.05 进行统计分析和灰色关联度分析，依据邓聚龙等方法[6,8-9]，将 55 个新组合的产量作为参考数列 X_0，其余 14 个农艺性状为比较数列，分别为：株高（X_1）、穗位高（X_2）、茎粗（X_3）、穗长（X_4）、穗粗（X_5）、秃尖长（X_6）、穗行数（X_7）、行粒数（X_8）、百粒重（X_9）、出籽率（X_{10}）、籽粒含水量（X_{11}）、单株粒重（X_{12}）、倒折率（X_{13}）、空秆率（X_{14}），对数据进行标准化处理，并计算关联系数和关联度进行比较分析。此外，以产量最高组合为理想品种，计算不同组合的关联度并进行分析比较。

2 结果分析

2.1 玉米不同新组合农艺性状和产量表现

从表 2 可以看出，55 个不同玉米新组合的各个农艺性状和产量表现不同。参试组合的株高范围在 229.30~329.10 cm，株高低于两个对照平均值的组合有 22 个。穗位高的范围在 86.60~148.70 cm，高于两个对照平均值的组合有 7 个。穗长的范围在 14.90~19.95 cm，比两个对照平均值长的组合有 32 个，大于 19 cm 的有 5 个组合，分别是南 T928、荣玉 1756、渝单 703、荣玉 1754 和渝单 896。穗粗的范围在 45.05~55.68 mm，高于对照平均值的有 37 个组合，穗粗最大的组合为青青 733。秃尖长的范围在 0.13~2.00 cm，秃尖长在 1.5 cm 以上的组合有 5 个，分别是青青 733、绵 1308、青青 711、HZ1722 和恩 1010。穗行数的范围在 13.00~20.20 行，低于对照平均值的有 4 个组合，分别是成 1702、云瑞 1207、HZ1721 和德 3239。行粒数在 32.80~44.90 粒，大于对照平均值的有 11 个组合，南 T928 的行粒数最高。百粒重在 23.52~36.25 g，大于对照平均值的有 25 个组合。单株粒重在 0.094~0.197 kg，大于对照平均值的有 33 个组合，单株粒重大于 0.180 kg 的组合分别是川单 1752、1703、成单 388、绵 1308 和渝单 836。倒折率在 0~76.0%，未发生倒折的组合有 25 个，倒折率大于 20% 的组合有云瑞 007、川单 1753、荣玉 1754 和靖 2017-2。空秆率在 0~18.31%，空秆率为 0 的组合有 14 个，空秆率大于 15% 的组合有云瑞 506 和靖 2017-2。其中，倒折率、空秆率和秃尖长的变异系数较大，分别达到了 202.05%、93.91% 和 47.18%。对于产量而言，产量表现高于两个对照的有 28 个组合，介于对照之间的有 9 个，低于两个对照的有 18 个。其中，1703 的产量最高为 11 472.19 kg/hm^2，靖 2017-2 的产量最低为 5 495.12 kg/hm^2。

从变异系数大小来看，15 个农艺性状的变异系数范围在 1.78%~260.71%。其中，变异系数大于 45% 的性状有 3 个，从大到小依次为倒折率、空秆率和秃尖长；变异系数小于 5% 的性状仅有 1 个，即为出籽率；变异系数在 10%~15% 的性状有单株粒重、产量和穗位高；其余性状则介于 5%~10%。

表2 不同组合的产量和农艺性状

组合	X_1 (cm)	X_2 (cm)	X_3 (mm)	X_4 (cm)	X_5 (mm)	X_6 (cm)	X_7	X_8	X_9 (g)	X_{10} (%)	X_{11} (%)	X_{12} (kg)	X_{13} (%)	X_{14} (%)	X_0 (kg/hm²)
T1	307.60	146.50	19.15	19.10	46.04	0.80	15.6	39.6	29.34	86.68	12.85	0.14	22.37	7.89	8 165.04
T2	326.10	148.70	18.60	18.55	46.82	0.57	16.4	39.0	31.04	87.76	14.40	0.15	10.81	5.41	8 665.09
T3	287.00	134.90	21.31	17.75	46.75	1.30	15.6	34.7	30.40	86.53	12.55	0.09	20.27	16.22	5 495.12
T4	265.50	119.40	17.11	16.60	48.92	0.70	15.8	36.5	29.33	88.20	14.50	0.15	2.74	2.74	8 877.95
T5	268.30	111.00	17.30	17.05	48.34	0.55	14.6	38.6	33.97	85.99	12.90	0.17	5.88	5.88	8 938.54
T6	271.80	120.50	19.07	18.65	49.61	0.35	16.4	39.3	28.41	88.52	15.05	0.18	1.45	7.25	9 969.00
T7	229.30	99.00	18.27	18.90	45.05	0.35	15.0	39.1	28.84	87.32	12.75	0.14	5.26	0.00	8 329.52
T8	280.80	103.50	19.59	16.95	54.98	0.70	20.2	35.0	32.22	89.33	13.10	0.18	2.67	1.33	10 570.57
T9	277.60	112.70	20.16	17.25	49.37	1.35	15.0	36.8	35.56	87.14	14.25	0.15	1.33	4.00	8 753.77
T10	286.90	123.60	20.02	17.45	47.32	0.80	16.6	42.3	26.46	87.30	13.50	0.16	8.11	5.41	9 257.45
T11	299.80	104.30	18.06	17.90	46.62	0.63	15.4	34.6	32.82	85.48	14.35	0.13	0.00	7.89	7 887.46
T12	264.60	117.30	17.22	16.75	46.87	0.93	16.0	37.3	29.26	87.44	16.15	0.14	0.00	0.00	8 273.99
T13	276.20	125.20	18.50	16.05	46.30	0.85	15.2	37.4	28.90	86.74	12.80	0.13	1.35	4.05	7 642.75
T14	296.40	124.60	18.81	17.15	50.22	0.75	16.4	35.5	28.64	86.49	13.20	0.14	2.74	4.11	7 822.85
T15	285.20	125.20	19.82	17.15	50.68	0.67	18.8	38.0	30.05	85.20	14.65	0.16	0.00	8.82	8 873.76
T16	291.60	115.10	18.48	17.30	46.47	0.52	16.2	38.9	23.77	85.47	14.10	0.13	2.70	0.00	7 684.36
T17	244.80	97.50	18.26	18.25	45.11	1.32	13.0	36.0	30.13	86.58	13.05	0.12	2.70	1.35	7 006.76
T18	263.50	98.20	16.36	16.80	48.41	0.80	15.0	35.4	28.31	87.71	13.50	0.12	0.00	6.25	7 839.99
T19	273.10	100.40	18.49	17.45	48.61	0.65	15.0	36.3	35.26	87.01	14.20	0.17	0.00	2.67	10 005.75
T20	258.90	116.50	19.19	19.00	46.68	1.05	15.6	42.0	29.64	86.53	15.30	0.17	0.00	6.76	9 755.02
T21	263.30	108.70	18.47	17.15	46.28	0.60	16.0	34.5	33.22	86.99	13.40	0.16	0.00	0.00	9 664.01
T22	286.20	117.20	19.17	17.60	49.18	1.00	16.0	33.5	34.58	86.29	13.20	0.17	9.33	1.33	10 387.11
T23	279.80	122.10	19.62	17.10	52.92	1.50	16.6	35.9	35.72	86.86	15.05	0.17	0.00	6.76	9 968.41
T24	286.90	123.00	20.24	17.85	49.36	0.35	15.6	42.1	34.50	87.18	13.55	0.20	5.48	0.00	11 394.58
T25	274.30	103.20	20.46	17.35	50.62	0.95	19.0	37.8	30.44	88.09	14.55	0.18	0.00	1.30	11 472.19
T26	245.90	121.10	18.22	17.15	49.17	0.35	16.2	35.7	33.43	88.17	14.90	0.18	0.00	0.00	10 580.64
T27	248.10	119.20	21.26	16.25	45.25	1.25	15.2	37.3	31.62	86.84	12.80	0.11	76.00	1.33	6 398.66
T28	283.80	125.80	20.19	19.45	49.30	0.95	15.6	41.2	31.41	88.30	14.75	0.18	5.19	6.49	10 769.43
T29	239.60	97.90	20.39	16.90	45.98	1.25	15.8	40.9	31.93	87.29	12.25	0.14	1.41	18.31	7 638.49
T30	248.70	109.80	19.69	16.95	47.93	0.40	15.8	41.2	28.81	88.26	13.15	0.14	0.00	0.00	8 708.21
T31	271.00	135.00	17.14	16.10	52.56	0.75	16.0	36.5	35.49	82.96	14.75	0.17	5.63	1.41	9 629.65
T32	320.80	137.20	20.60	17.95	51.27	0.65	15.0	36.3	36.25	87.25	12.30	0.16	26.32	3.95	9 604.18
T33	258.00	110.50	19.27	18.25	48.88	1.40	15.6	40.6	28.85	86.32	14.20	0.14	5.63	8.45	7 772.61
T34	281.40	118.10	17.04	16.50	49.93	0.95	16.4	35.5	32.26	84.26	14.05	0.15	0.00	1.33	8 835.87
T35	234.70	122.50	20.00	15.90	47.21	0.90	14.4	32.8	33.95	83.11	13.00	0.12	1.35	6.76	6 979.28
T36	259.70	98.50	18.34	18.35	50.81	0.38	17.6	40.9	30.75	88.06	13.95	0.17	0.00	10.14	9 296.14
T37	247.90	98.90	18.58	19.95	52.09	1.25	16.0	44.9	26.78	86.97	12.85	0.13	0.00	10.00	7 449.14
T38	272.30	111.20	20.79	16.40	48.97	0.13	16.4	39.1	30.24	89.69	14.45	0.16	0.00	1.32	9 794.59
T39	285.40	125.90	17.70	17.50	45.41	1.02	15.4	37.3	28.58	86.68	12.80	0.14	0.00	0.00	8 244.72

(续表)

组合	X_1 (cm)	X_2 (cm)	X_3 (mm)	X_4 (cm)	X_5 (mm)	X_6 (cm)	X_7	X_8	X_9 (g)	X_{10} (%)	X_{11} (%)	X_{12} (kg)	X_{13} (%)	X_{14} (%)	X_0 (kg/hm²)
T40	275.40	103.70	18.61	18.35	50.74	1.85	18.6	36.3	30.43	89.54	15.95	0.18	0.00	1.33	10 842.15
T41	262.60	120.50	17.66	17.20	55.68	2.00	18.4	39.0	33.55	86.46	14.55	0.16	0.00	0.00	9 163.70
T42	318.00	144.00	19.05	18.05	47.50	0.75	15.4	39.5	29.81	88.21	17.15	0.17	0.00	4.11	9 632.69
T43	254.40	104.10	16.66	19.30	48.62	0.65	16.0	40.9	29.67	89.39	13.90	0.15	0.00	1.33	9 113.90
T44	312.90	122.60	19.41	17.45	49.74	0.40	15.4	39.4	30.38	88.75	14.50	0.14	0.00	5.19	8 797.48
T45	284.10	123.00	16.03	16.90	47.37	0.40	18.6	33.4	27.94	86.47	13.35	0.16	1.32	0.00	9 350.56
T46	283.70	110.00	17.80	17.10	52.30	1.75	16.4	35.7	34.62	86.06	15.95	0.18	0.00	0.00	10 876.78
T47	278.80	126.70	16.59	14.90	45.75	0.60	14.8	34.3	27.62	87.71	13.75	0.12	0.00	1.27	7 675.31
T48	259.90	108.00	17.62	16.25	46.81	0.90	14.2	37.6	31.36	87.64	14.55	0.16	0.00	2.70	9 547.39
T49	276.90	112.10	15.65	18.35	49.70	0.80	16.6	38.3	33.35	86.96	14.50	0.16	0.00	0.00	9 942.36
T50	302.80	135.10	17.41	18.85	46.33	1.35	14.8	39.5	32.18	82.08	14.25	0.14	4.05	9.46	8 105.73
T51	268.10	126.10	19.52	18.15	54.36	0.65	18.8	38.5	30.02	87.75	15.15	0.19	0.00	4.05	11 001.51
T52	283.30	117.00	18.52	19.05	47.62	1.00	16.8	40.4	23.52	85.81	13.65	0.13	0.00	7.69	6 633.15
T53	251.90	86.60	17.16	18.25	48.33	0.65	16.8	37.8	30.68	86.87	15.25	0.18	0.00	0.00	10 359.32
T54	303.50	113.00	18.79	17.30	47.43	1.70	15.4	34.4	32.75	85.54	14.90	0.14	0.00	6.58	8 411.72
T55	246.00	94.10	18.05	18.00	46.63	0.85	16.2	38.4	31.69	90.06	12.50	0.18	5.41	0.00	10 331.02
CK1	281.68	142.62	19.86	16.94	47.17	0.34	14.1	40.0	32.54	85.39	14.15	0.15	3.73	4.93	8 909.69
CK2	257.88	113.85	18.94	18.36	46.74	0.57	15.5	39.5	29.29	87.09	13.59	0.14	1.58	6.08	8 299.01
平均值	275.36	117.27	18.72	17.61	48.74	0.88	16.17	38.01	31.01	87.01	14.07	0.15	5.50	4.31	9 014.97
标准差	21.52	13.72	1.32	1.00	2.49	0.41	1.35	2.57	2.79	1.55	1.04	0.02	11.11	4.05	1 306.44
变异系数	7.82	11.70	7.05	5.68	5.11	47.18	8.38	6.77	8.99	1.78	7.42	14.27	202.05	93.91	14.49

2.2 玉米不同新组合产量与农艺性状的关联系数比较

不同玉米组合各农艺性状与产量的灰色关联系数列于表3。不同组合之间同一性状的关联系数表现各不相同。其中，秃尖长的关联系数范围在0.333 4~0.995 9，变化幅度最大；单株粒重的关联系数范围在0.863 9~0.999 3，变化幅度最小。此外，同一组合不同性状间的关联系数变化明显。其中，恩932不同性状的关联系数变化范围在0.842 0~0.982 0，青青733的关联系数变化范围在0.333 4~0.988 0。

表3 产量与主要农艺性状的关联系数

组合	X_1	X_2	X_3	X_4	X_5	X_6	X_7	X_8	X_9	X_{10}	X_{11}	X_{12}	X_{13}	X_{14}
T1	0.756 8	0.655 2	0.850 2	0.787 6	0.946 6	0.966 3	0.917 8	0.829 1	0.944 4	0.883 3	0.989 5	0.966 2	0.868 6	0.929 2
T2	0.746 8	0.680 2	0.955 8	0.879 6	0.995 8	0.685 3	0.925 6	0.911 7	0.945 2	0.938 0	0.913 8	0.998 4	0.950 4	0.970 1
T3	0.602 1	0.546 8	0.553 5	0.622 5	0.652 5	0.420 5	0.647 8	0.683 7	0.638 8	0.631 5	0.698 4	0.999 0	0.748 5	0.715 2
T4	0.967 3	0.951 5	0.899 6	0.936 0	0.975 2	0.791 0	0.987 6	0.961 7	0.941 1	0.963 0	0.935 4	0.978 1	0.961 4	0.964 3

(续表)

组合	X_1	X_2	X_3	X_4	X_5	X_6	X_7	X_8	X_9	X_{10}	X_{11}	X_{12}	X_{13}	X_{14}
T5	0.9722	0.9354	0.9039	0.9618	0.9971	0.6488	0.8799	0.9669	0.8651	0.9888	0.8966	0.8803	0.9803	0.9773
T6	0.8444	0.8928	0.8795	0.9294	0.8786	0.4826	0.8763	0.8988	0.7731	0.8756	0.9465	0.8890	0.8891	0.8196
T7	0.8758	0.8910	0.9291	0.8167	0.9974	0.5576	0.9952	0.8636	0.9935	0.8965	0.9726	0.9656	0.9148	0.8519
T8	0.8087	0.6923	0.8356	0.7542	0.9328	0.6444	0.8957	0.7204	0.8281	0.8128	0.7294	0.9786	0.8023	0.8141
T9	0.9484	0.9848	0.8631	0.9911	0.9428	0.5227	0.9370	0.9934	0.7900	0.9614	0.9408	0.9822	0.9228	0.9633
T10	0.9799	0.9603	0.9420	0.9441	0.9182	0.8690	0.9987	0.8857	0.7883	0.9591	0.9056	0.9982	0.9010	0.9347
T11	0.7546	0.9785	0.8816	0.8246	0.8915	0.8195	0.8946	0.9506	0.7826	0.8633	0.8189	0.9674	0.7986	0.8901
T12	0.9401	0.8879	1.0000	0.9549	0.9400	0.8015	0.9017	0.9133	0.9647	0.8873	0.7400	0.9832	0.8428	0.8451
T13	0.8093	0.7481	0.8251	0.9143	0.8671	0.8247	0.8768	0.8290	0.8881	0.8183	0.9141	0.9986	0.7860	0.8156
T14	0.7592	0.7702	0.8288	0.8631	0.8027	0.9959	0.8173	0.9098	0.9234	0.8424	0.9034	0.9807	0.8200	0.8371
T15	0.9291	0.8869	0.9005	0.9804	0.9248	0.7594	0.7860	0.9793	0.9744	0.9860	0.9207	0.8811	0.9221	0.9443
T16	0.7610	0.8349	0.8311	0.8367	0.8683	0.7239	0.8142	0.7939	0.8820	0.8385	0.8140	0.9986	0.8041	0.7791
T17	0.8553	0.9235	0.7692	0.7180	0.8169	0.4636	0.9613	0.7949	0.7722	0.7534	0.8134	0.9987	0.7359	0.7261
T18	0.8838	0.9528	0.9963	0.8886	0.8430	0.9173	0.7287	0.9158	0.9400	0.8295	0.8798	0.9103	0.7935	0.8638
T19	0.8451	0.7199	0.8402	0.8429	0.8505	0.6473	0.7812	0.8067	0.9631	0.8512	0.8656	0.9797	0.9026	0.8669
T20	0.8199	0.8803	0.9165	0.9909	0.8375	0.8269	0.8470	0.9687	0.8357	0.8767	0.9930	0.9981	0.9388	0.8504
T21	0.8477	0.8180	0.8818	0.8663	0.8397	0.6346	0.8871	0.7974	0.9957	0.8952	0.8444	0.9804	0.9526	0.9497
T22	0.8508	0.8102	0.8336	0.8076	0.8177	0.9873	0.7999	0.7056	0.9432	0.7985	0.7525	0.9790	0.7568	0.8353
T23	0.8774	0.9099	0.9156	0.8261	0.9669	0.5067	0.8911	0.8004	0.9367	0.8539	0.9466	0.9981	0.9078	0.8249
T24	0.7447	0.7520	0.7790	0.7206	0.7202	0.4320	0.6852	0.8050	0.8096	0.7102	0.6839	0.9721	0.6987	0.7423
T25	0.7013	0.6243	0.7819	0.6923	0.7341	0.7933	0.8692	0.7000	0.6903	0.7116	0.7318	0.9313	0.7368	0.7241
T26	0.6982	0.8221	0.7631	0.7632	0.7960	0.4595	0.7680	0.7344	0.8698	0.7985	0.8498	0.9786	0.8295	0.8272
T27	0.7746	0.6803	0.6063	0.7561	0.7508	0.4682	0.7396	0.7073	0.6794	0.6967	0.7660	0.9870	0.5863	0.6753
T28	0.7976	0.8425	0.8462	0.8752	0.7786	0.8764	0.7390	0.8531	0.7804	0.7800	0.8162	0.9352	0.7573	0.7433
T29	0.9682	0.9812	0.7315	0.8560	0.8741	0.5195	0.8347	0.7419	0.7836	0.8114	0.9663	0.9466	0.7861	0.9999
T30	0.9105	0.9564	0.8866	0.9909	0.9772	0.5650	0.9840	0.8488	0.9441	0.9367	0.9533	0.9641	0.8988	0.9014
T31	0.8841	0.8873	0.8087	0.8067	0.9883	0.7675	0.8917	0.8564	0.8982	0.8460	0.9695	0.9336	0.8821	0.9350
T32	0.8696	0.8620	0.9526	0.9305	0.9763	0.6772	0.8252	0.8537	0.8657	0.9072	0.7729	0.9605	0.6855	0.9045
T33	0.8990	0.8908	0.7982	0.7920	0.8249	0.4610	0.8648	0.7616	0.9079	0.8385	0.8168	0.9457	0.8460	0.8818
T34	0.9420	0.9605	0.9010	0.9346	0.9396	0.8420	0.9512	0.9426	0.9184	0.9764	0.9735	0.9820	0.9166	0.9377
T35	0.8947	0.7071	0.6908	0.8379	0.7725	0.7070	0.8492	0.8819	0.6720	0.7866	0.8140	0.9987	0.7220	0.7716
T36	0.8794	0.7733	0.9238	0.9879	0.9865	0.5251	0.9201	0.9380	0.9401	0.9655	0.9418	0.8957	0.9875	0.8682
T37	0.8999	0.9747	0.7991	0.6825	0.7311	0.5110	0.8005	0.6489	0.9484	0.7942	0.8831	0.9312	0.7537	0.8588
T38	0.8681	0.8251	0.9679	0.8054	0.8859	0.4108	0.8996	0.9166	0.8522	0.9158	0.9156	0.9598	0.9329	0.9123
T39	0.8444	0.8041	0.9575	0.8950	0.9775	0.7079	0.9460	0.9092	0.9919	0.8940	0.9918	0.9832	0.8393	0.8416
T40	0.7617	0.6720	0.7560	0.7997	0.7992	0.4086	0.9249	0.7237	0.7451	0.7857	0.9034	0.9781	0.8000	0.7846
T41	0.9103	0.9835	0.8965	0.9390	0.8411	0.3334	0.8438	0.9880	0.9115	0.9603	0.9748	0.9569	0.9660	0.9690
T42	0.8853	0.8038	0.9247	0.9338	0.8717	0.7672	0.8483	0.9547	0.8568	0.9171	0.8135	0.9763	0.9575	0.8984
T43	0.8806	0.8410	0.8414	0.8883	0.9765	0.7178	0.9674	0.9115	0.9209	0.9819	0.9650	0.9815	0.9582	0.9811
T44	0.8044	0.9038	0.9175	0.9814	0.9390	0.5605	0.9645	0.9170	0.9971	0.9425	0.9236	0.9465	0.9112	0.9885

(续表)

组合	X_1	X_2	X_3	X_4	X_5	X_6	X_7	X_8	X_9	X_{10}	X_{11}	X_{12}	X_{13}	X_{14}
T45	0.989 3	0.982 7	0.781 1	0.890 7	0.906 3	0.532 4	0.853 0	0.803 2	0.825 6	0.932 0	0.880 0	0.961 5	0.982 9	0.999 8
T46	0.785 8	0.708 4	0.716 8	0.732 7	0.827 6	0.441 7	0.771 9	0.708 3	0.876 0	0.746 5	0.898 6	0.915 1	0.796 2	0.794 1
T47	0.803 6	0.740 3	0.952 1	0.988 6	0.884 6	0.808 0	0.911 5	0.929 4	0.945 6	0.810 6	0.839 5	0.924 9	0.776 2	0.790 6
T48	0.848 3	0.825 2	0.844 8	0.824 8	0.866 2	0.978 3	0.782 5	0.901 4	0.929 5	0.920 9	0.962 5	0.998 2	0.970 9	0.929 2
T49	0.868 5	0.816 1	0.708 3	0.911 2	0.884 2	0.789 1	0.894 7	0.870 8	0.956 5	0.858 3	0.899 4	0.939 9	0.911 5	0.908 8
T50	0.766 5	0.720 7	0.957 7	0.794 9	0.929 3	0.494 2	0.976 2	0.825 0	0.826 9	0.942 0	0.852 4	0.998 5	0.869 2	0.942 1
T51	0.724 4	0.818 0	0.783 7	0.772 2	0.858 5	0.583 4	0.917 9	0.757 3	0.719 7	0.751 2	0.818 9	0.997 9	0.783 0	0.743 3
T52	0.691 2	0.713 6	0.721 9	0.655 4	0.731 9	0.605 6	0.683 2	0.667 3	0.968 8	0.726 0	0.736 3	0.863 9	0.682 4	0.746 3
T53	0.734 6	0.613 7	0.735 6	0.849 6	0.803 2	0.623 1	0.854 9	0.806 8	0.801 5	0.808 1	0.908 2	0.998 5	0.856 2	0.853 8
T54	0.795 8	0.955 5	0.905 1	0.933 3	0.945 2	0.385 4	0.972 1	0.960 5	0.843 5	0.934 0	0.839 5	0.965 3	0.859 8	0.947 7
T55	0.719 8	0.655 0	0.780 1	0.849 9	0.773 1	0.803 6	0.818 5	0.826 2	0.838 2	0.849 8	0.716 5	0.998 0	0.800 5	0.857 3
CK1	0.951 8	0.741 3	0.902 5	0.957 8	0.966 5	0.522 9	0.845 9	0.912 3	0.917 1	0.983 4	0.975 1	0.999 3	0.981 4	0.996 8
CK2	0.978 3	0.928 6	0.879 8	0.845 5	0.947 4	0.714 0	0.944 1	0.848 0	0.967 5	0.895 6	0.936 3	0.977 1	0.864 3	0.924 0

2.3 玉米不同新组合产量与农艺性状的关联度比较

玉米不同组合各农艺状与产量的灰色关联度结果列于表4。按照灰色关联度大小顺序依次为单株粒重>穗粗>籽粒含水量>百粒重>空秆率>穗行数>出籽率>穗长>倒折率>行粒数>茎粗>株高>穗位高>秃尖长。关联度大于0.90有一个性状,即单株粒重,其关联度为0.964 4,表明它与玉米产量关系最为密切,对产量的影响最大;关联度小于0.7也仅有一个性状,即秃尖长,其关联度为0.654 2,表明它与产量关系最远,对产量的影响最小。而其余性状的关联度则在0.825 9~0.876 7,表明这些性状均对产量有较大影响。因此,在西南区玉米育种中,应将单株粒重放在首位,统筹考虑其他性状,而秃尖长可适当调整。

表4 不同农艺性状与产量的关联度及排序

性状	关联度	排序	性状	关联度	排序
株高	0.840 6	12	行粒数	0.847 2	10
穗位高	0.825 9	13	百粒重	0.872 1	4
茎粗	0.846 2	11	出籽率	0.861 8	7
穗长	0.856 4	8	籽粒含水量	0.874 1	3
穗粗	0.876 7	2	单株粒重	0.964 4	1
秃尖长	0.654 2	14	倒折率	0.850 6	9
穗行数	0.866 5	6	空秆率	0.869 3	5

2.4 玉米不同新组合与理想品种的关联度比较

为了进一步分析不同组合的优劣,本研究选取产量最高的组合1703为理想品种。依据灰色理论的分析原理,关联度越大则与理想品种越相似。从表5中可以看出,不同组合与理想品种的关联度在0.795 1~0.922 0。其中,关联度大于0.9且排在前五位的组合,按照关联度从大到小依次是成单388、绵1308、川单1751、荣玉1756和渝单701，前四个组合的产量均排在前十位以内，而渝单701的关联度与产量位次差别明显，主要是由于渝单701的出籽率和空秆率与理想品种相差较大所致。而新组合川单1752虽然产量排名第一位，但关联度排名第17位，主要是由于穗行数与理想品种相差较大所致。新组合青青711的产量排第三位，关联度则排在第20位，主要是由于秃尖长与理想品种差异较大所致。关联度排在两个对照前面的新组合共有28个，与产量排序基本一致，有个别组合排序不同。与采用产量单一性状相比，关联度分析可较为全面地比较不同组合的优劣。

表5 不同组合与产量最高品种的关联度

组合	品种名称	关联度	关联度排序	产量排序	组合	品种名称	关联度	关联度排序	产量排序
T51	成单388	0.9220	1	2	CK1	桂单162	0.8607	29	28
T40	绵1308	0.9133	2	4	T31	桂A1150	0.8604	30	19
T8	川单1751	0.9099	3	7	T44	HZ1720	0.8603	31	32
T28	荣玉1756	0.9088	4	5	T41	青青733	0.8597	32	25
T15	渝单701	0.9024	5	30	T30	CS17-2	0.8547	33	34
T48	HZ1721	0.8987	6	21	T14	南T605	0.8546	34	45
T49	恩912	0.8942	7	14	T42	S183	0.8535	35	18
T53	青青722	0.8911	8	9	T13	靖2017-1	0.8525	36	49
T38	黔1702	0.8904	9	15	T5	云瑞1207	0.8501	37	27
T19	黔1705	0.8903	10	11	T45	成单658	0.8500	38	22
T22	黔1701	0.8899	11	8	CK2	渝单8号	0.8483	39	38
T34	恩932	0.8895	12	31	T11	恩1077	0.8460	40	43
T55	苏科玉1604	0.8860	13	10	T52	渝单896	0.8457	41	54
T6	渝单836	0.8794	14	12	T33	南T931	0.8448	42	46
T36	GD1676	0.8793	15	23	T16	绵TX1701	0.8375	43	47
T20	渝单841	0.8785	16	16	T54	HZ1722	0.8365	44	36
T24	川单1752	0.8771	17	1	T29	云瑞506	0.8357	45	50
T12	1701	0.8762	18	39	T2	绵732	0.8287	46	35
T26	黔1704	0.8754	19	6	T17	成1702	0.8273	47	52
T46	青青711	0.8739	20	3	T35	德3239	0.8252	48	53
T4	恩1124	0.8718	21	29	T32	川单1753	0.8249	49	20
T21	苏科1614	0.8711	22	17	T7	CS17-1	0.8222	50	37
T9	黔1703	0.8708	23	33	T37	南T928	0.8219	51	51
T23	恩1010	0.8704	24	13	T47	苏科1618	0.8183	52	48
T43	渝单703	0.8664	25	26	T1	荣玉1754	0.8115	53	41
T18	1702	0.8656	26	44	T50	德10	0.8103	54	42
T10	荣玉1755	0.8644	27	24	T27	云瑞007	0.7967	55	55
T39	靖2017-3	0.8608	28	40	T3	靖2017-2	0.7951	56	56

3 讨论与结论

玉米的产量与植株多个农艺性状密切相关。前人针对不同类型玉米已开展了较多研究，结果各有不同。陈荣丽等研究发现，与甜玉米鲜穗产量密切相关的性状依次是穗粗、穗长、行粒数，而出苗-采收对产量的影响最小[3]；李清超等研究认为，与玉米产量关联度较大的性状依次是单穗粒重、株高和穗位高，关联度最小的是穗长[7]；马全姿等发现，与糯玉米鲜穗产量密切关联的性状是穗粗、穗行数和株高，秃尖长的影响最小[10]；安治良认为穗粒重、百粒重、穗行数是影响夏玉米产量的较大因素[11]；陈灿等认为穗粗、百粒重和穗行数对普通玉米的产量影响较大[12]。不同的玉米类型，从农艺性状的关联度大小来看，表现各有差异。本研究中发现，与玉米产量关联度最大的性状是单株粒重，其次为穗粗、籽粒含水量、百粒重和空秆率，关联度最低的是秃尖长。这与韩学坤等[9]的研究有一定的一致性，均认为单株粒重与产量的关联度最大，除病害性状外，秃尖长与产量的关联度最小。

综上可知，玉米产量与哪个农艺性状密切相关，除品种本身不同之外，还与品种类型、种植环境等有很大关系。因此，在应用灰色关联度过程中，应具体问题具体分析，针对育种目标和生态区域做合理分析，指导育种实践。

参考文献

[1] 霍仕平，晏庆九，向振凡，等．我国西南地区的玉米育种实践与思考［J］．作物杂志，2017（1）：20-24.

[2] 刘录祥，孙其信，王士芸．灰色系统理论应用于作物新品种综合评估初探［J］．中国农业科学，1989，22（3）：22-27.

[3] 陈荣丽，宋文兰，周胜，等．甜玉米鲜穗产量与农艺性状灰色关联度分析［J］．种子，2018，37（9）：92-95.

[4] 杨俊伟，王建军，邵林生，等．糯玉米鲜穗产量与玉米大斑病病情指数及主要农艺性状的灰色关联度分析［J］．河北农业科学，2018，22（3）：61-64.

[5] 连晓荣，陈苍．高密植条件下玉米杂交种主要农艺性状与产量的灰色关联度分析［J］．甘肃农业科技，2018（3）：3-6.

[6] 陈静，沈生元，谢庆春，等．甜玉米鲜穗产量与主要农艺性状的灰色关联度分析［J］．江苏农业科学，2017，45（10）：48-51.

[7] 李清超，马浪浪，文琼，等．玉米杂交组合主要农艺性状与产量的灰色关联度分析［J］．中国农学通报，2015，31（30）：74-78.

[8] 邓聚龙．农业系统灰色理论与方法［M］．济南：山东科学技术出版社，1988.

[9] 韩学坤，薛国峰，王会军，等．西南地区玉米产量与农艺性状的灰色关联度分析［J］．浙江农业科学，2018，59（2）：212-214.

[10] 马全姿，莫云锦，侯定基，等．糯玉米鲜穗产量与主要农艺性状灰色关联度分析［J］．现代农业科技，2018（8）：16-21.

[11] 安治良．豫北地区夏玉米产量与农艺性状的灰色关联度分析［J］．农业科技通讯，2018（6）：184-188.

[12] 陈灿，林秀芳，陈勤平，等．普通玉米高产品种产量与主要农艺性状的灰色关联度分析［J］．湖南农业科学，2015（3）：15-17，20.

水分和施氮量对亚热带一年两作玉米灌浆及产量的影响

胡雨欣,王桂阳,梁修仁,毛祥敏,吴海燕,周勋波

(广西大学农学院,广西 南宁 530004)

摘要:探究水分和氮肥对玉米(Zea mays L.)灌浆规律及产量形成的影响,为提高亚热带一年两作区域玉米周年产量提供理论依据。试验采取裂区设计,以灌溉(滴灌保持土壤含水量不低于田间最大持水量的60%)和雨养(生育期无灌溉)为主因素,施纯氮 0 kg/hm² (N0)、150 kg/hm² (N1)、200 kg/hm² (N2)、250 kg/hm² (N3)、300 kg/hm² (N4)为副因素,在玉米开花期后对干物质的积累进行测定;成熟期时进行测产与考种。试验结果表明,与雨养和其他氮素水平(N0,N1,N2和N4)相比,灌溉和N3处理提高了花后百粒重、最大灌浆速率所需要的时间(T_{max})、灌浆速率最大时的生长量(W_{max})和灌浆活跃期(P)。与雨养相比,灌溉处理增产15.7%,穗重增重11.0%、千粒重提高1.8%、收获指数提高17.6%;与N0相比,N3显著增产68.2%,穗重增重57.1%、千粒重提高30.2%、收获指数提高8.8%;但施氮量超过250 kg/hm²时,收获指数不增反降。施氮量250 kg/hm²并在阶段性干旱条件下结合适量灌溉,可有效提高亚热带一年两作种植区域玉米周年产量。

关键词:水分;氮肥;玉米;籽粒灌浆;产量

Effects of water conditions and nitrogen amounts on grain filling characteristics and yield of double-cropped maize in subtropical region

Hu Yuxin, Wang Guiyang, Liang Xiuren, Zhou Xunbo, Wu Haiyan

(Agricultural College of Guangxi University, Nanning 530004, Guangxi, China)

Abstract: The effects of water and nitrogen fertilizer on grain filling and yield formation of maize (Zea mays L.) were studied in order to provide theoretical basis for increasing annual maize yield of double cropping system in subtropical region. The experiment was conducted in the farm of Guangxi University from march to December, 2018, and adopted the split-plot design. Main plot was water conditions: drip irrigation (maintain soil moisture content as the maximum water holding

---

基金项目:国家自然科学基金(31760354);广西自然科学基金(2017GXNSFAA198036)。

作者简介:胡雨欣(1997—),研究生,主要从事作物栽培研究。Tel:(0771)3235612;E-mail:18191599483@163.com。

通信作者:周勋波,教授,主要从事农田生态研究。Tel:(0771)3235612;E-mail:xunbozhou@163.com。

in the field 60%) and rainfed (there was no irrigation during the maize growing season); split plot was nitrogen levels: 0 kg/ha (N0), 150 kg/ha (N1), 200 kg/ha (N2), 250 kg/ha (N3) and 300 kg/ha (N4). The accumulation of dry matter was measured after flowering stage of maize, and the yield and seed test of maize were carried out at maturity stage. The results showed that compared with rainfed and other nitrogen levels (N0, N1, N2 and N4), irrigation and N3 increased 100-grain weight after anthesis, the time when the grain filling rate was maximum (T_{max}), the growth weight when the grain filling rate was maximum (W_{max}) and the active grain filling period (P). Compared with rainfed, grain yield, ear weight, thousand gain weight and harvest index of irrigation condition was increased by 15.7%, 11.0%, 1.8% and 17.6%, respectively. Nitrogen application is beneficial to increase harvest index, whereas, when N>250 kg/ha (N3 and N4), harvest index increased; Compared with N0, grain yield, ear weight, thousand gain weight and harvest index of N3 was increased by 68.2%, 57.1%, 30.2% and 8.8%, respectively. The annual maize yield of one-year two-cropping maize in subtropical region could be effectively increased by nitrogen application 250 kg/ha combined with proper irrigation under the phased drought condition.

Keywords: Water; Nitrogen; Grain filling; Yield

水分是影响作物生长发育的重要因子，并且已经成为很多地区农业发展限制的因素。广西年平均降水量为1 086~2 755 mm，大部分地区降水量在1 300~2 000 mm[1]，降水量充沛，基本上能满足玉米生育期的水分需求。但因降水日期和降水量的不确定性及地形地貌的特殊性所导致的水流失严重，使玉米大喇叭口期、灌浆期等时期经常发生阶段性水分胁迫，而抽雄-灌浆期是玉米需水和耗水的关键时期，若该时期受到水分胁迫，籽粒灌浆受阻，导致产量下降[2,3]。在玉米的关键生育阶段合理灌溉能显著提高产量[4]。在作物生育期内，若降水量不足以支持其生长及养分吸收，科学灌溉就显得日益重要[5]。玉米产量高低与籽粒发育密切相关，而在玉米灌浆期氮素对籽粒发育起关键作用[6,7]。氮肥投入量低于经济最佳施氮量或最高产量施氮量，则导致产量较低；过高则易引起作物倒伏或病虫害增加，导致产量不再增加或是有所下降[8,9]，同时在土壤中残留或损失到大气和水体中的氮肥会显著增加，污染环境[10]。所以在生产上通过合理的灌溉与施氮结合的农业措施来控制氮肥使用量，降低氮肥损失量，缓解粮食生产压力。

灌溉和施肥对玉米农艺性状、产量以及氮肥利用率均有影响，如何科学地灌溉和施肥尤为重要。氮肥的增产效应与灌水量相关，合理的水肥协同优化组合可以提高水分、养分利用率，是提高产量的关键[11,12]。水肥一体化能够有效改善春玉米灌浆特性，延长灌浆时间并且提高灌浆速率[13]，灌水量和施肥量对玉米株高、茎粗、叶面积指数都有显著影响[14]。刘志恒等研究发现，长期施用氮肥量为180 kg/hm²，玉米能获得较高产量和氮肥利用率，保持土壤肥力，减少污染[15]。甄城等研究发现，施纯氮在0~270 kg/hm²，拔节期淹水条件下施氮量增加时，春玉米大喇叭口至乳熟期叶面积指数、株高、穗长、穗行数、行粒数、千粒重和籽粒产量均增加，施氮量进一步增加时上述指标增加不明显[16]。高天平等研究发现，充足灌溉能够显著提高30 cm土层有机碳含量和土壤含水量，有利于玉米产量的提高[17]。王炎等研究发现，中浓度氮肥处理促进苦荞生长发育、灌浆进程，增加籽粒充实度及产量的提高[18]。郑斯尹等研究表明，玉米不同生长阶段、不同施肥处理，以及二者交互作用显著影响土壤微生物量碳、氮含量[19]。

国内外在水肥对玉米品种的籽粒灌浆及氮素利用效率等方面的影响进行了大量研究,但对亚热带一年两季玉米种植区,在雨养和灌溉条件下施氮对玉米籽粒灌浆影响的研究未见报道。本研究将雨养与灌溉结合,研究不同施氮量下玉米的灌浆规律,明确不同水氮条件下玉米籽粒灌浆规律及产量与水氮间的关系,为亚热带玉米一年两作种植区水肥管理提供必要的技术支撑和理论依据。

1 材料与方法

1.1 试验材料

供试玉米品种为万川1306是由广西万川种业有限公司利用自选自交系GY21B作父本、自选自交系GY2121作母本,杂交选育而成的玉米单交种,其春播生育期为113 d,秋播为102 d,需有效积温2 300℃左右。

1.2 试验地概况

本试验于2018年3—12月在广西大学农场(22°50′N,108°17′E)进行。该地区为湿润的亚热带季风气候,年平均气温21.6℃,年均降水量1 304.2 mm,2018年试验地降水情况如图1(A)所示。试验地土壤类型为黏土,0~0.2 m土层的有机质17.5 g/kg,碱解氮126.2 mg/kg,速效磷40.0 mg/kg,速效钾124.5 mg/kg;pH值为5.4,容重为1.50 g/cm^3,田间持水量为37.2%(V%)。

1.3 试验方法

将供试玉米按0.6 m等行距;0.28 m等株距;2~3株/穴,使用播种机点播,种植密度为52 500株/hm^2。试验小区面积为4 m×4.2 m。试验采用裂区设计,设计水分和氮肥两个因素。水分设置两个水平:分别是滴灌(滴灌保持土壤含水量不低于田间最大持水量的60%;采用软管滴灌)和雨养(生育期无灌溉,作物生长发育水分仅来源于自然降雨)。氮肥设置5个水平:纯氮0 kg/hm^2(N0)、150 kg/hm^2(N1)、200 kg/hm^2(N2)、250 kg/hm^2(N3)、300 kg/hm^2(N4);氮肥为尿素,播前基肥占总施氮量的2/3,大喇叭口期追施剩余1/3。P_2O_5和K_2O施用量均为100 kg/hm^2(均作为基肥一次性施入)。灌溉处理的具体灌溉时间和灌溉量见图1(B),共2×5=10个处理,每个处理3次重复。

两季玉米分别于2018年3月22日和8月11日播种;4月15日与8月26日间苗;7月12日和12月16日收获,全生育期分别为112 d和127 d。

1.4 测定项目及方法

植株花后干物质积累测定:从玉米开花期开始,各处理取有代表性的2株玉米的果穗,每7 d取样1次,直到成熟,重复3次。用镊子取下果穗中部籽粒,以100粒为单位称量鲜重,在105℃烘箱杀青30 min,然后75℃烘干至恒重,称重。并参考徐田军等的方法,用Richards模型$W=\dfrac{A}{[HexpCB-Ct]^{\frac{1}{D}}}$模拟灌浆过程,式中W为籽粒干重(g);A为最终粒重

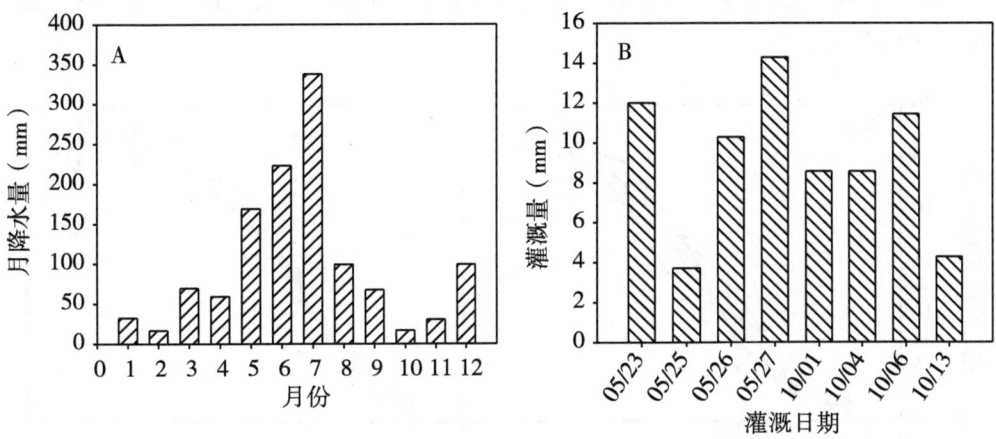

图 1　2018 年玉米生育期的月降水量（A）和灌溉量（B）

（g）；t 为授粉后天数（d）；B 为初值参数；C 为生长速率参数；D 为形状参数[20]。当 D=1 时，即为 Logistic 方程。R0（起始生长势）= C/D；达最大灌浆速率时的天数 T_{max} =（lnB-lnD）/C；灌浆速率最大时的生长量 W_{max}（g/100 粒）= A（D+1）$^{-1/D}$；灌浆活跃期（约完成总积累量的 90%）P = 2（D+2）/C。

测产与考种：在成熟期随机选取长势均匀的玉米 10 株，测定穗重和千粒重等，每个小区选取 2 m² 进行实收测产，自然风干至恒重后称重。

1.5　统计分析

试验数据利用 SPSS Statistics 21.0 进行统计分析（LSD 法），用 CurveExpert 1.4 拟合曲线，用 SigmaPlot 10.0 软件作图。

2　结果与分析

2.1　花后籽粒干物质积累

由图 2 可见，随着玉米的生长，籽粒干重逐渐增加，在花后 7 d 各处理间差异较小，随后增大。从花后 14 d 到 35 d 是籽粒干重快速增长的时期，灌浆速率快，在花后 42 d 灌浆基本完成。此外 N0 处理的干物质积累速率及积累量均显著低于其他氮素水平。春玉米，雨养和灌溉处理的百粒重均值分别为 14.6 g/100 粒和 16.3 g/100 粒，灌溉的百粒重较雨养提高了 11.9%（$P<0.05$）；秋玉米，雨养和灌溉处理的百粒重均值分别为 11.5 g/100 粒和 13.2 g/100 粒，灌溉的百粒重较雨养提高了 14.4%（$P<0.05$）。春玉米的百粒重较秋玉米提高了 26.4%（雨养）和 23.6%（灌溉），差异显著（$P<0.05$）。雨养和灌溉处理的百粒重均值分别为 13.1 g/100 粒和 14.8 g/100 粒，灌溉的百粒重较雨养提高了 13.0%（$P<0.05$）。

春玉米，N0、N1、N2、N3 和 N4 的百粒重均值分别为 13.2 g/100 粒、15.7 g/100 粒、16.2 g/100 粒、16.4 g/100 粒和 15.8 g/100 粒，即 N3>N2>N4>N1>N0。秋玉米，N0、N1、N2、N3 和 N4 的百粒重均值分别为 9.3 g/100 粒、12.2 g/100 粒、12.7 g/100 粒、12.8 g/100 粒和 14.8 g/100 粒，即玉米籽粒干物质随施氮量的增加而增加。春玉米氮素处理（N0、

N1、N2、N3 和 N4）的百粒重较秋玉米分别提高了 41.4%、28.1%、27.7%、28.3% 和 6.8%（$P<0.05$）。

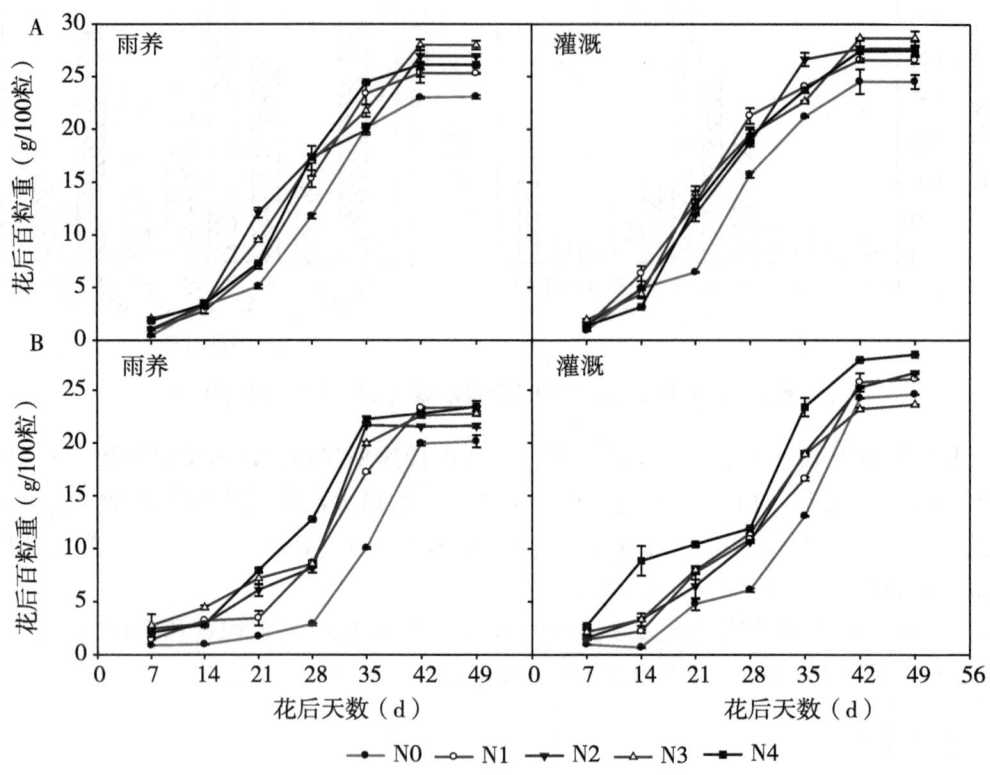

图 2　不同水分条件和施氮量对玉米花后百粒重的影响
（A：春玉米；B：秋玉米）

2.2　玉米籽粒灌浆特性

玉米籽千粒重 W 与授粉后天数 t 的变化用 Richards 方程进行模拟建立籽粒灌浆的动态方程（表1）。不同水分条件和施氮量处理 R^2 都在 0.99 以上，均达到极显著水平。雨养和灌溉的起始生长势（R0）均值分别为 0.099 6 和 0.096 9；N0、N1、N2、N3 和 N4 的起始生长势（R0）呈 N0>N1>N2>N4>N3，雨养和低氮素处理的起始生长势（R0）较高，表明雨养和低氮素处理的玉米进入灌浆盛期早。与雨养相比，灌溉的最大灌浆速率所需要的时间（T_{max}）提高了 1.1%（$P>0.05$）；N0、N1、N2、N3 和 N4 达到 T_{max} 呈 N0>N2>N3>N4>N1（雨养），N0>N3>N4>N2>N1（灌溉）；N0 的 T_{max} 显著高于其他氮素水平（$P<0.05$）。

雨养和灌溉条件下灌浆速率最大时的生长量（W_{max}）均值分别为 11.7 g/100 粒和 12.3 g/100 粒，灌溉的 W_{max} 较雨养显著提高了 4.5%（$P<0.05$）；N0、N1、N2、N3 和 N4 的 W_{max} 均值分别为 9.5 g/100 粒、12.6 g/100 粒、12.3 g/100 粒、13.0 g/100 粒和 12.6 g/100 粒，N3 的 W_{max} 显著高于其他氮素水平（$P<0.05$）；较 N0、N1、N2 和 N4 水平，N3 的 W_{max} 分别提高了 36.8%、3.2%、5.7% 和 3.2%，差异显著（$P<0.05$）。

雨养和灌溉条件下灌浆活跃期（P）均值分别为 29.3 d 和 31.4 d，灌溉的灌浆活跃期较雨养显著提高了 7.2%（$P<0.05$）；N0、N1、N2 和 N4 的灌浆活跃期均值分别为 27.7 d、

30.8 d、29.3 d、34.0 d 和 30.1 d，N3 的灌浆活跃期显著高于其他氮素水平，较其他氮素水平灌浆活跃期分别提高了 22.6%（N0）、10.2%（N1）、15.9%（N2）和 12.8%（N4）。综合表明，灌溉和 N3 处理有利于玉米灌浆。

表 1　不同水分条件和施氮量对玉米灌浆参数的影响

水分	施氮量	Richards 方程	R^2	R0	T_{max} (d)	W_{max} (g/100粒)	P (d)
雨养	N0	$W=17.38/(1+e^{(16.00-0.46t)})^{(1/4.05)}$	0.999 4	0.114 8	23.8	8.7	26.0
	N1	$W=25.01/(1+e^{(11.10-0.32t)})^{(1/2.90)}$	0.999 6	0.111 2	15.6	12.5	30.4
	N2	$W=24.24/(1+e^{(39.31-1.12t)})^{(1/12.88)}$	0.998 1	0.086 7	19.8	12.1	26.6
	N3	$W=26.13/(1+e^{(12.58-0.34t)})^{(1/4.32)}$	0.998 8	0.079 9	17.8	13.1	36.6
	N4	$W=24.51/(1+e^{(16.26-0.50t)})^{(1/4.76)}$	0.999 7	0.105 5	17.0	12.3	26.9
灌溉	N0	$W=20.68/(1+e^{(8.85-0.29t)})^{(1/2.22)}$	0.999 1	0.129 5	21.0	10.3	29.4
	N1	$W=25.50/(1+e^{(9.62-0.31t)})^{(1/2.82)}$	0.999 6	0.109 5	15.9	12.8	31.2
	N2	$W=24.75/(1+e^{(12.22-0.37t)})^{(1/3.97)}$	0.998 0	0.093 9	16.5	12.4	32.0
	N3	$W=25.98/(1+e^{(22.19-0.65t)})^{(1/8.22)}$	0.995 4	0.079 5	21.4	13.0	31.3
	N4	$W=25.69/(1+e^{(24.35-0.71t)})^{(1/9.84)}$	0.993 6	0.072 3	20.2	12.8	33.3

2.3　产量构成因素

两季试验结果表明，与雨养相比，灌溉处理增产 15.7%，穗重增重 11.0%、千粒重提高 1.8%、收获指数提高 17.6%（$P<0.05$），表明灌水利于产量提高（表 2 和表 3）。随施氮量的增加，穗重、千粒重、产量均表现为高氮素水平处理优于低氮素水平；N0、N1、N2、N3 和 N4 收获指数的均值分别为 0.34、0.39、0.39、0.37 和 0.36，即施氮有利于提高收获指数，但施氮量超过 250 kg/hm² 时，收获指数不增反降；与 N0 相比，N3 显著增产 68.2%，穗重增重 57.1%、千粒重提高 30.2%、收获指数提高 8.8%（$P<0.05$），表明施氮对玉米产量及产量构成有明显的影响。灌溉结合 N3 利于穗重、千粒重和产量提高，最终形成较高的收获指数。

表 2　不同水分条件和施氮量对穗重、千粒重、产量和收获指数的影响

处理	穗（g）			千粒重（g）			产量（kg/hm²）			收获指数 HI		
	春季	秋季	平均	春季	秋季	平均	春季	秋季	平均	春季	秋季	平均
雨养	118.9b	124.0a	121.4b	282b	270b	276b	5 165b	5 687a	5 426b	0.25b	0.34b	0.34b
灌溉	151.3a	118.2b	134.7a	289a	274a	281a	7 015a	5 555a	6 280a	0.32a	0.40a	0.40a
施氮												
N0	101.6e	80.2e	90.9e	235c	215d	225d	4 570e	3 448e	4 025e	0.24d	0.45c	0.34b
N1	132.4d	108.2d	120.3d	292b	276c	284c	5 778d	4 654d	5 216d	0.29c	0.48a	0.39a
N2	140.6c	131.0c	135.8c	300a	286b	293b	6 362c	6 195c	6 260c	0.31a	0.47b	0.39a
N3	146.9b	138.8b	142.8b	302a	289b	293ab	6 742b	6 798b	6 770b	0.30ab	0.44d	0.37b
N4	154.0a	147.1a	150.6a	298a	297a	293ab	7 014a	7 011a	7 005a	0.29bc	0.42e	0.36c

注：同一列不同小写字母表示差异显著（$P<0.05$）。

表 3　水分条件和施氮量对穗重、千粒重、产量和收获指数影响的显著性分析

处理	穗重（g）			千粒重（g）			产量（kg/hm²）			收获指数 HI		
	春季	秋季	平均	春季	秋季	平均	春季	秋季	平均	春季	秋季	平均
水分	0.000 1	0.001 6	0.000 2	0.013 2	0.075 0	0.013 0	0.000 3	0.015 1	0.000 6	0.002 0	0.000 5	0.001 1
施氮	0.000 1	0.000 1	0.000 1	0.000 1	0.000 1	0.000 1	0.000 1	0.000 1	0.000 1	0.000 1	0.000 1	0.000 1
水分×施氮	0.000 1	0.000 1	0.000 1	0.001 3	0.000 1	0.000 1	0.000 1	0.000 1	0.000 1	0.006 7	0.000 1	0.000 1

2.4　花后天数和籽粒干物质质量的回归分析

籽粒干重和花后天数的回归分析表明，随花后天数的增加籽粒干重增加；在相同时间下，N0 和 N1 的籽粒干物质积累较少。花后天数与籽粒干重之间符合一元一次的回归关系。春玉米，雨养条件下的回归方程是 y（g/100 粒）$= 0.752\ 3x$（花后天数）-5.520，$R^2 = 0.944\ 9$（$P<0.000\ 1$），灌溉条件下的回归方程是 $y = 0.756\ 5x-3.760$，$R^2 = 0.965\ 8$（$P<0.000\ 1$）；秋玉米，雨养条件下的回归方程是 $y = 0.572\ 5x-4.044$，$R^2 = 0.846\ 2$（$P<0.000\ 1$），灌溉条件下的回归方程是 $y = 0.668\ 7x-5.033$，$R^2 = 0.896\ 3$（$P<0.000\ 1$），试验结果表明，籽粒干重与花后天数呈极显著正相关（$P<0.01$）（表 4）。

表 4　不同水分条件和施氮量的授粉时间和籽粒干物质质量的回归分析

处理	水分	回归方程	方程参数	
			R^2	P
春玉米	雨养	$y = 0.752\ 3x-5.520$	0.944 9	0.000 1
	灌溉	$y = 0.756\ 5x-3.760$	0.965 8	0.000 1
秋玉米	雨养	$y = 0.572\ 5x-4.044$	0.846 2	0.000 1
	灌溉	$y = 0.668\ 7x-5.033$	0.896 3	0.000 1

2.5　灌浆参数与施氮量及产量的相关分析

如表 5 所示，施氮量与 R0 呈极显著负相关（$P<0.01$），与 W_{max} 和产量呈极显著正相关（$P<0.01$），与 T_{max} 无显著负相关（$P>0.05$），与 P 无显著正相关（$P>0.05$）。R0 与 T_{max} 和与 P 无显著负相关（$P>0.05$），与 W_{max} 呈显著负相关（$P<0.05$），与产量呈极显著负相关（$P<0.01$）。T_{max} 与 W_{max} 呈显著负相关（$P<0.05$），与 P 和产量无显著负相关（$P>0.05$）。W_{max} 与 P 无显著正相关（$P>0.05$），与产量呈极显著正相关（$P<0.01$）。P 与产量无显著正相关（$P>0.05$）。

表 5　灌浆参数与施氮量及产量之间的相关分析

参数	施氮	R0	T_{max}	W_{max}	P	产量
N	1.000	-0.798**	-0.401	0.838**	0.413	0.901**
R0		1.000	-0.009	-0.649*	-0.524	-0.790**

(续表)

参数	施氮	R0	T_{max}	W_{max}	P	产量
T_{max}			1.000	-0.660*	-0.338	-0.250
W_{max}				1.000	0.626	0.770**
P					1.000	0.427
产量						1.000

注：* 相关系数显著水平为 $P<0.05$；** 相关系数显著水平为 $P<0.01$。

* r values presented at $P<0.05$；** r values presented at $P<0.01$。

P 表示灌浆活跃天数。

3 讨论与结论

授粉后 10 d 内主要是玉米种皮干重增长时期，籽粒干重增长缓慢，干物质积累最大速率出现在授粉后 12~28 d，之后积累速率下降[21]。本试验结果表明，在花后 15~21 d 是籽粒干重快速增长时期，与前人研究基本一致，但快速增长期略有缩短，这可能与南北方气候差异有关。两季玉米结果表明，氮素水平为 0~250 kg/hm²，百粒重随氮素水平增加而呈上升趋势，说明氮素处理促进玉米干物质累积，但当施氮量超过一定值（250 kg/hm²）时，干物质积累将受到抑制，与一些学者的相关报道不一致[22,23]，可能是由于供试材料、土壤肥力、种植地区以及气候等差异所致，进一步表明不同区域、不同玉米品种的最适宜施氮量略有不同。本试验中与雨养相比，灌溉处理有效提高了玉米干物质积累，与杨明达研究认为滴灌能够增加吐丝后干物质积累量、氮素积累量及其对籽粒氮素的贡献率，最终增加产量的结果一致[24]。说明灌溉与施氮量的耦合效应可以增加玉米干物质积累，从而提高产量。

灌浆特性是影响玉米产量的主要因素，其受温、光、水、气和养分等因素影响较大[25]；灌浆持续期的长短和灌浆强度的高低决定籽粒库的充实程度和籽粒干物质积累量[26]，灌浆速率快、持续时间长的品种，籽粒饱满、千粒重高[27]。玉米籽粒灌浆期间灌浆速率呈"慢-快-慢"的"S"形单峰变化[28]；曹彩云等认为，随化肥用量增加，籽粒最大灌浆速率持续时间和灌浆进程延长，千粒重提高。本试验两季玉米均显示随施氮量增加籽粒干重呈上升趋势，与其研究一致[29]。本研究表明，灌溉条件下玉米的 T_{max}、W_{max}、P 和籽粒干重均显著高于雨养（$P<0.05$）。N3 的 W_{max} 和 P 较 N0 分别提高了 36.9% 和 22.6%，表明灌溉和 N3 利于玉米灌浆，主要原因在于灌溉和氮处理提高了玉米籽粒灌浆速率、延长了籽粒最大灌浆速率的持续时间。

氮是促进作物生长，提高作物产量的重要因素之一[30]。施氮能显著影响玉米千粒重和产量[31]，随施氮量增加单株粒重、穗粒数和 HI 显著增加[32]。不同时期不同程度的干旱胁迫均可能对玉米生长造成不同程度的影响[33]，比如穗期干旱会显著降低地上部干物质积累量和穗粒数，粒期干旱会明显降低粒重[34]。何海军等研究指出，在土壤水分缺乏的条件下，玉米的穗长、穗粒数、行粒数、行数、千粒重会显著降低，最终造成玉米减产[35]。合理的水氮互作可使玉米优质高产，产量随水氮的增加而增加，但增加到一定程度后趋势平缓，产量增加缓慢[12]。本研究表明，灌溉条件下穗重、千粒重、产量和收获指数均显著高于雨养；穗重、千粒重、产量均表现为高氮处理优于低氮处理；表明在一定范围内水氮耦合效应使玉米植株生长健壮，绿叶期延长，有效光合时间增加，保证了充足的"源"，为玉米籽粒灌浆提供了重要的物质基础，最终提高了产量。

本研究基于亚热带一年两作地区的水分条件和氮肥施用量对玉米产量、干物质累积和灌浆特性的影响，建立了不同水分条件下花后天数与籽粒干重的一元一次回归方程，对亚热带一年两作区玉米水肥管理提一定的供理论依据。但本研究因在雨养条件下，没有设置灌溉梯度，不同灌溉量和施氮量的水氮耦合效应对玉米产量、干物质累积以及灌浆特性的影响机制需要做进一步的深入研究。

阶段性干旱时适量灌溉可以促进玉米干物质累积、提升籽粒灌浆速率以及提高产量；不同氮肥水平处理下，施氮量为 250 kg/hm² 可以加快玉米干物质累积、延长籽粒灌浆时间且提高产量。

参考文献

[1] 覃卫坚，李栋梁. 近 50 年来广西各级降水气候变化特征分析 [J]. 自然资源学报，2014，29（4）：666-676.

[2] 何洁琳，谢敏，黄卓，等. 广西气候变化事实 [J]. 气象研究与应用，2016，37（3）：11-15.

[3] Liu Y, Yang H S, Li J S, et al. Estimation of irrigation requirements for drip-irrigated maize in a sub-humid climate [J]. Journal of Integrative Agriculture, 2018, 17（3）：677-692.

[4] 董平国，王增丽，温广贵，等. 不同灌溉制度对制种玉米产量和阶段耗水量的影响 [J]. 排灌机械工程学报，2014，32（9）：822-828.

[5] Trout T J, Bausch W C, Buchleiter G W. Water production functions for Central Plains crops [C]. ASABE-5th National Decennial Irrigation Conference Proceedings, 2010, Phoenix, A2. 5-8Dec.

[6] 吕新，胡昌浩，董树亭，等. 紧凑型玉米掖单 22 与 SC704 籽粒灌浆特性对比分析研究 [J]. 山东农业大学学报（自然科学版），2005，36（1）：70-74.

[7] 刘明，齐华，张卫建，等. 深松与施氮方式对春玉米子粒灌浆及产量和品质的影响 [J]. 玉米科学，2013，21（3）：115-119.

[8] 巨晓棠，谷保静. 我国农田氮肥施用现状、问题及趋势 [J]. 植物营养与肥料学报，2014，20（4）：783-795.

[9] 吕静瑶，申丽霞，晁晓乐. 不同氮效率玉米品种籽粒灌浆特性研究 [J]. 河南农业科学，2017，46（1）：7-12.

[10] Ju X T, Xing G X, Chen X P, et al. Reducing environmental risk by improving N management in intensive Chinese agricultural systems [J]. Proceedings of the National Academy of Sciences of the United States of America, 2009, 106（9）：3041-3046.

[11] 谭华，郑德波，邹成林，等. 水肥一体膜下滴灌对玉米产量与氮素利用的影响 [J]. 干旱地区农业研究，2015，33（3）：18-23.

[12] 尚文彬，张忠学，郑恩楠，等. 水氮耦合对膜下滴灌玉米产量和水氮利用的影响 [J]. 灌溉排水学报，2019，38（1）：49-55.

[13] 李晓龙，白云龙，闫东，等. 不同水肥管理方式对春玉米籽粒灌浆特性、氮素

利用及产量的影响[J].华北农学报,2017,32(3):182-187.

[14] 张富仓,严富来,范兴科,等.滴灌施肥水平对宁夏春玉米产量和水肥利用效率的影响[J].农业工程学报,2018,34(22):111-120.

[15] 刘志恒,徐开未,王科,等.不同施氮量对玉米产量及各器官养分积累的影响[J].浙江大学学报(农业与生命科学版),2018,44(5):573-579.

[16] 甄城,漆栋良,徐茵,等.拔节期淹水与施氮量互作对春玉米生长和产量的影响[J].灌溉排水学报,2019,38(S1):1-5.

[17] 高天平,张春,刘文涛,等.秸秆还田方式与灌溉量对土壤碳水环境和玉米产量的影响[J].山东农业科学,2019,51(6):108-112.

[18] 王炎,李振宙,黄凯丰.不同氮肥用量下苦荞灌浆特性及根系形态与充实度变化[J].热带作物学报,2019,40(6):1062-1067.

[19] 郑斯尹,陈莉莎,谢德晋.不同氮肥用量对玉米田土壤酶活性及微生物量碳、氮的影响[J].中国水土保持,2019(7):58-60+73.

[20] 徐田军,吕天放,赵久然,等.玉米籽粒灌浆特性对播期的响应[J].应用生态学报,2016,27(8):2513-2519.

[21] 高荣岐,董树亭,胡昌浩,等.夏玉米籽粒发育过程中淀粉积累与粒重的关系[J].山东农业大学学报,1993,24(1):42-48.

[22] 张家铜,彭正萍,李婷,等.不同供氮水平对玉米体内干物质和氮动态积累与分配的影响[J].河北农业大学学报,2009,32(2):1-5.

[23] 高遂,李春喜,周宝元,等.种植密度和施氮量耦合对夏玉米干物质积累及氮肥利用率的影响[J].玉米科学,2017,25(5):105-111.

[24] 杨明达,关小康,刘影,等.滴灌模式和水分调控对夏玉米干物质和氮素积累与分配及水分利用的影响[J].作物学报,2019,45(3):443-459.

[25] Valentinuz O R, Tollenaar M. Vertical profile of leaf senescence during the grain-filling period in older and newer maize hybrids[J]. Crop Science, 2004, 44(3): 827-834.

[26] 孙东升,刘成启.7个玉米自交系主要数量性状的配合力及遗传参数分析[J].江苏农业科学,2008(2):40-42.

[27] 刘宗华,张战辉.玉米籽粒灌浆速率研究进展[J].东北农业大学学报,2010,41(11):148-153.

[28] 徐田军,吕天放,赵久然,等.玉米生产上3个主推品种光合特性、干物质积累转运及灌浆特性[J].作物学报,2018,44(3):414-422.

[29] 曹彩云,李科江,马俊永,等.化肥施用水平对夏玉米籽粒灌浆进程的影响[J].河北农业科学,2007,11(1):57-59.

[30] Mokhele B, Zhan X J, Yang G Z, et al. Review: Nitrogen assimilation in crop plants and its affecting factors[J]. Canadian Journal of Plant Science, 2012, 92: 399-405.

[31] Khan A, Jan M T, Marwat K B, et al. Organic and inorganic nitrogen treatments effects on plant and yield attributes of maize in a different tillage systems[J]. Pakistan Journal of Botany, 2009, 41(1): 99-108.

[32] 魏淑丽, 王志刚, 于晓芳, 等. 施氮量和密度互作对玉米产量和氮肥利用效率的影响 [J]. 植物营养与肥料学报, 2019, 25 (3): 382-391.

[33] 李耕, 高辉远, 赵斌, 等. 灌浆期干旱胁迫对玉米叶片光系统活性的影响 [J]. 作物学报, 2009, 35 (10): 1916-1922.

[34] 宁芳, 张元红, 温鹏飞, 等. 不同降水状况下旱地玉米生长与产量对施氮量的响应 [J]. 作物学报, 2019, 45 (5): 777-791.

[35] 何海军, 寇思荣, 王晓娟. 干旱胁迫对不同株型玉米光合特性及产量性状的影响 [J]. 干旱地区农业研究, 2011, 29 (3): 63-66.

下 篇

下篇

Straw return and nitrogen fertilization regulate soil greenhouse gas emissions and global warming potential in dual maize cropping system

L. Yang[a], I. Muhammad[a], Y. X. Chi[a,b], Y. X. Liu[a],
G. Y. Wang[a], Y. Wang[a], X. B. Zhou[a]

([a] Guangxi Key Laboratory of Agro-environment and Agro-products Safety, Guangxi Colleges and Universities Key Laboratory of Crop Cultivation and Tillage, Agricultural College, Guangxi University, Nanning 530004, China; [b] Heilongjiang Bayi Agricultural University/Key Laboratory of Crop Germplasm Improvement and Cultivation in Cold Regions of Education Department, Daqing, China)

Abstract: Abundant nitrogen (N) fertilization is needed for maize (*Zea mays* L.) production in China because of its huge residual biomass return. However, excessive N fertilization has a negative impact on the soil ecosystem and environment, which contributes to climate change. Soil incorporation of maize residues is a well-known practice for reducing chemical N fertilization without compromising maize yield and soil fertility. Thus, residues incorporation has the capacity to minimize N fertilization uses and hence mitigate soil greenhouse gas emissions by improving plant N uptake and use efficiency. There is still a research gap regarding the effects of maize residues incorporation on maize yield, soil fertility, greenhouse gas emissions, and plant N and carbon (C) contents. Therefore, we conducted a field experiment during spring and autumn involving four different N fertilization rates (N0, N200, N250, and N300 kg N/ha), with and without maize residues incorporation, to evaluate grain yield, soil fertility, plant N and C contents, and greenhouse gas emissions (GHGs). Compared to N0, N fertilizer application at 300 kg N/ha with residues incorporation significantly increased area-scaled global warming potential (GWP) compared to other N fertilization rates in both spring and autumn seasons, but soil nutrient contents and plant N and C contents were not statistically different from the N250 treatment. In contrast, the N recovery use efficiency (NRUE), physiological N use efficiency (PNUE), and agronomic N use efficiency (ANUE) were significantly lower in the N300 treatment than in the lower N treatment groups. Nitrous oxide (N_2O) and carbon dioxide (CO_2) fluxes, area-scaled GWP, and greenhouse gas intensity (GHGI) were significantly lower in the N200 treatment with straw incorporation than the N250 and N300 treatments of the traditional planting system. Thus, we concluded that N200 treatment with residues incorporation is optimal for improving grain yield, soil fertility, plant

---

These authors contributed equally.
Corresponding author E-mail address: xunbozhou@gmail.com.

N uptake, and mitigating greenhouse gas emissions.

Keywords: Greenhouse gas emissions; Nitrogen fertilizer; Nitrogen use efficiency; Residues incorporation; Soil fertility

1 Introduction

Maize (*Zea mays* L.) is the most important crop in China, yielding over 200 million metric tons per year and accounting for 42.2% of cereal crop yields (Afreh et al., 2018; Li et al., 2021; Sorkhi and Fateh, 2014). Chemical fertilizers are the only way to sustain high crop production under an intensive cropping systems (Roozbeh and Rajaie, 2021). It has been estimated that more than half of the applied N is lost from agricultural land through denitrification, volatilization, leaching, and soil erosion from agricultural land (Abbasi et al., 2013; Afreh et al., 2018. Similarly, N fertilization may promote immobilization in various soil N pools, thus increasing soil GHGs and somewhat negating the benefits of improved soil quality (Luxhøi et al., 2007; Zhou et al., 2017). To improve soil fertility and reduce N losses, it is commonly understood that using chemical fertilizers along with straw is an effective strategy to increase N use efficiency (NUE) and soil fertility on a sustainable basis (Pan et al., 2017; Yang et al., 2022). To evaluate and implement the combined application of chemical N fertilization and straw incorporation, it is necessary to understand how N can be immobilized in soil N pools and GHGs emissions. Currently, increased global warming is causing concern owing to the increasing concentrations of GHGs in the atmosphere. Additionally, it threatens global food production and food supply security as it impacts agricultural productivity (Foley et al., 2011; Liu et al., 2022).

Current agricultural output needs to decrease GHGs emissions and increase NUE by implementing the proper fertilization strategies (Rochette et al., 2013; Wu et al., 2021). Numerous studies have shown that straw incorporation increases the available C and N content and stimulates soil microbial activity, resulting in higher CO_2 and N_2O emissions (Muhammad et al., 2021; Wang et al., 2018). In addition, researchers have demonstrated that straw decomposition increases C and N availability, which increases CO_2 emissions and tends to exacerbate or diminish CH_4 and N_2O emissions (Lenka and Lal, 2013; Yagioka et al., 2015). Therefore, a straw-based method must be devised to minimize soil GHGs emissions without compromising the maize yields. Therefore, combining commercial N fertilization and straw return is often an encouraging approach for increasing soil fertility, soil moisture content, and agricultural sustainability (Liu et al., 2022; Yang et al., 2022). Straw return has long been a widespread agronomic practice in Southern China, where the combined use of straw return and N fertilization is a commonly used method to achieve high crop yield, improve soil porosity, and mitigate GHGs emissions. Reducing N loss in agroecosystems using various methods, including modifying fertilization patterns and fertilizer types, is a top priority for researchers (Cheng et al., 2020; Sun et al., 2015). On the other hand, boosting NUE and reducing N fertilization are important for sustainable crop production and protecting soil health (Chi et al., 2022; Muhammad et al., 2018).

Agricultural management strategies, such as organic or synthetic N fertilization, are major contributors to GHG emissions from agricultural soils (Smith et al., 2008; Snyder et al., 2009). Straw return is a vital organic fertilizer that enhances the immobilization of inorganic N, and mulching protects the soil surface from direct raindrop strikes and even reduces wind and rainstorm erosion (Chen et al., 2017a; Luxhøi et al., 2007). Straw is a rich source of soil organic carbon (SOC), N, phosphorous, potassium, and trace elements, and its incorporation into the soil can increase its fertility and physicochemical properties (Hoang and Marschner, 2019; Zhao et al., 2020). Long-term straw return has been shown to enhance the predominant soil C and greatly boost CH_4 emissions (Xia et al., 2014). Straw mulching directly affects soil moisture content and temperature, which in turn affects GHGs emissions (Berger et al., 2013; Liu et al., 2014). The increase in soil moisture can be maximized by increasing SOM content due to straw return, which increases aggregate stability and porosity (Pulido-Fernández et al., 2013).

Straw and chemical fertilizers have varying impacts on soil N pools, although straw returning plays an important role in increasing NUE (Afreh et al., 2018), whereas chemical N fertilization generally increases soil N_2O emissions (Ju et al., 2009; Shcherbak et al., 2014). Agricultural practices, such as excessive N fertilization, have contributed to increased soil GHGs emissions, reducing NUE resulting in substantial N losses, predominantly through leaching and N_2O emissions (de Oliveira Silva et al., 2019; Shahbaz et al., 2019). Similarly, in the subtropical monsoon humid zone, straw incorporation and higher N application rates increased CO_2 and N_2O emissions owing to faster decomposition coupled with high rainfall and a hotter climate during the maize growth period (Li et al., 2019; Zhao et al., 2016). However, the impact of chemical fertilization and straw incorporation on GHGs emissions must be determined (Frimpong and Baggs, 2010; Wang et al., 2018).

The rapid increase in Chinese society has resulted in significant shifts in labor. Agricultural development is limited by a workforce shortages in remote rural areas, despite the need to increase food supply and economic benefits. To reduce the use of synthetic fertilizers and increase yield, the fertilization and planting methods must be improved (Wu et al., 2021). Straw incorporation can provide more available C and N, which stimulates the activity of soil microorganisms and significantly increases CO_2 and N_2O emissions in upland soils (Li et al., 2019; Ren et al., 2020; Zhao et al., 2020). The objectives of this study was to evaluate the effect of N fertilization and straw incorporation on maize yield and plant N uptake. This study aimed to investigate the combined effects of chemical fertilization and straw incorporation on soil fertility and sustainability. Finding the optimum combination to improve crop yield and mitigate GHGs emissions and their relationship with soil fertility.

2 Materials and Methods

2.1 Experimental site

The field experiment was performed at the Agronomy Research Farm of Guangxi University, Nanning, Guangxi, China (22°50′ N, 108°17′ E). This region is characterized by a subtropical climate, with a mean annual precipitation of 1,298.0 mm and a mean annual temperature of 21.7℃ (Muhammad et al., 2022). This region has two distinct seasons: the spring growing season from March to July and the autumn growing season from August to November. According to Chinese Soil Taxonomy, the soil was classified as clay loam with a pH of 6.5, soil organic C of 14.6 g/kg, and total N, available phosphorus, and potassium of 0.80 g/kg, 0.43 g/kg, and 0.89 g/kg, respectively.

2.2 Experimental design

The field experiment was carried out in a randomized complete block design with a split-plot arrangement and three replications. Straw management was allotted to the main plot, and N fertilizer dosages were assigned to the subplot. The straw management was the incorporation of spring and autumn maize straw after mechanically crushing it into 2-3 cm and thoroughly mixing it into the soil through a rotavator. Maize straw was removed from the field after harvest using a traditional planting system. Urea was used as an N source at four different rates: as no N fertilizer (N0), 200 kg N/ha (N200), 250 kg N/ha (N250), and 300 kg N/ha (N300).

Maize cultivar "Zhengda 619" was sown twice in the first week of March (4th) and in the second week of August (8th) with a row-row spacing of 60 cm and plant-plant spacing of 30 cm and harvested in the first week of July (5th) and last week of November (26th) 2021, respectively. Before planting, the recommended basal dosages of calcium magnesium phosphate (18% P_2O_5) and potassium chloride (60% K_2O) were in the soil at a rate of 100 kg/ha. Urea fertilizer was applied two-thirds before sowing and one-third during the large trumpet period. Other field management practices such as weeding and hoeing were consistently performed in all the plots.

2.3 Greenhouse gas fluxes, cumulative emissions

The static chamber method was used to measure the soil CO_2, N_2O, and CH_4 emissions (Ding et al., 2013). A set of iron boxes (30 cm × 30 cm × 15 cm in width, length, and height, respectively) were installed in each plot at a depth of 12 cm. A vented polyvinyl chloride (PVC) lid (30 cm × 30 cm × 30 cm in width, length, and height, respectively) was covered with reflective aluminum foil as the insulation layer to minimize temperature changes within the chamber during measurements, and the flux chamber was equipped with a vent tube and a sampling port. The flux chamber was anchored to a permanent anchor that was driven into the soil during each measurement. Gas samples were taken by using a (50 mL) polypropylene air-tight syringe at regular intervals (0, 10, 20, and 30 minutes) after chamber installation. The measurement time was limited to 9:00–11:00. throughout the study. Gas samples were collected at 10 different growth stages: including emergence (VE), third-leaf (V3), sixth-leaf (V6), twelfth-leaf (V12), tassel (VT), silking (R1), blister (R2), milk (R3), dough (R4), and maturity (R6). During gas sampling, soil moisture (%) and temperature were measured with a TDR 100 portable moisture meter, and a mercury thermometer was inserted into the hole on the top of each box to measure the inside temperature. The concentrations of CO_2, CH_4, and N_2O were determined using a gas chromatography (Agilent 7890A) equipped with a thermal conductivity detector (TCD), flame ionization detector (FID), and an electron capture detector (ECD) with high-purity N gas as a carrier. According to Wu et al. (2021) the gas flux was calculated using the following formula:

$$F = \rho \times h \times \frac{\Delta c}{\Delta t} \times \frac{273}{273 + T}$$

where F is the N_2O flux in $\mu g/(m^2 \cdot h)$, CO_2 and CH_4 fluxes in $mg/(m^2 \cdot h)$, and ρ ($CO_2 = 1.964$ kg/m^3) is the density of gas in normal conditions, h is the static box height in meter (m), $\Delta c/\Delta t$ is the differences in N_2O concentration per unit of time, and T is the temperature of the box (℃). Following the sampling date and emission rate, the cumulative emissions were calculated as previously described by Xu et al. (2020):

$$\text{Cumulative emission} = \sum_{i=1}^{n} (F_i + F_{i+1})/2 \times (t_{i+1} - t_i) \times 24$$

Where F is the flux of N_2O [$\mu g/(m^2 \cdot h)$] or CO_2 or CH_4 flux [$mg/(m^2 \cdot h)$], i is the ith measurement, and the term ($t_{i+1} - t_i$) is the number of days between two consecutive measurements, and n is the total number of measurements.

2.4 Global warming potential and emissions factor

The global warming potential of CO_2-equivalent (CO_2-eq) was calculated by multiplying the seasonal N_2O and CH_4 emissions by their respective radiative forcing potentials (Solomon et al., 2007). The area-scaled GWP is calculated as follows:

Area-scaled GWP = 25×CH_4 (kg/ha) + 298×N_2O (kg/ha) + 1×CO_2 (kg/ha)

The area-scaled GWP-to-grain yield ratio was used to determine the GHGI (Bayer et al., 2015).

$$\text{GHGI} = \frac{\text{Area-scaled GWP}}{\text{Yield}}$$

2.5 Sampling, nitrogen, and carbon analysis

Soil samples at a depth of 0-20 cm for each subplot were collected using augers at the maturity stage (R6) before maize harvesting. The samples were air-dried and sieved through a 0.15 mm-mesh sieve for the determination of soil total N (TN) using the semi-micro Kjeldahl method and SOC using the potassium dichromate external heating oxidation method. Organic forms of N were determined using the acid hydrolysis distillation method (Page et al., 1982).

Three representative maize plants from each subplot were harvested at the R6 growth stage and divided into stems, leaves, and grains to test the plant N uptake. After deactivating the enzymes by heating at 105℃ for 30 min, the above-ground plant parts were dried to a constant weight at 80℃, and the above-ground dry biomass was weighed for dry matter (Muhammad et al., 2022). The dry matter was then crushed and sieved using a 100 mesh sieve. The Kjeldahl method was used to determine leaf, stem, and grain N content (Page et al., 1982), and N uptake was calculated from the biomass N content.

Nitrogen use efficiency, including N apparent recovery use efficiency (NRUE), physiological

NUE (PNUE), and agronomic N use efficiency (ANUE), were calculated using the following formulas (Wang et al., 2021):

$$\text{NRUE}(\%) = \frac{U_N - U_0}{\text{Nitrogen applied}(F_N)} \times 100$$

Where F_N represents the amount of N fertilizer applied (kg/ha).

$$\text{PNUE}(kg/kg) = \frac{Y_N - Y_0}{U_N - U_0}$$

where Y_N represents grain yield in N-treated plots (kg/ha), Y_0 represents grain yield in the plot without N application (kg/ha), U_N represents N uptake by plants in N-treated plots (kg/ha), and U_0 represents N uptake by plants in the plot without N application (kg/ha).

$$\text{ANUE}(kg/kg) = \frac{Y_N - Y_0}{\text{N applied}(F_N)}$$

2.6 Grain yield

At the mature growth stage of maize, an area of 2 m^2 was randomly selected in each plot, and the grain yield was determined after maize threshing and drying.

$$\text{Grain yield}(kg/ha) = \text{maize grain weight}(kg/m^2) \times 10\,000\,(m^2)$$

2.7 Statistical analysis

The experimental data were analyzedusing SPSS version 21.0 (SPSS, Chicago, IL, USA). The Duncan's multiple range test (Duncan, $P<0.05$) was used to compare the mean values of the three replicates for each treatment. All data were expressed as means and standard errors. All graphs were constructed using Origin Pro, version 2021 (Origin Lab Corporation, Northampton, MA, USA).

3 Results

3.1 Soil organic carbon and nitrogen dynamics

Planting patterns and N fertilization significantly affected the SOC content (Figure 1). Straw incorporation with N300 treatment significantly increased SOC by 6.84% and 37.90% in spring and 5.15% and 32.71% in autumn seasons compared to the N200 and N0 treatments, respectively ($P<0.05$). Similarly, N fertilization increased SOC content in the traditional planting system compared to that in the N0 treatment, but the SOC content in the N250 and N300 treatments was statistically similar in both seasons. Our results showed that the average SOC content was 5.03% higher in spring than that in autumn.

Nitrogen fertilization, straw management, and their interactions significantly affected soil TN, NH_3-N (NHN), amino sugar N (ASN), amino acid N (AAN), hydrolyzable unknown organic N (HUN), total acid-hydrolyzable organic N (THAN), and acid-insoluble N (AIN; Table 1). In the form of chemical fertilizer and/or straw incorporation during the spring and autumn seasons,

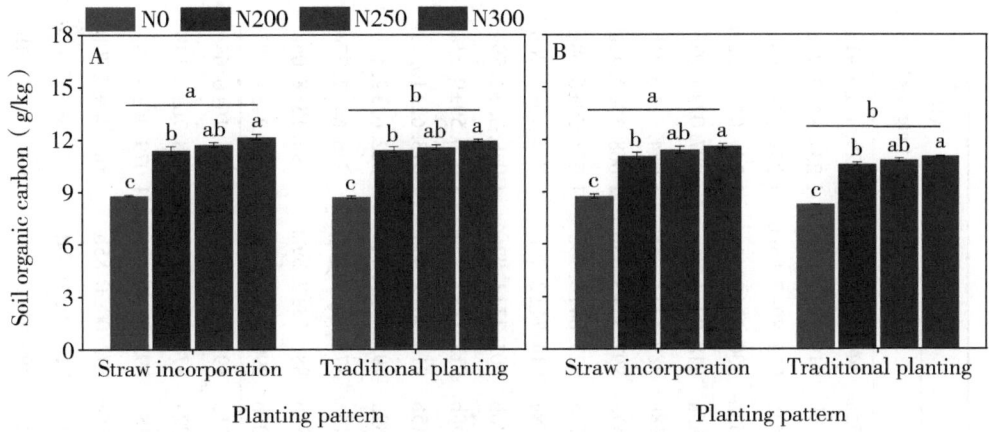

Figure 1 Effect of straw incorporation and nitrogen fertilization on soil organic carbon under dual-cropping system during spring (A) and autumn (B) seasons

Notes: Different letters indicate significant differences between samples ($P<0.05$).

the N input was reflected in the soil accumulation of TN, NHN, ASN, AAN, HUN, THAN, and AIN. Except for HUN, the levels of TN, NHN, ASN, AAN, HUN, THAN, and AIN were significantly higher during spring than autumn, and vice versa. In addition, straw incorporation significantly increased TN, NHN, ASN, AAN, THAN, and AIN by 12.57%, 2.70%, 13.00%, 12.80%, 2.70%, and 22.32%, but decreased HUN by 34.54% compared to traditional planting. Soil TN, NHN, ASN, AAN, THAN, and AIN were significantly higher in the N300 treatment than in the N0 treatment in both seasons and planting patterns (Table 1). In contrast, the HUN was substantially greater under the traditional planting system for N0 treatment during the autumn season. In both planting patterns, the N250 and N300 treatments had a statistically similar effects on TN, ASN, AAN, HUN, and AIN during the spring and on TN, NHN, ASN, AAN, HUN, and AIN during the autumn season. The interaction analysis showed that season, planting pattern, N fertilization, season × planting pattern, season × N fertilization, planting pattern × N fertilization, and season × planting pattern × N fertilization significantly and positively affected the soil TN, ASN, AAN, and HUN.

3.2 Carbon dioxide

The CO_2 fluxes in all the treatments showed an increasing trend and then gradually decreased throughout the entire experiment, and the lowest flux was observed at the R6 growth stage. During the experimental period, the CO_2 fluxes ranged from 4.27 to 125.64 mg C/($m^2 \cdot h$) in the spring season and 13.65 to 214.87 mg C/($m^2 \cdot h$) during the autumn season (Figure 2). Carbon dioxide fluxes increased after maize straw incorporation, reaching peaks at VT in spring and R1 in autumn, respectively. Additionally, the CO_2 fluxes in both seasons decreased with time following straw incorporation, although the fluxes in the autumn season were much higher, possibly because of increased microbial activity ($P<0.05$). Compared to straw incorporation, traditional planting resulted in 24.25% and 29.32% lower CO_2 fluxes during the spring and autumn seasons, respectively. The average CO_2 fluxes of N0 treatment [34.91 and 47.67 mg C/($m^2 \cdot h$)] were lower

Table 1 Changes in soil total and organic nitrogen with straw incorporation and nitrogen fertilization under dual-cropping system during spring and autumn in 2021

Seasons	Planting patterns	Nitrogen	TN (g/kg)	NHN	ASN	AAN	HUN	Total	AIN (mg/kg)
Spring	Straw incorporation	N0	1.87±0.01c	337.76±1.77c	50.03±0.18c	324.17±1.21c	132.44±3.55c	844.41±4.44c	1 029.42±9.91b
		N200	2.15±0.00b	367.75±0.63b	92.63±0.15b	409.30±0.68b	49.69±1.19b	919.39±1.56b	1 234.84±3.76a
		N250	2.17±0.00ab	370.36±0.83ab	93.11±0.052ab	411.40±0.23ab	51.02±1.43b	925.89±2.08ab	1 239.39±2.95a
		N300	2.18±0.01a	372.21±1.25a	93.72±0.31a	414.11±1.36a	50.48±3.53b	930.52±3.12a	1 249.01±10.27a
	Traditional planting	N0	1.72±0.01c	327.20±0.78c	46.06±0.24c	298.43±1.51c	146.32±0.85a	818.00±1.96c	907.02±7.03c
		N200	1.93±0.01b	355.65±0.54b	83.12±0.26b	367.30±3.15b	83.05±1.98b	889.12±1.34b	1 044.01±6.91b
		N250	1.97±0.01a	359.10±1.37b	84.53±0.29a	373.52±1.30a	80.60±0.47c	897.75±3.43b	1 068.15±3.42ab
		N300	1.98±0.01a	369.22±2.38a	84.99±0.51a	375.52±2.29a	93.32±0.77c	923.04±5.95a	1 053.39±6.08a
Autumn	Straw incorporation	N0	1.52±0.01c	325.88±1.52c	40.56±0.17c	288.63±1.22c	159.63±2.78a	814.71±3.79c	704.40±7.76b
		N200	1.95±0.00b	361.31±0.69b	83.92±0.14b	409.85±0.71b	48.19±0.30b	903.27±1.72b	1 048.38±1.79a
		N250	1.97±0.01ab	364.88±0.79a	84.66±0.33ab	413.45±1.60ab	49.21±2.98b	912.19±1.99a	1 056.59±9.31a
		N300	1.99±0.01a	365.56±0.97a	85.46±0.51a	417.38±2.48a	45.50±2.45b	913.90±2.43a	1 073.62±10.59a
	Traditional planting	N0	1.42±0.00c	318.25±1.37c	37.87±0.11c	269.47±0.83c	170.05±1.32a	795.63±3.44c	622.62±2.12b
		N200	1.65±0.01b	350.24±1.54b	71.05±0.22b	346.97±1.09b	107.34±3.15b	875.61±3.86b	776.64±7.77a
		N250	1.68±0.01a	354.45±0.95a	72.25±0.25a	352.85±1.23a	106.57±2.84b	886.13±2.39a	794.13±8.06a
		N300	1.68±0.01a	356.31±0.59a	72.43±0.52a	353.74±2.52a	108.28±3.01b	890.76±1.47a	793.74±11.85a
Spring	Straw incorporation		2.00±0.00a	357.41±0.23a	78.52±0.04a	371.72±0.13a	85.87±0.31b	893.51±0.57a	1 103.16±0.64a
Autumn			1.73±0.00b	349.61±0.29b	68.53±0.079b	356.54±0.39b	99.35±0.72a	874.02±0.72b	858.76±2.14b
Spring	Traditional planting		1.97±0.00a	358.21±0.38a	78.01±0.11a	386.04±0.58a	73.27±1.18b	895.53±0.94a	1 079.46±3.78a
Autumn			1.75±0.00b	348.80±0.34b	69.04±0.027b	342.22±0.15b	111.94±0.68a	872.00±0.85b	882.46±1.70b
	Traditional planting	N0	1.64±0.00c	327.28±0.56d	43.63±0.08c	295.17±0.55c	152.11±0.95a	818.19±1.40d	815.863±3.00c
		N200	1.92±0.01b	358.74±0.077c	82.68±0.20b	383.36±0.91b	72.07±1.21b	896.85±0.19c	1 025.97±4.74b
		N250	1.94±0.00a	362.19±0.29b	83.64±0.06a	387.81±0.26a	71.85±0.30b	905.49±0.73b	1 039.57±0.93ab
		N300	1.96±0.01a	365.82±0.50a	84.15±0.38a	390.19±1.77a	74.39±1.48b	914.55±1.25a	1 042.44±7.71a

(continued)

Seasons Planting patterns Nitrogen	TN (g/kg)	THAN (mg/kg)					AIN (mg/kg)
		NHN	ASN	AAN	HUN	Total	
Interactions							
Season	***	***	***	***	***	***	***
Planting pattern	***	***	***	***	***	***	***
Nitrogen	***	***	***	***	***	***	***
Season * Planting pattern	***	ns	***	***	***	ns	***
Season * Nitrogen	***	*	**	***	***	*	***
Planting pattern * Nitrogen	***	*	***	***	***	*	***
Planting pattern * Nitrogen * Season	***	ns	***	***	***	ns	***

Notes: Different letters indicate significant differences between samples ($P<0.05$). Values are means ± SE ($n=3$).

* Significant at $P<0.05$; ** Significant at $P<0.01$; *** Significant at $P<0.001$; ns, not significant.

TN, total nitrogen; THAN, total acid hydrolysable organic nitrogen; NHN, NH_3–N; ASN, amino sugar nitrogen; AAN, amino acid nitrogen; HUN, hydrolysable unknown organic nitrogen and AIN, acid insoluble nitrogen.

than N200 [44.64 and 65.06 mg C/(m² · h)], N250 [58.82 and 91.87 mg C/(m² · h)], and N300 [70.27 and 112.14 mg C/(m² · h)] treatments in spring and autumn seasons, respectively (Figure 2). Averaged across seasons, the CO_2 fluxes were significantly higher in the N300 treatment than in the other treatments. These results suggest that increasing N fertilization significantly increased CO_2 fluxes with and without straw incorporation ($P<0.05$). In contrast, the N200 treatment with straw incorporation significantly decreased the CO_2 fluxes by 12.26% and 36.71% compared to the N250 and N300 treatments of the traditional planting system, respectively.

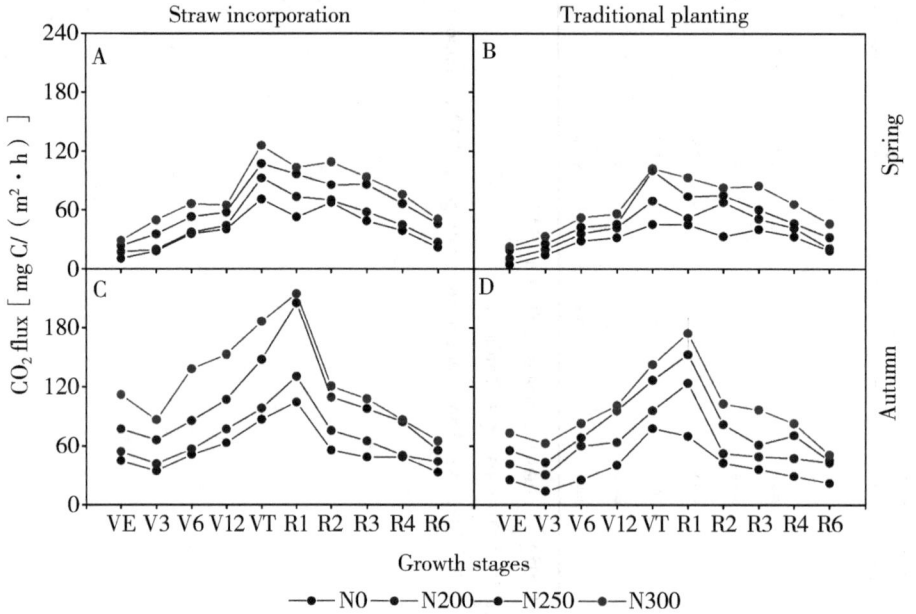

Figure 2 Effect of straw incorporation and nitrogen fertilization on CO_2 fluxes under dual-cropping system during spring with straw incorporation (A) and traditional planting (B), and in autumn seasons with straw incorporation (C) and traditional planting (D)

Notes: The VE, V3, V6, V12, VT, R1, R2, and R3 represents: emergence, third-leaf, sixth-leaf, twelfth-leaf, tassel, silking, blister, milk, dough, and maturity.

The averaged cumulative CO_2 emissions for N300 treatment were 1,222.16 kg C/ha (straw incorporation) and 1,015.14 kg C/ha (traditional planting) in spring and 1,987.91 kg C/ha (straw incorporation) and 1,526.53 kg C/ha (traditional planting) in autumn (Figure 5A, B). Compared to N0, the N300 treatment increased cumulative CO_2 emissions by 114.48% and 87.08% under straw incorporation and traditional planting systems in the spring season, respectively. Similarly, the N300 treatment resulted in higher cumulative CO_2 emissions in the autumn. Straw incorporation increased the cumulative CO_2 emissions by 26.86% compared to the traditional planting system. The combined application of N200 and straw incorporation significantly decreased the cumulative CO_2 emissions by 12.64% and 36.81% compared to the N250 and N300 treatments of traditional planting systems, respectively (Figure 5). Our results showed that straw incorporation significantly decreased cumulative CO_2 emissions in spring compared to autumn ($P<0.05$; Figure 5A, B). Regressions analysis demonstrated that the soil CO_2 fluxes polynomially increased with increasing soil total N content ($R^2 = 0.64$; $P<0.001$; Figure 6A) and SOC content ($R^2 = 0.84$; $P<0.001$; Figure 6D).

3.3 Nitrous oxide emissions

Except for a few small peaks following fertilization and straw incorporation in the early growing stages and compared to spring, the N_2O fluxes were relatively low during the autumn season (Figure 3). Straw incorporation rapidly increased N_2O fluxes, and pronounced peaks appeared at the start of the maize growing stages (VE, V3, and V6), with maximum values of 350.46 and 309.91 μg N/($m^2 \cdot h$) during the spring and autumn seasons, respectively. During the spring season, the N_2O fluxes [112.69 and 96.66 μg N/($m^2 \cdot h$)] were significantly higher than the autumn season [58.97 and 47.92 μg N/($m^2 \cdot h$)] under straw incorporation and traditional planting system, respectively ($P<0.05$). The results showed that higher N fertilization combined with straw incorporation significantly increased the N_2O fluxes compared to traditional planting ($P<0.05$), which could be due to straw incorporation increasing soil N content and microbial diversity, which increased N_2O fluxes. In addition, N_2O fluxes were lower in the N0 and N200 treatments in the straw incorporated plot than in the N250 and N300 treatments of the traditional plot. These results suggest that the straw incorporation with low N fertilization rates significantly improves crop yield, soil fertility, and plant NUE, and mitigates GHGs emissions compared to higher N fertiliza-

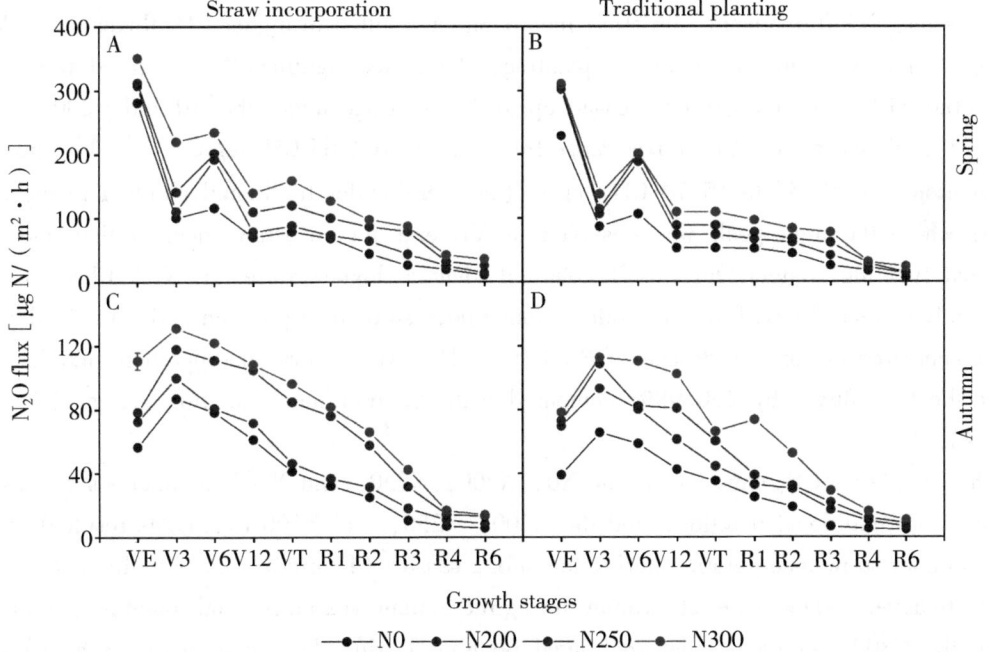

Figure 3 Effect of straw incorporation and nitrogen fertilization on N_2O fluxes under dual-cropping system during spring with straw incorporation (A) and traditional planting (B), and in autumn seasons with straw incorporation (C) and traditional planting (D)

Notes: The VE, V3, V6, V12, VT, R1, R2, and R3 represents: emergence, third-leaf, sixth-leaf, twelfth-leaf, tassel, silking, blister, milk, dough, and maturity.

tion rates in traditional planting systems.

Cumulative N_2O emissions were significantly affected by N fertilization and straw incorporation in both seasons. Averaged across straw and N fertilization, the spring season resulted in 131.80% higher cumulative N_2O emissions than the autumn season. The cumulative N_2O emissions were 1.41 and 0.58 kg N/ha for N0 treatment, 1.89 and 0.75 kg N/ha for the N200 treatment, 2.23 and 1.01 kg N/ha for N250 treatment, and 2.73 and 1.22 kg N/ha for N300 treatment during spring and autumn seasons, respectively (Figure 5C and D). The results of the current study results demonstrated that the cumulative N_2O emissions increased with higher N fertilization rates under both straw incorporation and traditional planting systems. Furthermore, straw incorporation resulted in higher cumulative N_2O emissions than traditional planting under the same N fertilization rate. However, N200 treatment in the straw-incorporated plot had significantly lower cumulative N_2O emissions during autumns compared to the N250 and N300 treatments under traditional planting (Figure 5D). Regressions analysis showed a significant and positive relationship between soil total N and SOC content and soil N_2O fluxes, these results suggesting that N_2O fluxes increased polynomially with soil total N content $R^2 = 0.67$; $P < 0.001$ and SOC content $R^2 = 0.87$; $P < 0.001$ (Figure 6B and E).

3.4　Methane emissions

In all the treatments, the CH_4 fluxes ed throughout the growing season (Figure 4). In the spring season, N300 treatment with straw incorporation resulted in higher CH_4 fluxes at the V6 and VT stages; however, under traditional planting, the fluxes significantly decreased from the VE stage to the V12 stage and again increased up to the R3 stage under the N0 and N200 treatments ($P<0.05$). Similarly, the CH_4 fluxes range from 4.59 to 104.65 CH_4 mg C/($m^2 \cdot h$) under straw incorporation and -50.57 to 15.74 CH_4 mg C/($m^2 \cdot h$) under traditional planting in the autumn season, where the high peak was observed at VE and R1 in straw incorporation and VE in traditional planting system. Our results showed that a higher dose of N fertilizer (N300) significantly increased CH_4 fluxes in both seasons under straw incorporation and traditional planting systems compared to other N dosages ($P<0.05$). However, straw incorporation significantly increased the CH_4 fluxes by 230.98% compared with the traditional planting system ($P<0.05$; Figure 4).

The cumulative CH_4 emissions of the N0, N200, N250, and N300 treatments showed net absorption under traditional planting, and the N200, N250, and N300 treatments resulted in net emissions under straw incorporation during the spring season (Figure 5E and F). In contrast, all N fertilizer treatments showed net absorption during the autumn season for both planting patterns, except for the N300 treatment under traditional planting. During the spring season, the cumulative CH_4 emissions for N200, N250, and N300 treatments were 0.14, 0.30, and 0.62 kg C/ha under straw incorporation and absorption were 0.07, 0.20, and 0.65 kg C/ha for N0, N200, and N250 treatments under traditional planting, respectively (Figure 5E). Moreover, cumulative CH_4 absorption decreased with a higher N fertilization rate under the traditional planting system in both seasons. However, increased N fertilization significantly increased cumulative CH_4 emissions under

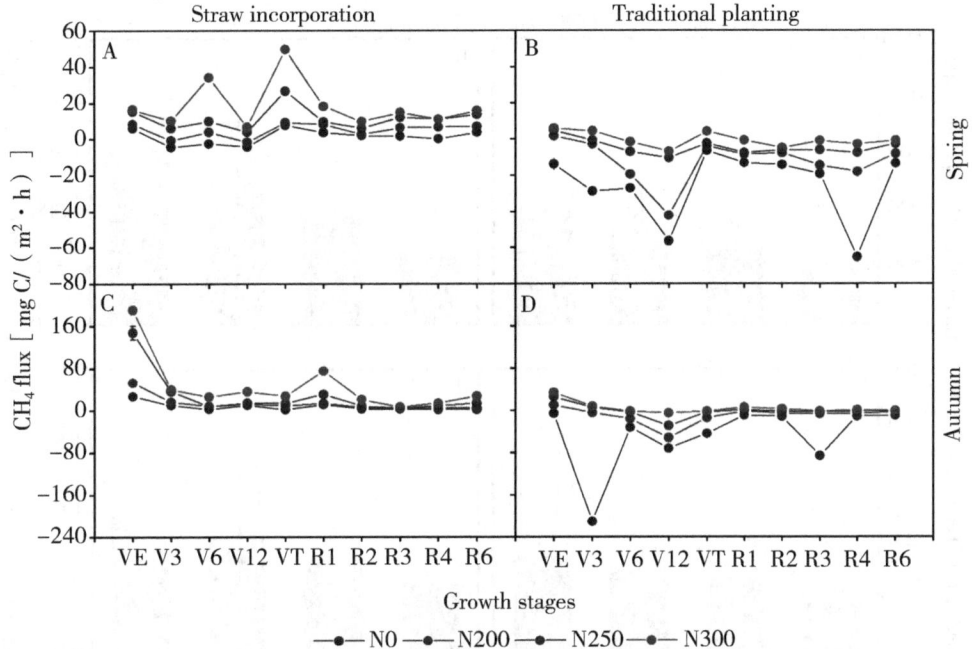

Figure 4 Effect of straw incorporation and nitrogen fertilization on CH$_4$ fluxes under dual-cropping system during spring with straw incorporation (A) and traditional planting (B), and in autumn seasons with straw incorporation (C) and traditional planting (D)

Notes: The VE, V3, V6, V12, VT, R1, R2, and R3 represents: emergence, third-leaf, sixth-leaf, twelfth-leaf, tassel, silking, blister, milk, dough, and maturity.

straw incorporation during spring season ($P<0.05$). Regression analysis showed a polynomially positive relationship between soil total N, SOC content, and soil CH$_4$ fluxes (Figure 6C and F). Soil CH$_4$ fluxes polynomially increased with increasing total soil N content $R^2=0.77$; $P<0.001$ and SOC content ($R^2=0.54$; $P<0.001$).

3.5 Maize grain yield, area-scaled GWP, and GHGI

Nitrogen fertilization, straw management, and season significantly affected maize yield ($P<0.001$, Table 2). The highest maize grain yield was observed in the straw-incorporated plots compared to the traditional planting patterns in the spring and autumn seasons. On average, maize yield in spring was significantly higher than that in autumn. The grain yield was significantly higher in the N300 treatment (6,479 kg/ha) than in the N0 treatment (2,425 kg/ha; $P<0.05$), but was not statistically different from the N200 and N250 treatments ($P>0.05$).

The area-scaled GWP showed considerable variation among seasons, straw management, and N treatments (Table 2). The N300 treatment resulted in a significantly higher area-scaled GWP followed by N250 compared to the N0 treatment. On average, the area-scaled GWP of straw-incorporated plots increased by 25.56% and 30.39% compared to traditional planting during the spring and autumn, respectively. Averaged across N fertilization and straw management, the spring season resulted in a 32.75% lower area-scaled GWP than autumn. These results suggest that a higher dose

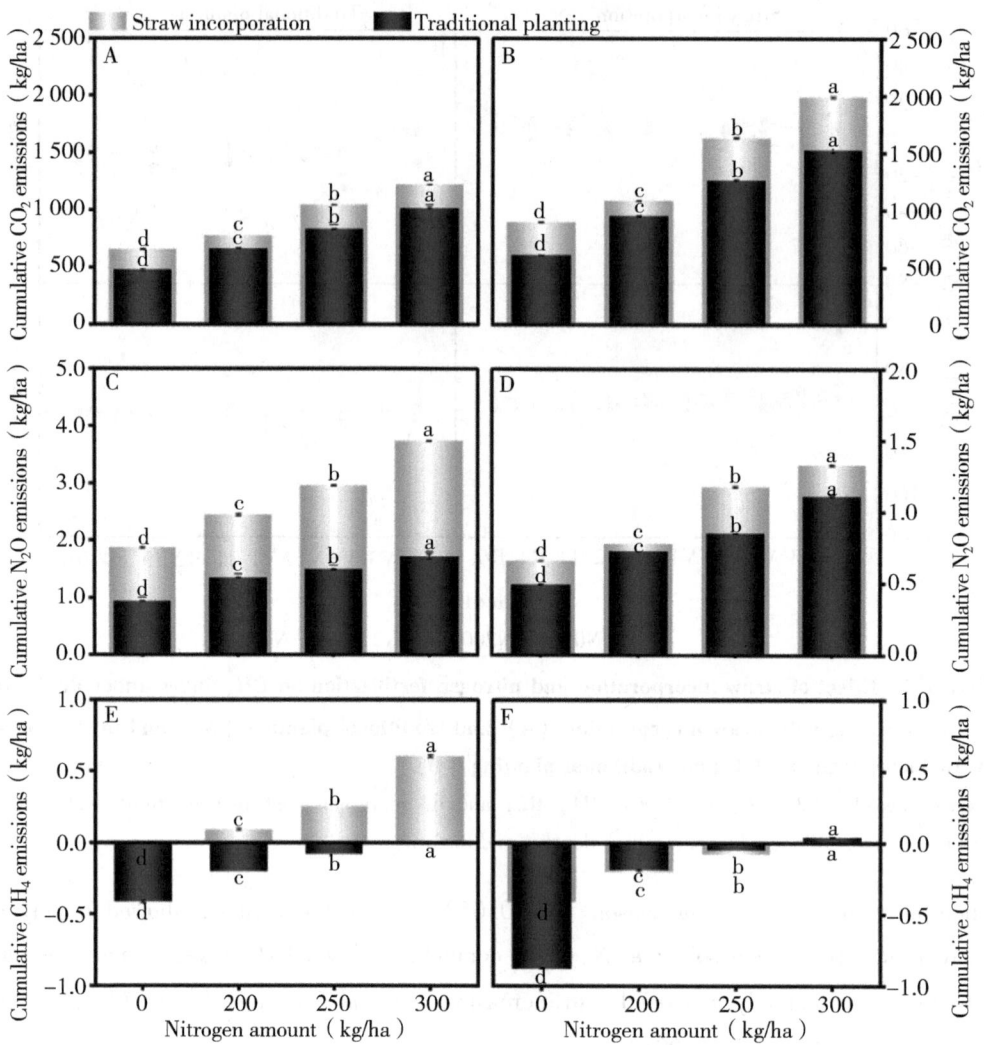

Figure 5 Cumulative emissions of CO_2 during spring (A) and autumn (B), N_2O emissions during spring (C) and autumn (D) and CH_4 emissions and uptake during spring (E) and autumn (F) seasons with planting pattern and nitrogen fertilization under dual-cropping system

Notes: Different letters indicate significant differences between samples ($P<0.05$).

of N fertilizer with straw incorporation during the spring season significantly increased the area-scaled GWP. Our results showed that N200 with straw incorporation resulted in a lower area-scaled GWP compared to the N250 and N300 treatments of traditional planting. Nitrous oxide emissions contributed at least 0.04% and CO_2 emissions contributed 99.89% to area-scaled GWP, while CH_4 (0.07%) uptake had a marginal impact. The GHGI was significantly lower in the N200 treatment, resulting in a higher grain yield than in N0, and higher GHGI in the N0 treatment resulted in a lower grain yield. In addition, GHGI was significantly higher during autumn under straw incorporation and decreased during spring under the traditional planting system ($P<0.05$; Table 2). The interactions between season, straw management, and N fertilizer treatments were found to be significant ($P<0.001$; Table 2).

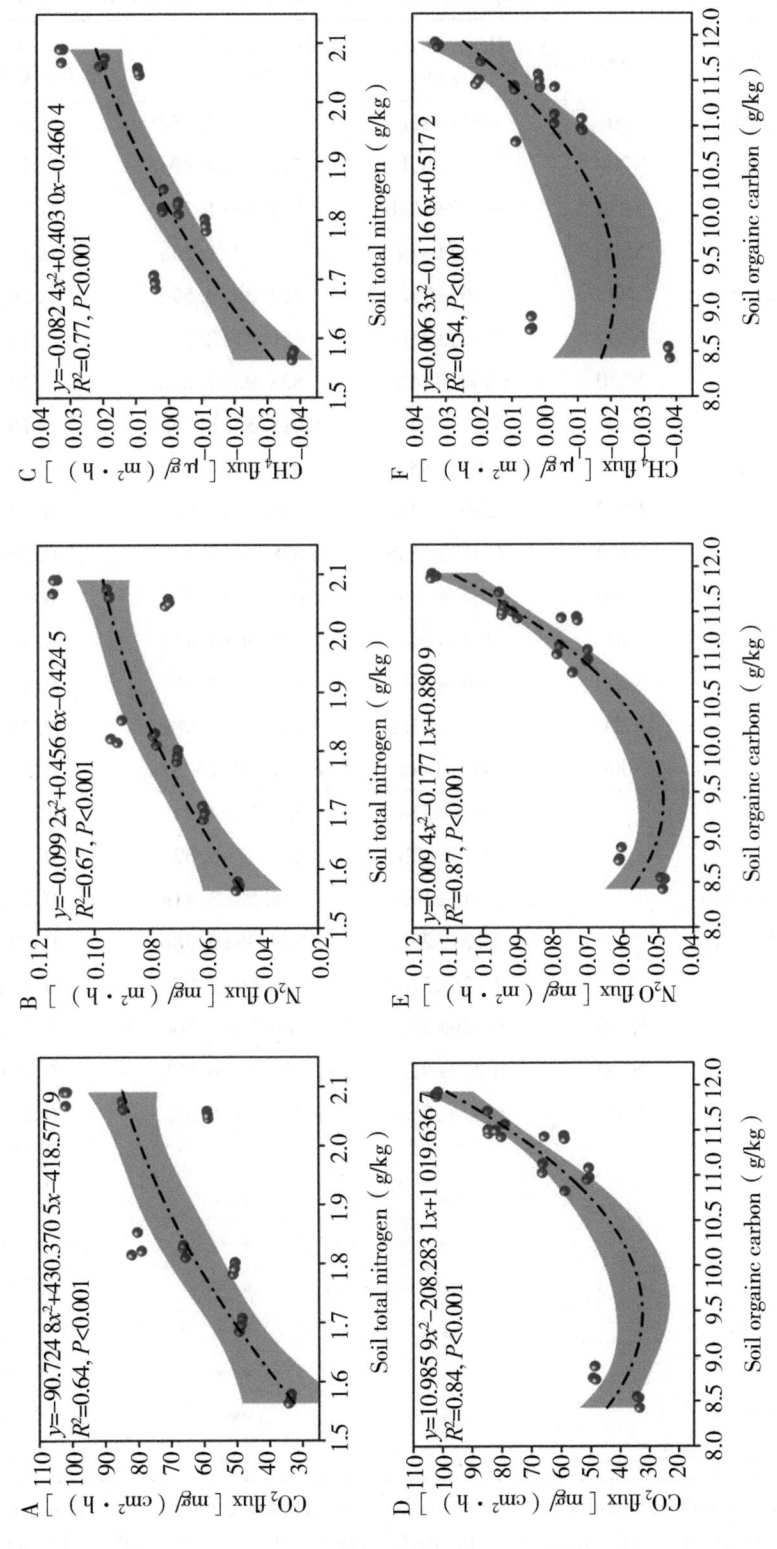

Figure 6　Relationship of soil total nitrogen with CO_2 flux(A), N_2O flux(B), CH_4 flux(C), and relationship of soil orgainc carbon with CO_2 flux(D), N_2O flux(E), CH_4 flux(F) under dual-cropping system

Table 2 Effect of straw incorporation and nitrogen fertilization on grain yield, area-scaled GWP, and GHGI under dual-cropping system during spring and autumn in 2021

Seasons	Planting patterns	Nitrogen	Grain yield (kg/ha)	Area-scaled GWP (kg/ha)	GHGI (kg CO_2 eq/kg)
Spring	Straw Incorporation	N0	2 917±83c	653.75±3.55d	0.22±0.01a
		N200	6 417±83b	778.00±4.89c	0.12±0.00d
		N250	6 667±83ab	1 051.76±1.30b	0.16±0.00c
		N300	6 917±83a	1 238.37±1.26a	0.18±0.00b
	Traditional Planting	N0	2 367±73c	463.31±9.59d	0.20±0.01a
		N200	5 900±76b	656.71±2.31c	0.11±0.00d
		N250	6 000±144b	828.93±2.60b	0.14±0.00c
		N300	6 333±83a	1 015.35±8.76a	0.16±0.00b
Autumn	Straw Incorporation	N0	2 667±83c	896.28±4.94d	0.34±0.01a
		N200	6 250±144b	1 087.67±1.08c	0.17±0.00d
		N250	6 417±83ab	1 638.43±4.16b	0.26±0.00c
		N300	6 667±83a	2 004.43±6.58a	0.30±0.00b
	Traditional Planting	N0	1 750±0c	581.30±1.47d	0.33±0.00a
		N200	5 667±83b	945.07±4.26c	0.17±0.00d
		N250	5 833±83ab	1 261.18±5.98b	0.22±0.00c
		N300	6 000±144a	1 527.80±26.32a	0.25±0.00b
Spring			5 440±21a	835.77±1.64b	0.16±0.00b
Autumn			5 156±0b	1 242.77±3.62a	0.25±0.00a
	Straw Incorporation		5 615±10a	1 168.58±3.41a	0.22±0.00a
	Traditional Planting		4 981±25b	909.96±6.38b	0.20±0.00b
		N0	2 425±57c	648.66±2.07d	0.27±0.01a
		N200	6 059±33b	866.86±0.56c	0.14±0.00d
		N250	6 229±91b	1 195.08±0.87b	0.19±0.00c
		N300	6 479±75a	1 446.49±6.86a	0.22±0.00b
Interactions					
Season			**	***	***
Planting pattern			***	***	**
Nitrogen			***	***	***
Season * Planting pattern			ns	***	ns
Season * Nitrogen			ns	***	***
Planting pattern * Nitrogen			ns	***	ns
Planting pattern * Nitrogen * Season			ns	***	***

Notes: Different letters indicate significant differences between samples ($P<0.05$). Values are means ± SE ($n=3$). * Significant at $P<0.05$; ** Significant at $P<0.01$; *** Significant at $P<0.001$; ns, not significant.

3.6 Plant nitrogen and carbon content

3.6.1 Nitrogen content

The results showed that straw incorporation and N fertilization significantly increased the leaf, stem, and grain N content in both seasons ($P<0.05$; Figure 7). The nitrogen contents in maize crops showed a wide range relative to N fertilization, with 0.42%–1.14% for leaf, 0.38%–0.67% for the stem, and 1.24%–2.51% for grain N content. Averaged across seasons and straw management, the N200, N250, and N300 treatments significantly increased leaf N content by 33.33%, 35.22%, and 36.75%; stem N content by 38.10%, 40.08%, and 41.67%; and grain N content by 44.76%, 45.69%, and 46.51% compared to N0 treatment. However, the N250 and N300 treatments had statistically similar effects on leaf, stem, and grain N content. Compared to traditional planting, straw incorporation significantly increased leaf, stem, and grain N content by 16.21%, 11.33%, and 15.29% during the spring, and by 41.31%, 11.98%, and 21.00% during autumn. These results suggest that leaf, stem, and grain N content were significantly higher during spring than in autumn under both planting patterns. In general, the N contents in leaves and

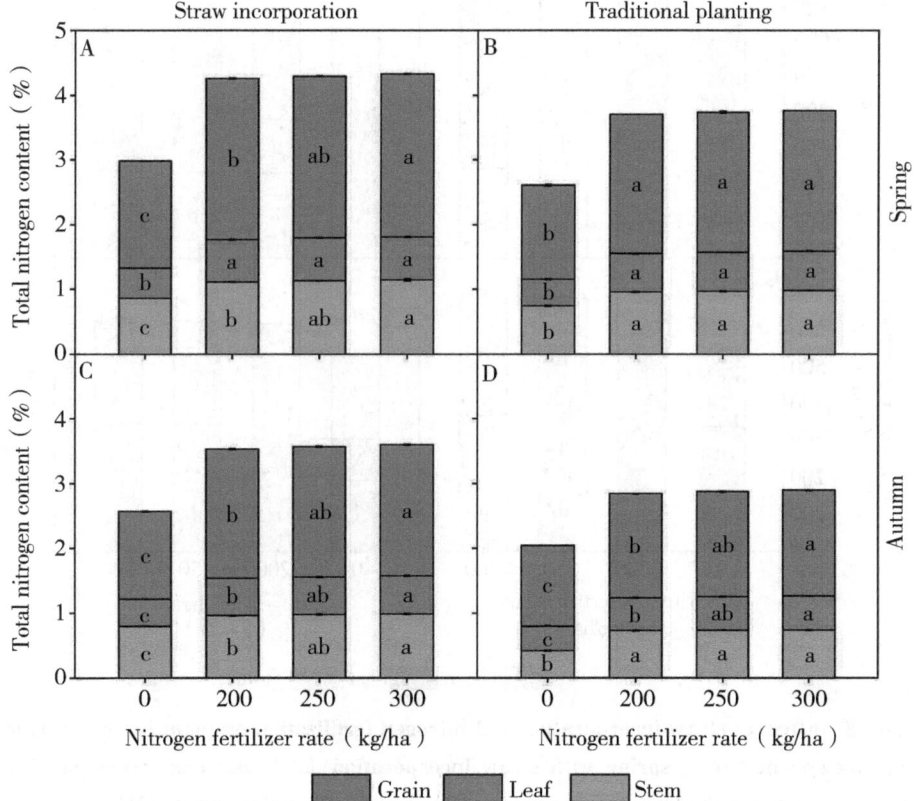

Figure 7 Effect of straw incorporation and nitrogen fertilization on plant total nitrogen content under dual-cropping system during spring with straw incorporation (A) and traditional planting (B), and in autumn seasons with straw incorporation (C) and traditional planting (D)

Notes: Different letters indicate significant differences at $P=0.05$; the vertical bar represent SE (3).

stems is lower than that in grain.

3.6.2 Carbon content

Carbon content showed a large variation among N fertilizer dosages and ranged from 113.07 to 158.16 g/kg for leaf, 193.39 to 243.35 g/kg for the stem, and 245.19 to 326.29 g/kg for grain (Figure 8). The highest C content in leaves, stems, and grain was found for the N300 treatment, followed by N250, and the lowest was observed for N0 treatment in both seasons and straw management. However, no significant differences were observed between the N250 and N300 groups. The spring season significantly increased the C contents in leaves, stems, and grains compared to autumn by 13.54%, 7.34%, and 4.69%, respectively. In addition, straw incorporation considerably increased the leaf, stem, and grain C content by 8.72%, 4.00%, and 10.05% compared to traditional planting. On average, the C content of leaf (139.59 g/kg) and stem (224.69 g/kg) was much lower than that of grain (302.19 g/kg). These results suggest that straw incorporation with the N250 treatment significantly increased leaf, stem, and grain C content, which was not statistically different from the N300 treatment during both seasons.

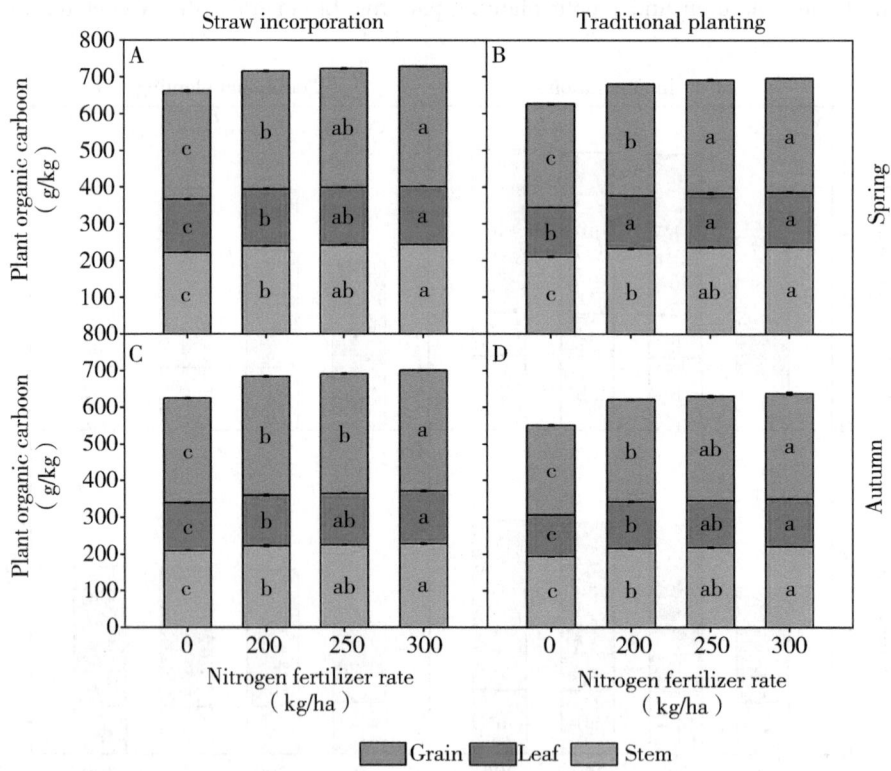

Figure 8 Effect of straw incorporation and nitrogen fertilization on plant organic carbon under dual-cropping system during spring with straw incorporation (A) and traditional planting (B), and in autumn seasons with straw incorporation (C) and traditional planting (D)

Notes: Different letters indicate significant differences at $P=0.05$; the vertical bar represent SE (3).

3.7 Plant N uptake and nitrogen use efficiency

Straw management, N fertilization, planting pattern, and seasons significantly affected plant

N uptakes, NRUE, and PNUE ($P<0.05$; Table 3). The results showed that ANUE was significantly decreased with increasing N fertilization rates in both planting patterns and seasons, and there were no significant differences between seasons or planting patterns. Straw-incorporated plots resulted in higher N uptake than traditional planting. Compared with N0, the N200 treatment significantly increased N uptake by 55.63% in straw-incorporated and 60.64% in traditional planting patterns during the spring season. During the autumn season, the N300 treatment significantly increased N uptake by 1.10%, 0.15%, and 0.02% compared with the N0, N200, and N250 treatments, respectively (Table 3). On average, the N uptakes were 194.96 kg/ha and 142.54 kg/ha during the spring and autumn seasons, respectively. These results showed that N uptake by the plants was 36.77% higher during spring than during autumn.

In spring, the NRUE for the N250 treatment was 1.62% and 13.00% higher than that of the N200 and N300 treatments in straw-incorporated plots and 1.59% and 15.31% higher in the traditional planting system, respectively (Table 3). However, the NRUE for N200 and N250 was not significantly different from each other, but was significantly higher than that for N300 in both planting patterns and seasons. Our results showed that average N uptake and NRUE were significantly higher during spring than during autumn, which might be due to a higher temperature and decomposition of residues. Furthermore, the PNUE and ANUE were substantially higher in the N200 treatment than in the N250 and N300 treatments (Table 3). The PNUE for the N250 and N300 treatments was statistically similar in both seasons and planting patterns ($P>0.05$). These results further revealed that the PNUE was significantly higher in the traditional planting system during autumn compared to residue incorporation during spring ($P<0.05$; Table 3), whereas neither the season nor the planting pattern had a significant impact on ANUE.

Table 3　Effect of straw incorporation and nitrogen fertilization on N uptake, NRUE, PNUE, and ANUE under dual-cropping system during spring and autumn in 2021

Seasons	Planting patterns	Nitrogen	N uptake (kg/ha)	NRUE (%)	PNUE (kg/kg)	ANUE (kg/kg)
Spring	Straw Incorporation	N0	142.00±1.00d			
		N200	221.00±1.00c	39.62±0.52a	44.20±1.40a	17.50±0.42a
		N250	242.67±0.33b	40.26±0.15a	37.25±0.69b	15.00±0.33b
		N300	248.33±0.67a	35.63±0.23b	37.43±0.94b	13.33±0.28c
	Traditional Planting	N0	114.33±0.67d			
		N200	183.67±0.33c	34.59±0.27a	51.08±1.15a	17.67±0.38a
		N250	202.00±1.73b	35.14±0.75a	41.46±2.53b	14.53±0.58b
		N300	205.67±0.67a	30.48±0.13b	43.39±1.00b	13.22±0.28b
Autumn	Straw Incorporation	N0	94.67±0.33d			
		N200	168.67±0.88c	36.76±0.47a	48.77±2.25a	17.92±0.72a
		N250	188.33±0.33b	37.39±0.13a	40.12±1.01b	15.00±0.33b
		N300	192.67±0.33a	32.70±0.14b	40.78±0.95b	13.33±0.28b
	Traditional Planting	N0	68.67±0.67d			
		N200	130.00±0.58c	30.73±0.23a	63.74±1.49a	19.58±0.42a
		N250	147.33±0.88b	31.54±0.38a	51.79±0.52b	16.33±0.33b
		N300	150.00±0.00a	27.03±0.04b	52.42±1.86b	14.17±0.48c

(continued)

Seasons	Planting patterns	Nitrogen	N uptake (kg/ha)	NRUE (%)	PNUE (kg/kg)	ANUE (kg/kg)
Spring			195.00±0.00a	36.00±0.00a	42.67±0.33b	15.00±0.00a
Autumn			142.67±0.33b	33.00±00.00b	49.67±0.33a	16.00±0.00a
	Straw incorporation		187.33±0.67a	37.00±0.00a	41.33±0.67b	15.33±0.33a
	Traditional planting		150.33±0.33b	31.67±0.33b	50.33±0.33a	16.00±0.00a
		N0	105.00±0.00d			
		N200	176.00±0.58c	35.33±0.33a	52.00±0.58a	18.00±0.00a
		N250	195.33±0.33b	36.00±0.00a	42.67±0.88b	15.33±0.33b
		N300	199.67±0.33a	31.67±0.33b	43.33±0.88b	13.33±0.33c
Interactions						
Season			***	***	**	*
Planting pattern			***	***	***	ns
Nitrogen			***	***	***	***
Season * Planting pattern			ns	***	**	*
Season * Nitrogen			***	**	ns	ns
Planting pattern * Nitrogen			***	**	ns	ns
Planting pattern * Nitrogen * Season			ns	**	ns	ns

Notes: Different letters indicate significant differences between samples ($P<0.05$). Values are means ± SE ($n=3$).

* Significant at $P<0.05$; ** Significant at $P<0.01$; *** Significant at $P<0.001$; ns, not significant.

4 Discussion

Nitrogen fertilization and residue incorporation generally increased SOC, TN, NHN, AAN, THAN, and AIN in both seasons and planting patterns. However, HUN was significantly higher in the traditional planting system in the N0 treatment than in the other fertilization treatments. The extent of these increases varied with the N fertilization rate and residue management. Soil organic carbon increased with higher N fertilization treatments, and more SOC accumulated in the plot with residue incorporation and higher N fertilization. Similarly, previous studies have demonstrated that crop residue incorporation and N fertilization significantly increase SOC and SOM and improve soil quality and characteristics (Hu et al., 2019; Muhammad et al., 2021). In fact, a high amount of crop residues incorporation or surface mulch considerably increases SOC (Casado-Murillo and Abril, 2013); this high SOC content might probably be attributed to the huge amounts of residues mulching, which affect soil temperature and consequently increases biological activity (Roozbeh and Rajaie, 2021). The higher SOC content in spring might be due to the higher soil moisture and temperature. These results are supported by Han et al. (2012), who reported that optimum soil temperature and moisture content increased the residue decompositions. The incorporated residues may serve as a source of energy for soil microbes and promote further decomposition (Mangalassery et al., 2015; Nachimuthu and Hulugalle, 2016), and subsequently increasing the SOC accumulation (Figure 1). The higher SOC in the residue-incorporated plot might be due to the maintenance of soil moisture, stabilization of SOC, and prevention of C from mineralization in the combination of N fertilization and residue incorporation (Roozbeh and Rajaie, 2021; Yang et al., 2022).

The findings of the current study revealed that residue management and N fertilization had a substantial effect on TN, NHN, ASN, AAN, HUN, THAN, and AIN (Table 1). The N300 treatment with straw incorporation had the highest TN, NHN, ASN, AAN, HUN, THAN, and AIN during spring, whereas the N0 treatment with traditional planting had the lowest during the autumn season (Table 1). It may be necessary to use an inorganic N fertilizer to meet the N needs of soil microorganisms to speed up the decomposition of crop residues such as maize (Singh and Rengel, 2007). These results are supported by Chen et al. (2017b), who reported that maize straw incorporation and N fertilization significantly increased soil N content, which might be due to the higher lignin (13.8 g/kg) and N content. The increase in TN, NHN, ASN, AAN, HUN, THAN, and AIN in the N300 treatment with straw incorporation might be attributed to increased nutrient release from crop residues, incredible biological processes, more N mineralization, and a consequent increase in soil N transformation (Roozbeh and Rajaie, 2021). Our results are in line with recent studies that have reported that plant residue decomposition rates are influenced by seasonal factors and residue management practices in arid and semi-arid regions (Roozbeh and Rajaie, 2021; Zhang et al., 2008). Maintaining agricultural productivity and soil fertility, minimizing N loss, and enhancing SOC sequestration can be achieved by incorporating residues into soil (Liu et al., 2014; Zhang et al., 2014b).

Different fertilization types showed a consistent trend for CO_2 fluxes; however, the treatments showed substantial variation. Straw incorporation with N fertilization resulted in much larger CO_2 emissions than the traditional planting system. In contrast, the N200 treatment with straw incorporation had significantly lower CO_2 fluxes than the N250 and N300 treatments in the traditional planting system. Seasonal CO_2 emissions were substantially higher in the autumn than in spring owing to temperature and rainfall variations. The ecosystem acts as a C sink in spring, but as a net C source in autumn due to the asymmetric response of ecosystem productivity and respiration in response to warming (Tang et al., 2022). Researchers have demonstrated that various fertilizers, SOC content, and soil microbial biomass and activity affect soil CO_2 emissions both directly and indirectly (Muhammad et al., 2018; Sauze et al., 2017). Compared to N0, the N300 treatment had significantly higher cumulative CO_2 emissions; however, this was much higher under residue incorporation with the same N fertilization rates. These results indicated that crop residues incorporated as manure are the primary source of CO_2 production (Muhammad et al., 2018). Furthermore, crop residues as substrate for soil respiration may provide available C and N for microbial decomposition (Bolinder et al., 2020; Mahal et al., 2019), which promotes the formation of soil aggregates, increase soil porosity, and promote CO_2 emissions (Muhammad et al., 2018; Shah et al., 2016). This might explain why the sole application of N fertilizer resulted in lower CO_2 emissions than the combined application of N fertilizer and maize straw (Figure 5A and Figure 5B). This could be related to greater breakdown, which promotes high respiration and CO_2 emissions (Bolinder et al., 2020; Mahal et al., 2019).

Maize residue incorporation with N300 treatment increased the N_2O fluxes and cumulative emissions by 24.02% and 79.45% relative to traditional planting with N300 treatment, respectively ($P<0.05$; Figure 3, Figure 6C and Figure 6D). The seasonal cumulative N_2O emissions ranged from 0.94–3.74 kg N/ha in spring and 0.49–1.33 kg N/ha in autumn. In addition, the residues incorporated during spring had significantly higher cumulative N_2O emissions than those from traditional planting. Soil microbial activities such as nitrification and denitrification are the primary sources of N_2O emissions (Zhu et al., 2013). Soil N_2O emissions are affected by soil C and N availability, water content, and residue management practices (Muhammad et al., 2018; Muhammad et al., 2021). The highest N_2O fluxes were realized with residue incorporation compared to traditional planting during the spring season. These findings clearly demonstrate that the incorporation of maize residues significantly affected both N_2O fluxes and cumulative emissions. Nitrogen sources from maize residues increased N_2O fluxes and cumulative N_2O emissions in both seasons. However, our results suggest that N200 treatment with straw incorporation is the best choice to increase soil fertility, crop yield, and diminish GHGs emissions compared to the N250 and N300 treatments of traditional planting systems. Previous studies have reported that the incorporation and surface mulching of residues greatly influences the N_2O emissions in long-term experiments (Hu et al., 2019; Muhammad et al., 2018). Thus, residue incorporation plays a vital role in the release of mineral N, which is the primary driver of N_2O emissions (Zhao et al., 2016). According to Afreh et al. (2018), increased N_2O emissions were observed in treatments with higher SOC and lower bulk density. Soil microbes play an essential role in limiting N_2O emissions in the field owing to the accessibility of SOC to soil microbes (Afreh et al., 2018; Saggar et al., 2013). Maize plants may also reduce soil N_2O emissions by increasing N absorption (Jiang et al., 2016). Nonetheless, our findings demonstrated that, compared to sole N fertilization, the incorporation of residues boosted plant N uptake while simultaneously increasing soil N_2O emissions, consistent with previously published results from long-term fertilization experiments (Afreh et al., 2018). Nitrogen fertilization and crop residues stimulate N_2O emissions (García-Ruiz et al., 2012; Shcherbak et al., 2014). Fertilizer N is most effective when applied to black soil in northeast China (Li et al., 2013). Bai et al. (2020) reported that chemical fertilization accounted for 64% of soil N_2O emissions, and chemical fertilization with residue incorporation treatment accounted for 84% of the soil N_2O emissions. The application of residues may alleviate the limitation on soil N_2O emissions by providing C and N sources (Bai et al., 2020; Chen et al., 2014).

The CH_4 fluxes mainly remained negative in the traditional planting system and were positive for straw incorporation during the spring and autumn seasons (Figure 4). These results are in line with the findings of Li et al. (2019) from cultivated land and Shimizu et al. (2013) from grassland; they reported that straw incorporation into the soil increased CH_4 fluxes, which might be due to increased SOC. Wang and Luo (2018) also reported that straw incorporation increases CH_4 emissions by 60%. Globally, the quantity of C deposited in the soil is more than twice that of vegetation or atmospheric C (Amundson, 2001; Zhang et al., 2014a). This is because they cover 35%–40% of the global land surface, which may play an essential role in the global C cycle (Lal, 2002; Zhang et al., 2014a). In our study, CH_4 fluxes fluctuated between emission and consumption, with small fluxes, but net absorption was observed for all N treatments in the traditional planting system. Lin et al. (2021) reported that N fertilization significantly decreased CH_4 absorption compared to the N0 treatment; this is consistent with the results that the CH_4 absorption decreased as the rate of N fertilization increased (Figure 4). In addition, the cumulative CH_4 emissions in the combined application of N fertilizer and residue incorporation were significantly higher than those in the sole N application (Lin et al., 2021). The effects of straw incorporation on CH_4 emissions have been studied mostly in maize cropping systems and showed that CH_4 emissions increased due to C and N released from straw decomposition, but varied with the season, residue management, and residue quantity and quality (Muhammad et al., 2018; Mutegi et al., 2010). However, we found a significant increase in CH_4 absorption under the traditional planting system in all N treatments in both seasons, similar to published investigations in upland areas, where straw mulching improved soil properties and could increase CH_4 uptake (da Silva et al., 2014; Gao et al., 2014). However, in our study, straw incorporation with all N treatments increased CH_4 emission compared to the traditional planting system (Figure 5E and Figure 5F), probably because of greater residues decomposition through soil microbes (Fan et al., 2020).

Our results showed that N fertilization and residue incorporation significantly increased maize grain yield, which is in agreement with previous studies (Sadeghi and Bahrani, 2009; Wu et al., 2022). Compared to the N0 and N200 treatments, the N300 treatment significantly increased maize grain yield by 167.18% and 6.95% respectively, but the N200 and N250 treatments were not statistically different (Table 2). These results are supported by Garrido-Lestache et al. (2005), who demonstrated that grain yield increased with increased N fertilization, which might be due to the soil having enough N to support root growth and development, allowing water and nutrients to be transported from the soil to the above-ground parts of the plant (Cheng et al., 2020; Wu et al., 2022). In this study, the area-scaled GWP was significantly higher in the N300 treatment, and the GHGI was higher for N0 treatment with residues incorporation since increased CO_2, N_2O, and CH_4 emissions (Table 2). The availability of C is required for both ammonia-oxidizing and denitrifying bacteria, and can be obtained through residues responsible for the N_2O production (Muhammad et al., 2018). In addition, methanogenic bacteria require available C, which produce CH_4 (Le Mer and Roger, 2001).

Nitrogen fertilization with and without residue incorporation had varying effects on the area-scale and GHGI per unit grain yield, while the N300 treatment resulted in significantly higher area-scale and GHGI (Table 2). However, residues incorporation stimulated higher GHGs emissions than traditional planting. Furthermore, straw incorporation increased the total area-scaled GWP and GHGI, mainly because of the enhanced CO_2, N_2O, and CH_4 emissions compared with traditional planting under the same amount of N fertilization (Table 2). These results are in line with the findings of a the previous study (Wang and Luo, 2018), which reported that residue incorporation enhanced SOC sequestration, which can only offset less than 7% of the total GWP of CH_4 and N_2O emissions. Other agronomic practices may alleviate poor NUE and excessive GHGs emissions. A growing cover crop can effectively decrease GHGs emissions, improve crop productivity and NUE, and increase GHGs emissions during residue incorporation (Muhammad et al., 2018; Wang and

Luo, 2018). The average GHGI was 0.14 kg CO_2-eq/ha for the traditional planting and 0.27 kg CO_2-eq/ha for the residues incorporation (Table 2). Our results showed that straw incorporation increased CO_2 and N_2O fluxes compared with traditional planting with the same N fertilization rates. However, the N200 treatment with straw incorporation had much lower CO_2 and N_2O fluxes, area-scaled GWP, and GHGI compared to the N250 and N300 of traditional planting. The net GWP for traditional planting was significantly lower than that for the residues incorporated with N fertilization which might be due to residue incorporation increasing SOC, which may also increase GHGs emissions (Lehtinen et al., 2014; Wang and Luo, 2018).

Nitrogen fertilization and residue incorporation into the soil stimulate leaf, stem, and grain N and C contents. Leaf, stem, and grain C and N contents were significantly higher in the fertilized plot than in the N0 plot; however, no significant differences were found among the N200, N250, and N300 treatments. These results are consistent with the findings of Bolinder et al. (2020), who reported that plant C and N content increased with increasing N fertilization, and too high N dosages resulted in lower plant C and N contents. Our results showed that plant C and N contents were much higher in residues incorporated plots than in traditional planting in both seasons. However, the spring season had 13.54%, 7.34%, and 4.69% higher C and 23.90%, 14.25%, and 26.74% higher N content in maize leaves, stems, and grains than in the autumn season, respectively (Figure 7 and Figure 8). The lower C and N contents of leaves, stems, and grains during the spring season might be due to higher nutrient losses through GHGs emissions or leachate (Duan et al., 2018; Muhammad et al., 2018). Afreh et al. (2018) reported that organic forming increased plant N content along with GHGs emissions compared to sole N fertilization, and also demonstrated that GHGs emissions were not significantly influenced by plant N uptake during long-term fertilization. Residue incorporation into soil increases the soil microbial growth owing to the direct interaction of soil microbes with crop residues, reducing nutrient leaching and runoff by reducing the direct erosion caused by heavy rainfall (Li et al., 2021; Yang et al., 2022), thus increasing leaf, stem, and grain C and N content (Figure 7 and Figure 8). Previous studies have reported that residue incorporation increases soil microbial growth and development due to increased soil moisture and nutrient content (Muhammad et al., 2021; Yang et al., 2022), improved soil fertility (Muhammad et al., 2018), and increased plant C and N content (Luxhøi et al., 2007).

Nitrogen fertilization resulted in higher plant N uptake in straw incorporated plots than in traditional planting, which might be due to higher above-ground biomass increasing N uptake (Afreh et al., 2018). Under the same N fertilization rate, straw-incorporated plots had considerably higher N uptake and NRUE than traditional planting ($P<0.05$, Table 3). Sole N fertilization dramatically decreases soil pH, and lower pH has a detrimental effect on plant N uptake, soil mineralization, and nitrification, making N available (Wang et al., 2013). In the current study, the increase and subsequent decrease in NRUE with increasing N fertilizer rates emphasizes the importance of balanced crop nutrition. Physiological N use efficiency and ANUE decrease were observed in our study because of higher N fertilization (N250 and N300). Advanced agronomic management with contemporary technology and environmentally friendly practices should be adopted to achieve the optimum utilization of N fertilizer (Wang et al., 2021).

5 Conclusions

The carbon and nitrogen contents in the soil and plant, nitrogen use efficiency, and maize grain yield were higher with N300, but statistically similar to the N250 treatment. Nitrogen fertilization significantly increased CO_2 and N_2O emissions and decreased CH_4 uptake in the residue incorporation plots compared with the N0 treatment. Residues incorporation resulted in higher soil nutrients, plant nitrogen and carbon content, and CO_2, N_2O, and CH_4 cumulative emissions with an equal amount of nitrogen fertilization when compared to the traditional planting system. Based on our findings, the spring season had significantly higher soil total nitrogen, NH_3-N, amino sugar nitrogen, amino acid nitrogen, total acid hydrolyzable organic nitrogen, acid insoluble nitrogen, plant nitrogen and carbon content, grain yield, CO_2, CH_4 uptake, and low N_2O emissions, area-scaled global warming potential, and greenhouse gas intensity than autumn. The N200 treatment with straw incorporation decreased CO_2 fluxes by 12.26% and 36.71%, N_2O fluxes by 6.20% and 24.22%, area-scaled global warming potential by 12.03% and 36.31%, and greenhouse gas intensity by 20.01% and 40.46% under both seasons compared to the N250 and N300 treatments of the traditional planting system. Therefore, residue incorporation could be an effective measure to improve soil fertility, grain yield, plant nutrient uptake, and nitrogen use efficiency and mitigate area-a-scaled global warming potential and greenhouse gas intensity during the spring season in the subtropical region of China. Furthermore, we concluded that crop yield, soil fertility, and plant nutrient uptake could be improved by mitigating greenhouse gas emissions in the N200 treatment in straw-incorporated plots.

Acknowledgments

This study was funded by the Natural Science Foundation of Guangxi Provinces (2019GXNSFAA185028) and National Natural Science Foundation of China (31760354). The authors express their special gratitude to Dr. Ihsan Muhammad for their strong support in revising the final version of the manuscript. We are also thankful to all funding sources and particularly to Guangxi University, for their financial assistance.

References

 Abbasi MK, Tahir MM, Rahim N, 2013. Effect of N fertilizer source and timing on yield and N use efficiency of rainfed maize (*Zea mays* L.) in Kashmir-Pakistan. Geoderma. 195: 87-93.

 Afreh D, Zhang J, Guan D, et al., 2018. Long-term fertilization on nitrogen use efficiency and greenhouse gas emissions in a double maize cropping system in subtropical China. Soil Tillage Res. 180: 259-267.

 Amundson R, 2001. The carbon budget in soils. Ann. Rev. Earth Planet. Sci. 29: 535-562.

 Bai J, Qiu S, Jin L, et al., 2020. Quantifying soil N pools and N_2O emissions after application of chemical fertilizer and straw to a typical chernozem soil. Biol. Fertil. Soils. 56: 319-329.

Bayer C, Gomes J, Zanatta JA, et al., 2015. Soil nitrous oxide emissions as affected by long-term tillage, cropping systems and nitrogen fertilization in Southern Brazil. Soil Tillage Res. 146: 213-222.

Berger S, Kim Y, Kettering J, et al., 2013. Plastic mulching in agriculture—friend or foe of N_2O emissions? Agric. Ecosyst. Environ. 167: 43-51.

Bolinder MA, Crotty F, Elsen A, et al., 2020. The effect of crop residues, cover crops, manures and nitrogen fertilization on soil organic carbon changes in agroecosystems: a synthesis of reviews. Mitig. Adapt. Strateg. Glob. Chang. 25: 929-952.

Casado-Murillo N, Abril A, 2013. Decomposition and carbon dynamics of crop residue mixtures in a semiarid long term no-till system: effects on soil organic carbon. Open Agric. 7: 11-21.

Cheng Y, Wang HQ, Liu P, et al., 2020. Nitrogen placement at sowing affects root growth, grain yield formation, N use efficiency in maize. Plant Soil. 457: 355-373.

Chen H, Liu J, Zhang A, et al., 2017. Effects of straw and plastic film mulching on greenhouse gas emissions in Loess Plateau, China: A field study of 2 consecutive wheat-maize rotation cycles. Sci. Total Environ. 579: 814-824.

Chen X, Mao A, Zhang Y, et al., 2017. Carbon and nitrogen forms in soil organic matter influenced by incorporated wheat and corn residues. Soil Sci. Plant Nutr. 63: 377-387.

Chen Z, Ding W, Luo Y, et al., 2014. Nitrous oxide emissions from cultivated black soil: a case study in Northeast China and global estimates using empirical model. Glob. Biogeochem. Cycles. 28: 1311-1326.

Chi YX, Gao F, Muhammad I, et al., 2022. Effect of water conditions and nitrogen application on maize growth, carbon accumulation and metabolism of maize plant in subtropical regions. Arch. Agron. Soil Sci. 1-15.

Da Silva AP, Babujia LC, Franchini JC, et al., 2014. Soil structure and its influence on microbial biomass in different soil and crop management systems. Soil Tillage Res. 142: 42-53.

De Oliveira Silva B, Moitinho MR, de Araujo Santos G A, et al., 2019. Soil CO_2 emission and short-term soil pore class distribution after tillage operations. Soil Tillage Res. 186: 224-232.

Ding W, Luo J, Li J, et al., 2013. Effect of long-term compost and inorganic fertilizer application on background N_2O and fertilizer-induced N_2O emissions from an intensively cultivated soil. Sci. Total Environ. 465: 115-124.

Duan YF, Hallin S, Jones CM, et al., 2018. Catch crop residues stimulate N_2O emissions during spring, without affecting the genetic potential for nitrite and N_2O reduction. Front. Microbiol. 9: 2629.

Fan D, Liu T, Sheng F, et al., 2020. Nitrogen deep placement mitigates methane emissions by regulating methanogens and methanotrophs in no-tillage paddy fields. Biol. Fertil.

Soils. 56: 711-727.

Foley JA, Ramankutty N, Brauman KA, et al., 2011. Solutions for a cultivated planet. Nature. 478: 337-342.

Frimpong KA, Baggs E, 2010. Do combined applications of crop residues and inorganic fertilizer lower emission of N_2O from soil? Soil Use Manag. 26: 412-424.

Gao X, Rajendran N, Tenuta M, et al., 2014. Greenhouse gas accumulation in the soil profile is not always related to surface emissions in a prairie pothole agricultural landscape. Soil Sci. Soc. Am. J. 78: 805-817.

García-Ruiz R, Gómez-Muñoz B, Hatch D, et al., 2012. Soil mineral N retention and N_2O emissions following combined application of ^{15}N-labelled fertiliser and weed residues. Rapid Commun. Mass Spectrom. 26: 2379-2385.

Garrido-Lestache E, López-Bellido R J, López-Bellido L, 2005. Durum wheat quality under Mediterranean conditions as affected by N rate, timing and splitting, N form and S fertilization. Eur. J. Agron. 23: 265-278.

Han X, Cheng Z, Meng H, 2012. Soil properties, nutrient dynamics, and soil enzyme activities associated with garlic stalk decomposition under various conditions. PLoS One. 7: e50868.

Hoang KTK, Marschner P, 2019. P pools after seven-year P fertiliser application are influenced by wheat straw addition and wheat growth. J. Plant Nutr. Soil Sci. 19: 603-610.

Hu W, Sui N, Yu C, et al., 2019. Comparative effects of crop residue incorporation and inorganic potassium fertilization on soil C and N characteristics and microbial activities in cotton field. J. Cotton Res. 2: 1-12.

Jiang Y, Huang X, Zhang X, et al., 2016. Optimizing rice plant photosynthate allocation reduces N_2O emissions from paddy fields. Sci. Rep. 6: 1-9.

Ju XT, Xing GX, Chen XP, et al., 2009. Reducing environmental risk by improving N management in intensive Chinese agricultural systems. PNAS. 106: 3041-3046.

Lal R, 2002. Carbon sequestration in dryland ecosystems of West Asia and North Africa. Land Degrad. Dev. 13: 45-59.

Lehtinen T, Schlatter N, Baumgarten A, et al., 2014. Effect of crop residue incorporation on soil organic carbon and greenhouse gas emissions in European agricultural soils. Soil Use Manag. 30: 524-538.

Le Mer J, Roger P, 2001. Production, oxidation, emission and consumption of methane by soils: a review. Eur. J. Soil Biol. 37: 25-50.

Lenka NK, Lal R, 2013. Soil aggregation and greenhouse gas flux after 15 years of wheat straw and fertilizer management in a no-till system. Soil Tillage Res. 126: 78-89.

Li J, Li H, Zhang Q, et al., 2019. Effects of fertilization and straw return methods on the soil carbon pool and CO_2 emission in a reclaimed mine spoil in Shanxi Province, China. Soil Tillage Res. 195: 104361.

Li L J, You MY, Shi HA, et al., 2013. Soil CO_2 emissions from a cultivated Mollisol: Effects of organic amendments, soil temperature, and moisture. Eur. J. Soil Biol. 55:

83-90.

Li N, Ma X, Bai J, et al., 2021. Plastic film mulching mitigates the straw-induced soil greenhouse gas emissions in summer maize field. Appl. Soil Ecol. 162: 103876.

Lin S, Zhang S, Shen G, et al., 2021. Effects of inorganic and organic fertilizers on CO_2 and CH_4 fluxes from tea plantation soil. Elem. Sci. Anth. 9.

Liu J, Zhu L, Luo S, et al., 2014. Response of nitrous oxide emission to soil mulching and nitrogen fertilization in semi-arid farmland. Agric. Ecosyst. Environ. 188: 20-28.

Liu Y, Pan Y, Yang L, et al., 2022. Stover return and nitrogen application affect soil organic carbon and nitrogen in a double-season maize field. Plant Biol. 24: 387-395.

Luxhøi J, Elsgaard L, Thomsen I, et al., 2007. Effects of long-term annual inputs of straw and organic manure on plant N uptake and soil N fluxes. Soil Use Manag. 23: 368-373.

Mahal NK, Osterholz WR, Miguez FE, et al., 2019. Nitrogen fertilizer suppresses mineralization of soil organic matter in maize agroecosystems. Front. Ecol. Evol. 7: 59.

Mangalassery S, Mooney SJ, Sparkes D, et al., 2015. Impacts of zero tillage on soil enzyme activities, microbial characteristics and organic matter functional chemistry in temperate soils. Eur. J. Soil Biol. 68: 9-17.

Muhammad I, Khan F, Khan A, et al., 2018. Soil fertility in response to urea and farmyard manure incorporation under different tillage systems in Peshawar, Pakistan. Int. J. Agric. Biol. 20: 1539-1547.

Muhammad I, Wang J, Sainju UM, et al., 2021. Cover cropping enhances soil microbial biomass and affects microbial community structure: A meta-analysis. Geoderma. 381: 114696.

Muhammad I, Yang L, Ahmad S, et al., 2022. Irrigation and nitrogen fertilization alter soil bacterial communities, soil enzyme activities, and nutrient availability in maize crop. Front. Microbiol. 13: 833758.

Mutegi JK, Munkholm LJ, Petersen BM, et al., 2010. Nitrous oxide emissions and controls as influenced by tillage and crop residue management strategy. Soil Biol. Biochem. 42: 1701-1711.

Nachimuthu G, Hulugalle N, 2016. On-farm gains and losses of soil organic carbon in terrestrial hydrological pathways: a review of empirical research. Int. Soil Water Conserv. Res. 4: 245-259.

Page A, Miller R, Keeney D, 1982. Methods of soil analysis, part 2: Chemical and microbiological properties. Madison WI: American Society of Agronomy. Soil Sci. Soc. Am. J. pp: 595-624.

Pan FF, Yu WT, Ma Q, et al., 2017. Influence of ^{15}N-labeled ammonium sulfate and straw on nitrogen retention and supply in different fertility soils. Biol. Fertil. Soils. 53: 303-313.

Pulido-Fernández M, Schnabel S, Lavado-Contador J F, et al., 2013. Soil organic matter of Iberian open woodland rangelands as influenced by vegetation cover and land management. Catena. 109: 13-24.

Ren H, Feng Y, Liu T, et al., 2020. Effects of different simulated seasonal temperatures on the fermentation characteristics and microbial community diversities of the maize straw and cabbage waste co-ensiling system. Sci. Total Environ. 708: 135113.

Rochette P, Angers DA, Chantigny M H, et al., 2013. Ammonia volatilization and nitrogen retention: how deep to incorporate urea? J. Environ. Qual. 42: 1635-1642.

Roozbeh M, Rajaie M, 2021. Effects of residue management and nitrogen fertilizer rates on accumulation of soil residual nitrate and wheat yield under no-tillage system in south-west of Iran. Int. Soil Water Conserv. Res. 9: 116-126.

Sadeghi H, Bahrani MJ, 2009. Effects of crop residue and nitrogen rates on yield and yield components of two dryland wheat (*Triticum aestivum* L.) cultivars. Plant Prod. Sci. 12: 497-502.

Saggar S, Jha N, Deslippe J, et al., 2013. Denitrification and N_2O: N_2 production in temperate grasslands: Processes, measurements, modelling and mitigating negative impacts. Sci. Total Environ. 465: 173-195.

Sauze J, Ogée J, Maron PA, et al., 2017. The interaction of soil phototrophs and fungi with pH and their impact on soil CO_2, $CO^{18}O$ and OCS exchange. Soil Biol. Biochem. 115: 371-382.

Shah A, Lamers M, Streck T, 2016. N_2O and CO_2 emissions from South German arable soil after amendment of manures and composts. Environ. Earth Sci. 75: 1-12.

Shahbaz M, Menichetti L, Kätterer T, et al., 2019. Impact of long-term N fertilisation on CO_2 evolution from old and young SOM pools measured during the maize cropping season. Sci. Total Environ. 658: 1539-1548.

Shcherbak I, Millar N, Robertson G P, 2014. Global metaanalysis of the nonlinear response of soil nitrous oxide (N_2O) emissions to fertilizer nitrogen. PNAS. 111: 9199-9204.

Singh B, Rengel Z, 2007. The role of crop residues in improving soil fertility. Nutr. Cycl. Agroecosystems. Springer, pp: 183-214.

Smith P, Martino D, Cai Z, et al., 2008. Greenhouse gas mitigation in agriculture. Philos. Trans. R. Soc. Lond., B, Biol. Sci. 363: 789-813.

Snyder CS, Bruulsema TW, Jensen TL, et al., 2009. Review of greenhouse gas emissions from crop production systems and fertilizer management effects. Agric Ecosyst Environ. 133: 247-266.

Solomon S, Manning M, Marquis M, et al., 2007. Climate change 2007 - the physical science basis: Working group I contribution to the fourth assessment report of the IPCC. Vol 4: CambridgeUniversity Press.

Sorkhi F, Fateh M, 2014. Effect of nitrogen fertilizer on yield component of maize. Int. J. Biosci. 5: 16-20.

Sun H, Zhang H, Powlson D, et al., 2015. Rice production, nitrous oxide emission and ammonia volatilization as impacted by the nitrification inhibitor 2-chloro-6-(trichloromethyl)-pyridine. Field Crops Res. 173: 1-7.

Tang R, He B, Chen HW, et al., 2022. Increasing terrestrial ecosystem carbon release in response to autumn cooling and warming. Nat. Clim. Change. 12: 380-385.

Wang L, Du H, Han Z, et al., 2013. Nitrous oxide emissions from black soils with different pH. Res. J. Environ. Sci. 25: 1071-1076.

Wang M, Pendall E, Fang C, et al., 2018. A global perspective on agroecosystem nitrogen cycles after returning crop residue. Agric. Ecosyst. Environ. 266: 49-54.

Wang XY, Luo Y, 2018. Crop residue incorporation and nitrogen fertilizer effects on greenhouse gas emissions from a subtropical rice system in Southwest China. J. Mt. Sci. 15: 1972-1986.

Wang Z, Wang Z, Ma L, et al., 2021. Straw returning coupled with nitrogen fertilization increases canopy photosynthetic capacity, yield and nitrogen use efficiency in cotton. Eur. J. Agron. 126: 126267.

Wu P, Liu F, Chen G., 2022. Can deep fertilizer application enhance maize productivity by delaying leaf senescence and decreasing nitrate residue levels? Field Crops Res. 277: 108417.

Wu P, Liu F, Li H, et al., 2021. Suitable fertilizer application depth can increase nitrogen use efficiency and maize yield by reducing gaseous nitrogen losses. Sci. Total Environ. 781: 146787.

Xia L, Wang S, Yan X, 2014. Effects of long-term straw incorporation on the net global warming potential and the net economic benefit in a rice-wheat cropping system in China. Agric. Ecosyst. Environ. 197: 118-127.

Xu Y, Wang Y, Ma X, et al., 2020. Ridge-furrow mulching system and supplementary irrigation can reduce the greenhouse gas emission intensity. Sci. Total Environ. 717: 137262.

Yagioka A, Komatsuzaki M, Kaneko N, et al., 2015. Effect of no-tillage with weed cover mulching versus conventional tillage on global warming potential and nitrate leaching. Agric. Ecosyst. Environ. 200: 42-53.

Yang L, Muhammad I, Chi YX, et al., 2022. Straw return and nitrogen fertilization to maize regulate soil properties, microbial community, and enzyme activities under dual cropping system. Front. Microbiol. 13: 823963.

Zhang D, Hui D, Luo Y, et al., 2008. Rates of litter decomposition in terrestrial ecosystems: global patterns and controlling factors. J. Plant Ecol. 1: 85-93.

Zhang J, Wang XJ, Wang J P, et al., 2014. Carbon and nitrogen contents in typical plants and soil profiles in Yanqi Basin of Northwest China. J. Integr. Agric. 13: 648-656.

Zhang X, Fan C, Ma Y, et al., 2014. Two approaches for net ecosystem carbon budgets and soil carbon sequestration in a rice-wheat rotation system in China. Nutr. Cycl. 100: 301-313.

Zhao X, He L, Zhang Z, et al., 2016. Simulation of accumulation and mineralization (CO_2 release) of organic carbon in chernozem under different straw return ways after corn harvesting. Soil Tillage Res. 156: 148-154.

Zhao X, Liu BY, Liu SL, et al., 2020. Sustaining crop production in China's cropland by crop residue retention: A meta-analysis. Land Degrad. Dev. 31: 694-709.

Zhou Y, Zhang Y, Tian D, et al., 2017. The influence of straw returning on N2O emissions from a maize-wheat field in the North China Plain. Sci. Total Environ. 584: 935-941.

Zhu X, Burger M, Doane TA, et al., 2013. Ammonia oxidation pathways and nitrifier denitrification are significant sources of N_2O and NO under low oxygen availability. PNAS 110: 6328-6333.

Straw return and nitrogen fertilization to maize regulate soil properties, microbial community, and enzyme activities under dual cropping system

L. Yang[1], I. Muhammad[1], Y. X. Chi[1,2], D. Wang[3] and X. B. Zhou[1]

([1] Guangxi Colleges and Universities Key Laboratory of Crop Cultivation and Tillage, Agricultural College, Guangxi University, Nanning 530004, China;

[2] The Key Laboratory of Germplasm Improvement and Cultivation in Cold Regions, College of Agronomy, Heilongjiang Bayi Agricultural University, Daqing 163319, China;

[3] College of Horticulture and Landscape, Tianjin Agricultural University, Tianjin 300392, China)

Abstract: Soil sustainability is based on soil microbial communities' abundance and its composition. Straw returning (SR) and nitrogen (N) fertilization influence soil fertility, enzyme activities, and the soil microbial community and structure. However, still remain unclear due to heterogeneous composition and varying decomposition rate of added straw. Therefore, the current study aimed to determine the effect of SR and N fertilizer application on soil organic carbon (SOC), total nitrogen (TN), urease (S-UE) activity, sucrase (S-SC) activity, cellulose (S-CL) activity, bacterial, fungal and nematode community composition from March to December 2020 at Guangxi University, China. Treatments included two planting patterns that is SR and traditional planting (TP) and six N fertilizer with 0, 100, 150, 200, 250 and 300 kg N/ha. Straw returning significantly increased soil fertility, enzymatic activities, community diversity and composition of bacterial and fungal compared to TP. Nitrogen fertilizer application increased soil fertility and enzymes, decreased richness of bacterial and fungal. In SR added plots, the dominated bacterial phyla were Proteobacteria, Acidobacterioia, Nitrospirae, Chloroflexi and Actinobacteriota; where fungal phyla were Ascomycota and Mortierellomycota and nematode genera were *Pratylenchus* and *Acrobeloides*. Co-occurrence network and Redundancy Analysis (RDA) analysis showed that TN, SOC and S-SC were closely correlated with bacterial community composition. It was concluded that the continuous SR and N fertilizer improved soil fertility and improved soil bacterial, fungal and nematode community composition.

Keywords: Straw return; Nitrogen fertilization; Soil enzymes; Soil microbes; Soil properties

---

These authors contributed equally.
Corresponding author E-mail address: xunbozhou@gmail.com.

1 Introduction

Carbon sequestration and long-term sustainability can be improved with residues return to the field (Su et al., 2020a; Lu et al., 2021). Straw mainly composed of carbon (C), nitrogen (N) and organic matter (OM) (Fan and Wu 2020), which can effectively enhance soil fertility and mitigate the negative impacts of excessive synthetic fertilizers uses (Zhao et al., 2019; Wang et al., 2021b). Numerous studies have demonstrated that SR reduced mineral fertilizer application by enhancing nutrient efficiency and improved organic C inputs (Wu et al., 2020; Wang et al., 2021a). Furthermore, SR has tremendous potential to improve soil health and micro-ecological environment (Su et al., 2020a). Straw decomposition is a complex process (Zhao and Zhang, 2018; Zhao et al., 2019), which predominantly mediated by soil microorganisms with specialized functions (Zhao and Zhang, 2018). A variety of microbial communities play significant roles in the crop residues decomposition, such as bacteria preferring to decompose labile compounds and dominating straw degradation at initial stage of decomposition (Mwafulirwa et al., 2021; Wu et al., 2021). In contrast, fungi decompose more abrasive materials principally in final stages of decomposition (Marschner et al., 2011). The energy and C derived from the crop residues incorporated in to the soil are distributed throughout the trophic levels, affecting different soil microorganisms such as soil bacteria, fungi and nematodes (Chen et al., 2021).

In addition to biotic component, the abiotic variables such as temperature and soil moisture, soil N, pH, and soil organic carbon (SOC) also affect the decomposition process and soil microbial activity, structure and community (Geisseler and Scow, 2014; Kamble and Bååth, 2016). Guangxi is located in south of China, which is typical subtropical monsoon humid region and committed to double-cropping system, where maize (*Zea mays* L.) is one of the main food crop (Huang et al., 2021). Straw decomposition and nutrient release are accelerated by the high temperature in this region (Ren et al., 2020). Residues incorporation into the soil decomposed more rapidly, which releases a variety of mineral nutrients that may be easily available and absorbed by plants (Muhammad et al., 2021). The straw C/N ratio is a factor that determines the rate of decomposition of the material, with lower C/N ratios favoring bacteria while higher ratios fungi, which have ability to degrade more complex organic molecules such as lignin.

The activity of soil enzymes has long been considered as a fundamental indication of soil quality (Borase et al., 2020). Soil enzymatic activities are generated and released by soil microbes, which are responsible for organic matter degradation (Burns et al., 2013). The activity of soil enzymes can provide insight into the processes of microbial sensitivity to added C and N (Zhao et al., 2016). The addition of C and N to soil and field management approaches may affect different enzymes in different ways (Gong et al., 2021; Huang et al., 2021). Soil sucrase activity can help plants and soil bacteria use sucrose as an energy source by hydrolyzing it into glucose and fructose (Yu et al., 2018; Wu et al., 2020). Similarly, cellulose (S-CL) activity is significantly correlated with sucrase activity when organic matter was added to the soil, showing that the individual substrate can affect an enzyme-catalyzed biological reaction (Salazar et al., 2011). Thus, understanding the relationship between the impact of SR and N application on soil microorganisms and the production of soil enzyme activities is crucial and will provide insight into the fundamental mechanics of SOC changes under long-term SR.

Therefor this study was conducted to investigate the effects of SR on the soil bacteria, fungi, and nematode communities, enzyme activities, and soil fertility. High-throughput DNA sequencing of PCR-amplified marker genes sequencing technology has recently provided significant insights into the diversity of the microbial community in various fertilizer management practices and SR decomposition (Zhao et al., 2016; Yuan et al., 2018). We conducted a field experiment in a subtropical region of China with two planting patterns (SR and TP) under N fertilizer application. The objectives of the study were (1) to investigate the impacts of N fertilization and SR on the soil microbial diversity and community composition; (2) to determine the response of soil enzyme activity under various N levels and SR systems; and (3) to understand how soil microbes and enzyme activities alter SOC and total N under SR and N application.

2 Materials and Methods

2.1 Experimental site

A filed experiment was conducted at Agronomy Research Farm of Guangxi University, Nanning, Guangxi, China (22°50′N, 108°17′E) from March to December 2020. The region belongs to a sub-tropical monsoon climate, with a mean annual temperature of 21.7℃ and mean annual precipitation of 1,298 mm. The soil of the experimental site was classified as clay loam, with initial properties of 14.6 g/kg SOC, 0.8 g/kg total N, 42.7 mg/kg available P, 88.5 mg/kg available K, and pH 6.5 when sampling at 0-20 cm depth.

2.2 Experimental design

The experiment was carried out in randomized complete block design in split-plot arrangement with three replications. The planting pattern was the main plot and N fertilization was the split-plot factors. The two planting patterns were straw return (SR; spring and autumn maize residue were mechanically crushed stalks in to 2-3 cm and mixed with soil in top 0-20 cm with rotary tillage) and traditional planting (TP; maize straw was removed from the field after harvested) since 2018. The six N fertilizer treatments were control (N0), 100 kg N/ha (N100), 150 kg N/ha (N150), 200 kg N/ha (N200), 250 kg N/ha (N250) and 300 kg N/ha (N300).

In 2020, the maize cultivar "Zhengda619" was sown twice, the first time on March 11th and harvested on July 9th for spring maize, and the second time on August 2nd and harvested on November 30th for autumn maize. The plot size was 4.2 m × 4.2 m, with a planting density of 55,556 plants/ha, having 60 cm row to row space and 30 cm planting space. The recommended basal doses of phosphate fertilizer (calcium magnesium phosphate, P_2O_5 content of 18%) and potash fertilizer (KCl, K_2O content of 60%) were incorporated at the rate of 100 kg/ha into the soil before sowing. Two-thirds of the N from urea was applied prior to sowing, with the remaining one-third applied during the large trumpet period. Other field management measures were consistent with typical farming procedures.

2.3 Soil sampling

Five random soil samples from 0-20 cm soil depth were obtained from each sub-plot at the jointing stage (V6) and mature stage (R6) before maize harvest. These soil samples were thoroughly mixed, sieved through a 2 mm mesh and then divided into two parts, one of which was immediately stored at-80℃ in the laboratory for molecular analysis. Additionally, the other portion of the soil samples were air-dried at room temperature and then sieved through 0.069 mm mesh for the analysis of enzyme activities, and/or sieved through 0.15 mm mesh for TN and SOC determination.

2.4 Soil fertility and enzyme activities analysis

Potassium dichromate volumetric and external heating techniques were used to quantify SOC (Chen et al., 2021), while soil TN was measured by the semi-micro kelvin method (Wang et al., 2021a). The activities of soil enzymes, soil urease (S-UE), soil sucrase (S-SC), and soil cellulase (S-CL), were evaluated using Solarbio analytical kits BC0125, BC0245, and BC0155S, respectively, (Science & Technology Co., Ltd. Beijing China) as per the procedure of the manufacturer.

2.5 DNA preparation, PCR amplification, and Highthroughput sequencing

Total genomic DNA was extracted through using the E.Z.N.A.® soil DNA Kit according to the manufacturer instructions (Omega Bio-tek Inc., Norcross, USA). The final DNA concentration and purity were determined using a Nano Drop 2000 UV-vis spectrophotometer (Thermo Scientific, Wilmington, USA), and the DNA quality was determined using 1% agarose gel electrophoresis. A Thermocycler PCR system was used to perform PCR amplification (Gene Amp 9700, ABI, USA). Bacterial primers 338F 5′-ACTCCTACGGGAGGCAGCAG-3′ and 806R 5′-GGACTACHVGGGTWTCTAAT-3′ were used to amplify the V3-V4 hypervariable sections of the 16S rRNA gene (Chen et al., 2018). The barcode primers ITS1F (5′-CTTGGTCATTTAGAG-GAAGTAA-3′) and ITS2R (5′-GCTGCGTTCTTCATCGATGC-3′) were used to amplify the fungal rRNA gene in the ITS1 sequence region (Dang et al., 2020). The fungal ITS1 sequence region was used for nematode DNA gene amplification using the barcode primers NF1 5′-GGTGGTG-CATGGCCGTTCTTAGTT-3′ and 18Sr2bR 5′-TACAAAGGGCAGGGACGTAAT-3′ (Xue et al., 2019). The PCR products from bacteria, fungi, and nematodes were extracted from a 2% agarose gel, purified using the AxyPrep DNA Gel Extraction Kit (Axygen Biosciences, Union City, CA, USA), and quantified using a QuantusTM Fluorometer (Promega, USA).

2.6 Processing of the sequencing data

All three PCR products (bacterial, fungal, and nematode) were purified, pooled in equimolar amounts and paired-end sequenced (2×300) on an Illumina MiSeq platform (Illumina, San Diego, USA) by Majorbio Bio-Pharm Technology Co., Ltd. (Shanghai, China; Zeng et al.,

2020). UPARSE version 7.1 was used to cluster the processed sequences into operational taxonomic units (OTUs) that had at least 97% similarity (Gdanetz et al., 2017). The RDP Classifier method was used to assess the taxonomy of the bacterial sequences against the SILVA database (version 128/16S-bacteria database) and fungal against the USA database (version 7.00; fungal-database) with a confidence level of 70% (Zheng et al., 2020). The taxonomy of nematode sequences was analysed by the RDP Classifier algorithm against the NCBI database (version NT/its-nematode database) using a confidence threshold of 70% (Xue et al., 2019).

2.7 Alpha and beta diversity analysis

An OTU-based analytical technique was used to assess the bacterial, fungal, and nematode diversity in each sample. The OTU richness and diversity of each sample were assessed using QIIME software version v1.8.0, with a sequencing depth of 3% to measure the diversity index and species richness (alpha diversity). Beta diversity analysis was used across all samples to estimate the community structure comparison index. The beta diversity of genotypes was calculated at the OTU level using weighted UniFrac distances and visualized using main coordinates analysis (PCoA). The QIIME tool was used to group and evaluate the weighted UniFrac distance matrices. They discovered evolutionary connections between various groups and the quantity of those samples.

2.8 Statistical analyses

The results of enzymatic activities and nutrient contents were analyzed using two-way ANOVA under two planting patterns and six nitrogen fertilizer application rates. The least significant difference (LSD) was used to separate means and interactions, and statistical significance was evaluated at $P \leqslant 0.05$. The alpha diversity was calculated utilizing the Chao1 and Shannon diversity indices. Soil nutrient, enzyme activity and alpha diversity were correlated using R package "pheatmap" (version 3.3.1). Beta diversity was estimated using the Bray-Curtis distance matrix and PCoA. Redundancy analysis (RDA) was used to examine the relationship between soil sample distribution and soil properties using R 4.0 package (https://cran.r-project.org/web/packages/rda/). These analyses were carried out to assess community compositions across all samples using OTU composition and to depict the link between bacterial, fungal, and nematode communities and soil properties.

3 Results

3.1 Soil Fertility and enzyme activities

Straw return had enhanced the soil SOC and TN under both spring and autumn maize cultivation (Figure 1). Compared to the TP treatment, SR significantly increased soil SOC content by 2.8%–9.0% and TN content by 6.0%–7.1% in both spring and autumn maize cultivated fields ($P < 0.05$). Averaged across SR, SOC and TN contents were 10.7 and 1.9 g/kg in spring maize field, and 10.2 and 1.6 g/kg in autumn maize field, respectively. Our results showed that SOC contents significantly increased by 7.40% and 2.98% at the V6 stage in spring and autumn maize field, respectively (Figure 1A, Figure 1B). In contrast, TN contents decreased by 5.13% in spring maize field and 6.6% in autumn maize field (Figure 1C, Figure 1D). In both spring and autumn maize fields, the SOC and TN contents generally increased with N fertilizer applications. The average SOC and TN contents of N100, N150, N200, N250 and N300 significantly increased compared to the N0 treatment ($P<0.05$). These results demonstrated 8.8%–32.1% changes in SOC and 3.1%–8.5% changes in soil TN, however no significant differences were observed between N200 and N250, and N250 and N300 treatments ($P>0.05$).

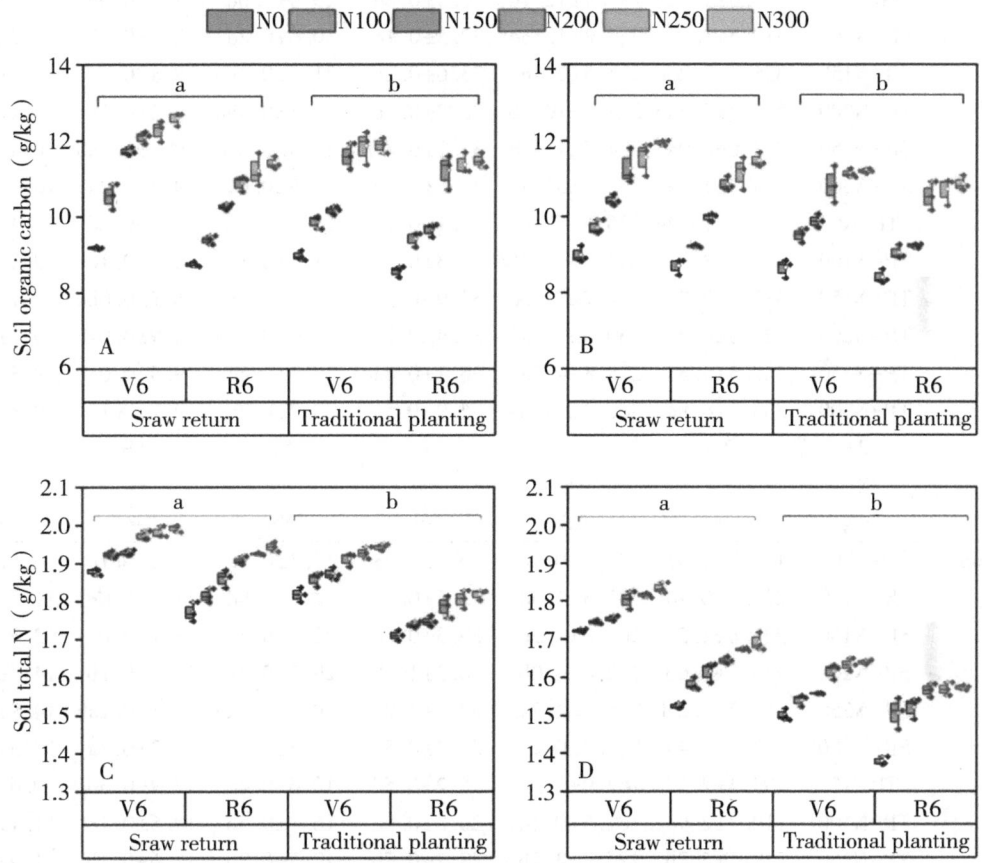

Figure 1 Changes in soil physicochemical properties in double-cropped maize of combined straw with nitrogen

The combined application of straw and nitrogenfertilizer (Table 1) improved soil enzyme activities for both seasons (S). Averaged across two stages, the SR increased S-UE, S-SC, and S-CL by 5.5%, 3.5%, and 5.8% in spring and 5.9%, 18.4%, and 12.2% in autumn, respectively (Table 1). Nitrogen fertilization significantly increased the soil enzyme activities in both planting patterns and seasons, suggesting that the S-UE, S-SC, and S-CL activities were significantly boosted in N300 compared to N0 treatment. However, the N250 treatment was not statistically different from N200 and N300, respectively. On average basis, S-UE, S-SC, and S-CL activities were 290.1, 32.7, and 8.99 U/g in spring, and 216.3, 24.3, and 12.24 U/g in autumn, respectively. These results demonstrated that the S-UE and S-SC activities were higher in spring than in autumn, but the S-CL activity was lower. Moreover, the S-UE and S-SC activities were higher and S-CL activity was lower at V6 stage than R6 in both seasons.

Table 1　Changes in soil enzyme activities in double-cropped maize of combined straw with nitrogen fertilizer applications

Season	Treatment	S-UE (U/g)		S-SC (U/g)		S-CL (U/g)	
		V6	R6	V6	R6	V6	R6
Spring	SR-N0	278.6±12.2d	237.4±12.1d	30.3±0.2d	29.0±0.6e	7.8±0.0d	8.1±0.0e
	SR-N100	310.6±2.9c	246.2±5.1cd	32.5±0.1c	30.6±0.2d	8.6±0.1c	8.7±0.0d
	SR-N150	326.4±2.2bc	263.5±2.4c	33.6±0.2c	31.3±0.4cd	8.8±0.1c	9.3±0.0c
	SR-N200	343.7±7.5ab	293.3±10.1b	35.7±0.9b	32.4±0.5bc	9.2±0.1b	10.1±0.1b
	SR-N250	353.6±6.0a	304.7±1.4ab	36.7±0.8ab	33.8±0.6ab	9.5±0.1ab	10.4±0.2ab
	SR-N300	354.3±1.2a	318.7±5.6a	38.5±0.7a	34.5±0.6a	9.7±0.2a	10.7±0.2a
	TP-N0	273.3±13.4c	226.5±5.1c	29.1±0.7e	26.3±0.7d	7.2±0.2d	7.4±0.0d
	TP-N100	301.5±3.9c	241.0±11.5bc	31.3±0.8d	28.8±0.8c	8.5±0.1c	9.0±0.1c
	TP-N150	317.7±2.2b	246.7±7.1bc	33.9±0.9c	30.4±0.2bc	8.7±0.0bc	9.1±0.0bc
	TP-N200	327.6±4.9a	260.6±3.9ab	34.4±0.2bc	31.6±0.9ab	8.9±0.0ab	9.2±0.0ab
	TP-N250	328.8±2.3a	278.8±4.5a	36.1±0.4ab	33.3±0.4a	9.0±0.0a	9.3±0.0a
	TP-N300	333.6±4.8a	306.3±11.8a	36.9±0.5a	33.2±0.7a	9.1±0.1a	9.3±0.0a
	P	NS	**	*	NS	NS	**
	N	**	**	**	**	**	**
	P×N	*	NS	NS	NS	**	**
Autumn	SR-N0	199.0±1.5e	174.0±0.9e	19.6±0.2e	17.3±0.5e	10.2±0.1e	10.6±0.3d
	SR-N100	215.9±0.4d	185.3±1.6d	22.5±0.3d	20.9±0.8d	11.4±0.0d	11.5±0.1c
	SR-N150	226.8±1.5c	206.0±1.3c	25.3±0.6c	22.5±0.4c	12.7±0.0c	12.3±0.1c
	SR-N200	243.7±7.6b	232.4±4.0b	30.3±1.0b	29.5±0.4b	13.7±0.1b	14.3±0.3bc
	SR-N250	249.2±2.2ab	245.2±4.2ab	32.2±1.0ab	30.8±0.1ab	14.1±0.1ab	14.8±0.2ab
	SR-N300	252.2±1.4a	247.3±0.2a	33.3±0.5a	31.7±0.3a	14.3±0.3a	15.5±0.5a
	TP-N0	193.4±2.5d	162.9±3.5d	15.2±0.5d	10.6±0.4c	7.9±0.3d	9.6±0.3d
	TP-N100	210.3±2.9c	182.5±0.4c	20.3±0.8c	16.1±0.5b	10.6±0.1c	11.1±0.1c
	TP-N150	219.5±2.2b	195.8±3.2b	23.2±0.7b	17.3±0.2b	11.6±0.1b	11.9±0.0b
	TP-N200	235.3±3.7a	214.6±0.8a	29.7±0.7a	23.1±0.4a	12.0±0.2ab	12.7±0.5a
	TP-N250	236.6±0.5a	216.6±0.5a	30.9±0.6a	24.4±0.4a	12.2±0.2a	13.1±0.0a
	TP-N300	242.0±0.5a	219.5±1.1a	31.4±0.4a	24.6±0.8a	12.3±0.1a	13.4±0.0a
	P	NS	*	NS	**	**	*
	N	**	**	**	**	**	**
	P×N	NS	**	*	NS	**	**

Notes: Different letters indicate significant differences between samples ($P<0.05$). Values are means ± SE ($n=3$).

*, significant at $P<0.05$; **, significant at $P<0.01$; NS, not significant.

S-UE, soil urease activity; S-SC, soil sucrase activity; S-CL, soil cellulase activity. SR, TP, P and N are straw return, traditional planting, planting pattern and nitrogen, respectively.

3.2 Alpha diversity of soil bacterial, fungal and nematode community

Base on the morphological traits, yield, soil organic carbon, nitrogen and soil enzyme activities, highly significant differences were observed in response to 200 kg N/ha under both traditional and straw returning pattern. Therefore, the soil from the plot receiving 200 kg N/ha of maize field was adopted for soil bacterial, fungal, and nematode analysis.

After quality filtering, a total of 2,201 bacterial, 984 fungal, and 395 nematode OUTs were identified from 1,415,768, 1,775,268 and 1,433,535 RNA gene sequences, respectively. The bacterial OTUs number exhibited the trend SR-N0>TP-N0>SR-N200>TP-N200 during spring and autumn seasons (Figure S1-A, Figure S1-B). Where the trend for soil fungal OTUs number was SR-N0>SR-N200>TP-N0>TP-N200 during spring, and SR-N0>TP-N0>SR-N200>TP-N200 during autumn (Figure S2-A, Figure S2-B), and the nematode OTUs number exhibited the trend TP-N200>TP-N0>SR-N200>SR-N during spring, and during the autumn the OTUs number for SR-N0 and TP-N200 were found the same; however, SR-N0 have higher OTUs than SR-N200 (Figure S3-A, Figure S3-B). The results indicated that SR plot resulted in higher bacterial diversity (Shannon index) in both autumn season, however not significantly affected by N fertilization. The bacterial richness (Chao and ACE index) were not affected by planting pattern nor N fertilization, except SR-N0 had significantly higher ACE compared to TP-N200 in autumn season (Table 2). The fungal richness (Chao and ACE index) were significantly affected by seasons, planting pattern, and N fertilization. During spring season neither planting pattern nor N fertilization had significant effect on fungal diversity and richness, except SR-N0 treatment had significantly lower diversity compared to other treatments. Similarly, the fungal diversity was significantly higher in SR treatment compared to TP during autumn season, whereas SR-N0 had noticeably higher fungal richness than TP-N200 (Table 2). In contrast, the bacterial, fungal and nematode richness were significantly higher in TP during spring season compared to SR treatments. However, SR-N200 treatment had significantly higher nematode Chao1 and ACE richness compared to SR-N0 treatment during spring season. Moreover, our results showed that the nematode diversity was not significantly affected by seasons, planting pattern, and N fertilization (Table 2). No variation in alpha-diversity of bacterial and nematode communities were observed during spring season ($P>0.05$). However, bacterial and fungal diversity was significantly higher in SR than TP treatments during autumn season in response to N fertilization application suggesting that bacterial and fungal are more sensitive to straw returning than nematode in autumn season ($P<0.05$). The multi-factor analysis revealed that the bacterial ($P<0.05$) and fungal alpha-diversity ($P<0.01$) were significantly affected by planting pattern and seasons, respectively (Table 2).

Table 2 Alpha-diversity of the soil bacterial, fungal and nematode communities with straw return and nitrogen application

Seasons	Treatments	Bacteria						Fungi						Nematode					
		Diversity	Richness					Diversity	Richness					Diversity	Richness				
		Shannon	Chao		ACE			Shannon	Chao		ACE			Shannon	Chao		ACE		
Spring	SR-N0	6.61±0.02a	2 101.6±27.6a		2 080.6±32.8a			4.17±0.08b	576.9±75.9a		567.8±69.3a			1.39±0.36a	6.33±2.31c		5.04±5.06c		
	SR-N200	6.65±0.04a	2 091.5±28.6a		2 073.4±27.7a			4.30±0.03a	565.8±79.0a		556.7±77.3a			1.48±0.47a	14.67±2.89b		15.04±3.54b		
	TP-N0	6.59±0.06a	2 096.4±17.0a		2 080.7±23.6a			4.37±0.06a	545.7±78.8a		535.4±77.1a			1.61±0.62a	22.83±8.31a		29.97±15.71a		
	TP-N200	6.54±0.06a	2 047.8±45.0a		2 038.7±28.0a			4.30±0.02a	460.6±86.1a		458.1±85.4a			1.92±0.27a	21.50±1.8a		22.46±1.33a		
Autumn	SR-N0	6.61±0.04a	2 116.1±21.7a		2 101.4±17.6a			4.60±0.09a	691.9±9.6a		682.7±13.6a			1.72±0.21a	19.33±4.04a		19.33±4.04a		
	SR-N200	6.64±0.02a	2 084.2±26.8a		2 067.3±16.9ab			4.56±0.07a	580.9±80.6ab		575.5±80.6ab			1.29±0.44a	15.17±6.45a		11.42±13.47a		
	TP-N0	6.62±0.06ab	2 111.6±17.3a		2 090.0±21.0a			4.41±0.03b	588.7±78.6ab		579.8±82.0ab			1.42±0.46a	20.33±3.51a		20.76±2.93a		
	TP-N200	6.51±0.03b	2 069.9±38.1a		2 038.3±26.0b			4.41±0.05b	493.9±108.3b		485.39±105.4b			1.81±0.16a	21.83±2.02a		22.61±1.93a		
Planting pattern (P)		*	NS		NS			NS	*		*			NS	NS		NS		
Nitrogen (N)		NS	*		*			NS	**		**			NS	NS		NS		
Season (S)		NS	NS		NS			**	*		**			NS	NS		NS		
P×N		*	NS		NS			NS	NS		NS			NS	NS		NS		
P×S		NS	NS		NS			**	NS		NS			NS	NS		NS		
N×S		NS	NS		NS			NS	NS		*			*	NS		NS		
P×N×S		NS	NS		NS			NS	NS		NS			*	NS		NS		

Notes: Shannon, Chao and ACE indexes were calculated based on phylogenetic distance at OTU level. Different letters indicate significant differences between samples ($P<0.05$). Values are means± SE ($n=3$).
*, significant at $P<0.05$; **, significant at $P<0.01$; NS, not significant.
SR, TP, P, N and S are straw return, traditional planting, planting pattern, nitrogen and season, respectively.

3.3 Beta diversity of soil bacterial, fungal and nematode community

The variations in bacterial, fungal, and nematode communities caused by SR and fertilization regime were explored using PCoA (Figure 2). The first two principal coordinates for the bacterial community represented 28.32% (PC1) and 21.38% (PC2) of total variation in spring (Figure 2A), and 35.07% (PC1) and 11.70% (PC2) in autumn maize fields (Figure 2B). These re-

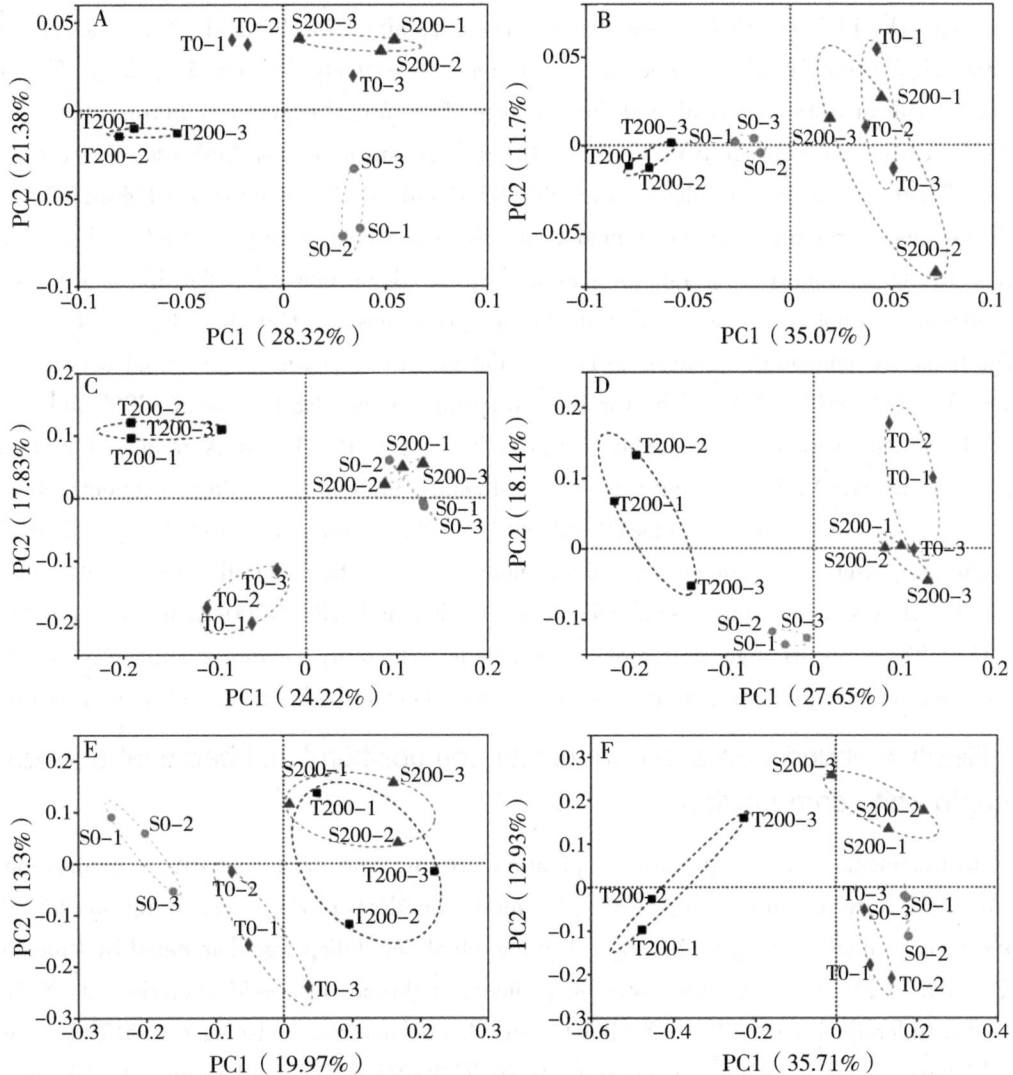

Figure 2 Beta diversities of soil bacterial (A, B), fungal (C, D) and nematode (E, F) communities in spring (A, C, E) and autumn (B, D, F) seasons were analyzed by principal coordinates analysis (PCoA) based on unweighted unifrac phylogenetic distance metrics at OTU level and displayed in scatter diagram, each treatment with 3 replications.

Notes: Abbreviations: S0 = straw returning without nitrogen fertilizer application; S200 = straw returning with 200 kg/ha nitrogen fertilizer application; T0 = traditional planting without nitrogen fertilizer application; T200 = traditional planting with 200 kg/ha nitrogen fertilizer application.

sults demonstrated that the beta-diversity of bacterial community structure was significantly affected by planting pattern and N fertilization during spring season; however, during the autumn season the SR-N200 and TP-N0 were overlapped and showed non-significant effect on soil bacterial community.

To assess the fungal beta diversity, the phylogenetic analysis of fungal composition were performed using unweighted UniFrac distances in spring (Figure 2C) and autumn (Figure 2D). The results of the current study revealed that autumn have more comparable community structure than spring season. The PC1 and PC2 principal coordinates explained 24.2% and 17.8% of variation in spring and 27.7% and 18.1% of variation in autumn, respectively (Figure 2C, 2D). The application of N fertilizer in the spring altered the structure of the fungal community in a positive direction of PC2. The correlation between SR-N200 and SR-N0 treatments during both seasons were closer, while significant separation were observed for TP-N200 and TP-N0 treatments (Figure 2C, Figure 2D). These results revealed that the fungal beta-diversity were strongly affected by both planting pattern and N fertilization during autumn season (Figure 2D); however, the SR-N200 and SR-N0 treatments were not significantly different during spring season ($P>0.05$; Figure 2C).

The nematode community analysis indicated that the first and second principal coordinates represented 34.36% and 25.74% of the variation in spring (Figure 2E), and 36.79% and 27.53% variation in autumn season, respectively (Figure 2F). The PCoA results indicated that during spring season the SR-N200 treatment was not significantly different from other treatments. However, SR-N0 was significantly different from TP-N0 and TP-N200 treatments ($P<0.05$; Figure 2E). During autumn season the treatment SR-N0 was non-significant from the other treatments, but SR-N200 treatment was significantly separated from TP-N0 and TP-N200 treatments ($P<0.05$; Figure 2F). These results suggesting that the nematode community was not significantly affected by N fertilization however planting pattern have significant effect in both spring and autumn seasons.

3.4 Relative abundance and community compositions soil bacterial abundance and community composition

In all treatments, including seasons, planting pattern, and nitrogen fertilizer application, 30 bacterial phyla were identified. In spring, 14 most abundant phyla, accounting for 96.51%–97.15% of all sequences (Figure S4-A). The microbial population was dominated by Proteobacteria (20.90%–25.89%) of the total sequences, followed by Acidobacterioia (18.56%–22.47%), Chloroflexi (13.32%–15.42%), and Actinobacteriota (16.41%–16.79%). In autumn, 15 most abundant phyla, accounting for 96.73%–97.64% of all sequences (Figure S4-B). Similar to spring, the dominant bacterial phyla were Proteobacteria (21.06%–23.26%), Acidobacterioia (13.68%–23.88%), Chloroflexi (13.72%–16.24%) and Actinobacteria (13.83%–17.74%) in the autumn. In the SR-N0, SR-N200, TP-N0, and TP-N200 soils, Proteobacteria, Acidobacteriota, Actinobacteriota, and Chloroflexi were the top four most abundant bacterial phyla irrespective of planting seasons (Figure 3A, Figure 3B). In SR-N200 soil, the Proteobacteria (22%), Acidobacteriota (22%), Actinobacteriota (16%), and Chloroflexi (14%) were the top four most abundant phyla in the spring planting pattern (Figure 3A). The abundance of Actinobacteriota and Chloroflexi were noticeably increased, where the abundance of Proteobacteria was markedly decreased in the TP-N200 soil compared to SR-N200 soil. In addition, the Proteobacteria and Actinobacteriota were potentally decreased, where the abundance of Acidobacteriota and Chloroflexi were increased in the SR-N0 soil compared to TP-N0 (Figure 3A). These results suggesting that soil with SR and optimum N fertilization (200 N kg/ha) present a significantly greater abundance of Proteobacteria, Acidobacteriota, and Chloroflexi in spring season. Similarly, in autumn season, the Proteobacteria, Acidobacteriota, Actinobacteriota, and Chloroflexi were the top four most abundant phyla in SR-N200 and TP-N200 soils (Figure 3B). The results of our current study suggesting that the Acidobacteriota (24%) and Chloroflexi (16%) were markedly increased in the TP-N200 soil, but the abundance of Actinobacteriota (14%) was decreased compared to SR-N200 soil (Figure 3B).

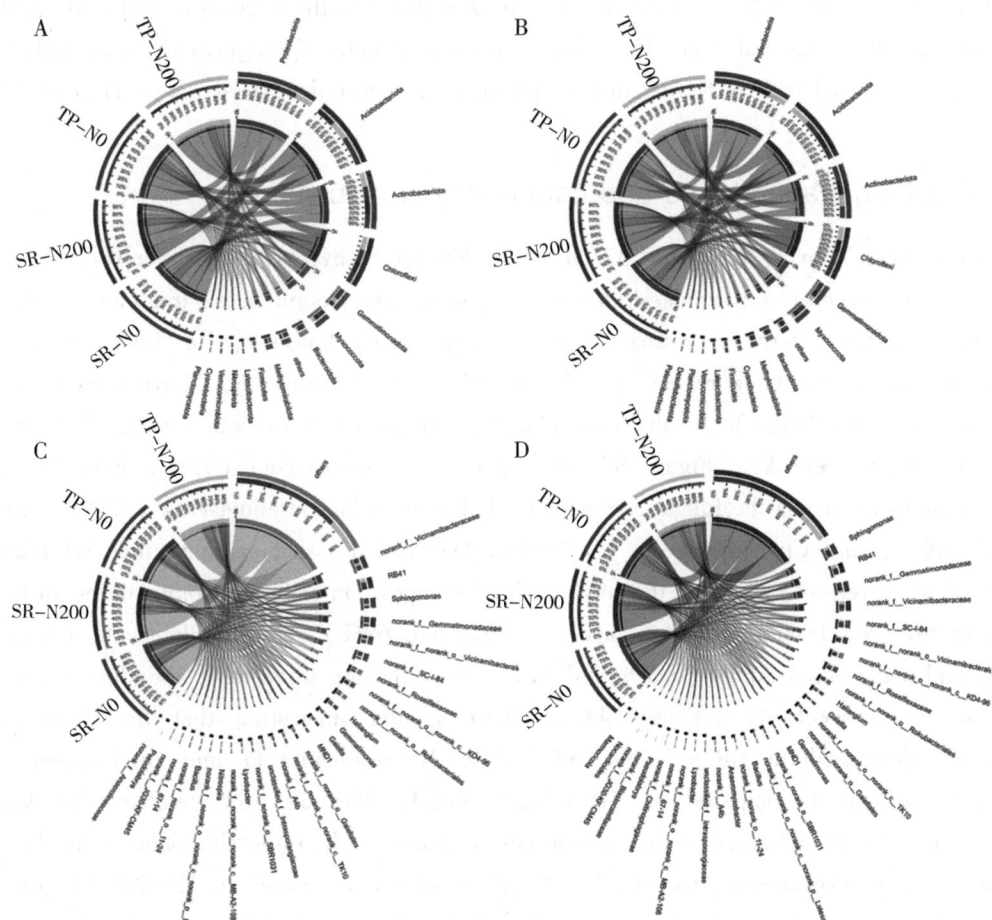

Figure 3 Relative abundances of bacterial in spring (A) and autumn (B) communities at the phylum level, and abundance at the genes level in spring (C) and autumn (D) seasons

Notes: SR-N0 = straw returning without N fertilizer application; SR-N200 = straw returning with 200 kg/ha N fertilizer application; TP-N0 = traditional planting without N fertilizer application; TP-N200 = traditional planting with 200 kg/ha nitrogen fertilizer application.

At the genus level, the bacterial community structures showed significant differences among the different treatment and seasons (Figure 3C, Figure 3D). For example, during the spring season in TP-N200 soil, the most dominant genera were norank_f_*Vicinamibacteraceae* (4.7%), RB41 (5.9%) and *Sphingomonas* (3.9%). Furthermore, suggesting that the TP-N0 soil significantly decreased the norank_f_*Vicinamibacteraceae* (3.3%) and RB41 (4.1%), however increased the abundance of *Sphingomonas* (5%). Likewise, in the SR-N200 and SR-N0 soils the dominant bacterial genus were norank_f_*Vicinamibacteraceae* (6.9% and 4.1%), RB41 (3.4% and 3.8%) and *Sphingomonas* (3.2% and 4.1%), respectively, suggesting that the RB41 and *Sphingomonas* were significantly dominant in SR-N0 soil compared to SR-N200 soil. During the autumn season, the TP-N200 soil have the highest abundance of genera *Sphingomonas* (4.8%) and RB41 (5.8%) compared to other treatments (Figure 3D). In addition, the abundance of *Sphingomonas* and RB41 were significantly increased in TP having 200 N/ha than soil with no fertilization, however, the SR-N200 soil decreased the abundance of these genera compared to SR-N0 soil. These results indicated that the abundances of genera *Sphingomonas* and RB41 were significantly decreased with N fertilization in SR plot, however decreased under TP plot (Figure 3D).

3.5 Soil fungal abundance and community composition

Various fungal phyla, including Ascomycota, Mortierellomycota, Chytridiomycota, Glomeromycota, Basidiomycota, and unclassified-fungal phyla, were found in all treatments, regardless of season, planting pattern, or nitrogen fertilizer application; however, Rozellomycota were only found in spring season (Figure S5-A, Figure S5-B). Ascomycota, Mortierellomycota, and Chytridiomycota were found to be the most abundant fungal phyla in both seasons (Figures 4A, Figure 4B, Figure S5-AC, Figure S5-B). The phyla Ascomycota (75%) have the highest relative abundance among the fungal community, followed by Mortierellomycota (12%), unclassified (5.39%), and Chytridiomycota (1.75%). Our findings demonstrated that SR treatments have variable degrees of impact on the phylum-level composition of fungal communities. In the TP-N200 and SR-N200 soils, the Ascomycota (79% and 69%), Mortierellomycota (8.2% and 13%), and unclassified-fungi (6.7% and 11%) were the top three most abundant phyla in the spring season, respectively (Figure 4A). These results suggesting that the abundance of Ascomycota decreased in SR-N200 soil, while Mortierellomycota and unclassified-fungi significantly increased compared to TP-N200 soil. In addition, the relative abundance of Ascomycota and Mortierellomycota were potentially increased, where the abundance of unclassified-fungi and Glomeromycota were decreased in the SR-N0 soil compared to TP-N0 (Figure 4A). These results demonstrated that SR increased the relative abundance of Ascomycota, Mortierellomycota, and Glomeromycota regardless of N fertilization. Similarly, in SR-N200 soil during autumn season, Ascomycota, Mortierellomycota, Chytridiomycota, and Basidiomycota were 76%, 5.5%, 9.4%, and 2.5%, respectively (Figure 4B). However, in TP-N200 soil the relative abundance of Ascomycota, Mortierellomycota, Chytridiomycota, and Basidiomycota were 81%, 7.3%, 4.4%, and 2.1%, respectively. The above results showed that the abundance of Ascomycota and Mortierellomycota were noticeably increased in TP-N220 soil where decreased the abundance of

Chytridiomycota and Basidiomycota (Figure 4B).

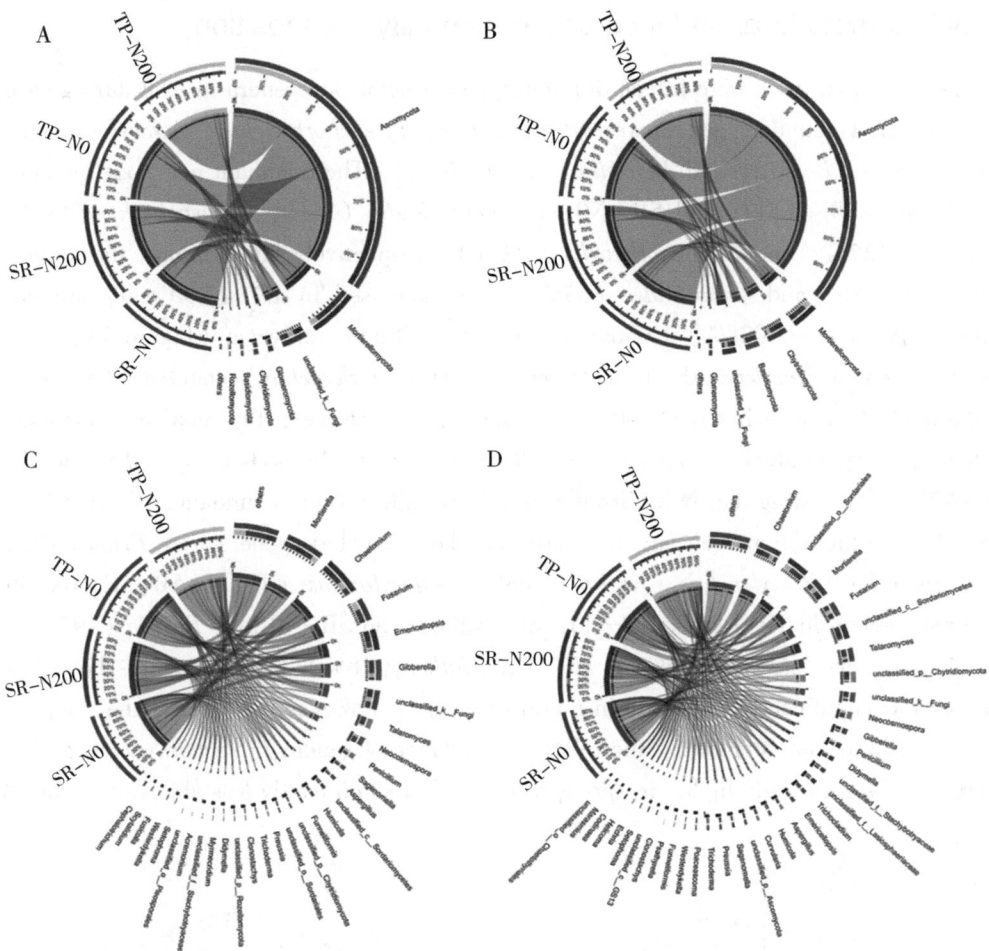

Figure 4 Relative abundances of fungal in spring (A) and autumn (B) communities at the phylum level, and abundance at the genes level in spring (C) and autumn (D) seasons

Notes: SR-N0 = straw returning without N fertilizer application; SR-N200 = straw returning with 200 kg/ha N fertilizer application; TP-N0 = traditional planting without N fertilizer application; TP-N200 = traditional planting with 200 kg/ha nitrogen fertilizer application.

At the genus level, the fungal community structure showed that *Mortierella*, *Chaetomium*, *Fusarium*, *Emericellopsis*, and *Gibberrella* were the most abundant genera in all the treatments (Figure 4C). The abundance of *Mortierella*, *Emericellopsis*, and *Gibberrella* were noticeable decreased in TP-N0 soil compared to TP-N200 soil, however *Chaetomium* and *Fusarium* were increased. Likewise, in SR-N200 and SR-N0 soils the relative abundance of top five genera were *Mortierella* (13% and 17%), *Chaetomium* (15% and 14%), *Fusarium* (3% and 6.4%), *Emericellopsis* (15% and 1%), and *Gibberrella* (3.5% and 12%), respectively. In autumn season, the highest relative abundance of Ascomycota was observed in TP-N200 soil (81%), TP-N0 soil (81%), SR-N200 soil (76%), and SR-N0 soil (75%). Furthermore, the abundance of genera *Ascomycota* and *Mortierellomycota* were increased in TP pattern, whereas the genre *Chytridio-*

mycot and *Basidiomycota* were decreased compared to SR (Figure 4D).

3.6 Soil nematode abundance and community composition

It was found that there were fifteen different types of nematode genera in total during spring season, and the most abundant genera were *Pratylenchus*, *Tylenchorhynchus*, *Acrobeloides*, unclassified, *Aphelenchus*, *Basiria*, and *Aporcella* (Figure S6-A). The dominant nematode genera in TP-N200, TP-N0, SR-N200, and SR-N0 were *Acrobeloides* (37%), *Tylenchorhynchus* (42%), *Pratylenchus* (42%), and *Tylenchorhynchus* (50%), respectively. In addition, the abundance of *Pratylenchus* (27%) and *Acrobeloides* (37%) were increased in TP-N200 soil, but decreased the *Tylenchorhynchus* (0.028%) compared to TP-N0 soil and vice versa (Figure 5A). Similarly, the nematode genera *Pratylenchus*, *Acrobeloides*, and *Tylenchorhynchus* accounted 42%, 19%, and 0% of the total abundance in SR-N200 soil, respectively. However, the relative abundance of *Tylenchorhynchus* was significantly higher in both TP-N0 and SR-N0 soils compared to the TP-N200 and SR-N200, suggesting that N fertilization negatively affected the abundance of *Tylenchorhynchus* (Figure 5A). In the autumn season, the most abundant nematode genera were *Prionchulus*, *Acrobeloides*, *Aporcella*, *Pratylenchus*, *Basiria*, and *Mesodorylaimus* (Figure 5B, Figure S6-B). The dominate genera in TP-N200, TP-N0, SR-N200, and SR-N0 were *Basiria* (19%), *Prionchulus* (40%), *Acrobeloides* (36%), and *Prionchulus* (28%). These results showed that the *Prionchulus* was significantly increased in unfertilized plot; however, decreased the *Aprocella* compared to N fertilized plot. Moreover, the relative abundance of genera *Tylenchorhynchus* and *Aphelenchus* were found significantly higher in spring seasons, but significantly less abundance was observed

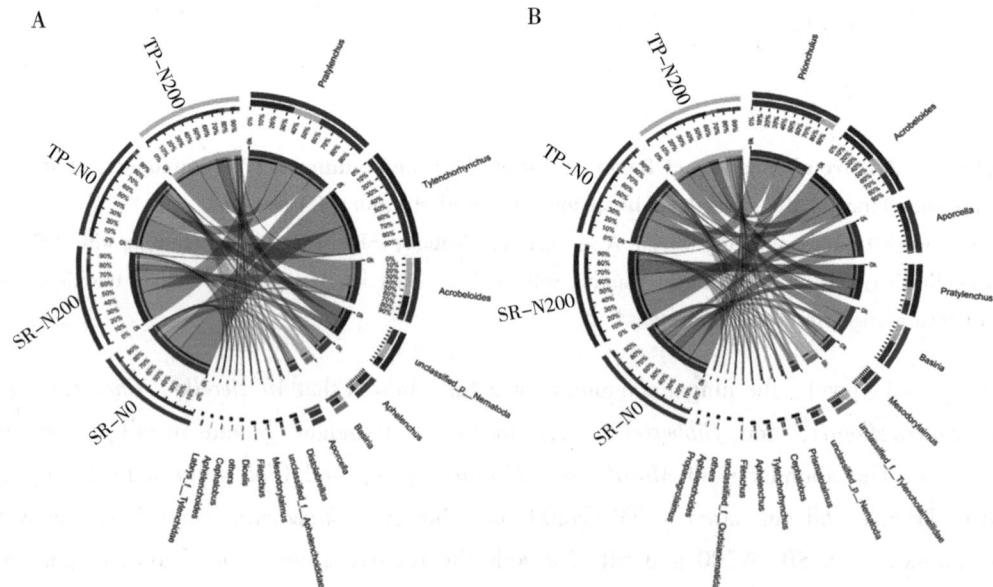

Figure 5 Relative abundances of nematode at the genes level in spring (A) and autumn (B) seasons

Notes: SR-N0 = straw returning without N fertilizer application; SR-N200 = straw returning with 200 kg/ha N fertilizer application; TP-N0 = traditional planting without N fertilizer application; TP-N200 = traditional planting with 200 kg/ha nitrogen fertilizer application.

in the autumn seasons (Figures 5A, Figure 5B).

3.7 Relationships between bacterial, fungal and nematode community and soil environmental properties

A multivariate RDA analysis indicated that soil environmental variables such as SOC, TN, S-UE, S-CL and S-SC contributed to the distribution of the bacterial, fungal and nematode OTUs (Figure 6). The first two RDA axes explained 39.17% and 27.11% variation in spring (Figure 6A) and autumn (Figure 6B), respectively. In addition, the relationship among the species and environmental factor, suggesting 25.52% and 13.65% variation in spring, and 17.82% and 9.29% variation in autumn for the first and second axes, respectively. The correlation analysis showed that the S-CS, TN, and S-UE significantly increased the bacterial phyla FCPU426 ($P<0.01$) and significantly decreased Cyanobacteria ($P<0.05$; Figure S7). Moreover, S-CL significantly decreased FCPU426 ($P<0.001$) and Nitrospirota ($P<0.05$), where SOC, S-CS, and S-UE increased Acidobacteriota. The relationship between changes in environmental factors and the relative abundances of fungal OTUs is shown in Figure 6C, and Figure 6D. In spring and autumn, the SOC, TN, S-UE, S-CL, and S-SC accounted for 45.51% and 55.88% of the overall variation, respectively. Furthermore, these results clearly demonstrated that the first and second axes account for 29.67% and 15.84% variation in spring, and 35.88% and 20.00% variation in autumn, respectively. Similarly, soil enzymes significantly affected the fungal phyla, the correlation analysis revealed that S-CL increased the abundance of Chytridiomycota and Basidiomycota ($P<0.001$), while decreased Mucoromycota and Rozellomycota. In contrast, the abundance of Mucoromycota and Rozellomycota increased with TN and S-UC (Figure S8). The first two axes suggested that environmental variables such as SOC, TN, S-UE, S-CL and S-SC strongly influenced the distribution of the nematode OTUs. Redundancy analysis was used to understand the relationship between nematode community and soil environmental characteristics (Figure 6E, Figure 6F). These results demonstrated that the first two axes of RDA accounted for 26.04% and 10.89% variation in spring season (Figure 6E), while 33.24% and 11.15% in autumn season (Figure 6F). Soil enzyme activities, SOC, and TN were much closely associated with nematode community distribution in TP-N200 and SR-N200 treatments during spring season. However, soil enzymes and SOC were associated to SR-N0 and SR-N200 treatment in autumn season, but TN was nearly correlated to TP-N0. The nematode genera *Pratylenchus* and unclassified *Aphelenchoididae* have positive, while unclassified *Tylencholaimellidae*, *Prismatolaimus*, *Prionchulus*, and *Mesodorylaimus* have negative correlation with S-CS, TN, and S-UE (Figure S9). The co-occurrence network diagram further clarified the specific changes in soil fertility and correlation with bacterial, fungal and nematode community structure under different planting pattern and N fertilizers (Figure 7). These results showed that S-SC have positive relationship with bacterial phyla Chlorflexi, S-CL with Proteobacteria, and TN with Chlorflexi, Acidobacteria, and Actinobacteriota in SR-N200 soil. However, SOC have positive relation with Chlorflexi and Actinobacteriota in TP-N200. The fungi and nematode have much more strong and positive correlation with soil enzymes, SOC and TN in TP-N200 compared to SR-N200.

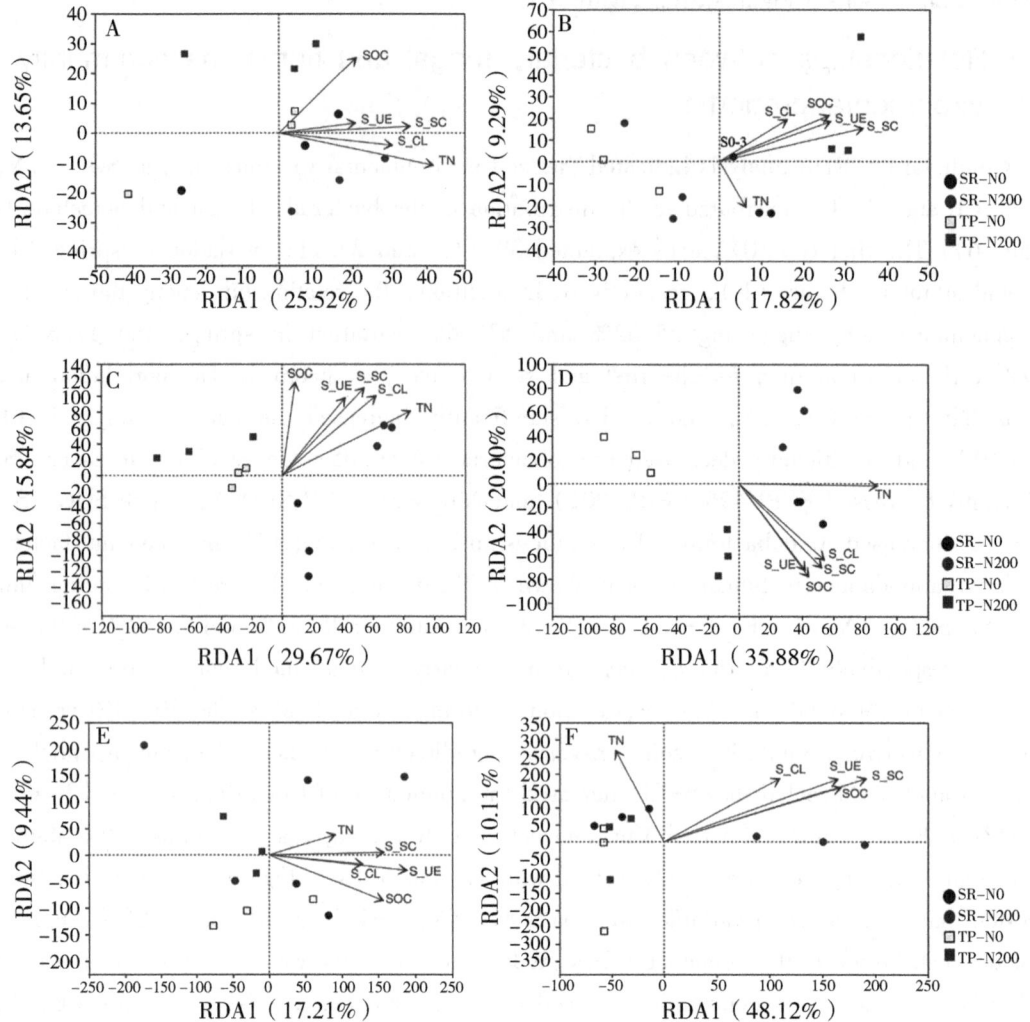

Figure 6 The redundancy analysis (RDA) to identify the relationship of bacterial fungal and nematode OTUs taxa and soil properties. Bacterial in spring (A) and autumn (B), fungal in spring (C) and autumn (D), and nematode in spring (E) and autumn (F) seasons

Notes: S0 = straw returning without nitrogen fertilizer application; S200 = straw returning with 200 kg/ha nitrogen fertilizer application; T0 = traditional planting without nitrogen fertilizer application; T200 = traditional planting with 200 kg/ha nitrogen fertilizer application.

4 Discussion

4.1 Changes in soil fertility and enzyme activities with planting pattern and N fertilization

Soil organic carbon and N are the primary sources of energy and nutrients for microbial growth (Zhao et al., 2016). The SOC is derived from incorporated crop residues and rhizodeposited organic matter (Banger et al., 2010; Yang et al., 2012). Incorporated residues are decomposed

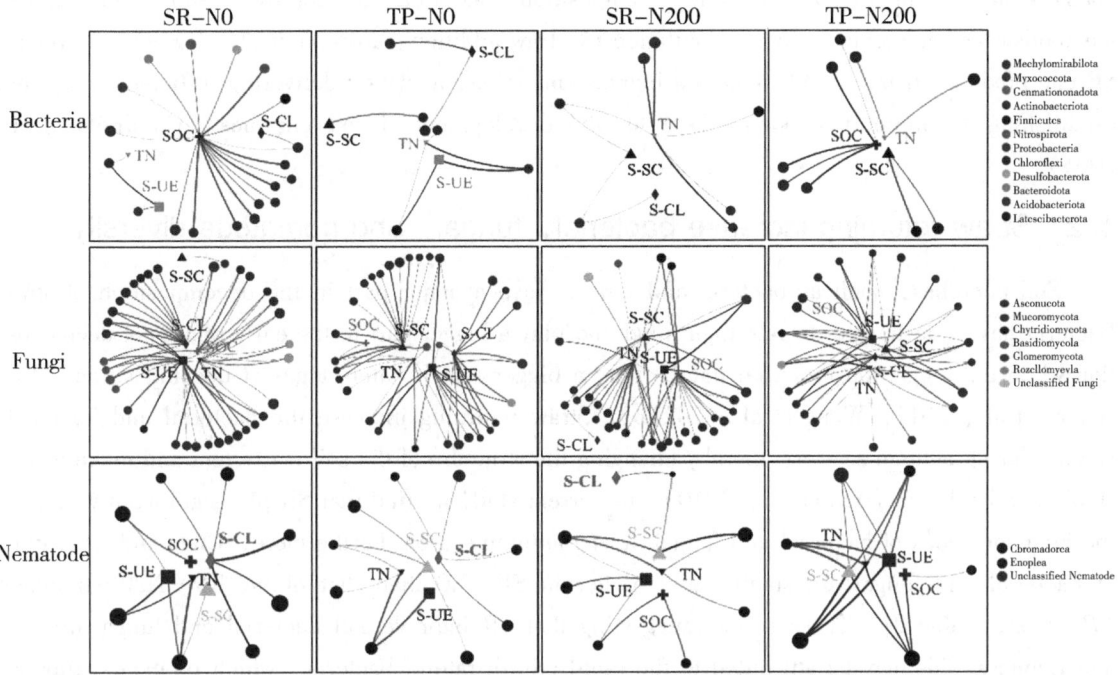

Figure 7 Co-occurrence network diagram between bacterial and fungal taxa at the phyla level and nematode on the class level with soil fertility and enzymes under different treatments

Notes: Dot size indicates the value of relative abundances. Positive correlations are labeled with red lines and negative correlations are colored in green, thick lines are high correlations and little correlations have thin lines. Treatments: no nitrogen addition + straw addition (SR-N0), no nitrogen addition + traditional planting (TP-N0), no nitrogen addition + straw addition (SR-200), and traditional planting with nitrogen fertilizer application (TP-N200). TN, SOC, S-UE, S-CL, and S-SC represent total nitrogen content, soil organic carbon content, soil urease activity, soil sucrase activity, and soil cellulase activity.

rapidly and returned large amounts of N and C into the field, which lead to improve soil quality and boost soil fertility (Liu et al., 2014). In addition, the application of N fertilizers increase soil TN and SOC compared to control treatment (N0). Averaged across fertilization, the SR significantly increase SOC and TN during both seasons compared to TP pattern (Figure 1). However, the spring season have much higher SOC and TN content than that of autumn season under both planting pattern. These findings showed that improved SOC and TN contents are mostly attributable to spring maize having more time to decompose the maize straw, however straw obtained after spring maize harvesting had less time to decompose for the autumn maize (Liao et al., 2018; Zhao et al., 2021). These findings are in line with findings from earlier research (Chao et al., 2019), they also demonstrated that spring maize have higher soil fertility compared to autumn maize, might be due to the straw incorporated into the soil have proper time to decompose before sowing.

The soil enzymatic activities boosted in response to chemical fertilizer application and SR due to greater C demand for microbial development under high N content (Liu et al., 2021). Soil enzyme activity is more important in evaluating soil quality and plays an important role in nutrient recycling (Demisie et al., 2014). The increase in soil enzyme activities (Table 1) could be explained by changes in the microbial community composition, as well as improved soil microorganism metabolism and microbial activity stimulation by straw addition (Iovieno et al., 2009). Therefore, SR can produce organic supplement and increase microbial biomass, delivering sufficient energy and establishing an appropriate environment for the development of soil enzymes (Gianfreda et al., 2005).

4.2 Straw returning increase bacterial, fungal, and nematode diversity

Soil microbes, such as bacteria and fungi, have a major part in the decomposition of straw. Bacteria have a faster growth rate than fungi and play a bigger part in the early phases of decomposition while the fungi, on the other hand, play a bigger role in later stages of decomposition (Marschner et al., 2011; Wang et al., 2021b). Straw returning increase the bacterial and fungal diversity during autumn season; thereby changing the structure of the soil ecosystem and make it more stable and healthier (Yin et al., 2010). In current study we find that SR plot significantly increase the Shannon index of bacterial and fungal during autumn season. Furthermore, Chao and ACE index of bacterial and fungal are significantly higher in SR-N0 than that of the TP-N200 treatment ($P<0.05$; Table 2). These results suggesting that SR improve soil bacterial and fungal diversity and richness. This may be attributed to the rapid growth rate of bacteria, which promotes straw decomposition in the early stages associated with larger number of nematodes diversity in the SR plot (Zhang et al., 2012; Wang et al., 2021a).

Our study find that SR treatments significantly reduce soil nematode richness and diversity compared to TP treatments in spring season, but SR-N200 have significantly higher soil nematode Chao and ACE indexes than that of the SR-N0 treatment (Table 2). Our findings are consistent with the findings of earlier researchers (Ferreira et al., 2021), who reported that fungal diversity and richness were unaffected by fertilization. Chemical fertilization, on the other hand, reduced bacterial diversity and richness, according to Feng et al. (2015). These results were also confirmed (Zhou et al., 2015), who found that long-term use of a single fertilizer reduces the diversity and richness of the soil bacterial community. In north China, the long-term uses of chemical and organic fertilizers increase the population of bacteria and fungi as reported by Li et al. (2017). As a result, straw addition with N fertilizer application have a key role in sustaining the diversity of soil bacteria, fungi, and nematode communities, which in turn may help to prevent the degradation of soil microbial community structure and function over time (Sun et al., 2015).

4.3 Effect of straw returning on bacterial, fungal, and nematode communities decomposition

The soil microbial community structure could be altered by SR and N fertilization (Li et al., 2017; Zhao et al., 2019). In this study, Proteobacteria, Acidobacteria, Nitrospirae, Chloroflexi and Actinobacteria are the important components of soil bacterial community (Figure S3-A, Figure S3-B), these results are supported by previous study (Liu et al., 2020). However, the current study found no significant difference in the relative abundance of major soil bacteria phyla between SR and TP plots. The high number of Ascomycota taxa could be attributed to their easy colonization and rapid growth (Liu et al., 2017; Wang et al., 2021b). In addition, SR have a considerable impact on the abundance of several fungi. Average abundances of fungal phyla Mucoromycota, Chytridiomycota, unclassified Fungi and Rozellomycota differ greatly in SR plot as compared to TP. Even though there were many different phyla in the soil, some of them weren't dominating and their abundance was extremely low. In the SR plot, the relative abundances of Rozelloomycoa varied, these results are in the line with the recent findings (Fan and Wu, 2020; Wang et al., 2021a). *Pratylenchus*, *Tylenchorhynchus*, *Prionchulus*, and *Acrobeloides* were dominant and important genera of soil nematode community (Figure 5A, Figure 5B).

The bacterial and fungal community compositions are changed with N fertilization, planting pattern, and seasons. The most dominant bacterial phyla are Proteobacteria and Acidobacteria, and dominant fungal phyla were Ascomycota, Mucoromycota and Chytridiomycota in both planting pattern and seasons. Furthermore, the phyla Chytridiomycota, unclassified Fungi, and Glomeromycota are significantly increase in N200 kg N/ha compared to N0 irrespective of the planting pattern. Similarly N fertilizer application have significantly positive effects on the nematode abundances (Figure S3). The abundance of genera *Pratylenchus* and *Acrobeloides* are noticeably increase in fertilized plots (N200 kg N/ha) compared to N0 treatment during both seasons. According to the finding of Allison et al. (2008), which reported that out of 38 publication 84% studies suggesting that the soil microbial communities are more sensitive to fertilization especially N.

4.4 Relationships of bacterial, fungal, and nematode community with soil fertility and enzyme activities

The Actinobacteria was the most dominant phylum of bacteria in SR-N0 compared to TP-N0, indicating that they are more prominent in straw decomposition than other microbial fractions (Chen et al., 2016; Li et al., 2017). Moreover, SR-N200 increase bacterial Acidobacteria at phyla level compared to TP-N200. Different bacterial community compositions suggested that bacteria dominates soil C decomposition, these changes in bacterial fractions suggested that bacteria are important in the degradation of straw and the activity of enzymes (Zhao et al., 2016). The microbial community composition varies in plant growth, but also in reaction to the soil environment in the field, including nitrogen, organic carbon and enzyme activity (Mouhamadou et al., 2013; Carrara et al., 2018; Hou et al., 2020). Our results show that the changes in the bacterial composition of the soil are identified in response to planting pattern and are closely related to N fertilizer application (Figure 6). Furthermore, the increase in diversity of bacterial composition in response to N fertilization is mostly attributable to changes in the relative abundance of certain bacterial phyla such as Chloroflexi, Actinobacteria, Acidobacteria, and Latescibacterota. The co-occurrence network analysis found that the abundance of Chloroflexi and Acidobacteria were closely related to TN in different planting patterns (Figure 7). These results are in the line with the finding of other researchers (Fierer et al., 2007), which reported that these bacterial phyla have been linked to the soil C and N cycles.

Changes in the soil fungal community in response to SR could be more directly linked to changes in SOC and N derived from straws. As several different types of fungi are saprotrophic in the soil, acquiring SOC, N, and energy from incorporated crop residues (Su et al., 2020b). This might be one reason for the results of RDA that soil enzymes, SOC, and TN were found to be significantly correlated with soil fungal community variation (Figure 6C, Figure 6D). These results further suggesting that the most of soil enzymes and soil nutrient are significantly correlate with SR-N200 during spring and with SR-N0 during autumn season. The current study results show that the fungal phylum Ascomycota are the most abundant phylum in SR-N0 compared to TP-N0, suggesting that Ascomycota predominated in the SR plot compared to other microbial segments (Ma et al., 2013; Yang et al., 2016). Additionally, at the phylum level, SR-N200 increase the fungal Chytridiomyeota division. Soil enzymes are essential for decomposing soil organic matter and unique to catalyze a specific biochemical reaction (Li et al., 2016). RDA results show differences in the enzymes affecting the changes of nematode community are mainly influenced by S-UE and S-SC in spring (Figure 6E) and autumn (Figure 6F), respectively. Moreover, S-UE and S-CL had a positive impact on nematode communities at genus level in SR-N0 and SR-N200 plots, and S-SC, S-UE, and SOC but have positive impact in TP-N200 plot (Figure 5).

5 Conclusion

This study investigates the impact of straw returning and N fertilizer on the composition and diversity of bacterial, fungal, and nematode communities, as well as their effects on soil enzymes, SOC, and TN of soil under maize plantation. The SOC, TN, S-UE, S-SC, and S-CL activities in the SR plot following N fertilization increase specifically at V6 stage than R6 in SR plot. The SR incorporation is an important determinant in influencing the composition of the bacterial and fungal communities and improved soil fertility compared to TP. The most dominant soil bacterial and fungal species in spring and autumn were Proteobacteria and Ascomycota, and the most dominant nematode genera were *Pratylenchus* and *Prionchulus* in spring and autumn, respectively. RDA and Co-occurrence network identified that TN, SOC and S-SC were closely correlated with bacterial and fungal community composition. These findings will be helpful in framing the straw returning and N fertilizer application for improving soil fertility, bacterial, and fungal structure and communities on sustainable way and minimizing the degradation of soil in sub-tropical regions of the China.

Acknowledgment

The authors express their special gratitude to all fundings. This study was financially supported by the National Natural Science Foundation of China (31760354) and the Natural Science Foundation of Guangxi Province (2019GXNSFAA185028).

References

Banger K, Toor G, Biswas A, et al., 2010. Soil organic carbon fractions after 16-years of applications of fertilizers and organic manure in a Typic Rhodalfs in semi-arid tropics. *Nutrient Cycling in Agroecosystems*. 86: 391-399.

Borase DN, Nath CP, Hazra KK, et al., 2020. Long-term impact of diversified crop rotations and nutrient management practices on soil microbial functions and soil enzymes activity. *Ecological Indicators*. 114.

Burns RG, Deforest JL, Marxsen J, et al., 2013. Soil enzymes in a changing environment: current knowledge and future directions. *Soil Biology and Biochemistry*. 58: 216-234.

Carrara JE, Walter CA, Hawkins JS, et al., 2018. Interactions among plants, bacteria, and fungi reduce extracellular enzyme activities under long-term N fertilization. *Global Change Biology*. 24: 2721-2734.

Chao P, Kan ZR, Peng L, et al., 2019. Residue management induced changes in soil organic carbon and total nitrogen under different tillage practices in the North China Plain. *Journal of Integrative Agriculture* 18: 1337-1347.

Chen B, Du K, Sun C, et al., 2018. Gut bacterial and fungal communities of the domesticated silkworm (Bombyx mori) and wild mulberry-feeding relatives. *The ISME Journal*. 12: 2252-2262.

Chen C, Zhang J, Lu M, et al., 2016. Microbial communities of an arable soil treated for 8 years with organic and inorganic fertilizers. *Biology and Fertility of Soils*, 52: 455-467.

Chen LF, He ZB, Wu XR, et al., 2021. Linkages between soil respiration and microbial communities following afforestation of alpine grasslands in the northeastern Tibetan Plateau. *Applied Soil Ecology*. 161: 103882.

Cusack DF, Silver WL, Torn MS, et al., 2011. Changes in microbial community characteristics and soil organic matter with nitrogen additions in two tropical forests. *Ecology*. 92: 621-632.

Dang K, Gong X, Zhao G, et al., 2020. Intercropping alters the soil microbial diversity and community to facilitate nitrogen assimilation: a potential mechanism for increasing proso millet grain yield. *Frontiers in Microbiology*. 11: 2975.

Demisie W, Liu Z, and Zhang M., 2014. Effect of biochar on carbon fractions and enzyme activity of red soil. *Catena*. 121: 214-221.

Fan W, and Wu J., 2020. Short-term effects of returning granulated straw on soil microbial community and organic carbon fractions in dryland farming. *Journal of Microbiology*. 58: 657-667.

Feng Y, Chen R, Hu J, et al., 2015. Bacillus asahii comes to the fore in organic manure fertilized alkaline soils. *Soil Biology and Biochemistry*. 81: 186-194.

Ferreira DA, Da Silva TF, Pylro VS, et al., 2021. Soil Microbial Diversity Affects the Plant-Root Colonization by Arbuscular Mycorrhizal Fungi. *Microbial Ecology*. 82: 100-103.

Fierer N, Bradford MA, and Jackson RB., 2007. Toward an ecological classification of soil bacteria. *Ecology*. 88: 1354-1364.

Gdanetz K, Benucci GMN, Pol NV, et al., 2017. CONSTAX: a tool for improved taxonomic resolution of environmental fungal ITS sequences. *BMC Bioinformatics*. 18: 1-9.

Geisseler D, and Scow KM., 2014. Long-term effects of mineral fertilizers on soil microorganisms-A review. *Soil Biology and Biochemistry*. 75: 54-63.

Gianfreda L, Rao MA, Piotrowska A, et al., 2005. Soil enzyme activities as affected by anthropogenic alterations: intensive agricultural practices and organic pollution. *Science of the Total Environment*. 341: 265-279.

Gong X, Dang K, Lv S, et al., 2021. Interspecific competition and nitrogen application alter soil ecoenzymatic stoichiometry, microbial nutrient status, and improve grain yield in broomcorn millet/mung bean intercropping systems. *Field Crops Research*. 270.

Hou Q, Wang W, Yang Y, et al., 2020. Rhizosphere microbial diversity and community dynamics during potato cultivation. *European Journal of Soil Biology*. 98: 103176.

Huang W, Wu JF, Pan XH, et al., 2021. Effects of long-term straw return on soil organic carbon fractions and enzyme activities in a double-cropped rice paddy in South China. *Journal of Integrative Agriculture*. 20: 236-247.

Iovieno P, Morra L, Leone A, et al., 2009. Effect of organic and mineral fertilizers on soil respiration and enzyme activities of two Mediterranean horticultural soils. *Biology and Fertility of Soils*. 45: 555-561.

Kamble PN, and Bååth E., 2016. Comparison of fungal and bacterial growth after alleviating induced N-limitation in soil. *Soil Biology and Biochemistry*. 103: 97-105.

Liao P, Huang S, Van Gestel NC, et al., 2018. Liming and straw retention interact to increase nitrogen uptake and grain yield in a double rice-cropping system. *Field Crops Research*. 216: 217-224.

Li S, Chen J, Shi J, et al., 2017. Impact of straw return on soil carbon indices, enzyme activity, and grain production. *Soil Science Society of America Journal*. 81: 1475-1485.

Li S, Zhang S, Pu Y, et al., 2016. Dynamics of soil labile organic carbon fractions and C-cycle enzyme activities under straw mulch in Chengdu Plain. *Soil and Tillage Research*. 155: 289-297.

Liu C, Gong X, Dang K, et al., 2020. Linkages between nutrient ratio and the microbial community in rhizosphere soil following fertilizer management. *Environmental research*. 184: 109261.

Liu C, Lu M, Cui J, et al., 2014. Effects of straw carbon input on carbon dynamics in agricultural soils: a meta-analysis. *Global Change Biology*. 20: 1366-1381.

Liu YX, Pan YQ, Yang L, et al., 2021. Stover return and nitrogen application affected soil organic carbon and nitrogen in a double-season maize field. *Plant Biology*.

Liu Z, He T, Lan Y, et al., 2017. Maize Stover Biochar Accelerated Urea Hydrolysis and Short-term Nitrogen Turnover in Soil. *Bioresources*. 12: 6024-6039.

Lu J, Li S, Wu X, et al., 2021. The dominant microorganisms vary with aggregates sizes in promoting soil carbon accumulation under straw application. *Archives of Agronomy and Soil Science*.

Ma A, Zhuang X, Wu J, et al., 2013. Ascomycota members dominate fungal communities during straw residue decomposition in arable soil. *PloS one*. 8: e66146.

Marschner P, Umar S, and Baumann K., 2011. The microbial community composition changes rapidly in the early stages of decomposition of wheat residue. *Soil Biology and Bio-*

chemistry. 43: 445-451.

Mouhamadou B, Puissant J, Personeni E, et al., 2013. Effects of two grass species on the composition of soil fungal communities. *Biology and Fertility of Soils*. 49: 1131-1139.

Muhammad I, Wang J, Sainju U M, et al., 2021. Cover cropping enhances soil microbial biomass and affects microbial community structure: A meta - analysis. *Geoderma*. 381: 114696.

Mwafulirwa L, Baggs EM, Russell J, et al., 2021. Identification of barley genetic regions influencing plant-microbe interactions and carbon cycling in soil. *Plant and Soil*.

Ren H, Feng Y, Liu T, et al., 2020. Effects of different simulated seasonal temperatures on the fermentation characteristics and microbial community diversities of the maize straw and cabbage waste co-ensiling system. *Science of the Total Environment*. 708.

Salazar S, Sánchez L, Alvarez J, et al., 2011. Correlation among soil enzyme activities under different forest system management practices. *Ecological Engineering*. 37: 1123-1131.

Schmidt JE, Kent AD, Brisson VL, et al., 2019. Agricultural management and plant selection interactively affect rhizosphere microbial community structure and nitrogen cycling. *Microbiome*. 7: 1-18.

Sun L, Xun W, Huang T, et al., 2016. Alteration of the soil bacterial community during parent material maturation driven by different fertilization treatments. *Soil Biology and Biochemistry*. 96: 207-215.

Sun R, Guo X, Wang D, et al., 2015. Effects of long-term application of chemical and organic fertilizers on the abundance of microbial communities involved in the nitrogen cycle. *Applied Soil Ecology*, 95: 171-178.

Su Y, He Z, Yang Y, et al., 2020a. Linking soil microbial community dynamics to straw-carbon distribution in soil organic carbon. *Scientific Reports*. 10.

Su Y, Lv J, Yu M, et al., 2020b. Long-term decomposed straw return positively affects the soil microbial community. *Journal of Applied Microbiology*. 128: 138-150.

Wang G, Bei S, Li J, et al., 2021a. Soil microbial legacy drives crop diversity advantage: Linking ecological plant-soil feedback with agricultural intercropping. *Journal of Applied Ecology*. 58: 496-506.

Wang L, Luo X, Liao H, et al., 2018. Ureolytic microbial community is modulated by fertilization regimes and particle-size fractions in a Black soil of Northeastern China. *Soil Biology and Biochemistry*. 116: 171-178.

Wang L, Wang C, Feng F, et al., 2021b. Effect of straw application time on soil properties and microbial community in the Northeast China Plain. *Journal of Soils and Sediments*. 21: 3137-3149.

Wu H, Cai A, Xing T, et al., 2021. Fertilization enhances mineralization of soil carbon and nitrogen pools by regulating the bacterial community and biomass. *Journal of Soils and Sediments*. 21: 1633-1643.

Wu L, Ma H, Zhao Q, et al., 2020. Changes in soil bacterial community and enzyme activity under five years straw returning in paddy soil. *European Journal of Soil Biology*.

100: 103215.

Xue B, Hou L, and Xue H., 2019. Research on the characteristics of soil nematode communities in alpine meadow in northern Tibet by using high-throughput sequencing. *Acta Ecologica Sinica*. 39: 4088-4095.

Yang H, Feng J, Zhai S, et al., 2016. Long-term ditch-buried straw return alters soil water potential, temperature, and microbial communities in a rice-wheat rotation system. *Soil and Tillage Research*. 163: 21-31.

Yang X, Ren W, Sun B, et al., 2012. Effects of contrasting soil management regimes on total and labile soil organic carbon fractions in a loess soil in China. *Geoderma*. 177: 49-56.

Yin X, Song B, Dong W, et al., 2010. A review on the eco-geography of soil fauna in China. *Journal of Geographical Sciences*. 20: 333-346.

Yuan Q, Hernández M, Dumont MG, et al., 2018. Soil bacterial community mediates the effect of plant material on methanogenic decomposition of soil organic matter. *Soil Biology and Biochemistry*. 116: 99-109.

Yu D, Wen Z, Li X, et al., 2018. Effects of straw return on bacterial communities in a wheat-maize rotation system in the North China Plain. *PloS one*. 13: e0198087.

Zeng XY, Li SW, Leng Y, et al., 2020. Structural and functional responses of bacterial and fungal communities to multiple heavy metal exposure in arid loess. *Science of The Total Environment*. 723: 138081.

Zhang X, Li Q, Zhu A, et al., 2012. Effects of tillage and residue management on soil nematode communities in North China. *Ecological Indicators*. 13: 75-81.

Zhao S, and Zhang S., 2018. Linkages between straw decomposition rate and the change in microbial fractions and extracellular enzyme activities in soils under different long-term fertilization treatments. *PloS one*. 13: e0202660.

Zhao S, Fan F, Qiu S, et al., 2021. Dynamic of fungal community composition during maize residue decomposition process in north-central China. *Applied Soil Ecology*. 167.

Zhao S, Li K, Zhou W, et al., 2016. Changes in soil microbial community, enzyme activities and organic matter fractions under long-term straw return in north-central China. *Agriculture, Ecosystems & Environment*. 216: 82-88.

Zhao S, Qiu S, Xu X, et al., 2019. Change in straw decomposition rate and soil microbial community composition after straw addition in different long-term fertilization soils. *Applied Soil Ecology*. 138: 123-133.

Zheng L, Chen H, Wang Y, et al., 2020. Responses of soil microbial resource limitation to multiple fertilization strategies. *Soil & Tillage Research*. 196.

Zhou J, Guan D, Zhou B, et al., 2015. Influence of 34-years of fertilization on bacterial communities in an intensively cultivated black soil in northeast China. *Soil Biology and Biochemistry*. 90: 42-51.

Effect of water conditions and nitrogen application on maize growth, carbon accumulation and metabolism of maize plant in subtropical regions

Y. X. Chi, F. Gao, I. Muhammad, J. H. Huang, X. B. Zhou

(Agricultural College of Guangxi University, Nanning 530004, China)

Abstract: Drought, water, and nitrogen losses have always been great challenges for agricultural production in subtropical regions. To study appropriate irrigation regimes and reasonable nitrogen applications in this area, a field experiment was conducted for summer maize (*Zea mays* L.) from 2018 to 2019. Two irrigation treatments, namely, rain-fed irrigation and supplementary irrigation, were designed, and five levels of nitrogen fertilizer were applied at 0, 150, 200, 250, and 300 kg N/ha. The differences in the growth period of biomass, leaf area index, agronomic traits, carbon accumulation, and carbon metabolism enzyme activity were measured. Findings revealed that the interaction of water and nitrogen has a significant impact on maize growth. Biomass, leaf area index, agronomic traits, carbon accumulation, and carbon metabolism enzyme activity showed a peak under N250 which higher than other treatments, and the differences were significant. In 2018, the agronomic traits and leaf area index were significantly higher than those in 2019, because the rainfall in 2018 increased by 170 mm compared with that in 2019. Meanwhile, additional irrigation could help improve agronomic traits and the leaf area index. Correlation analysis further revealed that carbon accumulation was positively correlated with carbon metabolism enzyme activity, although lower at maturity than flowering period. Overall, the findings suggested that supplementary irrigation in conjunction with N250 treatment is a worthy measure for sustainable agriculture in subtropical regions.

Keywords: Agronomic traits; Carbon accumulation; Irrigation conditions; Metabolism; Maize

1 Introduction

Maize (*Zea mays* L.) has become an important pillar of agricultural development and is the third primary crop in China (Zhou et al., 2021). One of the most important food crops in China is maize, which is grown primarily in subtropical regions. With the development of national green agri-

Corresponding author E-mail address: xunbozhou@gmail.com

culture, maize is playing an increasingly important role in crop production, and the pressure of increasing maize production in the future is high (Yao et al., 2021). Due to drought stress and low nitrogen use efficiency (NUE), summer maize production in subtropical regions is facing this huge challenge (Chen et al., 2014). However, due to natural factors, a series of problems, such as intermittent rainfall, high temperature, and high humidity, will affect the growth of maize (Porter et al., 2014, Welikhe et al., 2016). Water is an important limiting factor for maize production on arable land; at the same time, nitrogen is also an important factor for maize growth (Tilman et al., 2002). Drought and nitrogen loss are constantly increasing, and whether early or late in the growing season, can significantly reduce crop yields (Ge et al., 2012; Zhang et al., 2014). Urbanization and the development of high-water-consumption industries and agriculture must be restricted in water-scarce areas (Yuan et al., 2015). Therefore, there is an urgency to strengthen the management of water resources and NUE.

In the subtropical region, it is difficult to increase the irrigation area and amount of farmland. Long-term dependence on groundwater has caused severe environmental problems (Yang et al., 2015; Liang et al., 2019). Therefore, proper water storage, use, and effective water-saving irrigation have become critical issues (Ali et al., 2019). The use of different irrigation methods to increase the yield of maize and effectively use the limited water resources in the subtropical areas are essential.

According to some studies, nitrogen fertilizer is the most important input resource for agricultural production, as it improves crop yields and has a positive impact on biomass and maize grain yields (Vos et al., 2005; Javeed and Zamir, 2013). Today, the application amount of nitrogen fertilizer has far exceeded the suitable range in China (Fan et al., 2013). In many areas of China, the rate of nitrogen fertilizer has an impact on groundwater, air pollution, and NUE. (Cui et al., 2018). Therefore, increasing NUE rather than the nitrogen input is required for sustained increments in agriculture production, environmental protection, and resource constraints (Hedegaard et al., 2006), and fertilizer input and output balancing have received considerable attention. Increasing aboveground dry matter yield of summer maize by using multiple in-season N fertilizer applications can improve the utilization rate of N used in research methods (Shaviv, 2001; Zhou et al., 2017). However, regardless of which nitrogen application method is used, it will increase labor costs, thereby reducing NUE (Faostat, 2011). Previous studies indicated that the soil with a higher clay content has higher levels of transpiration, crop evapotranspiration, carbon accumulation, and yield due to the higher uptake of N (19). In addition to increased root biomass on clay loam soils also improve the crop uptake of N by diffusion, which may constitute an important part of the total uptake and labor cost (21).

Previous researchers demonstrated that the application of nitrogen fertilizer (Paponov and Engels, 2005; Nannen et al., 2011) is closely related to water management (Miao et al., 2011) and has a significant interaction effect (Gheysari et al., 2009; Huang et al., 2014; Jia et al., 2014). These factors affect and extend to carbon and nitrogen metabolism, accumulation, transit, and distribution. Therefore, the main purpose of this study is to determine the interaction effects of different irrigation methods and nitrogen fertilizers on spring maize yield, dry matter accumulation, agronomic traits, leaf area index, and carbon and nitrogen metabolism and accumulation in different growth periods. The goal is to investigate the impacts of different nitrogen application gradients on increasing maize yield and improving NUE in subtropical region under different water irrigation conditions.

One of China's most important maize-producing regions is the subtropical region. However, in this region, unsuitable water levels and excessive N supplementation are common, thereby resulting in a decrease in yield and NUE. Therefore, we conducted a fixed-position experiment that examines the effect of different irrigation methods and nitrogen fertilizers on maize production under high-yield conditions in the subtropical areas in 2018–2019. Its objective was to determine the interactive effects of different irrigation methods and nitrogen fertiliers on yield, dry matter accumulation, agronomic traits, leaf area index, and the interaction of carbon and nitrogen metabolism and its accumulation. The goal is to investigate the effect of different nitrogen application gradients on maize yield and NUE under different irrigation methods.

2 Materials and methods

2.1 Experimental site description

The field experiment was conducted at the farmland of Guangxi University in Nanning, Guangxi, China (altitude is 79 meters.) from 2018 to 2019. Maize growing season ranged from March to November, with a mean annual precipitation of 1,044.6 mm in 2018 and 874.6 mm in 2019 (Figure 1). According to Chinese Soil Taxonomy, the soil texture in the experimental field from 0–20 cm soil layer was loam, with a pH of 5.4; field capacity of 37.2%; soil bulk density of 1.50 g/cm^3; soil organic matter of 17.5 g/kg; and available nitrogen, phosphorus, and potassium of 126.2, 40.0, and 124.5 g/kg, respectively.

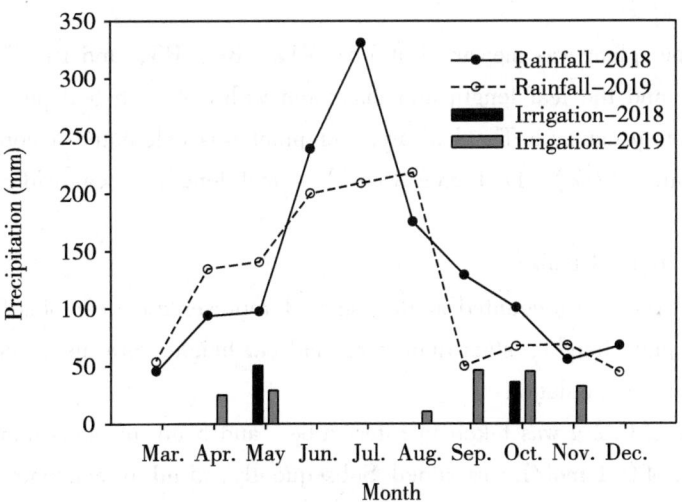

Figure 1 Cumulative monthly irrigation amount and precipitation during the growing season in 2018–2019

2.2 Experimental design

A randomized complete block design with split-plot arrangement was used in the experiment. The water treatments were the main plot while the nitrogen treatments were the split-plot factor. These treatments were replicated three times. The water treatments included supplementary irrigation and rain-fed treatment without irrigation. The five nitrogen treatments included control (N0), 150 (N150), 200 (N200), 250 (N250), 300 (N300) kg N/ha. Water was supplied to the plots using drip irrigation when the soil volumetric water content in the supplementary irrigation treatment, which was detected by soil moisture meter of TDR 100 (Spectrum Technologies Inc., Aurora, USA), was less than 28.6% or equivalently 60% of field capacity. The total irrigation volume during the whole maize growing season measured by a flow meter was 86.9 mm and 189.4 mm in 2018 and 2019, respectively (Figure 1). The 67% of total N fertilizers were applied at sowing and the 33% were used as top dressing at V12 stage. Furthermore, all plots received 100 kg/ha P_2O_5 and 100 kg/ha K_2O as base fertilizers before sowing. Each plot had an area of 4.20 m × 3.92 m, where 55,556 plants/ha were planted twice in a year with a row spacing of 0.60 m and plant spacing of 0.30 m on 22 March and 11 August 2018, and 27 February and 10 August 2019, and harvested on 12 July and 16 December 2018, and 7 July and 21 November 2019.

2.3 Sample collection and measurements

2.3.1 Biomass in stem, leaf, and ear of maize

For vegetative growth stages: 6th-leaf stage (V6), 12th-leaf stage (V12), silking stage (R0), milking stage (R3), and maturity (R6). The stem, leaf, and ear samples were separated and placed in heated at 105℃ for 30 min and then dried at 80℃ until a constant weight to achieve a dry matter weight (DW).

2.3.2 Leaf area index

The leaf area per plant was measured at V6, V12, R0, R3, and R6. Three plants per replicate were selected, and the leaf length and maximum width of each leaf per plant were measured manually using a measuring tape. The leaf area per plant was calculated according to the following formula (Wang et al., 2012). Leaf area (cm^2) = leaf length × leaf width × Correction factor (0.83)

2.3.3 Agronomic traits of maize

The agronomic traits were counted at R6 using 11 representative samples from area of 2 m^2 for each plot, and the plant height, stem diameter, and ear height were measured.

2.3.4 Plant carbon accumulation

A plant sample of 0.2 g was taken in a test tube, and 5 mL of potassium dichromate solution with a concentration of 0.1 mol/L was added. Subsequently, 5 mL of concentrated sulfuric acid was added to the test tube and was gently shaken to mix. The ventilation was opened, the temperature of the digester was controlled at 175℃, and heated for 5 min. A total of 50 mL of distilled water was used to wash the digested sample solution into the Erlenmeyer flask. A total of 5 drops of phenanthroline indicator were added, shaken, and titrated. The carbon accumulation was calculated by following Walkley and Black (1934) using the following formula:

Plant carbon accumulation = $(A-B) \times C_{Fe} - FeSO_4 \times 0.003/m \times 100\%$

Where A is the ferrous sulfate for blank titration (mL), B is the ferrous sulfate used for sample titration, C_{Fe} is the concentration of $FeSO_4$ standard solution, and m is the sample quality.

2.3.5 Phosphoenolpyruvate carboxylase (PEPCase)

From the flowering period to maturity, three fully expanded leaves or flag leaves were collected in each plot and stored in a refrigerator at $-80°C$ after quick freezing with liquid nitrogen. The leaf PEPCase content was measured using an enzyme-linked immune sorbent assay.

Phosphoenolpyruvate carboxylase enzymatic activity is defined as 1 nmol NADH consumed per minute in the reaction system per g of tissue, which is described as a unit of enzyme activity. Frozen plant tissues ($-80°C$, 0.2 g) were milled and homogenized with 1 mL extract solution and clarified by centrifugation at $8,000 \times g$ for 20 min. Absorbance was measured at 340 nm. According to the following formula:

PEPCase (U/g FW) = $\Delta A/(\varepsilon \times d) \times 10^9 \times V1/(V2 \times W/V3)/T$.

where ε is the NADH molar extinction coefficient (6.22×10^3 L/mol/cm), d is the cuvette light path (0.6 cm), V1 is the total volume of the reaction system ($2 \times 10^4/L$), V2 is the sample volume in the reaction system (0.02 mL), W is the sample quality (0.2 g), V3 is the extract volume (1 mL), T is the reaction time (5 min), and 10^9 is the unit conversion factor (1 mol = 10^9 nmol).

2.3.6 Sucrose phosphate synthase (SPS)

SPS enzymatic activity is defined as a unit of enzyme activity that catalyzes the production of 1 μg sucrose per minute per g of tissue. Frozen plant tissues ($-80°C$, 0.2 g) were milled and homogenized with 1 mL extract solution and clarified by centrifugation at $8,000 \times g$ for 10 min. Absorbance was measured at 480 nm. According to the following formula:

SPS (U/g FW) = $(C1 \times V1 \times \Delta A1/\Delta A2) \div (W \times V1/V2)/T$.

where C1 is the standard tube concentration (500 μg/mL), V1 is the sample volume to the reaction system (0.01 mL), $\Delta A1$ is the measure of the initial value at 340 nm after mixing well, $\Delta A2$ is the measure of the absorbance again after reacting at 25°C for 30 min, V2 is the extract volume (1 mL), W is the sample's fresh weight (0.2 g), and T is the reaction time (10 min).

2.3.7 Plant sampling and N content determination

The N contents were counted at R6 using 3 representative samples from area of 2 m^2 for each plot, and they were then dried at 80°C in a forced-air oven ( ) to a constant weight and weighted separately. After weighing, the samples were grounded using a cyclone sample mill with a fine mesh (0.5 mm). The following parameters were calculated: Agronomic NUE (ANUE, kg/kg) = [grain weight (with fertilizer) − grain weight (no fertilizer)] /N fertilizer applied.

Physiological NUE (PNUE, kg/kg) = [grain weight (fertilizer) − grain weight (no fertilizer)] / [plant N (fertilizer) − plant N (no fertilizer)].

2.3.8 Statistical analysis

All experimental data were analyzed using a two-way ANOVA in SPSS 16.0 (SPSS

Inc. Chicago, IL, USA). The least significant difference (LSD) was used to compare the mean value, and it was significant at $P<0.05$. All figures were created using Sigma Plot 10.0 (SPSS Inc., Chicago, IL, USA) and Origin 9.0 (Origin Lab Corporation, USA).

3 Results

3.1 Biomass in stem, leaf and ear of maize

With the growth of maize in 2018 and 2019, the stem, leaf, grain and maximum biomass gradually increased (Figure 2). The leaf biomass was higher than the stem before V6, as opposed to that greater biomass observed in stem after V6. When the ear occurred at R1, more and more biomasses continued to be accumulated in the ear, where the ear increased biomass by 72.55% compared with stem and by 264.17% relative to leaf at R6 in two years. Nitrogen-applied treatments played a significant role in enhancing plant biomass ($P<0.05$). In comparison with other N levels, N250 and N300 had greater stem, leaf, and grain biomass. Furthermore, in 2019, the biomass apportioned to the ear in N250 is not significantly different from that in N300 ($P>0.05$). Water treatment also had a significant effect on biomass after R1 ($P<0.05$). Supplementary irrigation contributed more towards biomass, especially the biomass of the ear that was significantly increased by 6.49% in 2018 and 51.39% in 2019 than those in rain-fed treatment ($P<0.05$). At R6, the ratio for the biomass of ear to total was 50.90% for rain-fed and 52.06% for supplementary irrigation. The total biomass in the whole plant of maize at R6 was 1,729.3 g/m^2 in 2018 and 1,356.7 g/m^2 in 2019.

3.2 Leaf area index

In 2018 and 2019, the leaf area index in the growth stage gradually increased, and the maximum leaf area was recorded at R0 (Figure 3). Averaged across the years the maximum leaf area index, which is 61.8%, 20.69%, 11.69% and 37.23% higher than V6, V12, R3 and R6, was observed at R0, respectively. N-applied treatments played a significant role in enhancing maize leaf area index ($P<0.05$). In the same growth stage, the LAI was increased with N levels. Our results showed that in the rain-fed and irrigation treatments, the LAI in N250 had no significant difference in 2018 ($P>0.05$). Rain-fed treatment also had a significant effect on LAI at R0 ($P<0.05$). Supplementary irrigation treatment significantly increased the LAI by 5.61% and 17.45% in 2018 and 2019 than those in rain-fed treatment, respectively ($P<0.05$).

3.3 Agronomic traits

Agronomic traits were affected by time, water, nitrogen, and their interaction for both years. The interactions were significant for plant height, ear height, and stem diameter ($P<0.05$). The existing difference between 2018 and 2019 could be due to climate change. Under rain-fed treatment, N150, N200, N250, and N300 were statistically similar and significantly higher than that of N0. Under supplementary irrigation treatment, plant height, ear height, and stem diameter

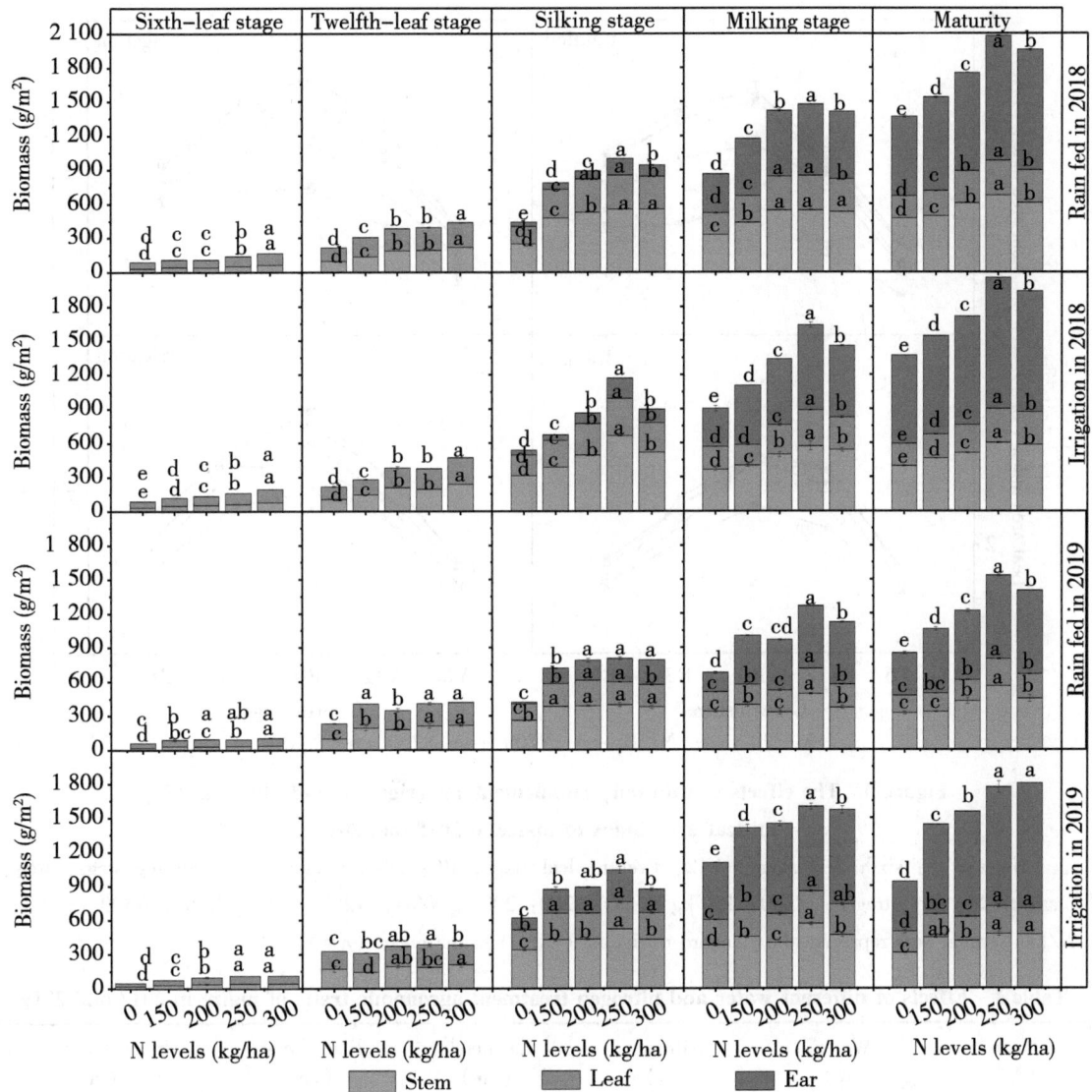

Figure 2　The effects of rain-fed, supplementary irrigation and nitrogen (N) on biomass in stem, leaf and ear of maize in 2018 and 2019

Notes: Error bars represent the standard deviation (± SD) of the mean (n=3).
Different letters indicate significant differences at $P<0.05$.

showed an increasing trend with increased amounts of N fertilizer. Compared with N0, the plant height for N250 was 11.41% and 8.61% higher in rain-fed in 2018 and 2019, respectively, similarly, 12.71% and 10.19% higher in irrigation treatment in 2018 and 2019. Agronomic traits of rain-fed and irrigation treatment were similar and show a gradual upward trend. Plant height, ear height, and stem diameter were not significantly different among N150, N200, and N300 in both years (Table 1).

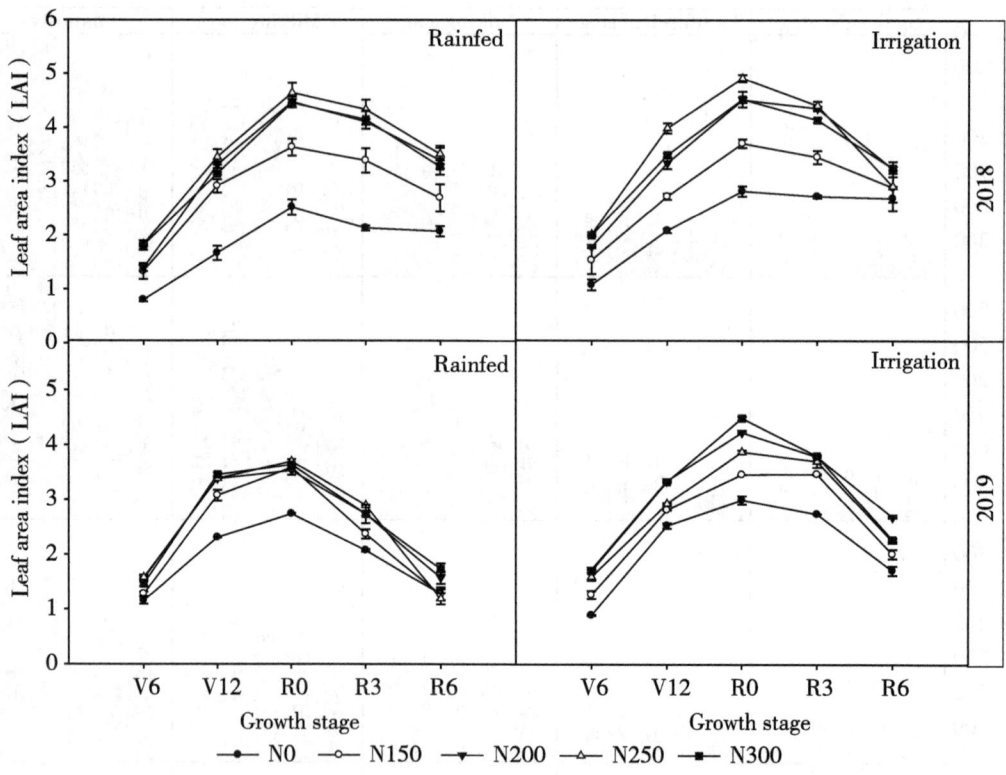

Figure 3 The effects of rain fed, supplementary irrigation and nitrogen (N) on leaf area index of maize in 2018 and 2019

Notes: V6: sixth-leaf stage; V12: twelfth-leaf stage; R0: silking stage; R3: milking stage; R6: maturity. N0: no nitrogen; N150: 150 kg N/ha; N200: 200 kg N/ha; N250: 250 kg N/ha; N300: 300 kg N/ha. Vertical bars represent the standard deviation (±SD) of the mean ($n = 3$).

Table 1 Effects of different water and nitrogen treatment agronomic traits of maize in 2018 and 2019

Year (Y)	Water (W)	Nitrogen (N)	Ear height (cm)	Plant height (cm)	Stem diameter (cm)
2018			153.45 a	293.55 a	27.25 a
2019			132.90 b	264.19 b	24.57 b
LSD (0.05)			0.463 9	1.816 8	0.217 6
	Irrigation		146.38 a	284.13 a	26.08 a
	Rained		139.97 b	273.61 b	25.75 b
	LSD (0.05)		0.529 2	0.405 1	0.099 3
		N0	130.09 e	263.22 d	24.02 d
		N150	139.73 d	275.15 c	25.19 c
		N200	145.32 c	281.17 b	26.04 b
		N250	152.56 a	291.94 a	27.39 a
		N300	148.18 b	282.17 b	26.93 a
		LSD (0.05)	0.567 0	1.217 4	0.151 9

(continued)

Year (Y)	Water (W)	Nitrogen (N)	Ear height (cm)	Plant height (cm)	Stem diameter (cm)
ANOVA					
Y			**	**	*
W			**	**	NS
N			**	**	**
Y×W			NS	*	NS
Y×N			**	*	*
W×N			NS	NS	NS
Y×W×N			NS	NS	NS

Notes: Different letters within the same column indicate significant differences among treatments at $P<0.05$. NS means not significant, * and ** indicate significant difference at the 0.05 and 0.01 levels of probability, respectively.

3.4 Carbon accumulation in stem, leaf, and ear of maize

Carbon accumulation was distributed in different parts of the maize, where stem, leaf, and ear were the main accumulation parts. Water and nitrogen treatment demonstrated a great impact on carbon accumulation during the flowering and maturity during stage (Figure 4). Compared with N0, all N treatments significantly increased carbon accumulation in all parts of the maize. Among the N treatment, the N250 resulted in higher carbon accumulation. Similarly, the plot receiving supplementary irrigation resulted in higher carbon accumulation compared with rain – fed plots. Averaged across fertilizer, the N250 treatment had significantly higher carbon content in stem, leaf, and grain compared with other N treatments. At the maturity stage in 2018, the carbon accumulation for N150 in maize stem, leaf, and ear was statistically similar to that in N200. Similarly, water treatment also had a significant effect on carbon accumulation. Supplement irrigation contributed much more to carbon accumulation, especially in the ear, which was 11.94% and 35.22% higher than those in rain-fed treatment in 2018 and 2019, respectively ($P<0.05$). At the flowering stage, the ratio for carbon accumulation of ear to total was 42.51% for rain-fed and 44.89% for supplementary irrigation, and at maturity, the ratio for carbon accumulation of ear to total was 50.74% for rain-fed and 53.29% for supplementary irrigation.

3.5 Carbon metabolism enzyme activity of maize

3.5.1 Phosphoenolpyruvate carboxylase (PEPCase) of maize

Different nitrogen and water treatments had significant effects on PEPCase at different growth stages of maize ($P<0.05$). During the flowering period, the PEPCase showed the same increasing trend, and the highest value was recorded for rain-fed and irrigation in 2019 (Figure 5). In N treatments, N300, N250, N200, N150, and N0 had 24.49%, 30.44%, 19.79%, 18.66%, and 11.84% higher PEPCase in flowering than in the maturity period ($P<0.05$). Our results

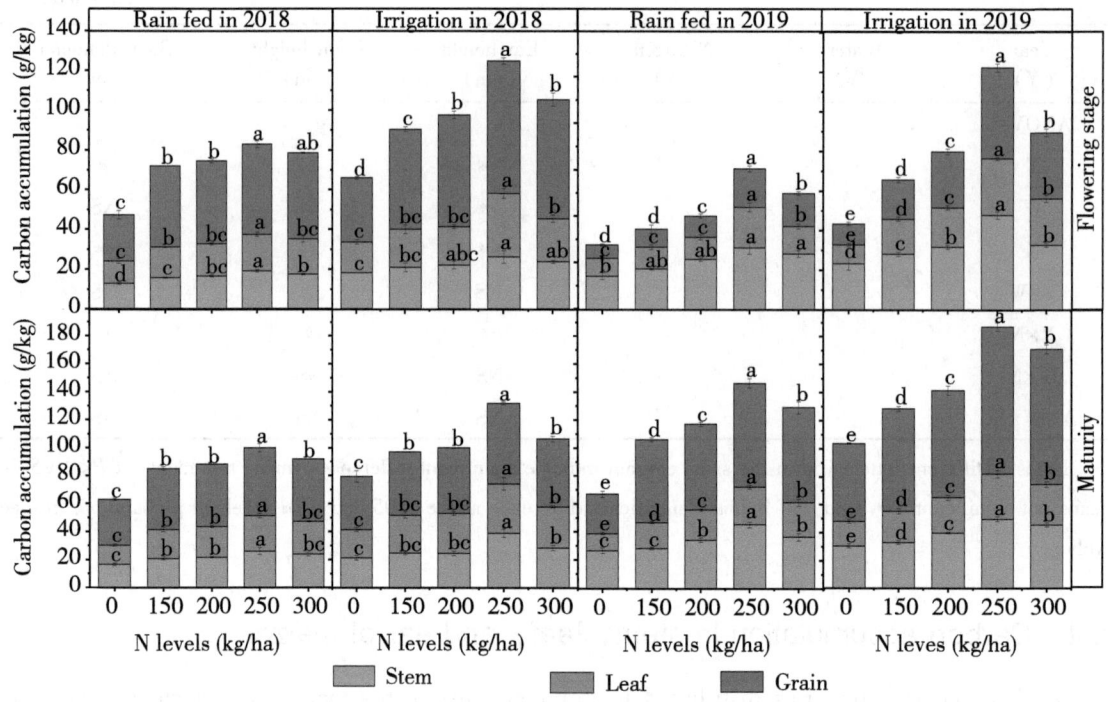

Figure 4 The effects of rain fed, supplementary irrigation and nitrogen (N) on carbon accumulation of maize in 2018 and 2019

Notes: Vertical bars represent the standard deviation (±SD) of the mean (n=3).
Different letters indicate significant differences at $P<0.05$.

showed that N250 had significantly higher PEPCase for the flowering and maturity periods. Likewise, the flowering period had higher PEPCase than the maturity period in rain-fed and irrigation treatments ($P<0.05$). The total of PEPCase content of N0, N150, N200, N250, and N300 were 1.53, 3.89, 7.16, 22.5, and 14.91 nmol NADH/(min·g) of fresh weight under rain-fed treatment, and 2.92, 6.11, 8.83, 19.62 and 12.52 nmol NADH/(min·g) of fresh weight in irrigation treatment, respectively. In both treatments, the PEPCase of N250 was significantly higher than those of N0, N150, N200, and N300.

3.5.2 Sucrose phosphate synthase (SPS) of maize

The results showed that both treatments (water and N fertilizer) had significant effect on SPS (Figure 6). Among the N fertilizer treatments, N250 had significantly higher SPS than other N treatments for rain-fed and irrigation. The SPS at the flowering period was significantly higher at rain-fed and irrigation treatments than in the maturity period. The SPS trend was N250>N300>N200>N150>N0 at irrigation, indicating that increasing nitrogen application could enhance SPS in the flowering and maturity periods under both rain-fed and irrigation treatments. In both treatments, the N250 had significantly higher SPS than other N treatments at rain-fed and irrigation ($P<0.05$).

3.6 Nitrogen utilization efficiencies

The presentexperiment revealed that the N utilization efficiencies in maize were significantly in-

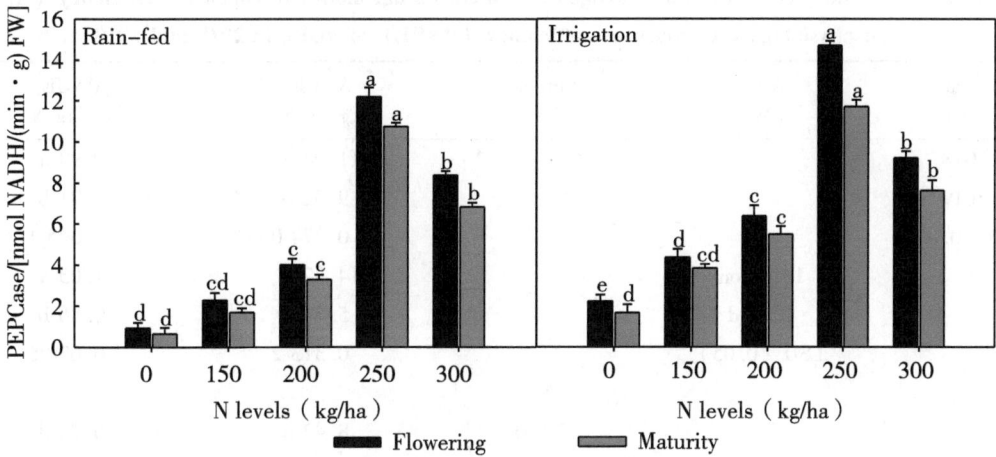

Figure 5   The effects of rain-fed, supplementary irrigation and nitrogen (N) on PEPCase of maize
Notes: Vertical bars represent the standard deviation (±SD) (n=3).
Different letters indicate significant differences at $P<0.05$.

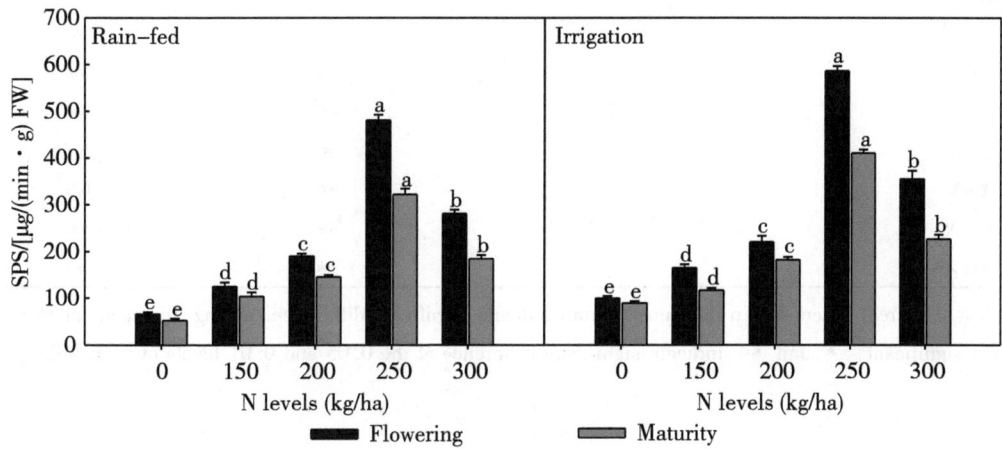

Figure 6   The effects of rain-fed, supplementary irrigation and nitrogen (N) on SPS of maize
Notes: Vertical bars represent the standard deviation (±SD) (n=3).
Different letters indicate significant differences at $P<0.05$.

fluenced by different irrigation methods and N fertilizer. A significance test showed that the effects of the interaction on ANUE, and PNUE reached a highly significant level in 2018 and 2019 (Table 2). The ANUE response to irrigation methods (Irrigation > Rained) was the same of PNUE response to irrigation methods (Irrigation > Rained) (Table 2); the irrigation was 41.9%, and 18.6% higher than rained respectively in 2018, and 67.6%, and 17.8% higher in 2019. Under the difference water conditions, both ANUE and PNUE showed increasing first and then decreased with a higher N application rate. Under rained and irrigation conditions, the ANUE and PNUE response to N were ranked N250>N300>N200>N150, and the differences were significant. With the irrigation treatment, the N250 was maintained higher N efficiency in combination.

Table 2 Effects of different water and nitrogen treatment on agronomic nitrogen use efficiency (ANUE), and physiological nitrogen use efficiency (PNUE) of maize in 2018 and 2019

Year (Y)	Water (W)	Nitrogen (N)	ANUE (kg/kg N)	PNUE (kg/kg N)
2018			11.04 a	0.99 a
2019			9.52 b	0.91 b
LSD (0.05)			0.374 0	0.045 1
	Irrigation		14.82 a	1.05 a
	Rained		5.74 b	0.86 b
	LSD (0.05)		0.313 2	0.038 5
		N0		
		N150	8.42 c	0.73 d
		N200	10.17 b	0.89 c
		N250	13.82 a	1.19 a
		N300	8.69 c	1.01 b
		LSD (0.05)	0.335 8	0.040 3
ANOVA				
Y			**	*
W			**	**
N			**	**
Y×W			*	NS
Y×N			**	NS
W×N			**	**
Y×W×N			**	*

Notes: Different letters within the same column indicate significant differences among treatments at $P<0.05$. NS means not significant, * and ** indicate significant difference at the 0.05 and 0.01 levels of probability, respectively.

4 Discussion

Generally, the final biomass and agronomic traits of maize are highly responsive to different water conditions (Jia et al., 2014; Fang et al., 2008). Irrigation at V6, V12, and R0 stages can significantly delay leaf senescence, increase the leaf area index, prolong the milking stage, and increase crop yield (Xu et al., 2018). Our results showed that biomass and agronomic traits in 2018 were significantly higher than those in 2019 (Figure 2 and Figure 4). The development of high-yield potential through increased dry matter accumulation and agronomic features is ensured because of the excessive rainfall in 2018 (Figure 1), which provided enough moisture to stimulate aboveground biomass in subtropical regions effectively. Under irrigation conditions, higher moisture availability also played a role in increasing the sink and source of N in maize. Gheysari et al. (2009) showed that the effects of N fertilizer on total aboveground biomass depended on the availability of water in the soil. The stunted plant growth in 2019 had adverse effects on dry matter accumulation than that in 2018 because rainfall is lesser in the former than in the latter. However, the total biomass and dry matter accumulation in irrigation treatments were not statistically different in N250 and N300 treatments, indicating that various nitrogen treatments had a greater effect on irrigation and rain-fed crops. These findings suggest that the application of nitrogen fertilizer can improve the photosynthetic capacity, nitrogen content, carbon content, total biomass, and agronomic traits but decrease the NUE (Peng et al., 2016; Hammad et al., 2017). Zhao et al. (2013) demonstrated that fertilizers could significantly increase biomass and agronomic traits in maize compared with common compound fertilizer. At the same water conditions, it increased with the levels of urea (CRU), but no significant difference was observed between N250 and N300 under adequate conditions.

Grain yield was based on dry matter. Thus, the first objective was to boost dry matter production as much as possible (Parija et al., 2013). LAI is a significant determinant of plant development, dry matter accumulation, and carbon fixation. The increase in LAI is possibly due to the growth of new leaves or the enlargement of expansion leaves (Mandal et al., 2005). Our results suggested that higher LAI with increased N application could be attributed to significant increases in leaf expansion (length and breadth) that resulted from cell division and cell enlargement at higher N rates. These results are in the line with the findings of Kar and Kumar (2015) and Shafi et al. (2011). Leaf senescence in the late stage of crop growth is an important reason for the decrease in LAI (Bandyopadhyay et al., 2010; Thomas 2013; Pradhan et al., 2013). The important sources of carbon in maize crop are stem and grain, whereas the interactive effects of different water regimes and nitrogen applications on stem, leaf, and grain in summer maize showed a significant increase (Figure 4). The reason for this increase is that optimum water-nitrogen interactions might effectively improve the photosynthetic capacity of ear leaves in summer maize, as well as plant production capacity and photosynthetic allocations (Srivastava et al., 2018). During heavy rainfall in 2018, carbon accumulation in each part of the maize increased with nitrogen input. Carbon accumulation was higher in the irrigation treatment than in the rain-fed treatment, especially in the maturity stage. Carbon accumulation could be at its peak in 2019 in irrigation treatment (Figure 4). Carbon accumulation was significantly higher in irrigation treatments with N250 nitrogen than in other N rates, indicating that this interaction has a favourable carbon accumulation environment.

Previous study reported that the management of water and nitrogen is closely related to the accumulation and distribution of carbon, and maize carbon metabolism enzyme activity plays a role in the entire growth process (Zhou et al., 2020). Researchers demonstrated that SPS is a multifunctional protein, a key enzyme sucrose in plant metabolism and performs various biological functions during plant growth. Sinclair and Rufty (2012) reported that moderate water conditions were required from the flowering to maturity stage to ensure the absorption and utilization of nitrogen in the soil. This condition improved plant growth and nitrogen usage by increasing carbon accumulation in the stems, leaves, and grains, as well as PEPc and SPS activities. Under irrigation and rain-fed conditions, the carbon metabolism enzyme activity increased with the nitrogen application, and the maximum activity was observed for to N250 (Figure 5 and Figure 6). The interactive effects of water and nitrogen are beneficial to carbon accumulation and carbon metabolism enzyme activities, which can provide more nutrients during the growth maize period. The carbon accumulation of each treatment in different parts exhibited clear differences because of the effects of the different water-nitrogen interactions. Low rainfall in 2019 limited nitrogen release capacity under low-moisture conditions in the flowering stage, probably due to a lack of mineral nitrogen availability and inhibited root activity in the soil (Shao et al., 2013). Nitrogen participates in morphological changes during the vegetative period, thereby affecting carbon accumulation. Despite the lack of water, carbon accumulation gradually increases in irrigation condition over the maturity stage, showing that the nitrogen in the soil has been fully absorbed (Teixeira et al., 2014).

Carbon accumulation and carbon metabolism enzyme activity increase with nitrogen application under the same water conditions; likewise, water content increases at the same nitrogen level (Guo et al., 2016). The activities of carbon accumulation and metabolic enzymes are much higher during the flowering stage than in the maturity stage possibly due to plant senescence, which may reduce the content in the maturity stage.

Furthermore, interactions between water and nitrogen fertilizer could increase biomass, agronomic traits, leaf area index, and plant growth. When spring maize was irrigated during the flowering season and treated with N250, it resulted in higher carbon accumulation and metabolic enzyme activity in plant maize. However, nitrogen migration mechanisms in the soil under different water conditions influenced the maize plant growth. Further study is needed to investigate these mechanisms in different climates and fertilizers.

5 Conclusions

Nitrogen and different irrigation treatments had significant interactive impacts on biomass, agronomic traits, leaf area index, carbon accumulation, and carbon metabolism enzyme activity in summer maize. Managing the optimal interactions of nitrogen and different irrigation treatments was beneficial to maize growth; the enhancement of summer maize under N250 and N300 treatments was excellent, regardless of irrigation treatments. Although the rain-fed treatment can boost summer maize growth and boost metabolites, it does not have the same impact as the supplementary irrigation treatment. This study found that the nitrogen fertilizer application rate of 250 kg N/ha was the best treatment for rain-fed and supplemental irrigation treatments. We suggest that an appropriate nitrogen fertilizer is more feasible under supplemental irrigation in subtropical areas. At the same time, compared with N300, N250 can reduce fertilizer costs and the risk of groundwater pollution.

Acknowledgments

The research was funded by the National Natural Science Foundation of China (31760354), Guangxi Natural Science Foundation (2019GXNSFAA185028) and China Postdoctoral Science Foundation Project (2020M683618XB). The authors are especially thankful to Peng Ju Shen for his valuable contributions.

References

Bandyopadhyay KK, Misra AK, Ghosh PK, et al., 2010. Effect of integrated use of farmyard manure and chemical fertilizers on soil physical properties and productivity of soybean. Soil Till Res. 110: 115-125.

Chen XP, Cui ZL, Fan MS, et al., 2014. Producing more grain with lower environmental costs. Nature. 514: 486-489.

Cui Z, Zhang FS, Dou ZX, et al., 2018. Pursuing sustainable productivity with millions of smallholder farmers. Nature. 555: 363-366.

Fang Q, Ma L, Yu Q, et al., 2008. Modeling nitrogen and water management effects in a wheat-maize double-cropping system. J Environ Qual. 37: 2232-2242.

Fan Z, Chai Q, Huang G, et al., 2013. Yield and water consumption characteristics of wheat/maize intercropping with reduced tillage in an oasis region. Eur J Agron. 45: 52-58.

Ge T, Sui F, Bai L, et al., 2012. Effects of water stress on growth, biomass partitioning and water use efficiency in summer maize (Zea mays L.) throughout the growth cycle. Acta Physiol Plant. 34: 1043-1053.

Gheysari M, Mirlatifi SM, Bannayan M, et al., 2009. Interaction of water and nitrogen on maize grown for silage. Agr Water Manage. 96: 809-821.

Guo LW, Ning TY, Nie LP, et al., 2016. Interaction of deep placed controlled-release urea and water retention agent on nitrogen and water use and maize yield. Eur J Agron. 75: 118-129.

Hammad HM, Farhad W, Abbas F, et al., 2017. Maize plant nitrogen uptake dynamics at limited irrigation water and nitrogen. Environ Sci and Pollut R. 24: 2549-2557.

Hedegaard J, Hauge M, Fage-Larsen J, et al., 2006. The effect of nitrogen fertilization on root distribution of winter wheat. Plant Soil Environ. 52: 308-313.

Huang P, Zhang J, Zhu A, et al., 2014. Coupled water and nitrogen (N) management as a key strategy for the mitigation of gaseous N losses in the Huang-Huai-Hai Plain. Biol Fert Soils. 51: 333-342.

Javeed HMR, Zamir MSI, 2013. Influence of tillage practices and poultry manure on grain physical properties and yield attributes of spring maize (Zea mays L.). Pak J Agr Sci. 50: 177-183.

Jia XC, Shao LJ, Liu P, et al., 2014. Effect of different nitrogen and irrigation treatments on yield and nitrate leaching of summer maize (Zea mays L.) under lysimeter conditions. Agr Water Manage. 137: 92-103.

Kar G, Kumar A, 2015. Effects of phenology based irrigation scheduling and nitrogen on light interception, water productivity and energy balance of maize (Zea mays L.). J Indian Soc Soil Sci. 63: 39-52.

Liang H, Qin W, Hu K, et al., 2019. Modelling groundwater level dynamics under different cropping systems and developing groundwater neutral systems in the North China Plain. Agr Water Manage. 213: 732-741.

Mandal KG, Hati KM, Misra AK, et al., 2005. Irrigation and nutrient effects on growth and water-yield relationship of wheat (*Triticum aestivum* L.) in Central India. J Agron Crop Sci. 191: 416-425.

Nannen DU, Herrmann A, Loges R, et al., 2011 Recovery of mineral fertiliser N and slurry N in continuous silage maize using the ^{15}N and difference methods. Nutr Cycl Agroecosys. 89: 269-280.

Paponov IA, Engels C, 2005. Effect of nitrogen supply on carbon and nitrogen partitioning after flowering in maize. J Plant Nutr Soil Sci. 168: 447-453.

Parija B, Kumar M, 2013. Dry matter partitioning and grain yield potential of maize (*Zea mays* L.) under different levels of farmyard manure and nitrogen. J Plant Sci Res. 29: 177-180.

Peng YF, Zeng XT, Houx JH, et al., 2016. Pre - and post - silking carbohydrate concentrations in maize ear-leaves and developing ears in response to nitrogen availability. Crop Sci. 56: 3218-3227.

Porter JR, Xie L, Challinor AJ, et al., 2014. Food security and food production systems, pp. 485-534. In Climate change 2014 - Impacts, adaptation and vulnerability: part A: global and sectoral aspects. Cambridge University Press, New York.

Pradhan S, Chopra UK, Bandyopadhyay KK, et al., 2013. Soil water dynamics, root growth and water and nitrogen use efficiency of rain-fed maize (*Zea mays* L.) in a semiarid environment. Indian J Agr Sci. 83: 542-548.

Shafi M, Bakht J, Khan MA, et al., 2011. Effects of nitrogen application on yield and yield components of barley (*Hordenum vulgare* L.). Pak J Bot. 43: 1471-1475.

Shao GQ, Li ZJ, Ning TY, et al., 2013. Responses of photosynthesis, chlorophyll fluorescence, and grain yield of maize to controlled-release urea and irrigation after anthesis. J Plant Nutri Soil Sci. 176: 595-602.

Shaviv A, 2001. Advances in controlled-release fertilizers. Adv Agron. 71: 1-49.

Sinclair TR, Rufty TW, 2012. Nitrogen and water resources commonly limit crop yield increases, not necessarily plant genetics. Glob Food Secur. 1: 94-98.

Srivastava RK, Panda RK, Chakraborty A, et al., 2018. Enhancing grain yield, biomass and nitrogen use efficiency of maize by varying sowing dates and nitrogen rate under rainfed and irrigated conditions. Field Crop Res. 221: 339-349.

Thomas P, 2013. Enhancing water productivity in wheat by optimization of irrigation management using AquaCrop model. M. Sc. thesis in agricultural physics. P. G. School IARI, New Delhi, India.

Tilman D, Cassman KG, Matson PA, et al., 2002. Agricultural sustainability and intensive production practices. Nature. 418: 671-677.

Vos J, Van Der Putten PEL, Brich CJ, 2005. Effect of nitrogen supply on leaf appearance, leaf growth, leaf nitrogen economy and photosynthetic capacity in maize (*Zea mays* L.). Field Crop Res. 93: 64-73.

Welikhe P, Essamuah-Quansah J, Boote K, et al., 2016. Impact of climate change on corn yield in Alabama. Prof Agr Workers J. 4: 1-14.

Xu XX, Zhang M, Li JP, et al., 2018. Improving water use efficiency and grain yield of winter wheat by optimizing irrigations in the North China Plain. Field Crop Res. 221: 219-227.

Yang XL, Chen YQ, Pacenka S, et al., 2015. Effect of diversified crop rotations on groundwater levels and crop water productivity in the North China Plain. J Hydrol. 522: 428-438.

Yao Z, Zhang W, Wang X, et al., 2021. Carbon footprint of maize production in tropical/subtropical region: a case study of southwest china. Environ Sci Pollu Res. 28: 28680-28691.

Yuan Z, Yan DH, Yang ZY, et al., 2015. Temporal and spatial 597 variability of drought in Huang-Huai-Hai River Basin, China. Theor Appl Climatol. 122: 755-769.

Zhang SL, Sadras V, Chen XP, et al., 2014. Water use efficiency of dryland maize in the Loess Plateau of China in response to crop management. Field Crop Res. 163: 55-63.

Zhao B, Dong ST, Zhang JW, et al., 2013. Effects of controlled-release fertilizer on nitrogen use efficiency in summer maize. Plos One, 8: e70569.

Zhou BY, Sun XF, Ding ZS, et al., 2017. Multisplit nitrogen application via drip irrigation improves maize grain yield and nitrogen use efficiency. Crop Sci. 57: 1687-1703.

Zhou XB, Wang GY, Yang L, et al., 2020. Double-double row planting mode at deficit irrigation regime increases winter wheat yield and water use efficiency in North China Plain. Agronomy 10: 1315.

Zhou XB, Yang L, Wang GY, et al., 2021. Effect of deficit irrigation scheduling and planting pattern on leaf water status and radiation use efficiency of winter wheat. J Agron Crop Sci. 207: 437-449.

Effects of nitrogen and water stress on the rehydration, endogenous hormonal regulation and yield of maize

Y. X. Chi[1,2], S. Ahmad[2], K. J. Yang[1], J. Fu[1], L. Yang[2], X. B. Zhou[2], H. D. Zhu[1]

([1]College of Agronomy, Heilongjiang Bayi Agricultural University/Key Laboratory of Modern Agricultural Cultivation and Crop Germplasm Improvement of Heilongjiang Province, Daqing 163319, China;
[2]Guangxi Colleges and Universities Key Laboratory of Crop Cultivation and Tillage, Agricultural College of Guangxi University, Nanning 530004, China)

Abstract: Water scarcity is known to be a strong limiting factor affecting maize growth and yield in cold semi-arid regions. Numerous studies have shown that rehydration improves maize growth. Our study aimed to explore the effects of rehydration treatments on maize growth and yield under water and nitrogen stress during different growth stages. We selected the drought-tolerant maize variety Nendan 19 (ND19) and subjected it to water stress during the V6 (sixth-leaf), R2 (filling), and R6 (maturity) growth stages, and a rehydration treatment after each stress stage. Our results indicated that N1 (N100 kg N/ha) and N3 (N300 kg N/ha) treatments significantly increased the leaf moisture status relative to water content (RWC), bound water content (BWC), free water content (FWC), and water potential (WP) at different growth stages. Similar trends were observed in the accumulation of plant leaf and root hormones (zeatin+zeatin riboside, indole-3-acetic acid, abscisic acid, and gibberellic acid), photosynthetic pigments, and chlorophyll fluorescence. However, under the same water stress conditions, they decreased as the N rate increased and reached a minimum value in the S3 (water stress for N3) treatments. In addition, with growth stage advancement and extension of the rehydration time, both showed a gradual upward trend. The results showed that to save water resources in the cold semiarid region, rehydration treatments (R2S1 and R2S3) significantly increased the photosynthetic pigments and chlorophyll fluorescence parameters, leaf moisture status, biomass, 100-grain weight, hormone content, ear characteristics, and grain yield of maize.

Keywords: Rehydration; Nitrogen; Hormones; Maize; Water stress

1 Introduction

North China is one of the main maize-producing areas, supplying more than 30% of the maize

---

These authors contributed equally to this work.
Corresponding author (XB. Z & HD. Z) E-mail address: xunbozhou@gmail.com; zhd495@163.com.

(*Zea mays* L.) production in China (Chi et al., 2021). Over the past 10 years, the annual average yield of maize in China has increased by 5% (Li et al., 2017). Climate change is an important factor limiting crop yield and quality and is also an issue of global food security (Ahmad et al., 2020). Recently, water shortages have increased globally due to the continuous increase in temperature and irregular rainfall (Ullah et al., 2018). Meanwhile, water supply has been affected by shortages and over-exploitation of groundwater due to the intensification of agriculture (Wei et al., 2008). According to Ahmad et al. (2019), rainfall in different regions (arid and semiarid regions) significantly affected maize growth. In semi-arid areas, water deficit is a significant challenge in crop improvement (Liu et al., 2016). Previous studies have shown that soil is not conducive to absorbing water in unique environments, resulting in constraints for increased maize yield during seasonal aridity (Ahmad et al., 2021; Gan et al., 2008). Water shortages during the flowering and grain-filling phases significantly affect grain yield (Guo et al., 2014). Maize hybrids require sufficient water to complete their growth cycle. Nevertheless, at particular intervals, plants are more susceptible to water scarcity, and any dryness at these times results in a considerable yield loss (Whitmore et al., 2009). Plaut et al. (2004) showed that grain yield, nitrogen absorption, and nitrogen-use efficiency (NUE) decreased under water-deficient conditions.

Long-term water deficits cause smaller leaves, earlier flowering, and prolonged flowering period and yield potential (Campos et al., 2004). These studies showed that leaf moisture status is an essential indicator of physiological tolerance and is closely related to soil water content (Sarker et al., 2010). Under water stress, the growth of maize seedlings shows several important physiological responses, including reduced cell swelling rate (Chen et al., 2015; Muhammad et al., 2022a), twisting of leaves (Asim et al., 2012), inhibition of CO_2 exchange, and reduced photosynthetic characteristics and chlorophyll fluorescence parameters (Mao et al., 2015). The reactions most sensitive to water constraints are photosynthetic, and gas exchange reactions, and maintaining a high level of photosynthetic activity can help plants withstand drought stress (Corina et al., 2009). Merah et al. (2001) indicated that the relative water content (RWC) of leaves is an important physiological tolerance characteristic that is directly related to soil water content. Therefore, RWC is an essential indicator of water stress in leaves, and Bai et al. (2016) and Khan et al. (2007) reported similar findings. In addition, chlorophyll fluorescence parameters may decrease with carbon exchange reduction under water stress (Fracheboud et al., 2004).

Hormones such as indole-3-acetic acid (IAA), abscisic acid (ABA), gibberellic acid (GA), ethylene (Eth), and jasmonic acid (JA) play a significant role in the grain filling process, plant growth, and development cycle in plants under water stress conditions (Yanping et al., 2018; Ahmad et al., 2022a). Auxin has been reported to play an essential role in plant growth, development, and responses to different abiotic stresses (Abid et al., 2018). Simultaneously with other hormones, auxin plays a role in meristem maintenance, leaf primordium, lateral root initiation, tropical response, vascular tissue development, root and shoot elongation, and the control of apical dominance (Quiroga et al., 2020). Kudoyarova et al. (2011) showed that regarding the effects of plant hormones on water transport, ABA generally enhances root water transport capacity by increasing root hydraulic conductivity (Lpr). The levels of various hormones, such as ABA, IAA, and Z + ZR, rapidly increase during the early growth stage of plant development and then progressively decline until the crop matures (Song et al., 2000). Previous studies have shown that high cytokinin levels are associated with rice spikelet development, early grain filling, and endosperm development (Yang et al., 2000).

Nitrogen (N) fertilizers play an essential role in modern agricultural production systems (Ahmad et al., 2022b; Muhammad et al., 2022b). The current nitrogen fertilizer application rate is relatively high, but the utilization rate in maize production is low, wasting resources and causing ecological and environmental problems (Zhang et al., 2016). In maize fields in northern China, insufficient soil moisture and drought stress are major issues that significantly limit plant yields in the early and late growth seasons (Zhang et al., 2014). Consequently, throughout agricultural production, reasonable and adequate water resource management is required in water-scarce areas, and rehydration treatment is performed after the critical stress phase. Water and nitrogen availability limits yield in many regions; therefore, careful management of optimum irrigation and nitrogen supply is vital to achieving high yields to fulfill agricultural demand (Muhammad et al., 2022c; Teixeira et al., 2014). The rehydration treatment strategy decreases the risk of environmental pollution caused by N losses and increases maize production and nitrogen use efficiency (Shao et al., 2013). However, rehydration treatment has not been widely investigated after different water and nitrogen interaction conditions. Only a few studies have investigated the consequences of rehydration on the leaf moisture status, photosynthetic pigments, chlorophyll fluorescence parameters, biomass, ear characteristics, grain yield, and endogenous hormones in maize. Therefore, it is necessary to explore the relationship between rehydration treatments at different growth periods under water and nitrogen stress.

This study evaluated the combined effects of rehydration treatment after water and nitrogen interaction on the leaf moisture status, photosynthetic pigments, chlorophyll fluorescence parameters, biomass, ear characteristics, endogenous hormones, and maize yield during the primary growth period. This study revealed the optimum management practices for water and nitrogen conservation, rehydration after stress in various development periods, and high yield production.

2 Material and Methods

2.1 Experimental location, design, and treatments

This study was conducted at the experimental station of Qiqihar Branch of Heilongjiang Academy of Agricultural Sciences (47°16′26″N, 123°41′46″ E, 143 m above sea level) in China in 2019 and 2020. The region is characterized by a semiarid climate in the western Songnen Plain. The maize hybrid cultivar Nendan 19 (ND19) was previously screened out for drought resistance and used in the current experiment. The experimental soil type was aeolian sandy soil with a pH of 7.82, and soil organic matter was 26.52 g/kg. The available nitrogen (N), phosphorous (P), and potassium (K) was 100.05 mg/kg, 16.91 mg/kg, and 134.03 mg/kg, respectively. The soil pH was determined using a pH meter. Soil field capacity and bulk density were determined using the Welcox method (Jiang et al., 2006). The study was conducted in pots (29.5 cm in diameter and 38 cm high). Each pot was filled with an equal amount of aeolian sandy soil from normal farmland.

There were three levels of water treatment 1) adequate water conditions, which kept the soil moisture at approximately 75%–85% of the soil field capacity (FC); 2) water stress [S], which maintained moisture content of 45%–50% of FC; 3) rehydration water [RS], which restored soil moisture to 75%–85% of FC. Three levels of nitrogen application (N0 with 0 kg N/ha, N1 with 100 kg N/ha, and N3 with 300 kg N/ha) were applied to all three water treatments. Previous studies demonstrated that maize yield results from water–nitrogen interaction (Bin et al., 2010). Therefore, we used three N application rates to determine the maize interactions under water stress, rehydration water, and well-watered conditions. All nitrogen applications were applied at sowing time, and all treatments received the same amount of phosphorus and potassium in the form of P_2O_5 165 kg/ha (P_2O_5 1.16 g per pot) and K_2O 105 kg/ha (K_2O 0.74 g per pot) as a basal dressing. Seeds were planted on May 12 and harvested on October 6, 2019, and 2020. There were three seeds in each pot, and during the third leaf stage, two plants were thinned, while the other continued to grow until maturity. We used a TDR-100 m (Spectrum Technologies Inc., California, USA) to measure the soil moisture daily. Water stress treatments were imposed when the soil moisture first fell below the designated standards during the maize season. Automatically operated triple-folding rain shelters were moved over the experimental plants before rainfall to prevent natural rainfall over the pots.

2.2 Plant sampling measurements

Plants were sampled from each treatment at the six-leaf stage. The samples were separated into the root, stem, leaf, ear, and grain sections. Agronomic traits were counted at the six-leaf, filling, and maturity stages using three representative samples from each pot. In addition, plant height, stem diameter, and ear height were measured. Half of the samples were dried at 80℃ in a forced-air oven to a constant weight and weighed separately, while the other half was stored at −80℃ until use in the hormone content assays.

2.3 Measurement of endogenous hormone contents

Extract samples of each treatment (each 0.5 g in leaves and roots) were used to measure the hormone contents (GA, IAA, ABA, and Z+ZR). Three replicates were prepared for each treatment. We measured endogenous hormone contents according to Ahmad et al. (2020) method.

2.4 Characteristics of maize ear and grain yield

At the physiological maturity stage (R6), all the remaining ears were harvested to determine yield (the length of the ear, ear diameter, rows per ear, and grain per row were counted manually from the selected cobs after measurement of the 100-grain weight, with a moisture content of approximately 14%).

2.5 Photosynthetic pigments and chlorophyll fluorescence parameter

Photosynthetic parameters of the second fully expanded leaf at different stages were measured using a Li-6800 photosynthesis instrument (LI-COR Inc., Lincoln, USA). The measurement was carried out at 9:00-11:00, using an artificial light source [1,500 μmol/($m^2 \cdot s$)], adjusting relative humidity to 60%, leaf chamber temperature to 30℃, airflow rate to 500 μmol/s, and CO_2 concentration to 400 μm/mol. The photosynthetic rate (Pn), stomatal conductance (Gs), intercellular CO_2 concentration (Ci), and transpiration rate (Tr) were derived from the photosynthetic instrument.

Chlorophyll (Chl) and carotenoid extraction was performed according to the method described by Hiscox and Israelstam (1979) with some modifications. A plant sample (0.2 g was taken in a test tube, and 50 mL of chlorophyll extract was added and extracted in the dark until the leaves were completely white and then measured at 470 nm, 646 nm, and 663 nm. The following formula was used to measure the chlorophyll (Chl) pigments:

$$Chl\ a\ (mg/g) = 12.7 \times A663 - 2.69 \times A646$$
$$Chl\ b\ (mg/g) = 22.9 \times A646 - 4.68 \times A663$$
$$Total\ chlorophyll\ (Chl\ t) = Chl\ a + Chl\ b$$

Carotenoid content (mg/g) = (4 × A470 nm × volume of sample) /fresh weight of the sample

2.6 Leaf moisture status

Relative water content (RWC), bound water content (BWC), free water content (FWC), and water potential (WP) were measured on fresh leaves. The RWC was measured following the method described by Machado and Paulsen (2001). FWC was measured according to the procedure described by Ming et al. (2011). Fresh leaves were quickly rinsed with distilled water, weighed, recorded as FW, dried, and immersed in a 60% sucrose solution for 6 h. This step was carried out in a dark environment at 4℃. The leaves were then quickly removed, washed with ultrapure water (at least three times), wiped off the surface moisture, weighed, and recorded as the sugar solution weight (SSW). Afterwards, the leaves were dried at 105℃ to a constant weight, weighed, and recorded as DW. WP was measured using the method described by Zhou et al. (2021). The leaf water potential (Lψ) was measured using the Psypro Water Potential System (Wescor, Inc., Logan, UT, USA) with eight model C-52-SF sample chambers. Before the measurement, the middle part of the leaf was wiped and punched to create a 0.6 cm diameter disk, which was placed into the sample chamber and balanced for 20 min before reading. RWC, FWC, and BWC with some modifications were calculated using the following formulas:

$$RWC\ (\%) = [(FW-DW)/(FTW-DW)] \times 100$$
$$FWC\ (\%) = (FW-SSW)/DW \times 100$$
$$BWC\ (\%) = TW-FWC$$

Where FW is fresh weigh, DW is dry weight; SSW is sugar solution weight, TW is turgid weight, and FTW is fully turgid weight.

2.7 Statistical analysis

Duncan's test was used for multiple comparisons and the analysis of differences. Statistical significance was set at $P<0.05$. All data in the tables are the average values of triplicate experiments. During the 2019 growing seasons, the endogenous hormone content, photosynthetic pigments, chlorophyll fluorescence parameters, and leaf moisture status were not significant; therefore, data from 2020 were used to measure the photosynthetic pigments and endogenous hormone content. LSD tests were conducted using the SPSS21.0 program (Ver. 21.0, SPSS, Chicago, IL, USA).

3 Results

3.1 Biomass in the stem, leaf, and ear of maize

The results demonstrated that the total dry matter accumulation (DMC) showed a significant difference between the various water stress conditions and N applications (Figure 1). Our findings show that the DMC of maize began to increase gradually during the V6 stage and reached a maximum value during the V6 stage. DMC showed N3>N1>N0 with different N applications (Figure 1). Normal water conditions increased more DMC per plant than water stress and rehydration treatments. During the R2 and R6 stages, the DMC per plant was improved by increasing the water stress to normal conditions. The extension of the rehydration time, DMC per plant, became more significant, especially from R2S3 to R6S3, after the R2 and R6 stages (Figure 1). The mean results during the 2019 and 2020 growing seasons showed that DMC per plant increased in N1 and N3 treatments by 32.87%, 44.35%, 49.07%, 56.39%, 50.67%, and 58.31%, whereas it increased by 27.85%, 37.52%, 38.81%, and 38.64% following R2S1 and R2S3 treatments compared with N0 under different growth stages. Under the R6 growth stage, DMC per ptant was relatively high at R6S1, and R6S3 was significantly higher than that at N0.

3.2 Ear characteristics and grain yield

Nitrogen application rates and different water stress conditions enhanced maize ear characteristics and grain yield during the 2019 and 2020 growing seasons. Table 1 shows that water conditions improved ear length (cm), row per ear, ear diameter (mm), grain per row, and grain yield (g/plant) compared to different N applications. The results showed that N3 treatment improved ear length by 37.18% and 44.27%, ear diameter by 13.08% and 14.54%, row per ear by 34.62% and 44.03%, grain per row by 35.28% and 31.03%, and grain yield by 73.80% and 75.13% during the 2019 and 2020 growing seasons, while N1 treatment increased ear length by 30.54% and 37.78%, ear diameter by 9.03% and 13.81%, row per ear by 32.03% and 39.13%, grain per row by 32.22% and 28.60%, and grain yield by 39.62% and 70.08% compared with N0. Under rehydration, ear characteristics and grain yield showed an upward trend with

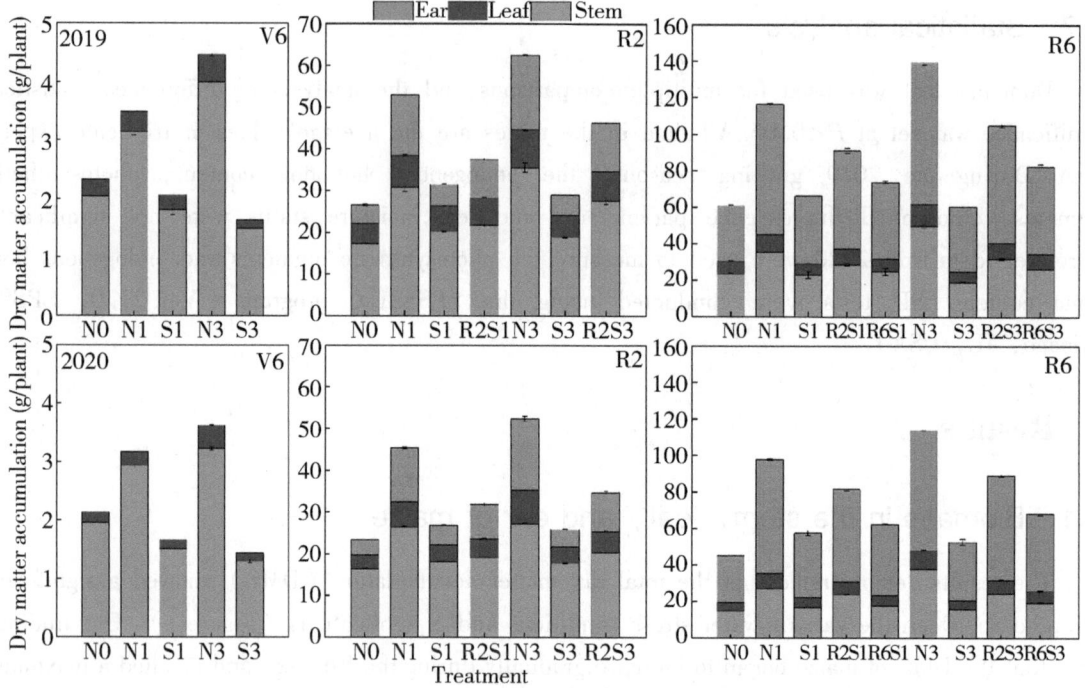

Figure 1 Effects of nitrogen and water stress on the rehydration of plant dry matter accumulation in maize during the 2019 and 2020 growing seasons

Notes：V6：sixth-leaf stage；R2：filling stage；R6：maturity stage；N0：N application of 0 kg/ha；N1：N application of 100 kg/ha；S1：N1 water stress；R2S1：N1 rehydration treatment under R2 stage；R6S1：N1 rehydration treatment under R6 stage；N3：N application of 300 kg/ha；S3：N3 water stress；R2S3：N3 rehydration treatment under R2 stage；R6S3：N3 rehydration treatment under R6 stage. Different lowercase letters indicate significant differences as determined by the LSD test （$P \leq 0.05$）.

increasing rehydration time and N application.

The yield of R2S3 was significantly higher than that of other rehydration treatments during the 2019 and 2020 growing seasons, and both were significantly higher than those of water stress treatments. Under the R2S3 treatment, yields of 66.71% (2019) and 66.44% (2020) were higher than N0; both other on-ear characteristic trends were similar to grain yield (Table 1). In the water stress treatment, S1 and S3 yields showed no significant difference between the 2019 and 2020 growing seasons. However, these values were significantly higher than those in the N0 treatment group. The ear characteristics and grain yield of N3 were the highest of all water conditions and N treatments during the maize 2019 and 2020 growing seasons.

Table 1　Effects of nitrogen and water stress on the rehydration of ear characteristics and grain yield in maize during the 2019 and 2020 growing seasons

Year (Y)	Treatment (T)	Ear length (cm)	Ear diameter (mm)	Rows per ear	Grains per row	Grain yield (g/plant)
2019	N0	10.37±0.21g	36.53±0.61e	11.33±0.67g	14.67±0.67f	37.17±0.92h
	N1	14.93±0.08b	40.16±0.23b	16.67±0.66ab	21.33±0.66ab	122.38±3.89b
	S1	12.21±0.06e	37.83±0.03de	13.33±0.67ef	17.33±0.67e	55.48±1.65g
	N3	16.51±0.17a	42.03±0.47a	17.33±0.67a	22.67±0.66a	141.89±3.94a
	S3	11.36±0.08f	37.26±0.61de	12.67±0.66fg	15.33±0.67f	47.74±1.39g
	R2S1	13.11±0.05d	39.13±0.32ab	15.33±0.66bcd	19.33±0.66cd	95.43±1.56d
	R2S3	13.73±0.08c	39.31±0.26ab	16.00±0.00abc	20.67±0.66bc	111.66±1.98c
	R6S1	12.23±0.13e	38.23±0.36cd	14.00±0.00deg	18.00±0.00e	67.62±1.81f
	R6S3	12.81±0.05d	38.96±0.99ab	14.67±0.66cde	18.66±0.67e	79.96±4.75e
2020	N0	8.81±0.43g	35.13±0.43d	9.33±0.67f	13.33±0.67f	23.86±2.99
	N1	14.16±0.03b	40.76±0.06ab	15.33±0.66ab	18.67±0.66ab	79.77±2.71
	S1	10.13±0.03c	37.01±1.72cd	11.33±0.67de	14.66±0.67ef	35.49±3.25
	N3	15.81±0.06a	41.11±0.25a	16.67±0.67a	19.33±0.66a	95.94±2.50
	S3	9.43±0.08f	36.86±1.27cd	10.67±0.67ef	14.00±0.00ef	30.42±1.95
	R2S1	12.71±0.05c	38.81±0.35bc	13.33±0.66cd	17.33±0.66bc	59.06±3.32
	R2S3	13.16±0.03c	39.51±0.20bc	14.67±0.66bc	18.00±0.00abc	71.10±4.54
	R6S1	11.11±0.06d	37.71±0.45cd	12.00±0.00de	15.33±0.67de	40.27±2.42
	R6S3	11.41±0.15d	38.33±0.88bc	12.66±0.67de	16.68±0.67cd	49.62±3.07
ANOVA						
Y		*	NS	**	*	*
N		NS	NS	NS	NS	*
W		**	**	**	**	**
Y×N		*	NS	NS	NS	NS
Y×W		**	NS	NS	NS	**
N×W		**	NS	NS	*	**
Y×N×W		*	NS	NS	NS	NS

Notes N0: N application of 0 kg/ha; N1: N application of 100 kg/ha; S1: N1 water stress; R2S1: N1 rehydration treatment under R2 stage; R6S1: N1 rehydration treatment under R6 stage; N3: N application of 300 kg/ha; S3: N3 water stress; R2S3: N3 rehydration treatment under R2 stage; R6S3: N3 rehydration treatment under R6 stage. In each data area, different letters within the same column indicate significant differences among treatments. NS means significance levels at $P \geqslant 0.05$, * means significance levels at $P<0.05$, ** means significance levels at $P<0.01$.

3.3 ABA and GA contents

Our results indicate that abscisic acid (ABA) content in maize leaves and roots increased remarkably at the sixth-leaf and filling stages and declined in the maturity stage. At different stages, water stress significantly inhibited ABA content. However, the ABA content gradually increased under the rehydration treatment in R2 and R6 compared with the water stress and N0 treatments. At the V6 stage, the ABA content of the leaves and roots was higher than that of the N0 treatment (Figure

2). In the R2 stage, ABA content was highest in the N3 treatment, followed by the N1, R2S3, and R2S1 treatments. Our results revealed that ABA content increased with increasing N application. Conversely, the reduction in N application and water content also inhibited ABA content.

The interaction between nitrogen and water significantly increased GA biosynthesis (Figure 3). As a result, the GA content gradually decreased in the leaves and roots during the maize growth stages (Figure 3). The V6 stage had a substantially higher GA content than R2 and R6. GA content was significantly higher in the N1 and N3 treatments than in the rehydration treatment. With increased rehydration time, the R2S3 and R2S1 treatments were better than the R6S3 and R6S1 treatments in the R6 stage. In all treatments, the GA content was lowest at S3.

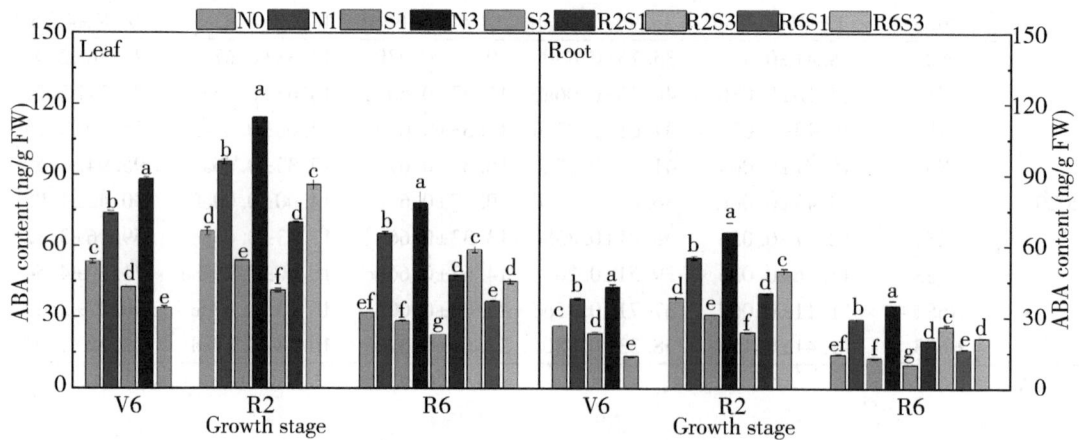

Figure 2 Effects of nitrogen and water stress on the rehydration of ABA content in maize during the 2020 growing season

Notes: The names of the treatments are the same as Figure 1. Different lowercase letters indicate significant differences as determined by the LSD test ($P \leq 0.05$). Vertical bars represent the standard error of means ($n=3$).

3.4 IAA and Z+ZR contents

Indole-3-acetic acid (IAA) plays an essential role in plants. Water and N application transiently improved IAA content at the R2 stage and reached the highest level of N3 treatment in the progression of plant growth and development. Our results showed that N3 and N1 treatments increased IAA content remarkably compared to the rehydration treatment. The IAA content of maize increased dramatically in the early stages of growth but gradually decreased with growth up to maturity (Figure 4). At the R2 and R6 growing stages, IAA content was higher in N3, N1, R2S3, and R2S1 treatments in the leaves and roots, whereas the R6S3 and R6S1 treatments increased slowly. During the R6 growth stage under the N3 treatment, the IAA content reached its peak in the leaves and roots, but the N1 treatment showed no significant difference compared with the rehydration treatment. Under the water stress treatment, the IAA content of the leaves was higher than that of the roots.

At all growth stages, the Z+ZR content did not significantly increase at the V6 stage,

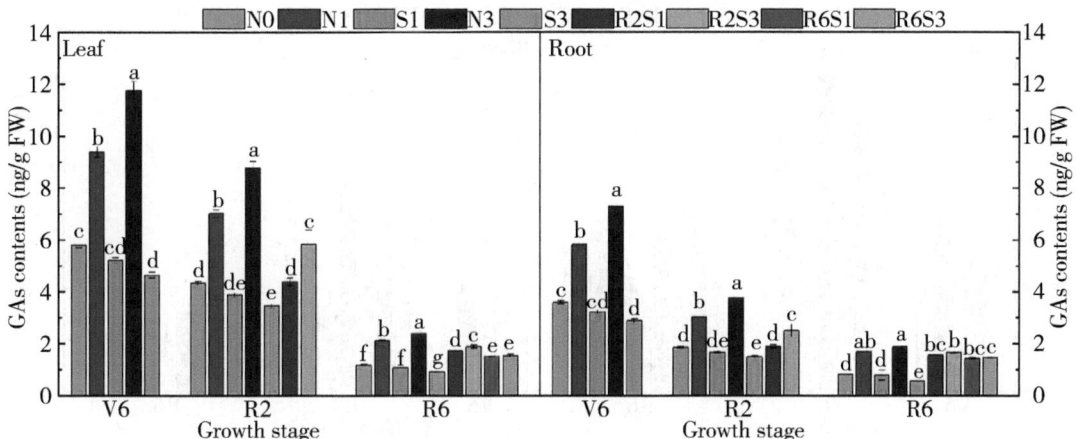

Figure 3 Effects of nitrogen and water stress on the rehydration of
GA content in maize during the 2020 growing season

Notes: The names of the treatments are the same as Figure 1. Different lowercase letters indicate significant differences as determined by the LSD test ($P \leq 0.05$). Vertical bars represent the standard error of means ($n=3$).

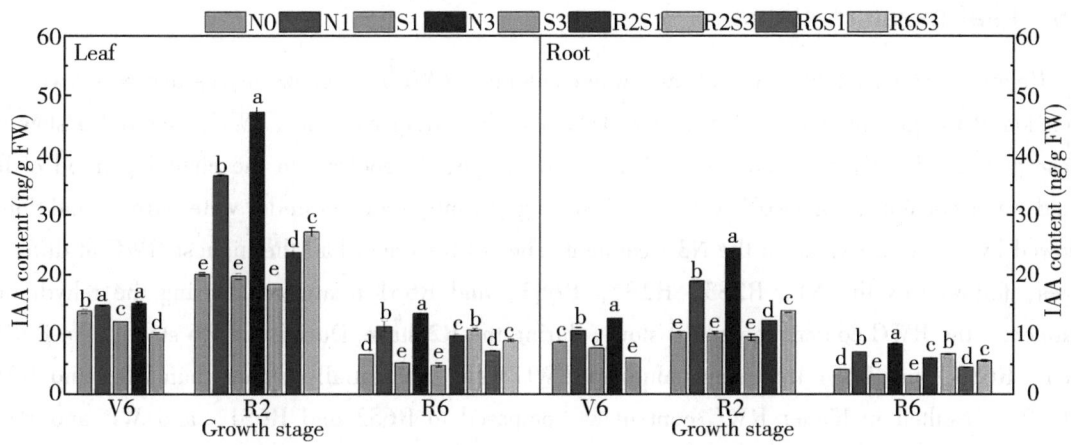

Figure 4 Effects of nitrogen and water stress on the
rehydration of IAA content in maize during the 2020 growing season

Notes: The names of the treatments are the same as Figure 1. Different lowercase letters indicate significant differences as determined by the LSD test ($P \leq 0.05$). Vertical bars represent the standard error of means ($n=3$).

increasing with the improvement of the R2 stage and reaching its lowest during the R6 stage (Figure 5). Under the V6 stage, the Z+ZR content were positively correlated with an increase in N application. Under water stress treatment, the Z+ZR content were reduced with increasing N application to the leaves and roots. Compared with rehydration, N3 and N1 considerably increased the Z+ZR content in the leaves and roots. In the R6 stage, the Z+ZR content declined gradually with increasing rehydration time. Among the various water conditions and N fertilizer treatments, the overall performance was N3>N1>R2S3>R2S1>R6S3>R6S1>N0>S1>S3 (Figure 5).

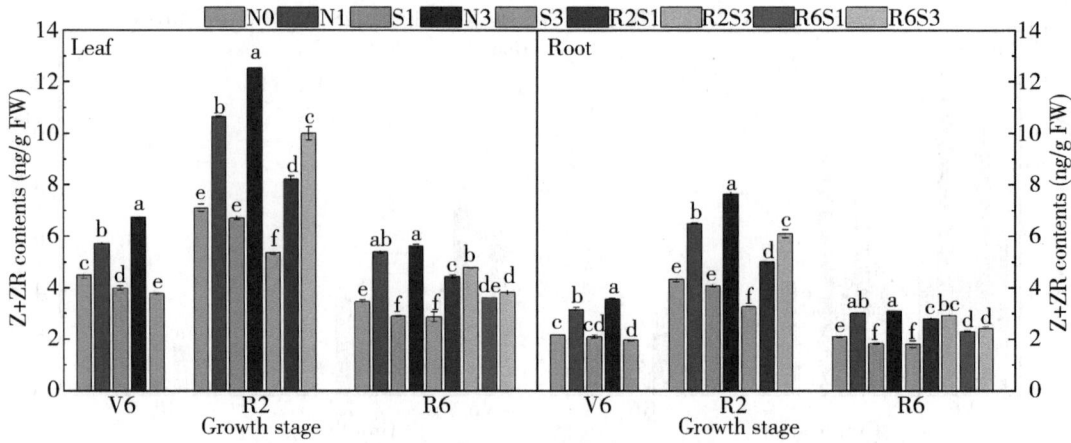

Figure 5 Effects of nitrogen and water stress on the rehydration of Z + ZR content in maize under different water conditions during the 2020 growing season

Notes: The names of the treatments are the same as Figure 1. Different lowercase letters indicate significant differences as determined by the LSD test ($P \leq 0.05$). Vertical bars represent the standard error of means ($n = 3$).

3.5 Leaf moisture status

Relative water content (RWC) and water potential (WP) of maize leaves decreased with the extension of the growth stages (Figure 6A–D). The free water content (FWC) showed a slow increase during the early stage and peaked at the R2 stage, in contrast to the changing trend of the bound water content (Figure 6B–C). RWC was significantly reduced under water stress conditions, followed by the lowest value in the N3 treatment. The N3 treatment had the highest RWC at different stages, followed by the N1, R2S3, R2S1, R6S3, and R6S1 treatments. During the rehydration treatment, the RWC content increased slowly during the R2 stage. During the R6 stage, the R2S3, R2S1, R6S3, and R6S1 treatment values of RWC were substantially higher than N0, and R2S3 and R2S1 resulted in higher RWC content as compared to R6S3 and R6S1, and WP and RWC showed a similar trend.

3.6 Chlorophyll and carotenoid contents

The chlorophyll content (Chl a, b, a+b) was significantly reduced under stress conditions in all treatments, and the S3 treatment was the lowest compared with the adequate water conditions (Figure 7). The water stress-induced decrease in Chl a content indicated the loss of the photosynthetic reaction under undesirable conditions (Figure 7A). From the R2 stage, the chlorophyll content (a, b, a+b) increased with the rehydration treatment. Under R6, with the extension of rehydration time, the chlorophyll content in R2S3 and R2S1 was slightly higher than that in R6S3 and R6S1, and both were significantly higher than that in the N0 treatment (Figure 7). Carotenoid content at different growth stages showed apparent differences due to the impacts of water and nitrogen interactions (Figure 7D). The N3 treatment yielded the highest levels of carotenoids during various maize growth stages.

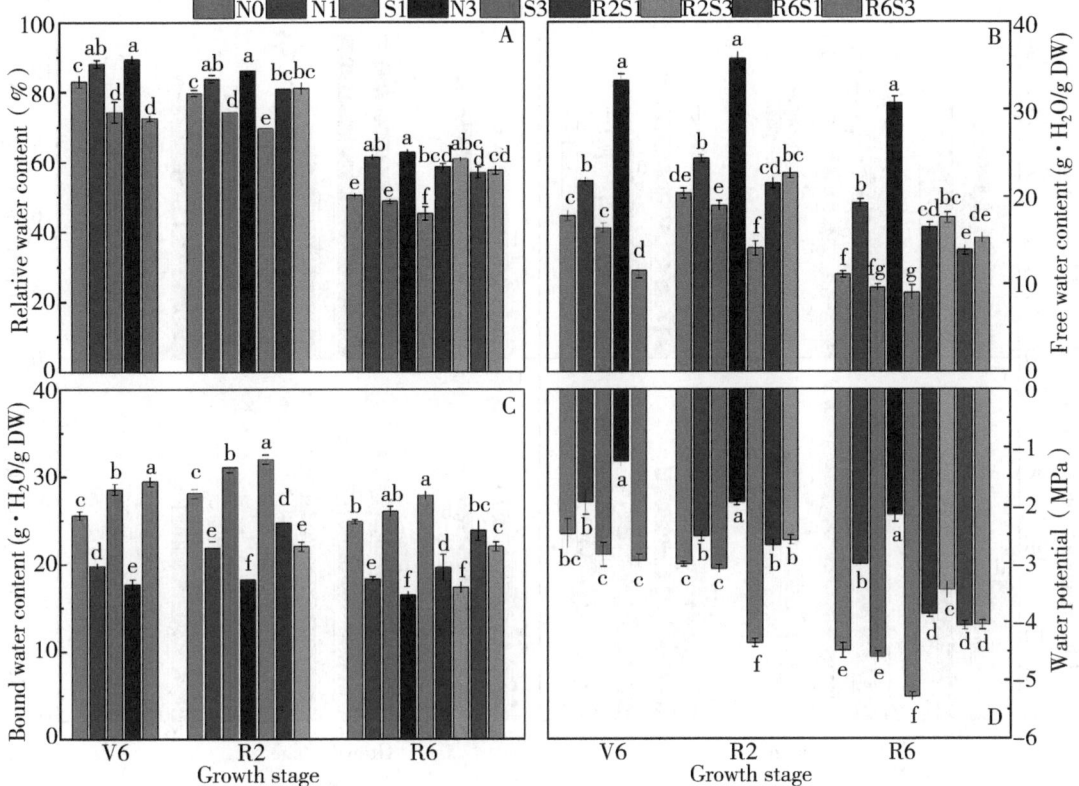

Figure 6 Effects of nitrogen and water stress on the rehydration of relative water content (A), free water content (B), bound water content (C), and water potential (D) in maize leaf during the 2020 growing season

Notes: The names of the treatments are the same as Figure 1. Different lowercase letters indicate significant differences as determined by the LSD test ($P \leq 0.05$). Vertical bars represent the standard error of means ($n=3$).

3.7 Photosynthetic gas exchange

The photosynthetic rate (Pn), transpiration rate (Tr), stomatal conductance (Gs), and intercellular CO_2 concentration (Ci) changed with each growth stage (Figure 8). The most substantial increase in photosynthetic pigment assimilation was exhibited by maize under N3 treatment. Under water stress, photosynthetic pigments were reduced with increasing amounts of N fertilizer. In contrast, maize under the S3 treatment showed the lowest decrease in photosynthetic pigments compared to N0. The most significant increase in photosynthetic pigments was observed in maize with the rehydration treatment. In contrast, the rehydration treatments considerably reduced photosynthetic pigments as the growth stage progressed. At the R6 stage, the photosynthetic pigments under different rehydration times showed that R2S3 > R2S1 > R6S3 > R6S1 and the differences were significant.

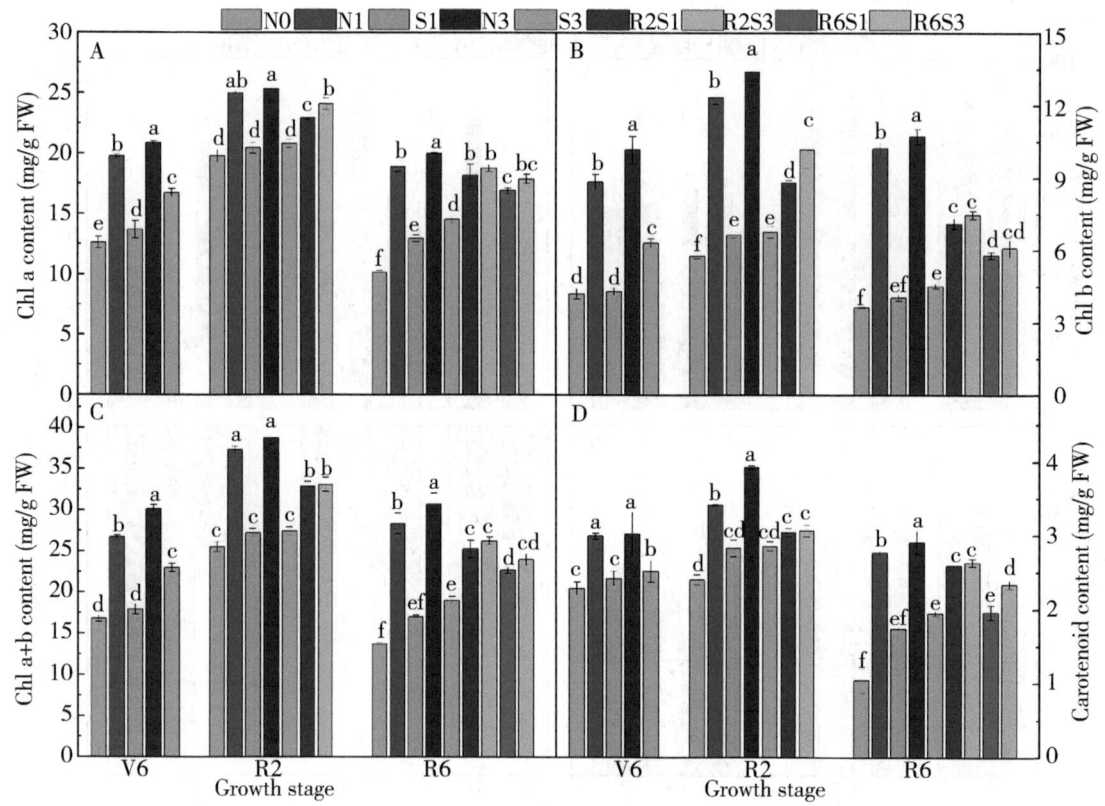

Figure 7 Effects of nitrogen and water stress on the rehydration of chlorophyll and carotenoid in maize leaf during the 2020 growing season

Notes: The names of the treatments are the same as Figure 1. Different lowercase letters indicate significant differences as determined by the LSD test ($P \leq 0.05$). Vertical bars represent the standard error of means ($n=3$).

3.8 Correlation analysis

Spearman's correlation results indicated that the photosynthetic pigments of maize were significantly ($P \leq 0.01$) positively correlated with leaf moisture status (FWC, RWC, WP) in maize (r-value ≤ 0.99). Simultaneously, a significant negative relationship was found between the BWC of the maize leaves (Figure 9). However, photosynthetic pigments in the leaves were negatively correlated with chlorophyll content (Chl a, b, a+b). However, these parameters were positively correlated with each other and with the hormonal content of maize leaves and roots. Hormonal contents, such as GA, Z+ZR, and ABA, were negatively correlated with BWC (r-value of -0.076, -0.003, and -0.168, respectively).

4 Discussion

Due to global climate change, warming, and increased temperature, many regions face drought stress (Shao et al., 2008). Crop productivity and dry land farming systems in semiarid areas depend on natural precipitation, and water shortages are important restrictions on crop yields

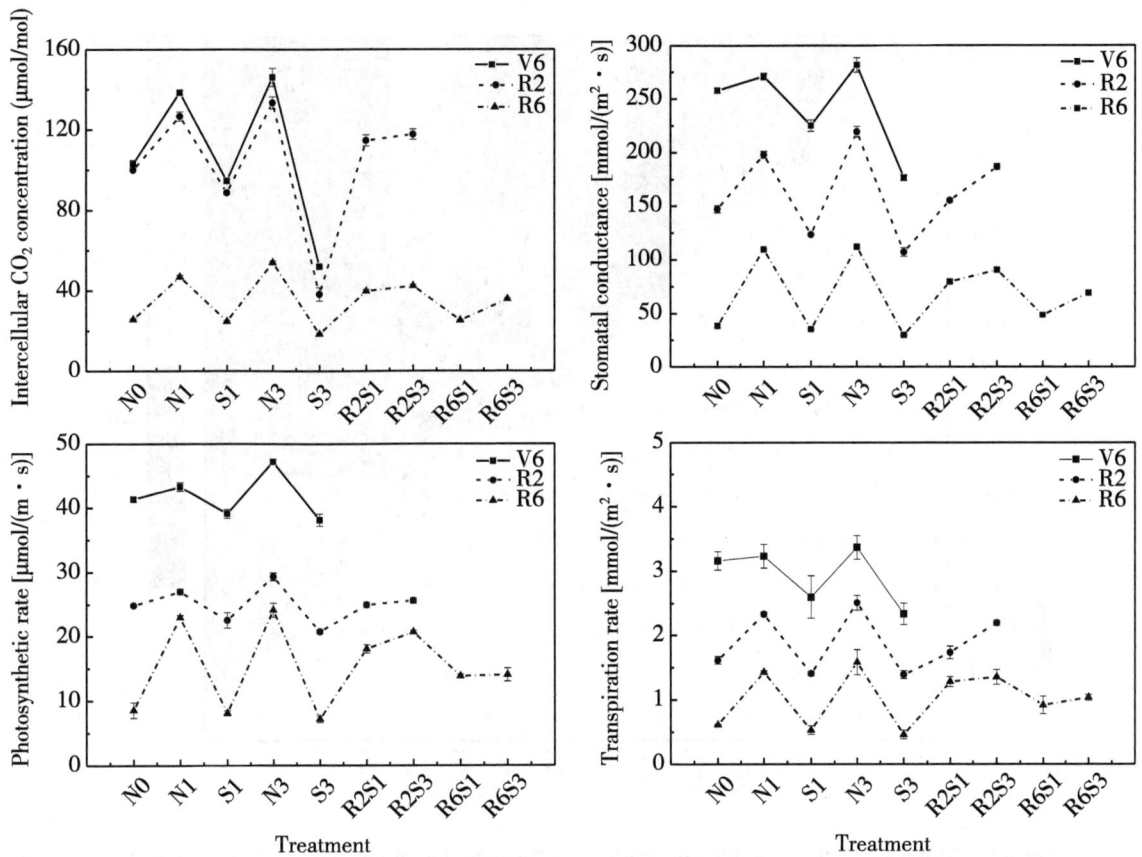

Figure 8 Effects of nitrogen and water stress on the rehydration of chlorophyll and carotenoid in maize leaf during the 2020 growing season

Notes: Transpiration rate (Tr), Photosynthetic rate (Pn), Stomatal conductance (Gs), and Intercellular CO_2 concentration (Ci) of maize leaf during the 2020 growing season. The names of the treatments are the same as Figure 1. Different lowercase letters indicate significant differences as determined by the LSD test ($P \leq 0.05$). Vertical bars represent the standard error of means ($n=3$).

(Ahmad et al., 2020; Ren et al., 2008). In the semiarid region of China, plant production is drastically reduced owing to water scarcity and irregular precipitation (Tang et al., 2011). Like other environmental stressors, water stress can affect many physiological, biochemical, and metabolic processes in plants (Ghahfarokhi et al., 2009). The results showed that plant growth was reduced under water stress conditions and was restored during the recovery period (Table 1). All maize hybrids exposed to water stress had lower biomass than the normal treatments, which occurs in many growth stages to water stress (Figure 1). To survive stress conditions, plants need a variety of metabolic pathways to reduce gas exchange and photosynthetic efficiency and ultimately prevent senescence (Zhang et al., 2018).

In the current study, N1 and N3 treatments increased maize yield in a cold and semi-arid region, and N3 was significantly higher than N1. Rehydration treatment can improve maize yield to avoid losses due to water stress. Other studies have found that controlled-release nitrogen fertilizer can decrease adverse effects under water stress conditions, such as water stress and decrease maize pro-

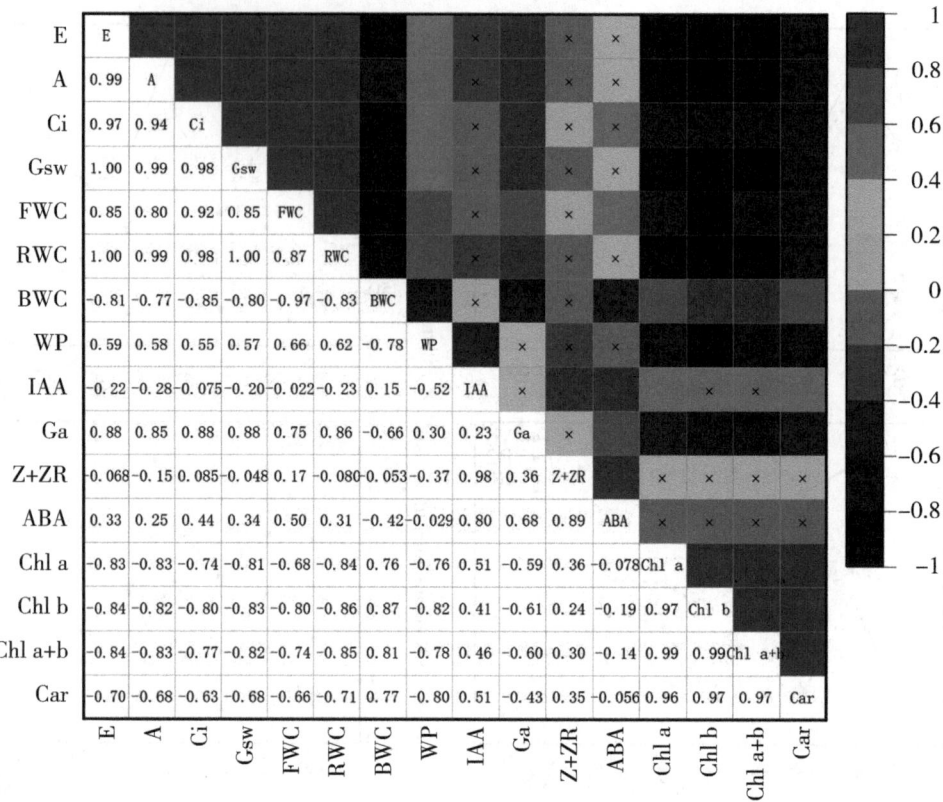

Figure 9 Matrices of Spearman's correlation coefficients (r values range -1.0 to $+1.0$) between leaves and root water conditions and N applications and different measured variables of maize at $P \leq 0.01$ ($n = 24$)

Notes: E: transpiration rate, A: Photosynthetic rate, Gs: stomatal conductance, Ci: intercellular CO_2 concentration, RWC: relative water content, FWC: free water content, BWC: bound water content, WP: water potential, Car: carotenoid, ABA: abscisic acid, GA: gibberellic acid, IAA: indole-3-acetic acid, Z+ZR: zeatin + zeatin riboside.

duction (Song et al., 2019). Previous studies found that the maize yield increased with nitrogen application rates under the same water content; and the longer rehydration time under the same N application (Guo et al., 2016). Our results showed that maize treated with N1 and N3 had higher ear characteristics and yields. As the growth period progressed, the rehydration treatment (R2S1, R2S3, R6S1, and R6S3) effectively alleviated the loss after stress. These findings are consistent with those reported by Nelissen et al. (2018) who reported that water stress causes oxidative membrane damage in maize crops. RWC and FWC content of maize leaves decreased because of water stress (S1 and S3). In contrast, the BWC content increased significantly (Figure 6). Previous reports have demonstrated that a decrease in RWC and FWC is triggered by water stress, which closes the stoma and may be caused by an accumulation of abscisic acid that preserves the cell structure and roots (Khan et al., 2007). Stomatal closure is a common mechanism used by plants to reduce the water runoff in response to drought stress. This can also reduce CO_2 assimilation and photosynthetic efficiency (Assmann and Jegla, 2016). According to previous studies, the inhibition of light-harvesting, multiplication of cells, and DMC decreases under water shortage (Corina et al., 2009).

Our results showed that the ratio of FWC to BWC increased, plant metabolism was vigorous, andplant growth was increased. Studies have shown that stomatal closure avoids water runoff defense mechanisms during water stress (Assmann and Jegla, 2016). Conversely, under water deficit conditions, the ratio of FWC to BWC decreases, metabolic activity weakens, and growth is reduced (Yongjin et al., 2015). FWC increased following the rehydration treatment, although not significantly during the R2 stage and FWC content decreased as the maize development stage increased during the R6 stage. Studies have also shown that with improved moisture, the leaves of plants change after drought stress (Corina et al., 2009). Study findings indicate that the response of water potential (WP) to water stress is more sensitive than that of FWC, BWC, and FWC, and the WP content of the leaves in maize was significantly decreased under water stress and led to the extension of the growth stage (Chaves et al., 2002). These results were consistent with our experimental results. Furthermore, the FWC, BWC, and WP decreased during water stress; however, these results can be improved by re-watering treatment (Figure 6A-B, Figure 6D), suggesting that dehydration is usually reversible. From a plant physiological perspective, chlorophyll content exists in different forms in the leaves. It is vital for the oxygenic exchange of light energy stored in chemical energy and as a source of energy (Liu et al., 2017). Our results showed that the chlorophyll and carotenoid content in the R2 stage reached a peak value in the N3 treatment and gradually declined during the R6 period (Figure 7). At different growth stages, the S1 and S3 treatments were higher than N0, but the effect was not significant. Our findings are consistent with previous studies where plants contain many carotenoids, which can act as non-enzymatic scavengers of ROS under stress conditions and remove surplus singlet oxygen and energy dissipation in the photosynthetic device (Wang et al., 2008; Jung et al., 2000).

Decreased chlorophyll fluorescence parameters may cause a decrease in the length and diameter of the ear (Pn, Tr, Gs, Ci, chlorophyll, and carotenoid content. In this study, the water-nitrogen interaction of the N1 and N3 treatments significantly increased the photosynthetic pigments at different growth stages (Figure 8). However, under the S1 and S3 treatments, the photosynthetic pigments were considerably reduced. Some studies have determined that chlorophyll a and b content is significantly reduced under water stress (Anjum et al., 2003). Our findings indicate that rehydration treatment is necessary for maize growth and influences each part of maize growth in the R2 and R6 stages. The photosynthetic pigments gradually increase under rehydration treatments (R2S1, R2S3, R6S1, and R6S3) with growth progress; however, photosynthesis, carbohydrate metabolism, and cell division are inhibited under water stress conditions. Thus, proper rehydration treatment is essential for maize growth (Chaves et al., 2002; Nelissen et al., 2018). Previous studies have demonstrated that many biological mechanisms are involved when plants are exposed to environmental challenges. However, the photosynthesis-related processes were affected mainly by water shortages and restorations (Osakabe et al., 2014). Thus, photosynthetic parameters have been used universally to evaluate maize growth characteristics (Scalabrin et al., 2016).

Raghavendra et al. (2010) reported that endogenous plant hormones are an important index for evaluating crop growth, development, yield, and presence under water stress conditions. Many studies have shown that drought stress is one way to increase the ABA content (Seki et al., 2002). Previous studies have shown that ABA is a stress hormone with multiple functions, and NCED3 genes are involved in water stress protection and stomatal closure (Ruggiero et al., 2004). Meanwhile, some studies have shown a widely used role of ABA in closing the stomata of drought-stressed plants (Zhang et al., 2006), which is essential for plants to quickly grow and restore water (Pustovoitova et al., 2004). In addition, ABA content plays a prominent role in the grain-filling process of maize under abiotic stress conditions (Ahmad et al., 2020). Correlation analysis showed that ABA content in maize leaves and roots was significantly correlated with the mean and early period stage (V6-R2), whereas it was negatively correlated with the late growth stage (R2-R6). During the maize growth stage, the ABA content of the roots was higher than that of the leaves. Under rehydration treatment, R2S1 and R2S3 could effectively alleviate the decrease in ABA content under water stress (S1 and S3), and the significant effect was longer in rehydration time (Figure 2). Water stress has been reported to cause a reduction in IAA content; however, in some instances, IAA content was enhanced in leaves, buds, and flowers (Wang et al., 2008b). In our experiments, the IAA content at different growth stages subjected to water stress decreased markedly; overall performance showed that R2>V6>R6. The total IAA accumulation in the leaves was significantly higher than that in the roots (Figure 4). Our findings showed that R2S1, R2S3, R6S1, and R6S3 significantly increased IAA content under water stress, and the effects of R2S3 and R2S1 were more significant than those of R6S1 and R6S3 during the late stage of maize growth. This result is similar to that of Xie et al. (2003). GAs and Z+ZR promote the growth of plants under osmotic stress in response to water stress and show a considerable decline in leaves and roots (Yang et al., 2006). The content of GAs gradually decreased with the change in the growth stage, and the minimum GAs was observed in the R6 growth stage (Figure 3). The Z+ZR content increased significantly from the V6 stage to the R2 stage and then decreased gradually to the R6 stage; the roots showed a similar trend of changes in Z+ZR content (Figure 5).

Our results showed that during the R2 stage, the R2S3 contents of GAs and Z+ZR were higher than that of R2S1 with prolonged rehydration time. For rehydration treatments during the R6 stage, GAs and Z+ZR contents with R2S3 and R2S1 were higher than those of R6S3 and R6S1, which simultaneously demonstrated the extension of the rehydration time. Our findings are consistent with those of previous studies in which the length of rehydration time affected maize growth hormone (Nelissen et al., 2018). Our findings revealed that N3 treatment increased total dry matter, maize ear characteristics, grain yield, leaf moisture status, photosynthetic pigments, chlorophyll fluorescence parameters, and hormonal content, whereas in the rehydration treatment, whether in the medium-term or later stage of maize growth, R2S3 treatment can have a recovery effect, increasing maize yield.

5 Conclusion

We herein describe the results of our comprehensive investigation regarding physiological responses and hormonal regulation in plants treated with nitrogen, water stress and rehydration. In conclusion, our results demonstrated that spring maize in cold semiarid areas under different growth stages enhanced the photosynthetic pigments and chlorophyll content, leaf moisture status, biomass, endogenous hormones, ear characteristics, and yield by increasing nitrogen application. Meanwhile, rehydration treatment improved IAA, ABA, and Z+ZR content, and it was significantly related to photosynthetic pigment, chlorophyll content, and leaf moisture status compared to the N0 treatment. Therefore, according to our results, N3 treatment can significantly improve the growth of maize and increase crop yields with a normal water supply. Under the rehydration treatment, R2S3 was better than other rehydration treatments, and it may be recommended to practice water conservation in the future.

Acknowledgment

The authors express their special gratitude to all funding sources, especially Heilongjiang Bayi Agricultural University, for their financial assistance. Supported by the Strategic Priority Research Program of the Chinese Academy of Science (Grant No. XDA28130103); Research and Development Program of Applied Technology in Heilongjiang Province (No. GA20B102)

References

Abid U, Hakim M, Muhammad S, et al., 2018. Phytohormones enhanced drought tolerance in plants: a coping strategy. *Environmental science and pollution research international*. 33: 33103-33118.

Ahmad S, Kamran M, Ding R, et al., 2019. Exogenous melatonin confers drought stress by promoting plant growth, photosynthetic capacity and antioxidant defense system of maize seedlings. *PeerJ*, 7, e7793.

Ahmad S, Kamran M, Zhou XB, et al., 2020. Melatonin improves the seed filling rate and endogenous hormonal mechanism in grains of summer maize. *Physiologia Plantarum*, 172: 1059-1072.

Ahmad S, Muhammad I, Wang GY, et al., 2021. Ameliorative effect of melatonin improves drought tolerance by regulating growth, photosynthetic traits and leaf ultrastructure of maize seedlings. *BMC Plant Biology*, 21 (1): 1-14.

Ahmad S, Wang GY, Muhammad I, et al., 2022a. Application of melatonin-mediated modulation of drought tolerance by regulating photosynthetic efficiency, chloroplast ultrastructure, and endogenous hormones in maize. *Chemical and Biological Technologies in Agriculture*, 9 (1): 1-14.

Ahmad S, Wang GY, Muhammad I, et al., 2022b. Interactive Effects of Melatonin and Nitrogen Improve Drought Tolerance of Maize Seedlings by Regulating Growth and Physiochemical Attributes. *Antioxidants*, 11 (2): 359.

Anjum F, Yaseen M, Rasool E, et al., 2003. Water stress in barley (Hordeum vulgare L.). II. Effect on chemical composition and chlorophyll contents. *Pakistan Journal of Agri-*

cultural Sciences. 26: 203-209.

Asim AK, Rabiye T, Neslihan S, et al., 2012. Current advances in the investigation of leaf rolling caused by biotic and abiotic stress factors. *Plant Science*, 182: 42-48.

Assmann SM, Jegla T, 2016. Guard cell sensory systems: recent insights on stomatal responses to light, abscisic acid, and CO_2. *Current Opinion in Plant Biology*, 33: 157-167.

Bai LR, Shi LR, Guo XL, et al., 2016. Effects of Water Stress on Physiological Characteristics of Different Genotypes of Triticale and Rye Seedlings. Crops.

Campos H, Cooper M, Habben JE, et al., 2004. Improving drought tolerance in maize: a view from industry. *Field Crops Research*, 90: 19-34.

Chaves M, Pereira JS, Maroco JP, et al., 2002. How Plants Cope with Water Stress in the Field? Photosynthesis and Growth. *Annals of Botany*, 89: 907-916.

Chen DQ, Wang SW, Cao BB, et al., 2015. Genotypic Variation in Growth and Physiological Response to Drought Stress and Re-Watering Reveals the Critical Role of Recovery in Drought Adaptation in Maize Seedlings. *Frontiers in Plant Science*, 6: 1241.

Chi YX, Yang L, Zhao CJ, et al., 2021. Effects of soaking seeds in exogenous vitamins on active oxygen metabolism and seedling growth under low-temperature stress. *Saudi Journal of Biological Sciences*, 28 (6): 3254-3261.

Corina HK, Carlos CV, Enrique IL, et al., 2009. Analysis of Gene Expression and Physiological Responses in Three Mexican Maize Landraces under Drought Stress and Recovery Irrigation. *Plos One*. 4: e7531.

Fracheboud Y, Jompuk C, Ribaut JM, et al., 2004. Genetic analysis of cold-tolerance of photosynthesis in maize. *Plant Molecular Biology*, 56: 241-253.

Gan Y, Campbell CA, Liu L, et al., 2008. Water use and distribution profile under pulse and oilseed crops in semi-arid northern high latitude areas. *Agricultural Water Management*, 96: 337-348.

Ghahfarokhi MG, Mansurifar S, Taghizadeh-Mehrjardi R, et al., 2009. Autosampler: Stanley A Stone assigned to Varian Associates Inc/Scientific Systems Inc-ScienceDirect. *Environment International*, 14: 127-131.

Guo LW, Ning TY, Nie LP, et al., 2016. Interaction of deep placed controlled-release urea and water retention agent on nitrogen and water use and maize yield. *European Journal of Agronomy*, 75: 118-129.

Guo ZJ, Zhang YL, Zhao JY, et al., 2014. Nitrogen use by winter wheat and changes in soil nitrate nitrogen levels with supplemental irrigation based on measurement of moisture content in various soil layers. *Field Crops Research*, 164: 117-125.

Hiscox JD, Israelstam GF, 1979. A method for the extraction of chlorophyll from leaf tissue without maceration. *Revue Canadienne De Botanique*, 57: 1332-1334.

Huo YJ, Wang PM, Wei YY, et al., 2015. Overexpression of the Maize psbA Gene Enhances Drought Tolerance Through Regulating Antioxidant System, Photosynthetic Capability, and Stress Defense Gene Expression in Tobacco. *Frontiers in Plant Science*,

6: 1223.

Jiang P, Lei T, Liu X, et al., 2006. Principles and experimental verification of capillary suction method for fast measurement of field capacity. *Nongye Gongcheng Xuebao/ Transactions of the Chinese Society of Agricultural Engineering*, 22: 1-5.

Jung S, Jin SK, Cho KY, et al., 2000. Antioxidant responses of cucumber (Cucumis sativus) to photoinhibition and oxidative stress induced by norflurazon under high and low PPFDs. *Plant Science*, 153: 145-154.

Khan UH, Link W, Hocking TJ, et al., 2007. Evaluation of physiological traits for improving drought tolerance in faba bean (Vicia faba L.). *Plant And Soil*, 292: 205-217.

Kudoyarova G, Veselova S, Hartung W, et al., 2011. Involvement of root ABA and hydraulic conductivity in the control of water relations in wheat plants exposed to increased evaporative demand. *Planta*, 233: 87-94.

Li GH, Zhao B, Dong S, et al., 2017. Interactive effects of water and controlled release urea on nitrogen metabolism, accumulation, translocation, and yield in summer maize. *Die Naturwissenschaften*, 104: 9-10.

Liu Y, Mingsheng MA, Guo X, 2017. Effects of Drought Stress and Rewatering on Plant Growth and Antioxidant Enzyme Activities at the Corn Seedlings. *Gansu Agricultural Science and Technology (in Chinese)*.

Liu Y, Wu W, Liao YC, et al., 2016. Towards the highly effective use of precipitation by ridge-furrow with plastic film mulching instead of relying on irrigation resources in a dry semi-humid area. *Field Crops Research*, 188: 62-73.

Machado S, Paulsen GM, 2001. Combined effects of drought and high temperature on water relations of wheat and sorghum. *Plant & Soil*, 233: 179-187.

Mao H, Wang H W, Liu SX, et al., 2015. A transposable element in a NAC gene is associated with drought tolerance in maize seedlings. *Nature Communications*, 6: 8326.

Merah O, 2001. Potential importance of water status traits for durum wheat improvement under Mediterranean conditions. *The Journal of Agricultural Science*, 137: 139-145.

Ming DF, Pei ZF, Naeem MS, et al., 2011. Silicon Alleviates PEG-Induced Water-Deficit Stress in Upland Rice Seedlings by Enhancing Osmotic Adjustment. *Journal of Agronomy and Crop Science*, 198: 14-26.

Muhammad I, Lv JZ, Yang L, et al., 2022a. Low irrigation water minimizes the nitrate nitrogen losses without compromising the soil fertility, enzymatic activities and maize growth. *BMC Plant Biology*, 22 (1): 1-13.

Muhammad I, Yang L, Ahmad S, et al., 2022b. Nitrogen fertilizer modulates plant growth, chlorophyll pigments and enzymatic activities under different irrigation regimes. *Agronomy*, 12 (4): 845.

Muhammad I, Yang L, Ahmad S, et al., 2022c. Irrigation and nitrogen fertilization alter soil bacterial communities, soil enzyme activities, and nutrient availability in maize crop. *Frontiers in microbiology*, 105.

Nelissen H, Sun XH, Rymen B, et al., 2018. The reduction in maize leaf growth under mild drought affects the transition between cell division and cell expansion and cannot be restored by elevated gibberellic acid levels. *Plant Biotechnology Journal*, 16: 615-627.

Osakabe Y, Osakabe K, Shinozaki K, et al., 2014. Response of plants to water stress. *Frontiers in Plant Science*, 5: 86.

Plaut Z, Butow BJ, Blumenthal CS, et al., 2004. Transport of dry matter into developing wheat kernels and its contribution to grain yield under post-anthesis water deficit and elevated temperature. *Field Crops Research*, 86: 185-198.

Pustovoitova TN, Zhdanova NE, Zholkevich VN, 2004. Changes in the Levels of IAA and ABA in Cucumber Leaves under Progressive Soil Drought. *Russian Journal of Plant Physiology*, 51: 513-517.

Quiroga G, Erice G, Aroca R, et al., 2020. Radial water transport in arbuscular mycorrhizal maize plants under drought stress conditions is affected by indole-acetic acid (IAA) application. *Journal of Plant Physiology*, 153115: 246-247.

Raghavendra AS, Gonugunta VK, Christmann A, et al., 2010. ABA perception and signalling. *Trends in Plant ence*, 15: 395-410.

Ren X, Jia Z, Chen X, 2008. Rainfall concentration for increasing corn production under semi-arid climate. *Agricultural Water Management*, 95: 1293-1302.

Ruggiero B, Koiwa H, Manabe Y, et al., 2004. Uncoupling the Effects of Abscisic Acid on Plant Growth and Water Relations. Analysis of sto1/nced3, an Abscisic Acid-Deficient but Salt Stress-Tolerant Mutant in Arabidopsis. *Plant Physiology*, 136: 3134.

Sarker AM, Rahman MS, Paul NK, 2010. Effect of Soil Moisture on Relative Leaf Water Content, Chlorophyll, Proline and Sugar Accumulation in Wheat. *Journal of Agronomy & Crop Science*, 183: 225-229.

Scalabrin E, Radaelli M, Capodaglio G, 2016. Simultaneous determination of shikimic acid, salicylic acid and jasmonic acid in wild and transgenic Nicotiana langsdorffii plants exposed to abiotic stresses. *Plant Physiology and Biochemistry*, 103: 53-60.

Seki M, Ishida J, Narusaka M, et al., 2002. Monitoring the expression pattern of around 7,000 Arabidopsis genes under ABA treatments using a full-length cDNA microarray. *Funct Integr Genomics*, 2: 301.

Shao G, Li Z, Ning T, et al., 2013. Responses of photosynthesis, chlorophyll fluorescence, and grain yield of maize to controlled-release urea and irrigation after anthesis. *Journal of Plant Nutrition and Soil Science*, 176: 595-602.

Shao HB, Chu LY, Jaleel CA, et al., 2008. Water-deficit stress-induced anatomical changes in higher plants. *Comptes rendus-Biologies*, 331: 215-225.

Song G, Wen W, Guo X, 2000. The changing rules of content of inner GA 3, ABA in big kernel wheat variety. Journal of Henan Agricultural University.

Song L, Jin J, He J, 2019. Effects of Severe Water Stress on Maize Growth Processes in the Field. *Sustainability*, 11: 1-18.

Tang QX, Li SK, Xie RZ, et al., 2011. Effects of Conservation Tillage on Crop Yield: a

Case Study in the Part of Typical Ecological Zones in China. *Agricultural Sciences in China*, 10: 860-866.

Teixeira EI, George M, Herreman T, et al., 2014. The impact of water and nitrogen limitation on maize biomass and resource-use efficiencies for radiation, water and nitrogen. *Field Crops Research*, 168: 109-118.

Ullah A, Akbar A, Luo Q, et al., 2018. Microbiome Diversity in Cotton Rhizosphere Under Normal and Drought Conditions. *Microbial Ecology*, 77: 429-439.

Wang C, Yang A, Yin H, et al., 2008a. Influence of Water Stress on Endogenous Hormone Contents and Cell Damage of Maize Seedlings. *Journal of Integrative Plant Biology*, 50: 427-434.

Wang YP, Russel JR, Chan ZL, 2018. Phytomelatonin: a universal abiotic stress regulator. *Journal of Experimental Botany*, 5: 963-974.

Wei FZ, Li JC, Wang CY, et al., 2008. Effects of nitrogenous fertilizer application model on culm lodging resistance in winter wheat. *Acta Agronomica Sinica*, 34: 1080-1085.

Whitmore AP, Richard WW, 2009. Physical effects of soil drying on roots and crop growth. *Journal of Experimental Botany*, 60: 2845-2857.

Xie Z, Dong J, Cao W, et al., 2003. Relationships of endogenous plant hormones to accumulation of grain protein and starch in winter wheat under different post-anthesis soil water statusses. *Plant Growth Regulation*, 41: 117-127.

Yang JC, Peng SB, Remeo MV, et al., 2000. Grain filling pattern and cytokinin content in the grains and roots of rice plants. *Journal of plant growth regulation*, 30: 261-270.

Yang JC, Zhang JH, Liu K, et al., 2006. Abscisic acid and ethylene interact in wheat grains in response to soil drying during grain filling. *New Phytologist*, 171: 293-303.

Zhang JH, Jia, W S, Yang JC, et al., 2006. Role of ABA in integrating plant responses to drought and salt stresses. *Field Crops Research*, 97: 111-119.

Zhang S, Sadras V, Chen X, et al., 2014. Water use efficiency of dryland maize in the Loess Plateau of China in response to crop management. *Field Crops Research*, 163: 55-63.

Zhang WF, Cao GX, Li XL, et al., 2016. Closing yield gaps in China by empowering smallholder farmers. *Nature*, 537: 671-674.

Zhang XB, Lei L, Lai JS, et al., 2018. Effects of drought stress and water recovery on physiological responses and gene expression in maize seedlings. *BMC Plant Biology*, 18: 68.

Zhao B, Dong ST, Zhang JW, 2010. Effects of Controlled-Release Fertilizer on Yield and Nitrogen Accumulation and Distribution in Summer Maize. *Acta Agronomica Sinica*, 10: 301-304.

Zhou XB, Yang L, Wang GY, et al., 2021. Effect of deficit irrigation scheduling and planting pattern on leaf water status and radiation use efficiency of winter wheat. *Journal of Agronomy and Crop Science*. 207: 437-449.

Low irrigation water minimizes the nitrate nitrogen losses without compromising the soil fertility, enzymatic activities and maize growth

I. Muhammad[a], J. Z. Lv[b], L. Yang[a], S. Ahmad[a],
S. Farooq[a], A. Khan[c], M. Zeeshan[a], X. B. Zhou[a]

([a] Guangxi Colleges and Universities Key Laboratory of Crop Cultivation and Tillage,
Agricultural College, Guangxi University, Nanning 530004, China;
[b] Maize Research Institute of Guangxi Academy of Agricultural Sciences, Nanning 530007, China;
[c] Department of Agronomy, The University of Agriculture, Peshawar, Pakistan)

Abstract: Nitrate nitrogen ($NO_3^- - N$) leaching increased with nitrogen (N) fertilization under high water supply to the field negatively affected the maize growth and performance. This study aimed to understand the mechanisms of nitrate N leaching on biochemical basis and its relationship with plant performance under various N fertilizers under low and high irrigation water supply. The experiment included two irrigation regimes i.e, 1) high irrigation 80% water holding capacity (HW) and 2) low irrigation 60% water holding capacity (LW) relative to the field capacity of soil and five N fertilizer levels i.e, control (no N application), N200 (200 kg N/ha), N250 (250 kg N/ha), N300 (300 kg N/ha), and N350 (350 kg N/ha). Soil and plant enzymes were observed at different growth stages of the maize, whereas the leachates were collected at ten-day intervals from date of sowing. The LW regime had 10.15% lower $NO_3^- - N$ leachate than HW, with correspondence increases in grain yield (25.57%), shoot (17.57%) and root (28.67%) dry matter. Irrespective of the irrigation water, the RubisCo, glutamine synthase (GS), nitrate reductase (NR), nitrite reductase (NiR), and glutamate synthase (GOGAT) activities increased with increasing N fertilizer up to the V9 growth stage and decreased with approaching the maturity stage in maize. In HW irrigation, soil total N, GOGAT, soil nitrate ($NO_3^- - N$), leached nitrate ($LNO_3^- - N$), root N (RN), leaf N (LN) were positively correlated with N factors suggesting the higher losses of N through leaching (11.3%) compared to LW irrigation. However, the malondialdehyde (MDA), hydrogen peroxide (H_2O_2), superoxide (O_2), and proline were negatively correlated with the other enzymatic activities both under LW and HW irrigation regimes. Thus, minimizing the $NO_3^- - N$ leaching is possible with LW and N300 combination, without compromising the yield benefit and with improving enzyme activities.

These authors contributed equally.
Corresponding author E-mail address: xunbozhou@gmail.com.

Keywords: Nitrogen leaching; Irrigation water; N fertilizer; Biochemical processes; Enzyme; Maize growth

1 Introduction

While protecting the environment's quality, the global demand for higher food production is a serious challenge, especially in the maize cropping system[1]. Irrigation and N fertilizer are the two vital leading inputs affecting the agricultural cropping systems[2]. Nitrogen fertilizer is one of the key element influencing crop growth. High water and fertilizer inputs are commonly seen to achieve high yields[3]. However, excessive fertilization and poorly planned irrigation regimes are common methods causing soil environmental problems[4], and resulting in 20% N contribution to environment as a result of farming[5]. Asia is the largest in N fertilizer consumption (58%), followed by the United States (22%), Europe (11%), and Africa (8%).

To achieve optimum yields especially in high-yield crops like maize (*Zea mays* L.), the regular N fertilization is needed. When fertilizer input exceeds crop requirements, contamination of water resources occurs[2,6,7]. The overuse of N increases the risks of nitrate leaching, and causes water pollution, an international problem (Huang et al., 2017b). Nitrate pollution of groundwater is most common in areas with higher rainfall, light-textured soils, and high inorganic N fertilization[8,9], and seriously damages the sustainability of food and agricultural soils[10,11]. On the other side the demand for food forced new agricultural practices to be developed, disturbing the N cycle in an agroecosystem[12]. There is a concern about optimizing the fertilization strategy, because only a minute fraction of applied mineral fertilizer is absorbed by plants, and almost 30% to 50% of these nutrients can be lost in different ways[13].

Nitrogen in the form of nitrate is highly mobile in soil and is primarily influenced by soil water conditions[14]. However, the N leaching due to greater water availability degrades the local groundwater, which causes water pollution[15,16]. The high water is needed for improved crop growth and development, however, it causes the potential N losses in the form of N leaching[6]. Thus, both N fertilization and water management are necessary for decreasing the potential nitrate losses, without affecting the growth and performance of maize crop. The N balancing technique considers all of the N transformation activities in the experimental field, including urea volatilization, hydrolysis, mineralization, nitrification, initial soil N, plant N uptake, residual soil N, and N leaching[17,18]. Consequently, both scientific and public attention has recently concentrated on the preserving of water against contamination caused by mineral N from different agricultural practices[19].

Maize crop is growing globally due to its multiple applications in food, feed, and industrial sectors. Average maize yields in China increased from 1 t (1949) to 6 t/ha (2013), demonstrating a 1,633% increase in total maize production[20]. Maize was grown on more than 36 million hectares in 2013, with greater production than any other crop in China[21] especially, in North China Plain having 39% China's maize. The most of farmers use their past experience in fertilization and the economic situation rather than assessing the potential consequences of their decisions[22]. Ju et al.[23] stated that some farmers apply 500–600 kg N/ha for better maize and wheat yield in crop rotation, with 60% application to maize. In North China Plain region, researchers have discovered that the average N application for the winter wheat-summer maize crop rotation system has increased fivefold in the last 30 years[6,24], and thus cause 70% losses to the environment. Therefore, the N fertilization and irrigation optimization is the need of the day, for possible increases in N-use efficiency, reducing $NO_3^- - N$ losses through leaching (8.43 kg N/ha for 142 kg N/ha fertilization) under 60 cm soil depth[2], and developing strategic management practices for increasing maize crop productivity and sustaining the environment[2,25]. Therefore, this study was designed to investigate the variation in N transport and leaching with varying N fertilization levels under two irrigation regimes (60% and 80% of field capacity) in maize crops. The objectives of this study are 1) to assess the feasibility of growing maize at different irrigation and N levels, and the impact of this approach on soil physicochemical properties, grain yield, plant N content, and mineral N leaching 2) to determine the effects of irrigation regimes coupled with synthetic N fertilizer on soil N accumulation, and to develop a practical N management scheduling approach that integrates local factors like soil moisture, maize water use, and rainfall and 3) to quantified the impact of higher N application rate on nitrate leaching under low and high irrigation water, and its relationship with soil plant N metabolism enzyme and maize growth.

2 Material and methods

2.1 Experimental design

A pot trial for 135 days in greenhouse was carried out at Guangxi University (22°50′ 24.6″N, 108°17′ 2.25″E) in the subtropical region of China. The average annual temperature and rainfall are 21.8℃ and 1,298 mm, respectively, and the difference is 227 mm of precipitation between the driest and wettest months. The pots were filled with soil of clay loam soil texture (according to Chinese Soil Taxonomy), with a pH of 5.6, a field capacity of 44%, soil bulk density of 1.40 g/cm^3, soil organic matter of 20 g/kg, and available N, phosphorus, and potassium of 127 mg/kg, 40 mg/kg, and 126 mg/kg, respectively. Five seeds of maize hybrid (Zhengda 619) were planted per pot (length of 32.5 cm and height of 29 cm) on September 28, 2020, which is the most commonly grown variety in the subtropical areas of China. The seeds were obtained from CP seed industry Yunnan Zhengda seed Co. Ltd., China. The selected seeds permission was granted from the respective authority. Before sowing, basal fertilizers like phosphorus (P_2O_5) and potash (K_2O) at the rate of 100 kg/ha each, and half of the respective N fertilizers (as urea) were mixed with soil in a pot. However, the remaining half of the N dose was applied as a top dressing at the nine-leaf stage (V9).

The experiment consisted of two factors (i.e., irrigation regimes and N treatments) conducted in control conditions. A 2 × 5 factorial experiment using four replication was carried out in a completely randomized design. The water regimes were 1) high irrigation 80% water holding capacity (HW) and 2) low irrigation 60% water holding capacity (LW) relative to the field capacity of soil. The five N treatments were 0, 200, 250, 300, and 350 kg N/ha application represented as control, N200, N250, N300, and N350, respectively. The pot positions were changed at a 10 days interval during irrigation inside the greenhouse to reduce the temperature effect.

Plants were irrigated with the tap water during the entire growth period, and both HW and LW water levels were maintained. Micro-tensiometers were used to measure the water of the soil in each pot (Nanjing Institute of Soil Science, Chinese Academy of Sciences). The soil moisture content was corrected at five-day intervals by weighting and adding additional water to maintain the required HW and LW water contents. In addition, six tiny holes were made at the bottom of each pot used to collet leachate in plastic trays placed below the pots. After each irrigation, a syringe was used to collect the leachate from the plastic plate and store it at 4℃ for lab examination.

2.2 Experimental analytical procedures

The full expended leaves at four different growth stages (V9, R1, R3, and R6), were collected for measuring the N metabolism enzyme activities. The activities of nitrate reductase (NR; D799304 – 0100), nitrite reductase (NiR; D799133 – 0100), glutamine synthetize (GS; D799578 – 0100), glutamate synthase (GOGAT; D799302 – 0100), RubisCo (D799834 – 0100), and glutamate dehydrogenase (GDH; D799834-0100) were determined using the plant enzyme kit from Sangon Biotech Co. Ltd (Beijing, China) following the appropriate manual supplied with the kit. Similarly, the total N contestants in leaves, stems, roots, and grains of maize were determined following the Kjeldahl procedure.

The NO_3^--N in the leached water was determined after first (sowing time) and second dose (V9) of N application. The leachates collected from a plastic tray (1L) every 10 days were stored at 4℃ in a 50 mL polyethylene tube until nitrate analysis. The UV-spectrophotometer was used to measure the nitrate contents in the leached water[26]. Similarly, at the end of the experiment the homogenized soil sample was taken from each pot, roots and other debris were removed. Half of the soil samples were air dried, sieved and used for determination of soil total N using the Kjeldahl procedure[27] and organic carbon ($K_2Cr_2O_2$ extraction method), while the half of the soil samples were kept in a refrigerator at −80℃ until the determination of soil NH_4^+-N and NO_3^--N using the procedure of KCl extraction dilatation method proposed by Paramasivam and Alva[28]. The N use efficiency was calculated by using the following formula:

$$ANUE(kg/kg) = \frac{[\text{grain weight(fertilized pot)} - \text{grain weight(control)}]}{\text{N fertilizer applied}}$$

Where, the ANUE is the agronomic N use efficiency.

The maize was harvested in April 2021, and thoroughly washed with tap water. The plants were separated into its components i.e., root, stem, leaves, stem and kernel. Kernel yield and kernel per ear were recorded after threshing to determine grain yield. The constituent parts of each plant were then separated into kernels, leaves, stalks, and roots. After that, the plant biomass was sundried for three days before being oven-dried for 72 hours at 70℃ to evaluate the dry matter content.

2.3 Statistical analysis

All the statistical analysis were performed using SPSS v25 (IBM SPSS Statistics; Chicago,

NY, USA). A mixed-model analysis of variance (ANOVA) was used to calculate the effects of N rates and irrigation regimes on N metabolism enzyme activities, nitrate leachate, and soil nutrient content. The differences between the means were determined using Tukey's test at $P<0.05$. The Mantel test was used in R v. 3.63 to evaluate the relationship between the N content (soil N, leachate and plant N), plant enzyme activities and irrigation regimes.

3 Results

3.1 Nitrogen metabolism enzyme

Remarkable differences in the N metabolism enzymes activities were noted in response to irrigation regimes, N fertilizer treatments and their interactions (Figure 1) across the growth stages. The NR, NiR, and GOGAT activities increased with increasing N fertilizer up to N300 treatment at R3

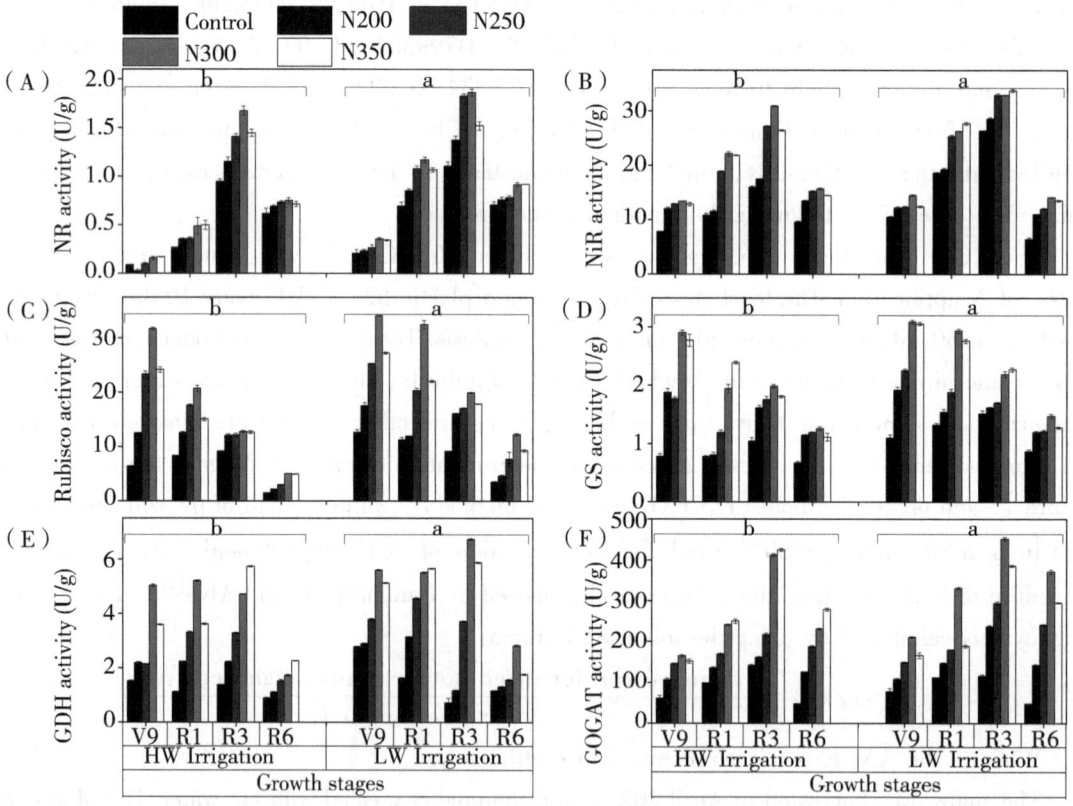

Figure 1 Effect of nitrogen dosages and irrigation regimes on the activities of nitrate reductase (NR), nitrite reductase (NiR), RubisCoRubisco, glutamine synthase (GS), glutamate dehydrogenase (GDH), and glutamate synthase (GOGAT). The bar represents the standard error

Notes: The nitrogen treatments control, N200, N250, N300 and N350 represent the application of nitrogen at the rate of 0, 200, 250, 300 and 350 kg N/ha. HW and LW represent the irrigation water at the rate of 80% and 60% of field capacity. The V9, R1, R3, and R6 stages represent the maize with 9 leaves, silking, milking, and maturity stages, respectively.

growth (milking stage) both under LW and HW irrigations (Figure 1A - B, and Figure 1D). However, the higher N fertilizer (N350) had decreased the enzyme activities. The growth stages had positive effects on NR, NiR, and GOGAT activities, and showed increasing trends up to the R3 growth stage, but lower activities were found at the R6 stage. These enzyme activities were significantly higher in LW irrigation than in the HW irrigation at all growth stages ($P<0.001$). In contrast, RubisCo and GS activities were highest in plants at the V9 growth stage and decreased when approaching the maturity in maize plants (R6) both under LW and HW (Figure 1C and Figure 1D). The maximum RubisCo and GS activities were found for N300 treatment at V9 stage under LW irrigation compared to HW irrigation. However, the lowest activities were observed for control treatment at the R6 growth stage both under LW and HW irrigation regimes. The RubisCo and GOGAT activities in the maize leaves kept decreasing with crop maturity at all N treatments. An irregular change of increases in GDH activities (a key enzyme of glutamate pathways for biosynthesis) was detected for maize across the growth stages (Figure 1E). However, at R3 stage, the GDH activities were significantly higher in N300 and N350 treatments with LW and HW irrigation regimes, respectively. At R3 stage with N300 treatment, the activities of NR, NiR, GDH, and GOGAT increased by 71.87%, 50.60%, 446.36%, and 235.65%, respectively, compared to control treatment ($P<0.001$). The regressions analysis showed that the N metabolism enzymes are highly dependent on irrigation water and N fertilization. In current study, the linear regression showed that NR ($r^2=0.88$ and 0.90), NiR ($r^2=0.95$ and 0.90), GOGAT ($r^2=0.86$ and 0.92), GS ($r^2=0.91$ and 0.96), GDH ($r^2=0.97$ and 0.89), and RuBisCo ($r^2=0.84$ and 0.88) activities were positively correlated to N fertilization under both LW and HW irrigation, respectively (Figure 2). Furthermore, the linear regression confirmed that the enzyme activities were strongly and positively correlated to N fertilization.

3.2 Soil nitrogen and carbon status

The soil NH_4^+-N, NO_3^--N, total N, and soil organic carbon (SOC) contents were significantly affected by irrigation regimes, N fertilizers, and their interaction (Figure 3, $P<0.01$). Soil total N, SNH_4^+-N, SNO_3^--N and SOC contents were increased by 29%, 17.7%, 243%, and 24.7% in N300 treatment compared to N0, respectively (Figure 3). The NH_4^+-N, NO_3^--N, and SOC contents were highest in the N300 treatment followed by the N350 treatment in LW irrigation. However, under HW irrigation, these parameters had resulted in greater values with N350 treatment as compared to other N treatments. In contrast, the soil total N content was higher in the N350 treatment under both LW (19.7 g/kg) and HW (18.7 g/kg) irrigation regimes as compared to the rest of N treatments (Figure 3A). Our results indicate that the LW irrigation regimes significantly increase soil total N, SNH_4^+-N, SNO_3^--N, and SOC content by 6%, 9.5%, 52%, and 4% compared with HW irrigation.

3.3 Leachate potential of nitrate

The leachate potential of NO_3^--N in response to irrigation regimes and N fertilization across the

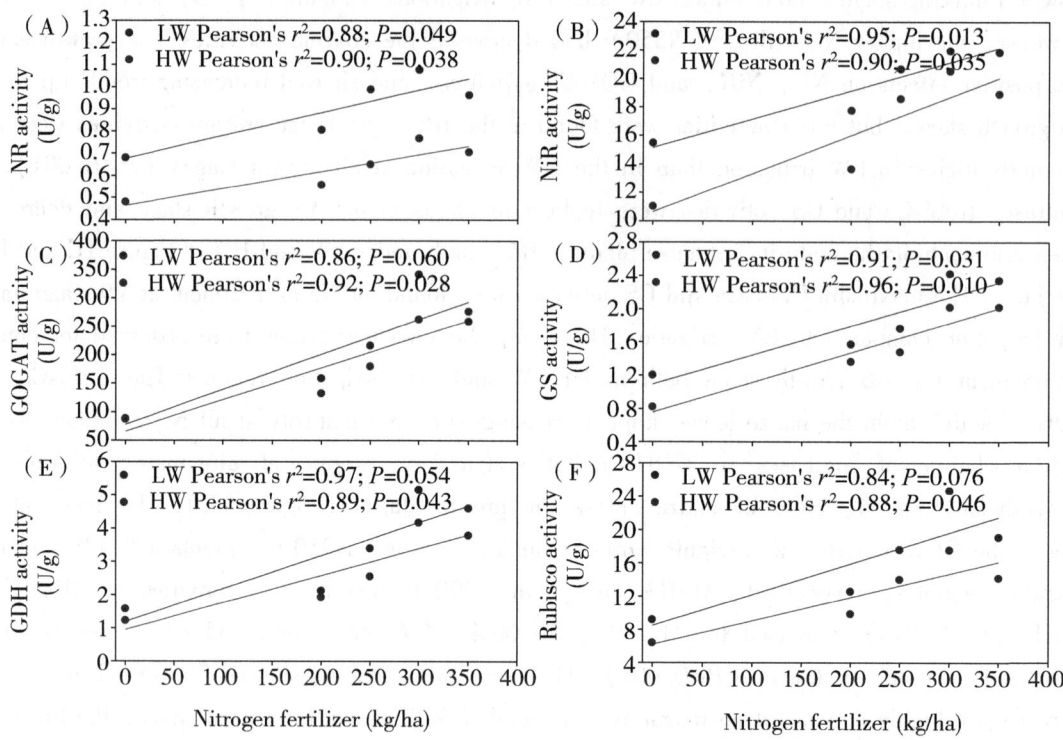

Figure 2 Effect of the interaction between nitrogen dosages and irrigation regimes on the activities of (A) nitrate reductase (NR) NR, (B) nitrite reductase (NiR) NiR, (C) Rubisco, (D) glutamine synthase (GS) GS, (E) glutamate dehydrogenase (GDH) GDH, and (F) glutamate synthase (GOGAT)

Notes: The nitrogen treatments are control (0), 200, 250, 300 and 350 kg N/ha. The number of observation ($n=16$) and HW and LW represent the irrigation water at the rate of 80% and 60% of field capacity.

growth stages (Figure 4) showed variable responses to the treatments. The $NO_3^- - N$ content in leachate was maximum on 10 days after planting, concurrent with HW irrigation and first dose of N fertilization (Figure 4A). It was further noted that the maximum $NO_3^- - N$ leaching was observed for the N350, followed by the N300 treatment, and least for the control treatment. These results suggest that during plant growing period the $NO_3^- - N$ leachate increased with increasing N fertilizer dosages under HW compared to LW irrigation (Figure 5A). Regardless of irrigation, the maximum leached $NO_3^- - N$ content was observed for N350 treatment on 10 days after maize planting and gradually decreased with increasing days after planting up to 60 days (Figure 4). Furthermore, the leached $NO_3^- - N$ increased again with 2nd dose of N fertilization on 70 days and 80 days after maize planting. Averaged across fertilization, LW irrigation significantly decreased $NO_3^- - N$ in leachate by 10.15% compared to HW irrigation (Figure 5A). The maximum $NO_3^- - N$ (5.36 mg/L) in leached water was observed for N350 treatment under HW irrigation, which is 8.84% higher than that of LW irrigation (Figure 5A). The relationship between irrigation and N fertilizer, showed that $NO_3^- - N$ in leached water linearly increased with increasing N fertilizer dosages in both irrigation regimes, but significantly lower leachate was observed under LW irrigation ($r = 0.99$; $P <$

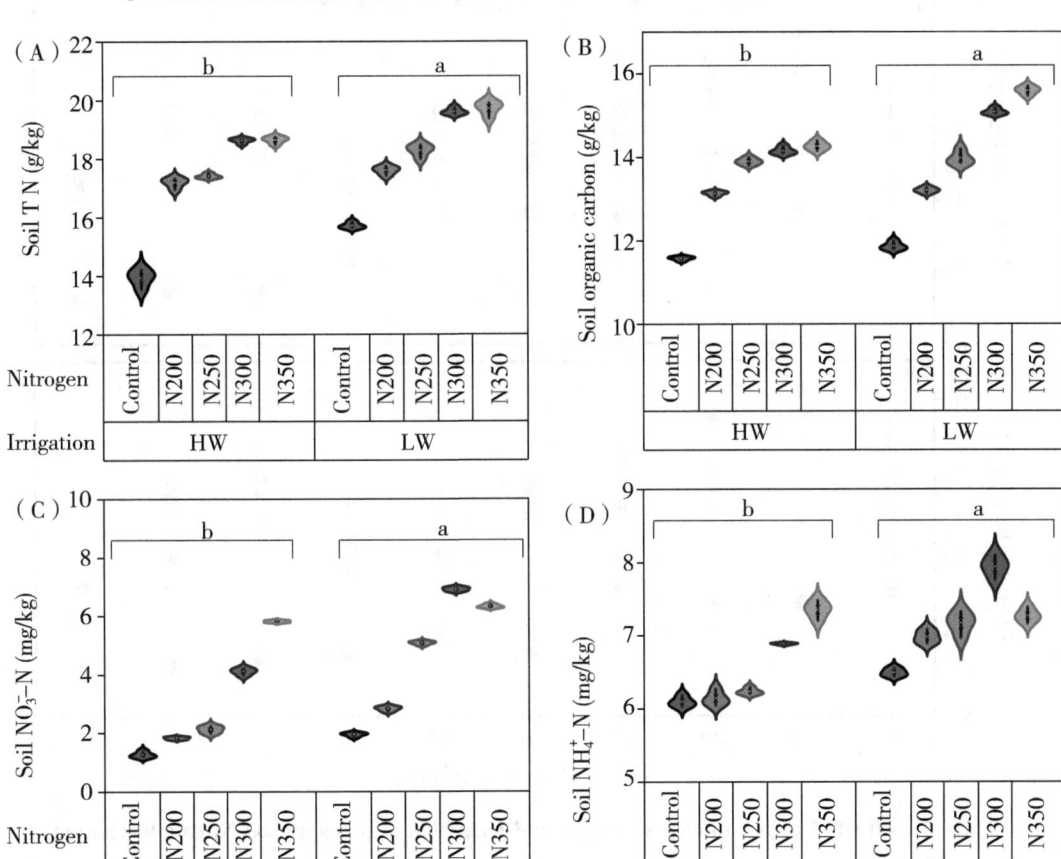

Figure 3 Effect of nitrogen dosages and irrigation regimes on soil total N, soil organic carbon, soil NO_3^--N, and NH_4^+-N content

Notes: The bar represents the standard error ($n=4$). The nitrogen treatments control, N200, N250, N300 and N350 represent the application of nitrogen at the rate of 0, 200, 250, 300 and 350 kg N/ha. HW and LW represent the irrigation water at the rate of 80% and 60% of field capacity. The V9, R1, R3, and R6 stages represent the maize with 9 leaves, silking, milking, and maturity stages, respectively.

0.001) compared to HW irrigation ($r=0.99$; $P<0.001$) as shown in Figure 5B.

3.4 Grain yield, dry matter and nitrogen contents

The irrigation and N fertilization had varied the grain yield, shoot, and root dry matter of maize (Table 1). However, the interaction between these two factors were non-significant for root dry matter ($P=0.328$). It was observed that N300 treatment under LW irrigation had higher shoot dry matter, root dry matter, and grain yield than the rest of the treatments. However, the shoot dry matter, root dry matter and grain yield were the highest at N350 treatment under HW irrigation than the other treatments. Averaged across fertilization, the LW irrigation significantly increased grain yield, shoot and root dry matter by 25.57%, 17.44%, and 28.66%, respectively compared to HW irrigation. However, when averaged over irrigations regimes, the data showed that the N300

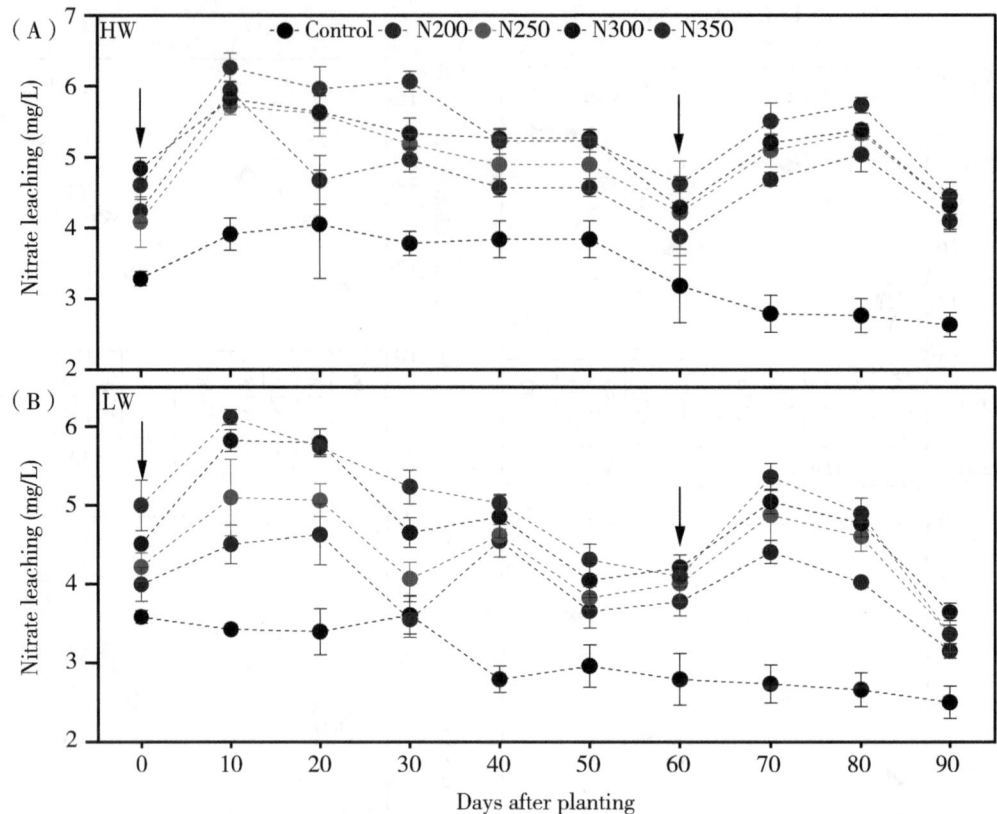

Figure 4　Effect of days, nitrogen dosages and irrigation regimes on soil Nitrate leachating

Notes: The bar represents the standard error ($n=4$). The nitrogen treatments control, N200, N250, N300 and N350 represent the application of nitrogen at the rate of 0, 200, 250, 300 and 350 kg N/ha. HW and LW represent the irrigation water at the rate of 80% and 60% of field capacity.

treatment significantly increased the grain yield by 82.13% and 2.06% compared to the control and N350 treatments.

　　Irrigation regimes, N fertilizer and their interaction, significantly affected the kernel, leaf, root, and stem N contents (Figure 6). Increasing the N fertilizer up to 300 kg N/ha significantly increased leaf, root, and stem N contents, which was not statistically different from the N350 treatment (350 kg N/ha). However, the kernel N contents were significantly higher in the N350 as compared to the other N fertilizer treatments. The plants N content in response to irrigation regime showed that kernel, leaf, root, and stem N content were significantly higher under LW irrigation than that of HW irrigation. Compared to LW irrigation, the kernel, leaf, root, and stem N content (averaged over N fertilizers) decreased by 14.33%, 11.81%, 13.65%, and 14.02% in HW irrigation, respectively. The regression analysis showed that the stem [$r=0.98$ ($P<0.01$) and $r=0.99$ ($P<0.01$)], root [$r=0.98$ ($P<0.01$) and $r=0.99$ ($P<0.01$)], leaf [$r=0.98$ ($P<0.01$) and $r=0.96$ ($P<0.05$)], and kernel N [$r=0.98$ ($P<0.01$) and $r=0.96$ ($P<0.01$)] content linearly increased with increasing N fertilizers rate both under LW and HW irrigation regimes, respectively (Figure 7).

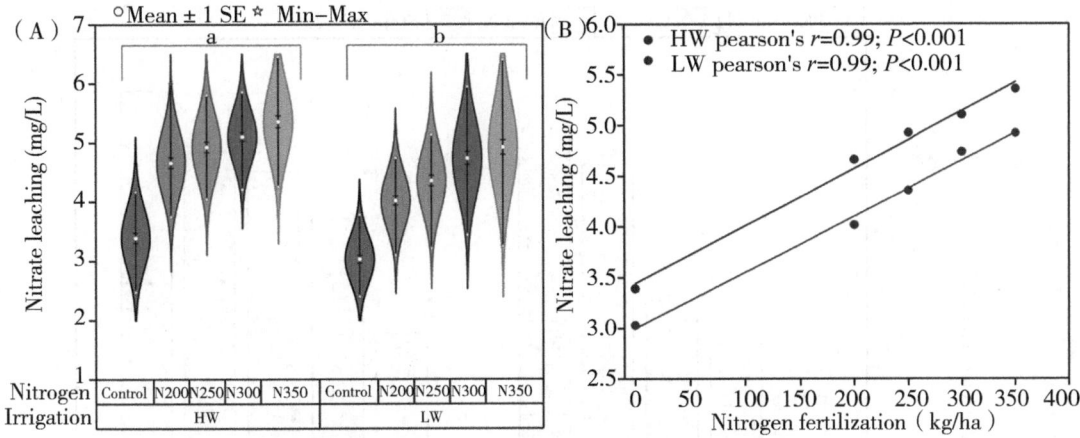

Figure 5 Effect of nitrogen dosages and irrigation regimes on soil Nitrate leachating, (B) the interaction between nitrogen dosages and irrigation regimes on Nitrate leachating

Notes: The bar represents the standard error ($n=16$). The nitrogen treatments control, N200, N250, N300 and N350 represent the application of nitrogen at the rate of 0, 200, 250, 300 and 350 kg N/ha. HW and LW represent the irrigation water at the rate of 80% and 60% of field capacity. The V9, R1, R3, and R6 stages represent the maize with 9 leaves, silking, milking, and maturity stages, respectively.

3.5 Nitrogen use efficiency

The present experiment revealed that the N use efficiencies in maize were significantly influenced by different irrigation methods and N fertilizer. The NUE was remarkably higher in LW compared to HW irrigation (Table 1). Our results showed that the NUE was significantly higher in the N200 treatment compared to other N dosages, these results suggesting NUE was 22.83% higher in LW irrigation compared to HW. Averaged across irrigation, N200 treatment resulted 68.6% and 134.3% higher NUE compared to N300 and N350 treatments, respectively. These results suggesting that NUE was significantly higher under LW irrigation with N200 treatment than other N treatments.

3.6 Mantel test correlation analysis

The Mantel test was run to explain the relationship between the plant enzyme activities and the N factors of soil and plants (i.e, soil total N, mineral N, stem N, root N, leaf N, and grain N). The results revealed that the correlation strength for LW irrigation was higher than in HW irrigation (Figure 8). The soil NH_4^+-N and NO_3^--N of LW irrigation were strongly positive correlated with NiR, GOGAT, GDH, and RubisCo activities. Similarly, the soil NO_3^--N also had a strong correlation with proline and APX activities (Figure 8). All the N factors were negatively correlated with MDA, O_2, and POD activities, except SN, which had a lighter positive correlation with O_2 activity. In HW irrigation, STN is positively correlated with all the plant enzyme activities except O_2 activity. Likewise, the GOGAT activity was also positively correlated with all the N factors and the correlation between GOGAT and NO_3^--N, LNO_3^--N, SN, RN, LN and GN were relatively stronger than others. Among the plant enzymes, all the enzymes were positive correlated with each other, except MDA, H_2O_2, O_2, and proline, which were negatively correlated with the other en-

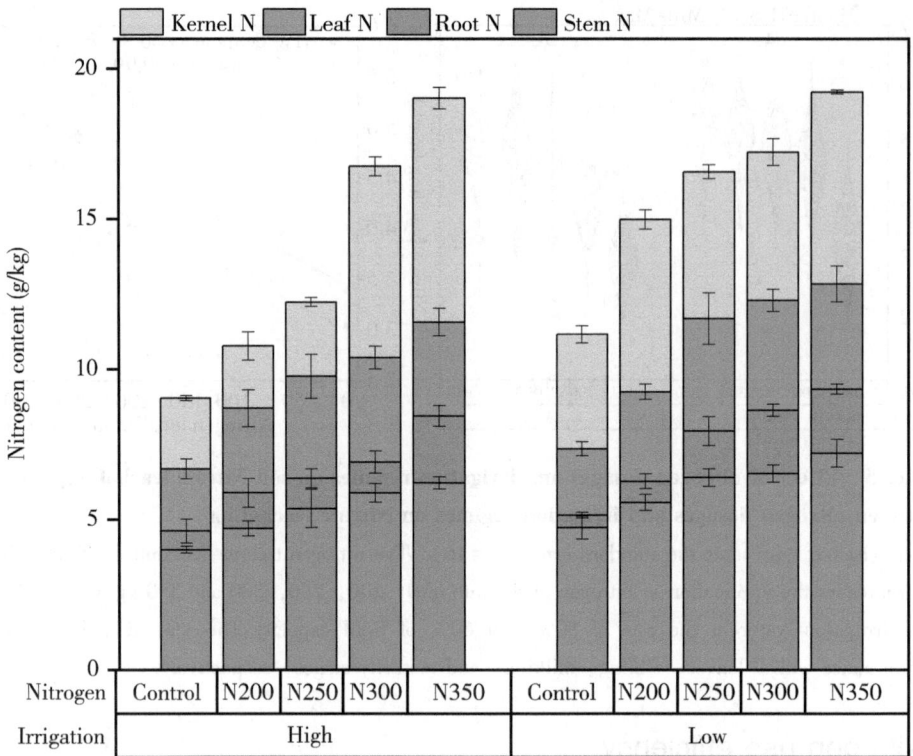

Figure 6 Effect of nitrogen dosages and irrigation regimes on kernel N, leaf N, root N and stem N content

Notes: The bar represents the standard error ($n=4$). The nitrogen treatments control, N200, N250, N300 and N350 represent the application of nitrogen at the rate of 0, 200, 250, 300 and 350 kg/ha. HW and LW represent the irrigation water at the rate of 80% and 60% of field capacity. The V9, R1, R3, and R6 stages represent the maize with 9 leaves, silking, milking, and maturity stages, respectively.

zymatic activities both under LW and HW irrigation regimes.

4 Discussion

The high rainfall>1,600 mm in subtropical region's agro-ecosystem is at risk due to greater nitrate N leaching, if applied with heavy N fertilization. In order to achieve high production without compromising the soil fertility and ecosystem, the farmers have to use the more feasible irrigation water for avoiding nitrate losses. The enhanced irrigation water and excessive N fertilization cause leaching into groundwater or through drainage ditches, which causes a number of environmental issues, including decline of arable land, contamination of groundwater, eutrophication, and the swamping of wetlands downstream[6,29,30]. Our results showed that LW irrigation with N300 treatment significantly increased the grain yield, shoot and root dry matter compared to HW irrigation. The possible reasons for this could be associated with lower N leaching in low irrigation water supply. The findings are consistent with previous researchers[6], who documented that NO_3^--N leakage increases as a result of a huge amount of irrigation while using the same N fertilizer application rate. Irrigation

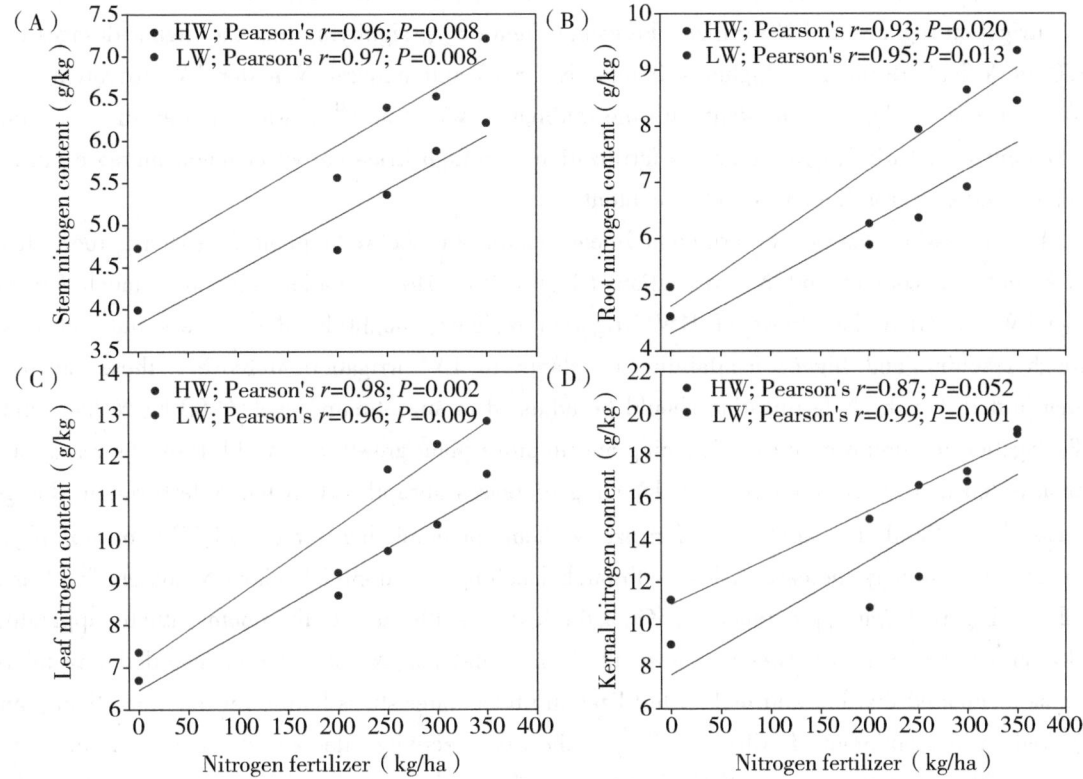

Figure 7 Effect of interaction between nitrogen dosages and irrigation regimes on kernel N, leaf N, root N and stem N content

Notes: The bar represents the standard error. The nitrogen treatments control, N200, N250, N300 and N350 represent the application of nitrogen at the rate of 0, 200, 250, 300 and 350 kg/ha. The number of observation ($n=4$) and HW and LW represent the irrigation water at the rate of 80% and 60% of field capacity, respectively.

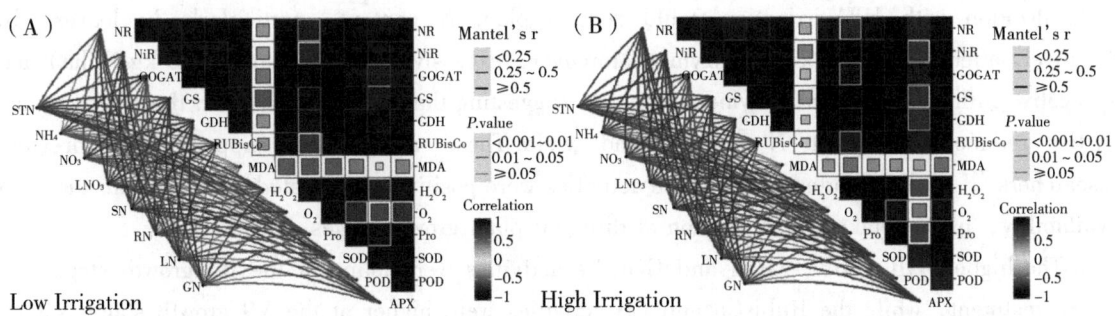

Figure 8 Correlation between nitrogen content (soil N, NO_3^--N leachate, and plant N) and plant enzyme activities under HW irrigation (a) and LW irrigation (b)

Notes: HW and LW represent the irrigation water at the rate of 80% and 60% of field capacity and GOGAT.

and N fertilizers are the key limiting factors in plant growth. The soil total N, SOC, and mineral N mainly vary with irrigation and N fertilization[31,32]. The results of the current study indicated that the soil total N, SOC, soil NH_4^+-N and NO_3^--N content and distribution significantly differed under various irrigation regimes and N fertilizer dosages (Figure 3), signifying that the higher dosages of N (300 kg N/ha) resulted in higher soil total N, SOC and mineral N under LW irrigation ($P <$ 0.05). These results are consistent with the findings of Wu et al.[31], who reported that soil nutrient content varied with irrigation and N fertilization. The high irrigation water might increase nutrient leaching and decrease the soil nutrient content[31].

The regression analysis described a linear relationship between plant N (stem, root, leaf, and kernel N) content and N fertilization (Figure 7). These relationships were much stranger under LW irrigation than those of HW irrigation regimes, might be due to less nutrient losses through leaching and higher availability to plants in LW irrigation. Moreover, these outcomes demonstrated that the N fertilization should be adjusted to an optimum level (300 kg N/ha) under LW irrigation to minimize nutrient leaching and improve plant growth and yield. Previous researchers demonstrated that the residual NO_3^--N leaching increases abruptly when the N fertilization dosages increase[6,33]. Similarly, applying N fertilizer without contemplating water availability or crop physiological demands may increase N losses through leaching and diminish plant N content[6]. Garcia et al.[34] reported that approximately 75% of the leaf N is allocated to the photosynthetic apparatus, and a strong relationship between photosynthetic rate and leaf N content was recorded. Several researchers reported that low and high N fertilization under water stress had lower tolerance than plants fertilized with optimum fertilization[34,35]. The GS activity decreased under water stress condition. Thus, an increase in NH_4^+-N concentration and an increase in protein and chlorophyll loss in detached rice leaves[36] better explain this scenarios.

In terms of nitrogenous substances and enzymes involved in N metabolism are significantly influenced by N fertilization and irrigation regimens. An adequate level of nitrate accumulation facilitates the N metabolism in the cells and excessive nitrate disrupts physiological processes[37]. We found that the N300 treatment (averaged irrigation and stages) significantly increased NR, NiR, RubisCo, GS, GDH and GOGAT activities compared to control. The current findings showed that lower N fertilization with HW irrigation could reduce plant N content, particularly in leaves, by negatively affecting N metabolism enzymes. Moreover, excessive N fertilization (350 kg N/ha) has a negative influence on plant enzyme activities, suggesting that a slow increase in these enzyme activities reduces the plant N content and yield. These results are supported by previous researchers[38], who reported that enzyme activities were positively correlated with N fertilization, N availability, plant uptake and utilization at different plant growth stages.

The highest NR, NiR, GDH and GOGAT activities were found at the R3 growth stage with N300 treatment, while the RubisCo and GS activities were higher at the V9 growth stage. Natywa et al.[39] also demonstrated that the activity of enzymes is affected by N fertilization and crop growth stages[40,41]. On the other hand, improvements in soil fertility have been shown to increase the activity of NR and ammonia-assimilating enzymes such as GOGAT and GDH activities[42,43]. High N doses did not increase enzyme activities. However, they can lead to the accumulation of toxic substances, such as ammonia, which is harmful to plants and inhibits the growth of certain microbes, and resulted in lower soil pH, which is crucial for enzyme performance[37,44]. After flowering, the leaf is a chief source of N for grain, and the photosynthetic rate and Rubisco activity increase with increasing leaf N content[42,45]. Moreover, they reported that water stress also caused metabolic imbalances which resulted in amino acid accumulation, diminished ATP and also adversely affected NR activities[42,45].

The $NO_3^- -N$ leaching differed with varied N fertilization dosages and irrigation regimes. The higher $NO_3^- -N$ was found under HW irrigation and with a higher N dose (350 N kg/ha) at the early growing stage after the first 10 days of N fertilization application. Additionally, the $NO_3^- -N$ losses (4.70 mg/L) through leaching were significantly higher under HW irrigation, compared to that of the LW irrigation regime (4.22 mg/L). The increase in N fertilization and irrigation water level led to the overall N losses through leaching[6]. These results were also supported by Gholamhoseini et al.[19], who reported that fertilizer N application from 0 to 450 kg N/ha resulted in a ten and six folds increase in $NO_3^- -N$ leaching in the HW and LW irrigation, respectively. In addition, they reported that 450 kg N/ha increased the grain yield by 6% compared to 300 kg N/ha, but increased $NO_3^- -N$ leaching by 67%. In contrast, other studies reported that higher N dosage does not increase crop yield, but significantly increases $NO_3^- -N$ leaching into groundwater[19,46]. Even with LW irrigation regimes, application of over 150 kg N/ha did not significantly improve crop yield, but $NO_3^- -N$ leaching was notably increased. Our findings indicated that $NO_3^- -N$ leaching increased with increasing N fertilization and high irrigation water (Figure 5A), which means the N350 treatment under HW irrigation had the highest $NO_3^- -N$ leaching compared to LW irrigation regimes. The overall NUE statistically higher in LW irrigation compared to HW irrigation, however the N200 treatment under HW had significantly higher NUE compared to LW irrigation (Table 1). Previous study reported that irrigation water and N fertilization could improve the N absorption and utilization[47], also play an important role in summer maize's growth and development, to boost N accumulation in various organs and then promote N accumulation in the ears[48].

Table 1 Effect of different water and N treatments on kernel yield, root and shoot dry matter components of maize

N-fertilizer	Shoot dry matter (g)		Root dry matter (g)		Grain yield (g/plant)		Nitrogen use efficiency (NUE)	
	HW	LW	HW	LW	HW	LW	HW	LW
Control	25.0+2.4d	26.2+1.9d	3.5+0.1c	5.2+0.1c	32.2±0.6d	27.9±1.7d	0.0	0.0
N200	34.3+1.5c	38.4+2.2c	5.6+0.2b	6.7+0.2b	47.44±1.32abc	37.6±1.8c	17.32±0.5a	14.75±0.7a
N250	36.7+4.1bc	45.4+2.0b	6.0+0.2b	7.5+0.4b	38.5±0.2bc	57.2±1.0b	9.65±0.1b	14.38±0.3a
N300	41.0+1.0ab	54.9+1.8a	6.8+0.4a	9.2+0.5a	39.4±2.2abbc	70.0±1.2a	6.88±0.4c	12.18±0.2b
N350	43.4+1.8a	47.0+2.2b	7.1+0.3a	8.7+0.3a	44.24±2.21a	60.1±0.9b	6.03±0.3c	7.7±0.1c

Notes: Means followed by different lowercase letters within each column indicate significant differences ($P < 0.05$) using LSD test. The nitrogen treatments control, N200, N250, N300, and N350 represent the application of nitrogen at the rate of 0, 200, 250, 300 and 350 kg/ha. HW and LW represent the irrigation water at a rate of 80% and 60% of field capacity.

In China, farmers apply about 500-600 kg N/(ha·a) to maximize the production in a maize-wheat rotation system. Thus, controlling the N application dosages is key in order to control nitrate concentrations in drainage waters, especially in south subtropical regions of China. To minimize leaching losses, it is important to reduce the $NO_3^- - N$ content in the soil, especially during the rainy season[6]. Land managers considered using optimal dosages of N fertilizer and the proper irrigation water for minimizing the leaching volume of $NO_3^- - N$. Tarkalson et al.[49] found that extremely heavy and deep seepage of surplus water and $NO_3^- - N$ could be reduced using proper irrigation scheduling techniques. Overall, we found a huge variation in $NO_3^- - N$ leaching with different irrigation and N treatments, indicating that $NO_3^- - N$ leaching increases with higher doses of N fertilizer application and irrigation water. Similarly, Gholamhoseini et al.[19] concluded that grain yield was not significantly increased by applying more than 300 kg N/ha in both irrigation regimes, but $NO_3^- - N$ leaching was dramatically increased.

5 Conclusions

Higher N fertilization under limited irrigation supply had positive effects on maize crop yield, plant enzyme activities, plant N content, $NO_3^- - N$ leachate and soil nutrient contents. The LW regime had 10.15% lower $NO_3^- - N$ leachate than HW, with correspondence increases in grain yield (25.57%), shoot (17.57%) and root (28.67%) dry matter. The Rubisco, GS, nitrate reductase, NiR, and GOGAT activities increased with increasing N fertilizer. In HW irrigation, soil total N, GOGAT, $NO_3^- - N$, $LNO_3^- - N$, RN, LN were positively correlated with N factors suggesting the higher losses of N. However, the MDA, H_2O_2, O_2, and proline were negatively correlated with the other enzymatic activities both under LW and HW irrigation regimes. Thus, minimizing the $NO_3^- - N$ leaching is possible with LW and N300 combination, without compromising the yield benefit and with improvement in enzyme activities. The findings suggested that optimizing the irrigation regimes and N fertilization effectively enhances soil fertility, plant N content, enzyme activities and yield.

Acknowledgments

The authors express their special gratitude to Prof. Dr. Xun Bo Zhou and Dr. Ahmad Khan for strong support revising the final version of the manuscript. We are also thankful to all the funding sources and especially to Guangxi University for the financial assistance.

References

[1] Cassman KG, Dobermann A, Walters DT. Agroecosystems, nitrogen-use efficiency, and nitrogen management. J Human environ. 2002, 31 (2): 132-140.

[2] Gheysari M, Mirlatifi SM, Homaee M, et al. Nitrate leaching in a silage maize field under different irrigation and nitrogen fertilizer rates. Agricultural Water Management. 2009, 96 (6): 946-954.

[3] Wang X, Fan J, Xing Y, et al. The effects of mulch and nitrogen fertilizer on the soil environment of crop plants. Advances in agronomy. 2019, 153: 121-173.

[4] Mack UD, Feger KH, Gong Y, et al. Soil water balance and nitrate leaching in winter wheat – summer maize double – cropping systems with different irrigation and N fertilization in the North China Plain. Journal of Plant Nutrition and Soil Science. 2005, 168 (4): 454-460.

[5] Jones JR, Downing, J. A. Encyclopedia of Inland Waters. 2009: 225-233.

[6] Jia X, Shao L, Liu P, et al. Effect of different nitrogen and irrigation treatments on yield and nitrate leaching of summer maize (*Zea mays* L.) under lysimeter conditions. Agricultural Water Management. 2014, 137: 92-103.

[7] Huang T, Ju X, Yang H. Nitrate leaching in a winter wheat-summer maize rotation on a calcareous soil as affected by nitrogen and straw management. Scientific reports. 2017, 7 (1): 1-11.

[8] Mas-Pla J, Menció A. Groundwater nitrate pollution and climate change: learnings from a water balance-based analysis of several aquifers in a western Mediterranean region (Catalonia). Environmental Science and Pollution Research. 2019, 26 (3): 2184-2202.

[9] Perez JMS, Antiguedad I, Arrate I, et al. The influence of nitrate leaching through unsaturated soil on groundwater pollution in an agricultural area of the Basque country: a case study. Science of the total environment. 2003, 317 (1-3): 173-187.

[10] Kong L, Xie Y, Hu L, et al. Excessive nitrogen application dampens antioxidant capacity and grain filling in wheat as revealed by metabolic and physiological analyses. Scientific Reports. 2017, 7 (1): 1-14.

[11] Li J, He Z, Du J, et al. Regional variability of agriculturally-derived nitrate-nitrogen in shallow groundwater in China, 2004-2014. Sustainability. 2018, 10 (5): 1393.

[12] Pelzer E, Bazot M, Makowski D, et al. Pea-wheat intercrops in low-input conditions combine high economic performances and low environmental impacts. European Journal of Agronomy. 2012, 40: 39-53.

[13] Jensen ES, Peoples MB, Boddey RM, et al. Legumes for mitigation of climate change and the provision of feedstock for biofuels and biorefineries. A review. Agronomy for sustainable development. 2012, 32 (2): 329-364.

[14] Sanchez-Martín L, Meijide A, Garcia-Torres L, et al. Combination of drip irrigation and organic fertilizer for mitigating emissions of nitrogen oxides in semiarid climate. Agriculture, ecosystems & environment. 2010, 137 (1): 99-107.

[15] Barton L, Colmer TD. Irrigation and fertiliser strategies for minimising nitrogen leaching from turfgrass. Agric Water Manage. 2006, 80 (1-3): 160-175.

[16] Wang GY, Hu YX, Liu YX, et al. Effects of Supplement Irrigation and Nitrogen Application Levels on Soil Carbon-Nitrogen Content and Yield of One-Year Double Cropping Maize in Subtropical Region. Water. 2021, 13 (9): 1180.

[17] Chowdary V, Rao N, Sarma P. A coupled soil water and nitrogen balance model for

flooded rice fields in India. Agriculture, Ecosystems & Environment. 2004, 103 (3): 425-441.

[18] Ju XT, Kou CL, Zhang F, et al. Nitrogen balance and groundwater nitrate contamination: Comparison among three intensive cropping systems on the North China Plain. Environmental pollution. 2006, 143 (1): 117-125.

[19] Gholamhoseini M, AghaAlikhani M, Sanavy SM, et al. Interactions of irrigation, weed and nitrogen on corn yield, nitrogen use efficiency and nitrate leaching. Agric Water Manage. 2013, 126: 9-18.

[20] Qin X, Feng F, Li Y, et al. Maize yield improvements in China: past trends and future directions. Plant Breeding. 2016, 135 (2): 166-176.

[21] Yearbook CCS: and 2015: edited by City Social and Economic Investigation Department, National Statistical Bureau. In.: Published by China Statistic Press, Beijing, China; 2014.

[22] Wei Y, Chen D, Hu K, et al. Policy incentives for reducing nitrate leaching from intensive agriculture in desert oases of Alxa, Inner Mongolia, China. Agric Water Manage. 2009, 96 (7): 1114-1119.

[23] Ju XT, Xing GX, Chen XP, et al. Reducing environmental risk by improving N management in intensive Chinese agricultural systems. Proceedings of the National Academy of Sciences. 2009, 106 (9): 3041-3046.

[24] Jin X, Xu Q, Huang C. Current status and future tendency of lake eutrophication in China. Science in China Series C: Life Sciences. 2005, 48 (2): 948-954.

[25] Ma Z, Lian X, Jiang Y, et al. Nitrogen transport and transformation in the saturated-unsaturated zone under recharge, runoff, and discharge conditions. Environmental Science and Pollution Research. 2016, 23 (9): 8741-8748.

[26] American Public Health Association. Standard methods for the examination of water and wastewater. Vol. 6. American Public Health Association., 1926.

[27] Paramasivam S, Alva A. Leaching of nitrogen forms from controlled-release nitrogen fertilizers. Communications in Soil Science and Plant Analysis. 1997, 28 (17-18): 1663-1674.

[28] Kjeldahl J. A new method for the estimation of nitrogen in organic compounds. Fresenius Z Anal Chem. 1883, 22 (1): 366.

[29] Chang F, Gao F, Hong M, et al. Effects of fertilization regimes on nitrogen leaching and maize yield in Hetao Irrigation Area. Chinese Journal of Ecology. 2018, 37 (10): 2951-2958.

[30] Tafteh A, Sepaskhah AR. Yield and nitrogen leaching in maize field under different nitrogen rates and partial root drying irrigation. International Journal of Plant Production. 2012, 6 (1): 93-113.

[31] Wu H, Du S, Zhang Y, et al. Effects of irrigation and nitrogen fertilization on greenhouse soil organic nitrogen fractions and soil-soluble nitrogen pools. Agric Water Manage. 2019, 216: 415-424.

[32] Muhammad I, Khan F, Khan A, et al. Soil fertility in response to urea and farmyard manure incorporation under different tillage systems in Peshawar, Pakistan. Int J Agric Biol. 2018, 20: 1539-1547.

[33] Mu X, Chen F, Wu Q, et al. Genetic improvement of root growth increases maize yield via enhanced post-silking nitrogen uptake. European Journal of Agronomy. 2015, 63: 55-61.

[34] Garcia A, Marcelis L, García-Sánchez F, et al. Moderate water stress affects tomato leaf water relations in dependence on the nitrogen supply. Biologia plantarum. 2007, 51 (4): 707-712.

[35] Bahadur A, Lama T, Chaurasia S. Gas exchange, chlorophyll fluorescence, biomass production, water use and yield response of tomato (Solanum lycopersicum) grown under deficit irrigation and varying nitrogen levels. Indian Journal Of Agricultural Sciences. 2015, 85 (2): 224-228.

[36] Hsu YT, Kao CH. Cadmium toxicity is reduced by nitric oxide in rice leaves. Plant Growth Regulation. 2004, 42 (3): 227-238.

[37] Sun Y, Sun Y, Yan F, et al. Coordinating Postanthesis Carbon and Nitrogen Metabolism of Hybrid Rice through Different Irrigation and Nitrogen Regimes. Agronomy. 2020, 10 (8): 1187.

[38] Sun Y, Sun Y, Li X, et al. Relationship of activities of key enzymes involved in nitrogen metabolism with nitrogen utilization in rice under water-nitrogen interaction. Acta Agronomica Sinica. 2009, 35 (11): 2055-2063.

[39] Natywa M, Sawicka A, Wolna-Maruwka A. Microbial and enzymatic activity in the soil under maize crop in relation to differentiated nitrogen fertilisation. Water-Environment-Rural Areas. 2010, 10 (2): 111-120.

[40] Ashraf MN, Jusheng G, Lei W, et al. Soil microbial biomass and extracellular enzyme-mediated mineralization potentials of carbon and nitrogen under long-term fertilization (>30 years) in a rice-rice cropping system. Journal of Soils and Sediments. 2021, 21 (12): 3789-3800.

[41] Ashraf MN, Hu C, Wu L, et al. Soil and microbial biomass stoichiometry regulate soil organic carbon and nitrogen mineralization in rice-wheat rotation subjected to long-term fertilization. Journal of Soils and Sediments. 2020, 20 (8): 3103-3113.

[42] Xu ZZ, Yu ZW. Nitrogen metabolism in flag leaf and grain of wheat in response to irrigation regimes. J Plant Nutr Soil Sci. 2006, 169 (1): 118-126.

[43] Nathawat N, Kuhad M, Goswami C, et al. Nitrogen-metabolizing enzymes: effect of nitrogen sources and saline irrigation. Journal of plant nutrition. 2005, 28 (6): 1089-1101.

[44] Brzezińska M, Włodarczyk T. Enzymes of intracellular redox transformations (oxidoreductases). Acta Agrophys Rozpr Monogr. 2005, 3: 11-26.

[45] Xu ZZ, Zhou GS. Effects of water stress and high nocturnal temperature on photosynthesis and nitrogen level of a perennial grass Leymus chinensis. Plant and Soil. 2005,

269 (1): 131-139.

[46] Abouziena HF, El-Karmany M, Singh M, et al. Effect of nitrogen rates and weed control treatments on maize yield and associated weeds in sandy soils. Weed Technol. 2007, 21 (4): 1049-1053.

[47] Li G, Zhao B, Dong S, et al. Impact of controlled release urea on maize yield and nitrogen use efficiency under different water condition. Plos One. 2017, 12 (7): e0181774.

[48] Chi YX, Gao F, Muhammad I, et al. Effect of water conditions and nitrogen application on maize growth, carbon accumulation and metabolism of maize plant in subtropical regions. Archives of Agronomy and Soil Science. 2022, 1-15.

[49] Tarkalson D, Payero J, Ensley S, et al. Nitrate accumulation and movement under deficit irrigation in soil receiving cattle manure and commercial fertilizer. Agric Water Manage. 2006, 85 (1-2): 201-210.

Nitrogen Fertilizer Modulates Plant Growth, Chlorophyll Pigments and Enzymatic Activities under Different Irrigation Regimes

I. Muhammad[1], L. Yang[1], S. Ahmad[1], S. Farooq[1], A. A. Al-Ghamdi[2], A. Khan[3], M. Zeeshan[1], M. Elshikh[2], A. M. Abbasi[4,5], X. B. Zhou[1]

([1]Guangxi Colleges and Universities Key Laboratory of Crop Cultivation and Tillage, Agricultural College, Guangxi University, Nanning 530004, China; [2]Department of Botany and Microbiology, College of Science, King Saud University, P. O. Box 2455, Riyadh 11451, Saudi Arabia; [3]Department of Agronomy, The University of Agriculture Peshawar, Peshawar 25120, Pakistan; [4]University of Gastronomic Science, Pollenzo, Piazza V. Emanuele II, I-12042, Bra/Pollenzo, Italy)

Abstract: Nitrogen fertilization and irrigation patterns have been extensively studied for common maize (*Zea mays* L.), but there is limited published work for Zhengda 619, especially in subtropical areas. Nitrogen (N) fertilizer and irrigation play an important role in crop growth and yield improvements. The study aimed to investigate the yield, growth, chlorophyll content, reactive oxygen species (ROS) and enzyme activities of hybrid maize (Zhengda 619) under greenhouse conditions. Individual plants grown in plastic pots were subjected to two irrigation types—low irrigation (LW; 60% field capacity) and high irrigation water (HW; 80% field capacity) —and five N rates. Our results demonstrate that the LW irrigation increased dry matter, kernel yield, leaf chlorophyll, total root length, root diameter, root volume, and root surface area, as well as soil enzymes and plant antioxidant enzymes, while it lowered malondialdehyde (MDA), proline, and ROS. Moreover, most of the above parameters increased with increasing N application rates up to N3 under LW irrigation due to the increased N available to the plant and soil enzymes. It is concluded that increasing N rates could improve soil enzyme activities as well as plant antioxidant enzymes and decrease ROS, ultimately resulting in a higher kernel yield under LW irrigation.

Keywords: Enzymatic activity; Leaf chlorophyll; Irrigation; Nitrogen; Maize

1 Introduction

Maize (*Zea mays* L.) is the third most important primary crop in China and is considered an important pillar of agricultural development[1]. Water and fertilizers, particularly nitrogen (N),

---

These authors contributed equally to this work.
Correspondence: xunbozhou@ gxu. edu. cn.

are ideal for increasing productivity in regions with water scarcity or abundant rainfall. Nitrogen application increases root and shoot dry matter and kernel yield, as well as improving crop quality[2]. Sustainable intensification in modern agriculture requires an increased efficiency of resources while maintaining or increasing productivity and improving environmental quality[3]. Similarly, adopting suitable water management strategies, such as irrigation management, is critical for achieving high crop water use efficiency (WUE) and yield[4].

One of the most important research topics in agriculture is nitrogen regulation and irrigation management. Nitrogen is a key plant nutrient that promotes and inhibits plant growth[5]. High-yield crops like maize are regularly fertilized with a large amount of nitrogen fertilizer to achieve optimal yields. Agricultural yields are improved by increasing nutrients and water uptake with appropriate fertilization and irrigation techniques[6,7]. However, such strategies must be backed up with research-based knowledge that addresses key issues that could reduce yield[8]. Water stress has a negative impact on plant development, plant height, and leaf area[9]. Similarly, N deficiency restricts plant growth and decreases leaf area and biomass yield[10]. Water and nitrogen, particularly in terms of plant growth and crop production, are well known for their complex interaction. Maize yield decreases under limited water conditions with high N fertilizers[11]. In contrast, large amounts of N fertilizer are required when corn is cultivated in areas with no water stress[12].

Nitrogen fertilizers have been applied to maize immediately before planting and partially side-dressed during the V6 to V8 stages[13]. Delayed side-dressing could result in irreversible yield loss. Delaying N application until the V6 stage resulted in a near 12% loss in kernel yield, according to[14]. However, research was limited to normal maize hybrids. There is a growing interest in learning more about the effects of different irrigation and nitrogen applications during plant growth, when N is most important for maximum yield, especially in leafy and hybrid maize. During kernel filling, a number of annual cereals show genetic diversity in the degree and pace of leaf withering[13]. While it is critical to use the optimal amount of water and nitrogen, when the amount of fertilizer input exceeds the level of nutrient absorbed by crops, contamination of water resources occurs as a result of crop management[15]. As a result, scientific and public concern have finally increased, with an emphasis on water pollution caused by nitrogen from agricultural sources[16].

Inappropriate use of N and water may increase N nitrate losses through leachate, with negative environmental consequences[17]. Furthermore, the price of N fertilizer has risen rapidly during the last few decades[18]. As a result, in an irrigated agricultural system, it is critical to improve nitrogen management in order to maximize farm income and reduce environmental impact[19]. Irrigation water and N have been studied extensively in relation to maize production and WUE[20-22]. However, few studies have investigated the combined effect of water and nitrogen on maize biomass, kernel yield, and enzymatic activity in subtropical areas of China, particularly in high-precipitation areas. In addition, there are discrepancies in the results between the amounts of water used and the rates of nitrogen applied[23]. Maize for kernel yield requires different management approaches than silage maize. Compared to kernel corn, silage maize is harvested at an earlier stage of maturity, so it requires less water. To avoid decreasing the overall nutritional value of the maize, nitrogen management is essential during the early reproductive stages[24]. This research aimed to determine the combined effects of irrigation water and N on hybrid maize growth, biomass yield, kernel yield, WUE and enzymatic activity at different growth stages for maize cultivated in subtropical regions in controlled conditions.

2　Materials and Methods

2.1　Experimental site

A pot experiment was carried out at Guangxi University in the subtropical region of China (22°50′, 108°17′), in a greenhouse with a controlled environment and nutrition systems. The area is characterized as a warm and temperate region with a mean air temperature of 21.7℃ and mean annual rainfall of 1,298 mm. According to Chinese Soil Taxonomy, the texture of soil was clay loam, with a pH of 5.6, a field capacity of 44%, soil bulk density of 1.40 g/cm^3, soil organic matter of 20 g/kg, and available nitrogen, phosphorus, and potassium of 127.0 mg/kg, 40.0 mg/kg, and 126 mg/kg, respectively. A pot was filled with a mixture of soil collected from the greenhouse, which has not been in use for the last six years.

2.2　Experimental design and management

A 2 × 5 factorial experiment was carried out in a completely randomized design with four replications, with a total of 40 pots, in a controlled-environment greenhouse in Guangxi, China (Figure 1). The experimental treatments were two irrigation levels, i.e., low irrigation water (LW; 60%) and high irrigation water (HW; 80%), field capacity, and five nitrogen rates, i.e., control (N0), 200 kg N/ha (N1), 250 kg N/ha (N2), 300 kg N/ha (N3), and 350 kg N/ha (N4). On September 28, 2020, five uniformly sized hybrid maize seeds of the Zhengda 619 variety, which is the most commonly grown variety in the subtropical areas of China, were planted per pot (with a length of 32.5 cm and a height of 29.0 cm). The seeds were obtained from CP seed industry, Yunnan Zhengda Seed Co. Ltd., China. The selected seeds permission was granted from the respective authority. The base fertilizers (P and K) and 1/2 of N were thoroughly mixed with soil before sowing, and the remaining 1/2 of N was applied as a top dressing at the nine-leaf stage (V9). The phosphorus (P) and potash (K) fertilizers were used in accordance with local fertilization standards, at 100 kg P/ha, and 100 kg K/ha, to ensure that all experimental treatments had equal P and K concentrations. The fertilizer types used in our experiment were urea (46% N), phosphorus pentoxide P_2O_5 (18% P), and potassium oxide K_2O (60% K).

Maize crops were trimmed to four plants per pot at the three-leaf stage to facilitate better adaptability to the pot environment. Throughout the growth stage, plants were watered with tap water to maintain soil moisture at 60% and 80% of the field's water holding capacity. Micro-tensiometers were used to measure the temperature of the soil in each pot (Nanjing Institute of Soil Science, Chinese Academy of Sciences).

2.3 Sampling and measurements

2.3.1 Determination of yield and growth attributes

At four growth stages (V9, R1, R3, and R6), data on several physiological aspects of maize crops were collected. The plants, ear width, and ear length were measured at physiological maturity. The number of rows per ear and the number of seeds per row were manually counted after harvesting at full maturity. Following threshing, yield characteristics such as kernel yield and kernel per ear were recorded. Each plant's components were then divided into kernels, leaves, stalks, and roots. Following that, the plant dry matter was sun-dried for three days before being oven-dried at 70℃ for 72 h to determine plant dry matter.

For the measurement of root length, diameter and surface area, root samples were taken after harvesting, washed, scanned, and analyzed by root image analysing software. In each treatment, four plants were selected, and the leaf area was calculated using the following formula.

$$LA(\text{cm}^2) = L \times W \times factor(0.75)$$

where LA represent the leaf area (cm^2), L is the length (cm), W is the width (cm), and 0.75 is the constant-coefficient factor for the maize leaf area.

2.3.2 Determination of antioxidant enzyme activity

Four plants from each replicate were selected at four growth stages (V9, R1, R3 and R6) for antioxidant enzyme activity, and leaf samples were put in liquid N_2 for 1 min before being stored at 80℃ for biochemical and physiological analyses. The activity of superoxide dismutase (SOD) was measured with the nitro blue tetrazolium (NBT) illumination technique[65]. The 50% decrease in absorbance at 560 nm was used to represent one unit of SOD activity and was expressed as U/g fresh weight (FW). Peroxidase (POD) activity was analyzed according to MacAdam et al.[66]. The reaction mixture included a phosphate buffer (50 mM), guaiacol (16 mM), and 0.2 mL of enzyme extract, followed by H_2O_2 (10 mM). The absorbance at 470 nm was measured until 5 readings were taken at 30 s intervals. The activity of ascorbate peroxidase (APX) was determined by Nakano and Asada[67]. The reaction mixture included a 50 mM phosphate buffer, 0.1 mM EDTA, 0.5 mM AsA, and 1.0 mM H_2O_2, along with 0.2 mL crude enzyme extract. The following formula was used to calculate the change in absorbance of the mixture at 290 nm.

$$APXactivity(\text{U/mg}) = \frac{\Delta A_{290} \times Vt}{2.8 \times M \times V \times t}$$

2.3.3 Chlorophyll content

With slight modifications, the chlorophyll content was determined using the method described in[68]. To avoid light from altering the results, the leaf samples from each treatment were chopped up and immersed in a graduated tube with 80% acetone. When chlorophyll had been extracted, the supernatant was then removed and placed in a new tube, and the absorbance was recorded at wavelengths of 663 nm, 645 nm, and 470 nm to measure the content of chlorophyll a and b, and 80% acetone was utilized as a blank control.

2.3.4 Determination of ROS, MDA and proline content

The superoxide anion (O_2^-) content was determined by following the method of[69]. Briefly, a fresh sample of the leaf (500 mg) was homogenized with a 65 mM potassium phosphate buffer (pH 7.8), and centrifuged for 10 min at 8,000× g at 4℃. Next, 2 mL of supernatant was mixed with 0.5 mL of potassium phosphate-buffer (PBS; 65 mM, pH 7.8), and 0.1 mL of hydroxylamine hydrochloride solution (10 M) was mixed together and kept at 25℃ for 1 h. One mL of amino benzene sulfonic acid solution (58 M) and 1 mL of an α-naphthylamine solution (7 M) was added, and incubated at 25℃ for 20 min. The pigments were then extracted with 1 mL of chloroform. The mixture was centrifuged for 10 min at 10,000× g at 4℃. The absorbance at 532 nm was determined by collecting the upper pink supernatant, while the H_2O_2 content was assayed according to Ohto et al.[70].

Malondialdehyde, a product of lipid peroxidation, was measured in plant leaves using the method described by Weisany et al.[71]. Trichloroacetic acid (TCA; 0.1% w/v) was used to extract 200 mg of fresh leaf, and the extract was centrifuged at 12,000× g for 5 min at 4℃. Subsequently, 20% of TCA was added to the solution and thoroughly mixed with 0.5% of 4 mL thiobarbituric acid (TBA) and incubated for half an hour in a hot water bath at 90℃. The entire extract was kept on ice. The absorbance values were measured at 532 nm, and the nonspecific absorption at 600 nm was subtracted from the absorbance data.

The ninhydrin procedure, described by Bates et al.[72], was used to determine the free proline content. A fresh leaf sample (500 mg) was homogenized in 10 mL of 3% aqueous sulfosalicylic acid and centrifuged at 10,000× g for 15 min. Following the filtration of the homogenized solution, 2 mL of the filtered solution was transferred to test tubes and treated with acid ninhydrin (2 mL) and glacial acetic acid (2 mL). The tubes were kept warm at 80℃ for 1 h. To stop the process, the tubes were placed in an ice bath. The liquid was aggressively stirred for a few seconds with a mixer after adding 4 mL of toluene. After separating the toluene chromophore from the aqueous phase, the absorbance at 520 nm was measured.

2.3.5 Soil enzyme activity analysis

The activities of soil acid phosphatase (BC0145), acid invertase (BC3075), β-glucosides (BC0165), catalase (BC0105), cellulose (BC0155), and soil urease (BC0125) were determined using the soil enzyme kit from Solarbio Science & Technology Co. (Beijing, China). The methods of determination are described in detail in the manual.

2.4 Statistical analysis

A mixed-model analysis of variance (ANOVA) was used to calculate the effects of N rates and irrigation on yields, enzymatic activity, and kernel quality during the four growth seasons. The analysis of variance (ANOVA) was performed using SPSS 21.0 software (SPSS Inc., Chicago, IL, USA). The least significant difference (LSD) test was used to separate means and interactions. Statistical significance was evaluated at $P \leqslant 0.05$.

3 Results

3.1 Kernel yield and yield components

The yield and yield-related components from different N and irrigation treatments are shown in

Table 1. We found that a high dose of N input with low irrigation water (LW) resulted in a higher yield. Our results show that kernel yield was significantly affected by irrigation rates ($P < 0.001$) and input N rate ($P < 0.001$), and the interactions between irrigation and N level ($P<0.001$; Table 2) also affected kernel yield. There was no effect of irrigation treatment on the number of rows per ear, but N rate had a significant effect on the number of rows per ear, kernel per row, kernel per ear, ear length and ear diameter in both LW and HW irrigation. In contrast, N had no effect on ear diameter in HW irrigation, but the ear diameter was significantly increased in LW irrigation (Table 1). Regardless of N fertilizer input, LW irrigation significantly increases crop yield and yield components compared to HW irrigation.

Table 1　Effect of different water and N treatments on yield and yield components of maize

Irrigation	N-Fertilizer	Rows Number	Kernels Per Row	Kernels Per Ear	Ear Length (cm)	Ear Diameter (cm)	Kernels Yield (g/plant)
HW	N0	11.0 ± 0.6 b	10 ± 1.0 b	108 ± 13.6 c	7.5 ± 0.5 b	4.0 ± 0.2 a	32.2 ± 0.6 d
	N1	14 ± 0.5 a	13 ± 1.3 a	169 ± 19.3 b	8.4 ± 0.3 b	4.2 ± 0.1 a	47.1 ± 2.2 c
	N2	14 ± 0.5 a	13 ± 0.9 a	177 ± 3.1 b	8.6 ± 0.3 b	4.3 ± 0.1 a	38.5 ± 0.2 bc
	N3	14 ± 0.5 a	14 ± 0.5 a	199 ± 16.5 ab	8.8 ± 0.7 b	4.3 ± 0.1 a	39.4 ± 2.2 ab
	N4	15 ± 0.5 a	14 ± 0.8 a	220 ± 17.1 a	11.0 ± 0.7 a	4.3 ± 0.1 a	44.2 ± 1.3 a
LW	N0	12 ± 1.3 c	11 ± 1.5 b	119 ± 11.0 b	8.3 ± 1.0 b	4.1 ± 0.2 c	27.9 ± 1.7 d
	N1	13 ± 0.6 bc	15 ± 1.9 ab	197 ± 18.5 a	9.3 ± 0.9 b	4.2 ± 0.3 bc	37.6 ± 1.8 c
	N2	14 ± 1.0 abc	17 ± 1.6 a	222 ± 18.5 a	10.0 ± 0.5 b	4.3 ± 0.1 bc	57.2 ± 1.0 b
	N3	16 ± 0.8 a	18 ± 1.0 a	259 ± 22.7 a	12.3 ± 0.2 a	4.9 ± 0.1 a	70.0 ± 1.2 a
	N4	15 ± 1.0 ab	17 ± 1.1 a	234 ± 21.3 a	10.1 ± 0.7 ab	4.7 ± 0.1 ab	60.1 ± 0.9 b

Notes: Means followed by different lowercase letters within each column indicate significant differences ($P<0.05$) using the LSD test. The nitrogen treatments N0, N1, N2, N3 and N4 represent the application of nitrogen at the rate of 0, 200, 250, 300, and 350 kg/ha. HW and LW represent irrigation water at a rate of 80% and 60% of field capacity.

3.2　Plant dry matter

Nitrogen fertilizer significantly increased the growth characteristics of maize under both LW and HW irrigations. As shown in Figure 1, the root and shoot dry matter of maize was significantly increased with increasing N rate in both LW and HW conditions ($P<0.001$). LW irrigation had 29% and 17% higher root and shoot dry matter than HW irrigation, respectively ($P<0.01$). Compared to N0, the root dry matter in N1, N2, N3, and N4 was 64%, 73%, 98%, and 105% higher in LW, and 28%, 43%, 76%, and 66% higher in HW irrigation, respectively. In contrast, the root dry matter in N3 and N4 treatments was statistically similar under both LW and HW irrigation.

Under both irrigation treatments, the accumulation of shoot dry matter in N-treated plants was significantly higher than in N0 plants. Averaged across N fertilizer applications, N3 markedly im-

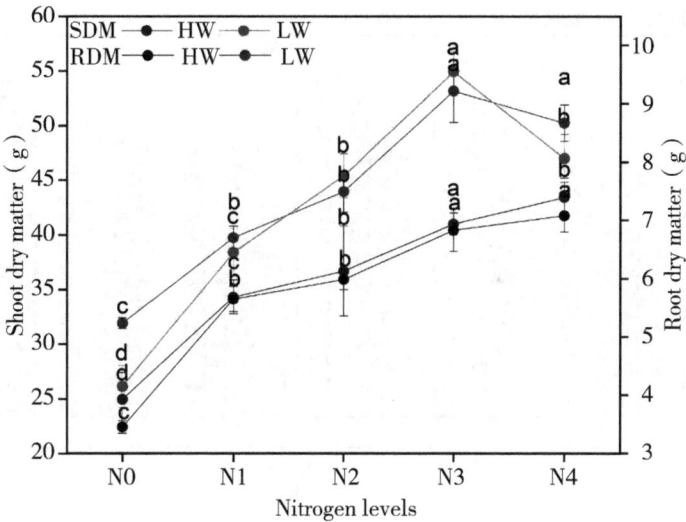

Figure 1 Effect of irrigation and nitrogen treatments on root and shoot dry matter

Notes: Means followed by different lowercase letters within each treatment indicate significant differences ($P<0.05$). Bars represents standard errors. The nitrogen treatments N0, N1, N2, N3 and N4 represent the application of nitrogen at the rates of 0, 200, 250, 300 and 350 kg/ha. HW and LW represent irrigation water at the rate of 80% and 60% of field capacity. RDM and SDM are root and shoots dry matter, respectively.

proved shoot dry matter by 88% compared to the N0 treatment (Figure 1). Similar to root dry matter, shoot dry matter was also significantly increased as the N level increased, but in the N2 and N4 treatments, the shoot dry matter was not statistically different. Our results show that the N rate had a better effect on shoot dry matter under LW compared to HW irrigation. The N3 treatment showed the most favorable influence on shoot dry matter, followed by the N4 and N2 treatments, indicating that N3 treatment significantly increased shoot dry matter by 110% and 64% compared to N0 in LW and HW irrigation, respectively.

3.3 Plant height and leaf area

Irrigation, N fertilizer, stages and their interactions had significant effects on plant height and leaf area (Figure 2 and Table 2). The results show that plant height and leaf area were significantly increased with increased N rate ($P<0.001$) under both irrigations treatments ($P<0.01$) at different growth stages ($P<0.001$). However, LW irrigation showed markedly higher plant height and leaf area than HW irrigation. LW irrigation showed 4.3% and 0.4% higher plant height and leaf area than HW irrigation, respectively. However, on average, the N4, N3, N2, and N1 treatments increased plant height by 10.7%, 16.5%, 9.5% and 5.7%, and leaf area by 24%, 37.5%, 13% and 6.5%, compared to N0. It is well understood that plant height and leaf area increase over time. Our findings reveal that the plant height at the R3 stage was statistically similar to the R6 stage, but significantly greater than at the V9 and R1 stages ($P<0.001$). The maximum leaf area was obtained at R3 followed by R6, and the lowest leaf area was observed at the V9 growth stage.

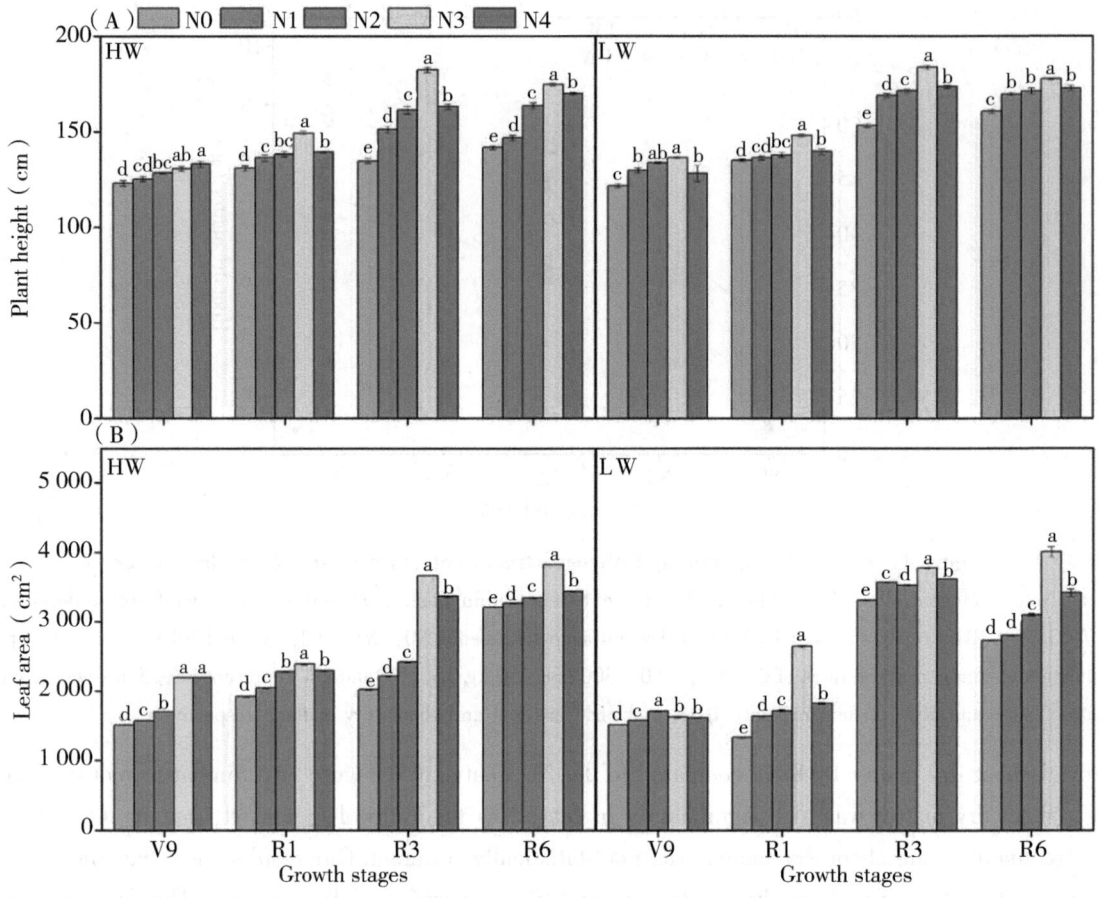

Figure 2　Effects of irrigation and nitrogen rates on plant height (a) and leaf area (b)

Notes: HW and LW represent irrigation water at the rate of 80% and 60% of field capacity. Means followed by different lowercase letters within each treatment indicate significant differences ($P<0.05$). Bars represents standard errors. The nitrogen treatments N0, N1, N2, N3 and N4 represent the application of nitrogen at the rates of 0, 200, 250, 300 and 350 kg/ha.

Table 2　Probability values (P values) of main and interaction effects for irrigation, nitrogen treatment and stage (V9, R1, R3, R6) for various maize parameters

Parameters	Irrigation	Nitrogen	Stage	Irrigation × Nitrogen	Stage × Irrigation	Stage × Nitrogen	Stage × Irrigation × Nitrogen
Number of rows/ear	0.367	<0.001	---	0.442	---	---	---
Kernels per row	0.021	<0.001	---	0.775	---	---	---
Kernels per ear	0.005	<0.001	---	0.578	---	---	---
Ear length	0.029	0.002	---	0.045	---	---	---
Ear diameter	0.043	0.007	---	0.248	---	---	---
Kernel weight	<0.001	<0.001	---	<0.001	---	---	---
Shoot dry matter	0.009	<0.001	---	0.035	---	---	---
Root dry matter	0.002	<0.001	---	0.328	---	---	---

(continued)

Parameters	Irrigation	Nitrogen	Stage	Irrigation × Nitrogen	Stage × Irrigation	Stage × Nitrogen	Stage × Irrigation × Nitrogen
Plant height	<0.001	<0.001	<0.001	<0.001	0.006	<0.001	<0.001
Leaf area	0.035	<0.001	<0.001	<0.001	<0.001	<0.001	<0.001
Malondialdehyde	<0.001	<0.001	<0.001	0.009	0.006	0.002	<0.001
Hydrogen peroxide	0.002	<0.001	<0.001	0.002	0.065	0.002	<0.001
Superoxide anion	<0.001	<0.001	<0.001	<0.001	0.043	<0.001	0.005
Proline	<0.001	<0.001	<0.001	0.004	0.018	<0.001	<0.001
Superoxide dismutase	<0.001	<0.001	<0.001	<0.001	0.003	<0.001	<0.001
Peroxidase	0.005	<0.001	<0.001	<0.001	0.040	<0.001	<0.001
Ascorbate peroxidase	<0.001	<0.001	<0.001	0.040	0.001	<0.001	<0.001
Chlorophyll a	<0.001	<0.001	<0.001	<0.001	0.024	<0.001	<0.001
Chlorophyll b	0.016	<0.001	<0.001	<0.001	0.019	0.002	0.017

3.4 Root growth and development

The number of roots, total root length, surface area, volume, and diameter were significantly affected by N treatment (Table 3; $P<0.001$). Compared with N0, the total root length in N1, N2, N3, and N4 for LW was increased by 77%, 141%, 375%, and 248%, while a 48%, 80%, 117%, and 175% increase was observed for HW irrigation (Table 3). On average, the total root length in LW was 13% longer than that in HW irrigation. Similarly, averaged across irrigation, the N3 treatment increased total root length by 229% over N0. The maximum root surface area was observed for the N3 treatment in LW (881.3 cm^2) and the N4 treatment in HW irrigation (720.1 cm^2), which was significantly higher than for the N0 treatment, which resulted in the lowest root surface area (Table 3). The maximum root surface area was observed for LW compared to HW irrigation, but these results were not significant. Similarly, the root volume and average diameter were not significantly affected by irrigation treatment ($P>0.05$), but they significantly increased with increasing N rate ($P<0.001$; Table 3). Compared to N0, the root volume and average diameter were increased by 183% and 19.6%, respectively, in the N3 treatment, regardless of irrigation.

Table 3 Effects of irrigation and nitrogen application on number of roots, total root length, acreage diameter and surface area

Irrigation	N-Fertilizer	Number of Roots	Total Root Length (cm)	Root Diameter (mm)	Root Surface Area (cm^2)	Root Volume (mm^3)
HW	N0	1 829 e	856.69 e	0.8 b	162.3 d	5 092.3 c
	N1	3 014 d	1 267.60 d	1.0 a	412.0 c	19 579.8 b
	N2	3 584 c	1 539.53 c	0.9 a	458.9 c	20 793.1 b
	N3	4 222 b	1 855.38 b	0.9 a	545.9 b	24 704.9 b
	N4	5 183 a	2 359.84 a	1.0 a	720.1 a	35 036.3 a

(continued)

Irrigation	N-Fertilizer	Number of Roots	Total Root Length (cm)	Root Diameter (mm)	Root Surface Area (cm^2)	Root Volume (mm^3)
LW	N0	2 323 c	663.45 e	0.7 c	160.0 c	20 197.8 c
	N1	2 868 c	1 172.20 d	1.1 a	398.4 b	21 697.8 c
	N2	3 699 b	1 599.53 c	0.9 b	472.4 b	21 571.4 c
	N3	6 256 a	3 148.81 a	0.9 b	881.3 a	47 005.9 a
	N4	4 397 b	2 306.67 b	1.1 a	818.6 a	37 606.5 b

Notes: Means followed by different lowercase letters within each column indicate significant differences ($P < 0.05$) using LSD tests. The nitrogen treatments N0, N1, N2, N3 and N4 represent the application of nitrogen at the rates of 0, 200, 250, 300 and 350 kg/ha. HW and LW represent irrigation water at a rate of 80% and 60% of field capacity.

3.5 Chlorophyll a and b contents

The application of N, irrigation treatment and their interaction significantly increased chlorophyll *a* and *b* content at different growth stages (Figure 3 and Table 2). The maximum content of chlorophyll *a* and *b* occurred at V9 and gradually decreased until R6. Under both HW and LW irrigation, the maximum chlorophyll *a* and *b* content was found at V9 with N3 treatment, and the minimum was found for N0. Our results show that chlorophyll *a* and *b* content increased with increasing N content up to the N3 treatment, but that higher N content (N4) decreased chlorophyll *a* and *b* contents. The application of N fertilization with LW irrigation resulted in higher chlorophyll *a* and *b* content compared with HW irrigation. The minimum chlorophyll *a* and *b* contents at all growth stages were obtained with HW in N0 treatment, and the maximum with LW in N3, followed by N2 treatment. The mean results for the four growth stages revealed that N1, N2, N3, and N4 significantly increased chlorophyll *a* content by 52%, 71%, 89%, and 62%, respectively, and chlorophyll *b* content by 33%, 66%, 96%, and 58%, respectively, compared to N0.

3.6 Malondialdehyde and proline content

The application of N and irrigation treatment significantly affected the MDA and proline content (Figure 4a, b). The application of N decreased the MDA content compared to the N0 treatment. The maximum MDA content was observed for N0 and the minimum was observed for N4 treatment in both HW and LW irrigations (Figure 4). This suggests that a higher N content has a negative impact on MDA content. For irrigation, the LW treatment decreased the MDA content by 29.72% compared to HW irrigation. The N0 treatment resulted in a higher MDA content compared to other N treatments. For fertilization, N4, N3, N2 and N1 significantly decreased the MDA content, by 51%, 43%, 26%, and 17%, respectively. The results show that N content and irrigation have a significant effect on proline content under both irrigation types. The proline content decreased with increasing N content at LW and HW irrigations. However, on average, the high N content (N4) has a 57.7% lower proline content than that the N0 treatment (Figure 4b). The minimum proline content was observed at V9 under LW treatment, and the maximum was observed at

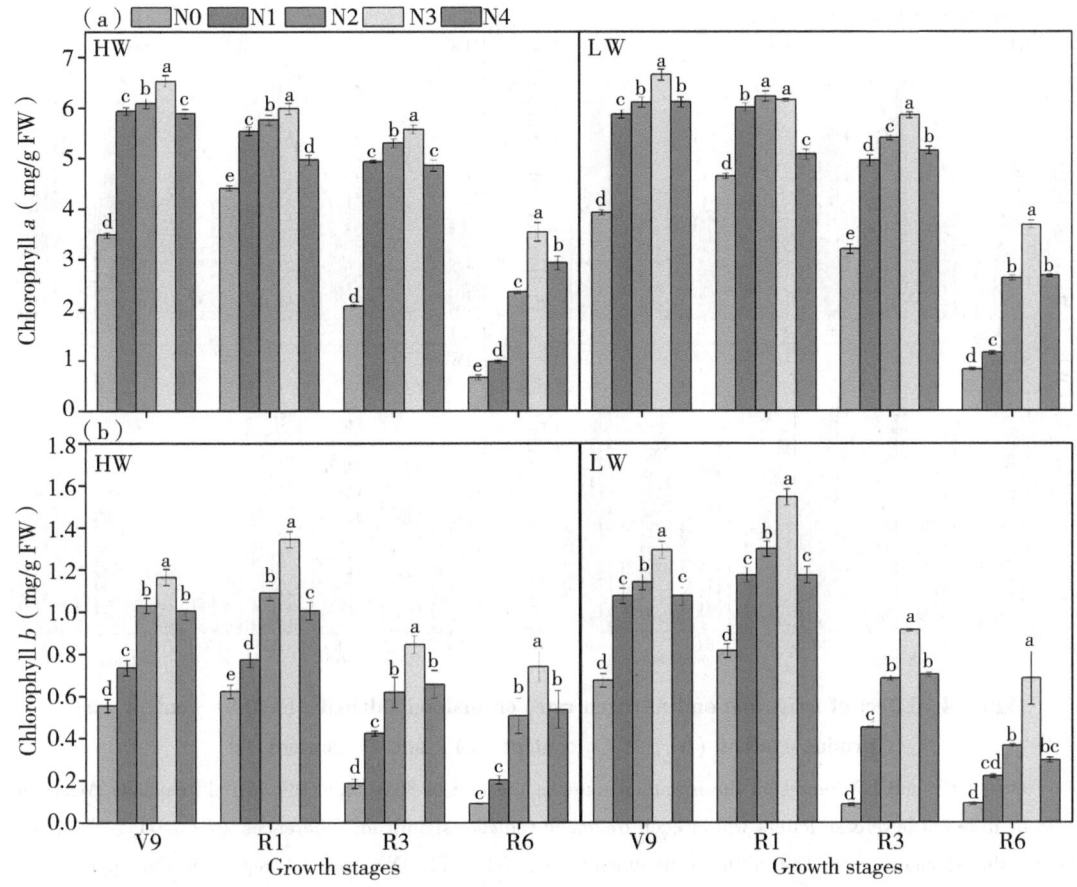

Figure 3 Effect of irrigation and nitrogen rates on chlorophyll *a* content (a) and chlorophyll *b* contents (b)

Notes: HW and LW represent irrigation water at the rate of 80% and 60% of field capacity. Means followed by different lowercase letters within each treatment indicate significant differences ($P<0.05$). Bars represent standard errors. The nitrogen treatments N0, N1, N2, N3 and N4 represent the application of nitrogen at the rates of 0, 200, 250, 300 and 350 kg/ha.

R6 under HW treatment. The LW treatment significantly decreased the proline content by 19% compared to HW irrigation.

3.7 Reactive oxygen species

Hydrogen peroxide and O_2^- content decreased significantly with N rates in HW and LW irrigation. In the present study, both H_2O_2 and O_2^- contents decreased with increasing N content, where the higher N content (N4) resulted in a lower ROS content compared to other N treatments (Figure 4c, d). Under both LW and HW, the minimum H_2O_2 and O_2^- contents were observed for N4, while N0 resulted in a higher ROS content. The H_2O_2 content was higher at R3 and R6 in HW and LW irrigations, respectively. Likewise, a lower O_2^- content was observed at V9 and gradually increased towards R6. On average, the LW irrigation resulted in an 8% lower H_2O_2 and a 43%

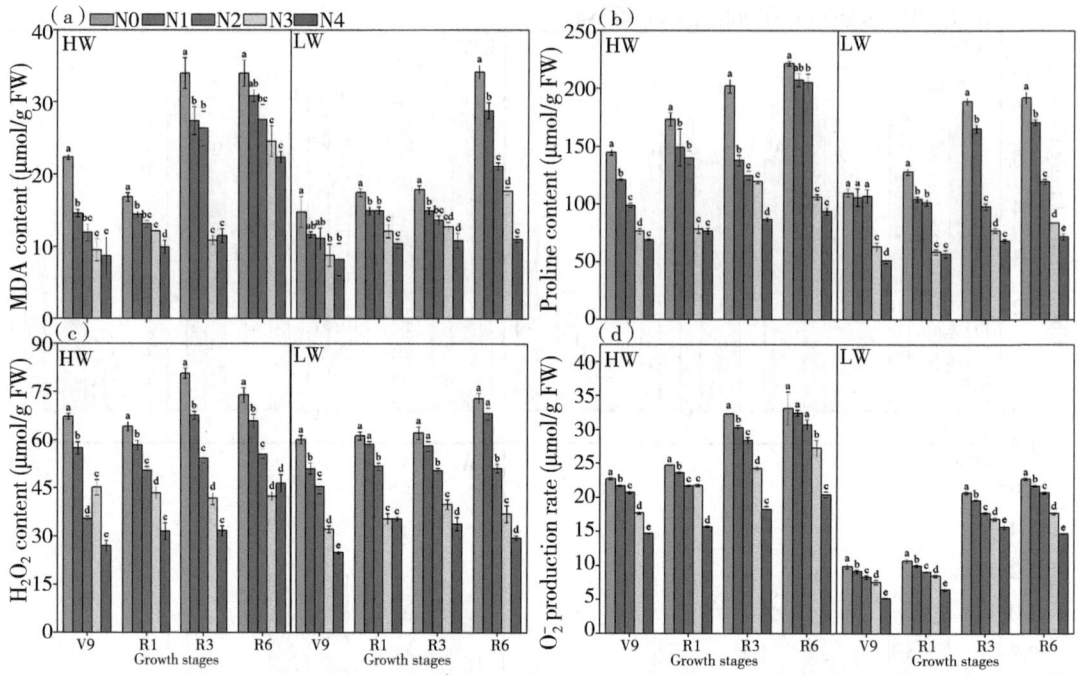

Figure 4 Effect of irrigation and nitrogen rates on malondialdehyde (MDA) content (a), proline content (b), H_2O_2 content (c), and O_2^- content (d)

Notes: HW and LW represent the irrigation water at the rate of 80% and 60% of field capacity. Means followed by different lowercase letters within each treatment indicate significant differences ($P<0.05$). The bar represents the standard error. The nitrogen treatments N0, N1, N2, N3 and N4 represent the application of nitrogen at the rate of 0, 200, 250, 300 and 350 kg/ha.

lower O_2^- content compared to the HW irrigation, indicating that N has a better response to ROS under HW irrigation. Similarly, compared to N0, the N1, N2, N3, and N4 treatments decreased the H_2O_2 content by 11%, 27%, 41% and 52% (Figure 4c), respectively, and the O_2^- content by 5%, 11%, 19% and 37% (Figure 4d).

3.8 Antioxidant enzymatic activity

The N application, stages, and irrigation treatments significantly affected the antioxidant enzymatic activity (Figure 5). The results show that the antioxidant enzymatic activities increased with increasing N content under LW and HW irrigation. On the other hand, LW irrigation showed significantly higher SOD, POD, and APX activity than HW irrigation. The activities of SOD, POD, and APX increased from V9 to R3 and then decreased towards R6. Increasing N content significantly increased the activity of the antioxidant enzyme, but N4 treatment had lower activity than N3. In comparison to N0, the N1, N2, N3, and N4 treatments increased SOD activity by 32%, 61%, 86%, and 65%, respectively, and HW decreased SOD activity by 5.7% compared to LW irrigation. The highest POD activity was observed for N3 treatment, which was 39% and 19% higher than the N0 and N4 treatments, respectively. Similarly, the APX activity was increased in the N3 treatment by 159% and 17% compared to the N0 and N4 treatments, respectively. The mean results based on four stages exhibited that the enzymatic activities (SOD, POD, and APX) were decreased in HW by 5.7%, 4.8%, and 12.2% compared to LW irrigation. The higher antioxidant enzymatic activities increase the yield by protecting the photosynthetic system of the maize crop.

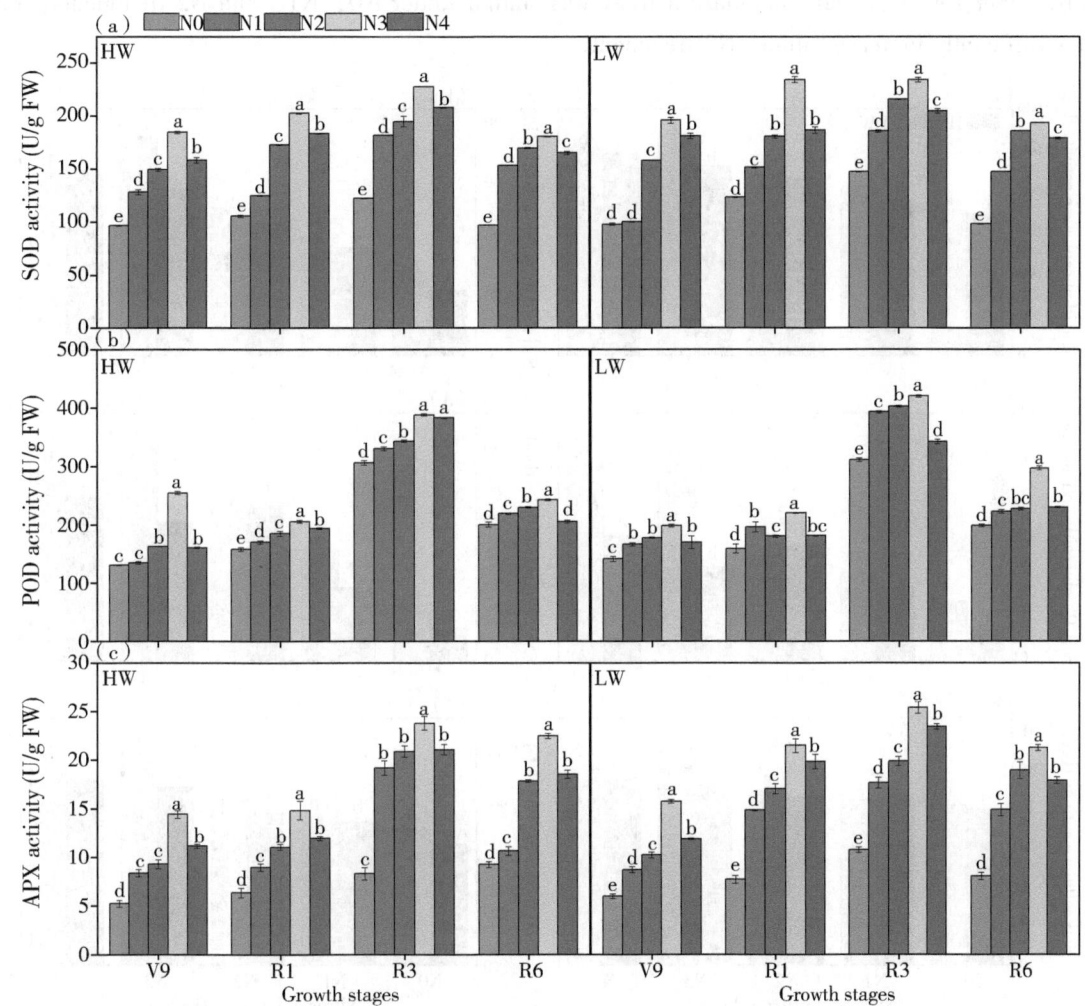

Figure 5　Effects of irrigation and nitrogen rates on SOD activity (a),
POD activity (b), and APX activity (c)

Notes: HW and LW represent irrigation water at the rate of 80% and 60% of field capacity. Means followed by different lowercase letters within each treatment indicate significant differences ($P<0.05$). Bars represents standard errors. The nitrogen treatments N0, N1, N2, N3 and N4 represent the application of nitrogen at the rates of 0, 200, 250, 300 and 350 kg N/ha.

3.9　Soil enzyme activity

As shown in Figure 6a-f, nitrogen fertilization significantly increased the activity of acid phosphatase, acid invertase, and urease, while the N4 treatment (350 kg N/ha) under LW irrigation resulted in lower enzyme activities than the N3 treatment (300 kg N/ha). Similarly, the activities of β-glucoside, catalase, and cellulase were significantly higher for the N3 treatment compared to control or other treatments of the group under the LW irrigation system. In addition, the N4 treatment under the HW irrigation system resulted in higher β-glucoside and urease activity, but β-glucoside and urease activity were not statistically different from N3 and N2 treatment, respectively. Under LW irrigation, cellulase activity was similar under N0, N1, and N2 treatments, but was significantly increased under N3 treatment.

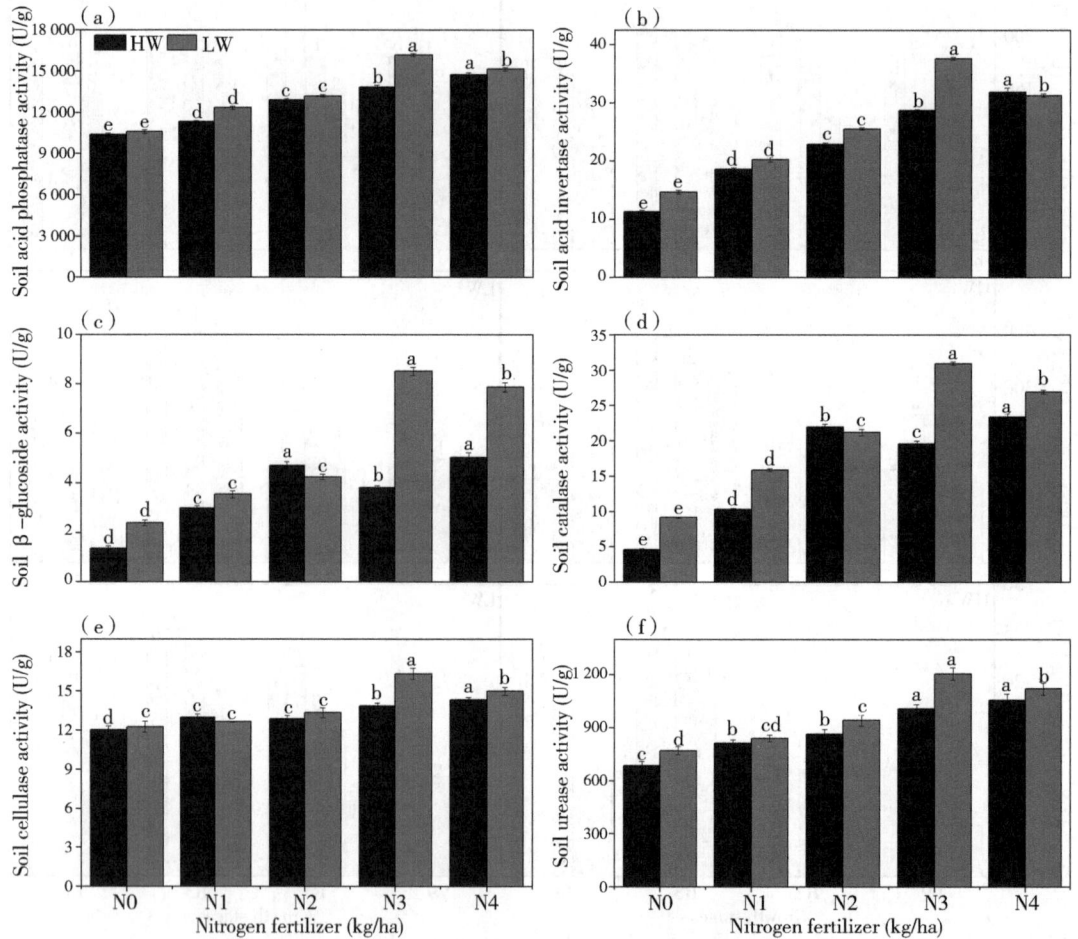

Figure 6 Effects of irrigation and nitrogen rates on soil acid phosphatase (a), soil acid invertase (b), β-glucosides (c), catalase (d), cellulose (e), and soil urease (f)

Notes: HW and LW represent irrigation water at the rate of 80% and 60% of field capacity. Means followed by different lowercase letters within each treatment indicate significant differences ($P<0.05$). Bars represent the standard errors. The nitrogen treatments N0, N1, N2, N3 and N4 represent the application of nitrogen at the rates of 0, 200, 250, 300 and 350 kg/ha.

3.10 Correlation analyses

Leaf enzymes had a positive correlation with kernel per row, kernel per ear, root dry matter, shoot dry matter, plant height, kernels yield, leaf area, root diameter, root surface area, root volume, and total root length. However, MDA, proline H_2O_2 and O_2^- had negative correlations with most of the above parameters. Moreover, a strong positive correlation was observed for chlorophyll a

and *b* content with plant enzymes, dry matter, and yield. However, a slight positive correlation was observed with root diameter, ear diameter, total root length, root surface area, and volume. Additionally, MDA content had a lower negative correlation with almost all of the above parameters, but had a strong negative correlation with root diameter (Figure 7).

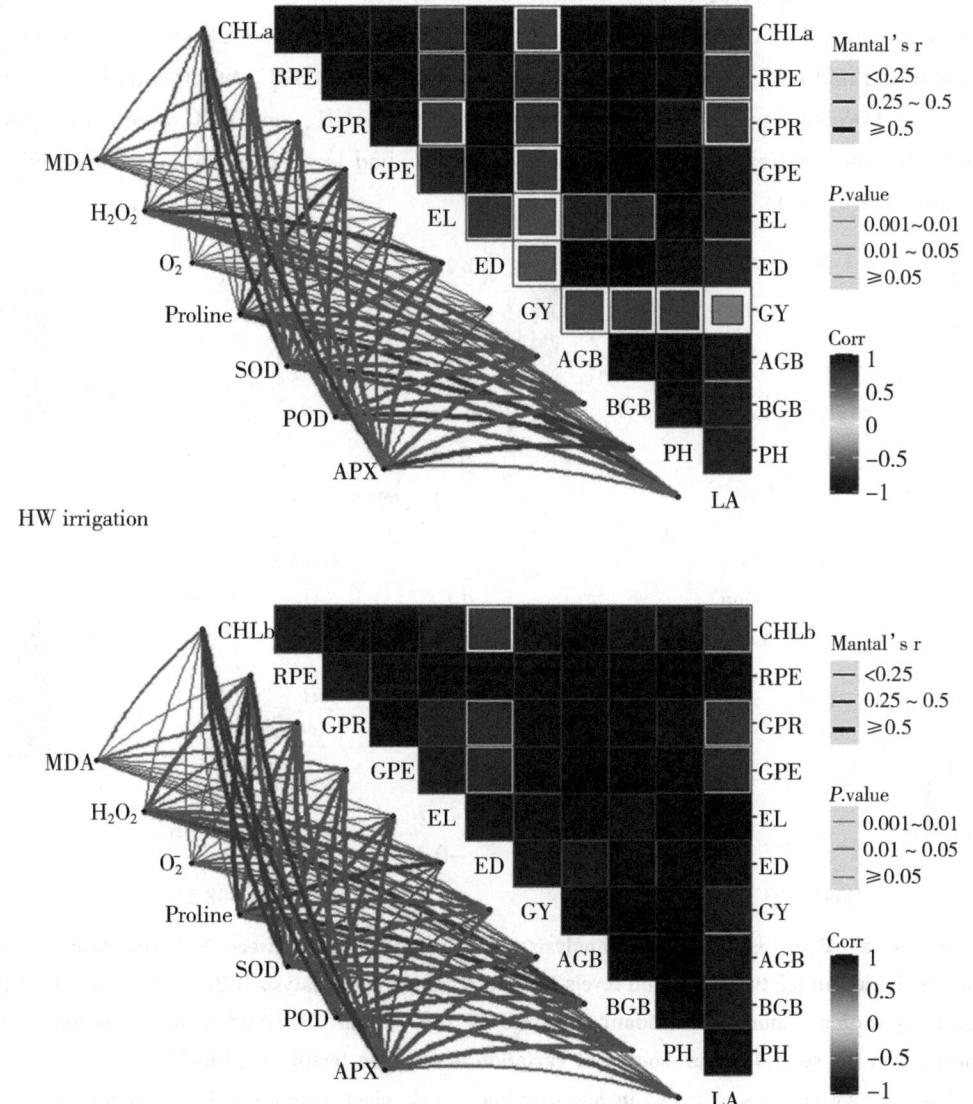

Figure 7 Correlation analysis of the yield, enzymes, chlorophyll, and ROS related parameters
Notes: CHLa, chlorophyll a; CHLb, chlorophyll b; GY, kernel yield; AGB, shoot dry matter, BGB; root dry matter; PH, plant height; LA, leaf area; MDA, malondialdehyde; H_2O_2, hydrogen peroxide; O_2^-, superoxide anion; SOD, superoxide dismutase; POD, peroxidase; and APX, ascorbate peroxidase.

To assess the relationship between N fertilization and enzyme activities, Principal Component Analysis (PCA) was performed (Figure 8). Dim1 separated the plant ROS and antioxidant enzyme activities under both low and high irrigation. These findings reveal that Dim1 contributed

75% and 77.8% of the overall variation under low and high irrigation, respectively (Figure 8a, b). Furthermore, the plant enzyme activities were strongly correlated with N3 treatment, while ROS and proline were correlated to N0 treatment under both irrigation types. However, SOD and POD have a much greater contribution under LW compared to HW irrigation. Similarly, Dim1 showed 93.2% and 98.5% variation in soil enzymes activities under low and high irrigation regimes, respectively (Figure 8c, d). These results suggest that UR, AI, and CL were more correlated to N3 treatment and had a higher contribution compared to ACP and BGC activities under low irrigation (Figure 8c). However, under high irrigation regimes, CL, UR, and BGC were significantly correlated with N3 and N4 treatments, whereas CL and BGC had lower contributions compared to other soil enzymes (Figure 8d).

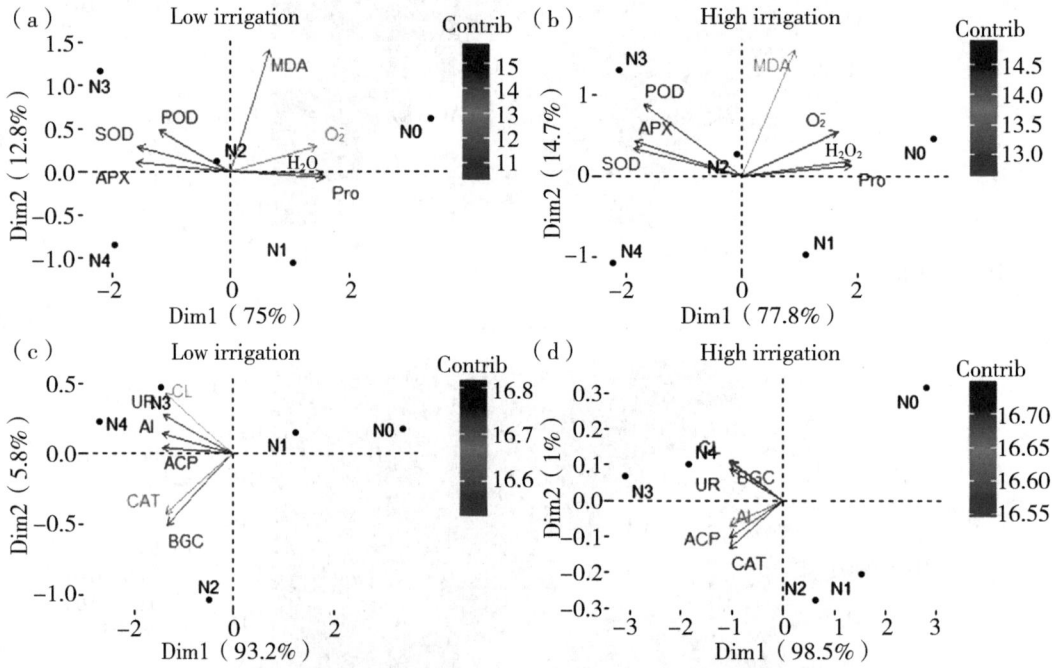

Figure 8 Principal component analysis of the relationships between N fertilization and enzymes activities under two irrigation levels Principal component analysis (PCA) indicates that the largest separation among antioxidant enzymes is the spatial distribution of irrigation water (Dim1) and the second largest source of variation is nitrogen fertilizer (Dim2)

Notes: Plant enzymes activities in low irrigation (a), plant enzymes activities in high irrigation (b), soil enzymes activities in low irrigation (c), and soil enzymes activities in low irrigation (d).

4 Discussion

Nitrogen (N) and water are the major governing factors in agricultural systems worldwide[15]. Nitrogen fertilizer is one of the key elements influencing crop growth. It is severely used in crop production due to its significant crop yield and the demands of the current population[25]. High water and fertilizer inputs are commonly seen to achieve high yields[26]. However, excessive fertilization and poorly planned irrigation systems are common for farmers, and these methods are substantial in causing environmental problems related to soil[27]. The application of N significantly increased the maize kernel yield, as well as root and shoot dry matter under both LW and HW irrigation. The higher yield and yield components under LW irrigation with N application are attributed to the higher chlorophyll content and an enhanced antioxidant enzymatic activity defense system. The higher N content of the LW irrigation system protected the photosynthetic system and enzymatic activity, resulting in higher yields. For better plant adaptation to stress conditions, it is important to maintain a high antioxidant enzymatic activity system[28]. Previous literature reported that N application increased the yield of different crops by reducing the loading rate and enhancing photosynthetic performance[16,29,30]. Ahmad et al.[31] reported that N application increased maize kernel yield with a medium plant population density by reducing leaf senescence at the bottom. In the present study, N3 under LW resulted in a higher kernel and dry matter yield, but N4 had a higher kernel yield under the HW irrigation system, which is not statistically different from the N3 treatment. These findings are consistent with previous findings[32,33], which show that high irrigation water causes oxidative membrane damage in maize crops. The ideal application rate reported is intended to maximize crop yield while conserving resources and the environment[34].

Root morphology is quite flexible; it can readily respond to the available mineral nutrients in the soil[35]. We found that the N3 treatment resulted in a higher root dry weight, implying that the association between the N fertilizer rates and root systems is not linear and positive. In fact, N input may be detrimental to root development and growth. In the current study, N fertilizer application promoted root growth in both the HW and LW irrigation systems and increased the proportion of roots in LW on average compared to HW, indicating that N application has a significant impact on root growth development under well irrigation systems. Root and shoot dry matter and proportion in the HW and LW irrigation systems were negatively affected by the N0 treatment, suggesting that an absence of N will be harmful for early plant growth is described as premature growth. Not only nutrient absorption and root morphology are inextricably linked, but crop development, yield, and crop growth are all also closely connected to the spatial distribution of root systems[36]. The root system is distributed efficiently in N4 under HW irrigation, with a higher root length density, diameter, surface area, and volume, compared to N3 under LW irrigation, resulting in larger and deeper infiltration scales. Nitrogen rate-induced increases in kernel yield are also aided by gentler root senescence in the N3 treatment. These results are supported by[37], which demonstrated the ideal root system distribution in N225 treatment with a greater total root length and surface area, which is the main contributor to the N-induced increase in kernel yield. According to previous studies, the root environment's relative stability is effective in promoting the root system's buffer capacity in detrimental soil environments, as well as resulting in a high kernel yield and utilization efficiency[37,38]. Our results for LW irrigation exhibit that disproportionate N (N4) adversely affects root growth as compared to N3, while in HW irrigation, N4 has better root growth compared to the other N treatments. The observed decrease in crop yield can be attributed to a variety of factors. Excessive N application may cause slight reductions in crop yield due to adverse effects on root development during the early stages of plant growth or different aging mechanisms leading to a relative nitrogen shortage during the reproductive stage[37]. Root thinning and longitudinal expansion are induced by N deficiency, which promotes root development in the soil, whereas high N inhibits root vertical expansion[37,39]. From the perspective of root morphology and development, our findings describe the role of excessive N and its deficiency on maize yield under two different irrigation regimes.

Reduced chlorophyll content and leaf area per plant are closely associated with leaf senescence[31,40], and the degradation of chlorophyll reduces photosynthetic efficiency[41]. For a high-density plant population, it is critical to reduce maize's accelerated leaf senescence and protect the photosynthetic apparatus[31]. Our results show that the chlorophyll a and b contents were significantly higher at V9, and the lowest value was observed at R6 under both HW and LW irrigation. The maximum chlorophyll a and b contents were observed for N3 treatment at all stages under LW irrigation compared to other N treatments and the control. Recently published studies have reported that the decrease in chlorophyll content in the later stages is due to an increase in leaf senescence in older plants[31,42]. The application of N with LW irrigation increased leaf area per plant and chlorophyll a and b content, suggesting that N had a crucial effect on enhancing the leaf area, plant growth and photosynthetic efficiency. The leaves stopped growing at the silking stage and became senescent as the plant grew older. The loss of leaf greenness due to chlorophyll loss, a result of chloroplast degradation, was the first symptom of senescence[43,44]. The primary element that maintains leaf photosynthesis is the leaf chlorophyll content[45], with leaves providing up to 50-80 percent of the photosynthesis required by kernels[43]. Lower dry matter and kernel yield in HW irrigation are primarily caused by a reduced chlorophyll content and enzymatic activities. This may be due to drought damage to the chloroplast structure[46], and a disorder of the N metabolism and down-regulation of the enzymatic activities[47,48]. All these adverse changes reduced plant photosynthesis and induced a decrease in dry matter accumulation, which ultimately decreased kernel yield[43,49].

Senescence increases the MDA content in leaves, which can be harmful to plant growth[50]. Our results show that MDA was lower at V9 and increased gradually with leaf senescence. The N-treated plants resulted in a lower MDA content in both HW and LW irrigation compared to N0. Averaged across irrigation, a lower MDA content was obtained in all growth stages under N4, and the maximum was found under N0 treatment. In addition, MDA accumulation has been found to be increased during the leaf senescence process in a variety of crops[41]. The MDA content is commonly used to assess lipid peroxidation. The MDA content was higher in the HW treatment than in the LW treatment in this study (Figure 4a), implying that the leaf cell membranes were damaged as a result of the high irrigation. Furthermore, MDA content was lower in N4 under HW than at other nitrogen application rates, implying that increasing nitrogen fertilizer supply rates (up to 350 kg N/ha) could reduce lipid peroxidation in maize leaves caused by high irrigation. Tian et al.[33] reported that the greatest reduction in lipid peroxidation in leaves was observed with N4, implying that a high nitrogen fertilizer supply can aid in the recovery of lipid peroxidation in leaves following high irrigation.

Under both HW and LW irrigation, proline accumulation was highest in N0-treated plants at all growth stages compared to other N-treated plants, while the lowest value was observed in N4 (Figure 4b). Plants treated with N have a reduced MDA content. This might be due to the accumulation of nitrogen-containing compounds (i. e., proline), which play an adaptive role by helping to stabilize sub-cellular structures, scavenge free radicals, and buffer cellular redox potential in stressful situations[51]. Previous studies have reported that proline metabolism has a significant impact on cellular redox potential, which could be important for stress tolerance signaling. Additionally, proline reduces equivalent N and carbon dioxide as an osmolyte and antioxidant as well as a source of energy[52].

In this study, we found that N application decreased the H_2O_2 and O_2^- contents in both HW and LW irrigation, indicating that N may be beneficial in maintaining aquaporin activity by reducing H_2O_2 accumulation. The maximum H_2O_2 and O_2^- contents were observed at R6 in N0 under HW irrigation, but the minimum was observed at V9 in N4 under LW irrigation, suggesting that high irrigation water may have caused damage to the leaf cell membranes. H_2O_2 is a ROS that is produced by cellular metabolism and is a measure of a plant's ability to scavenge ROS under stress. Previous research has largely demonstrated that the accumulation of H_2O_2 can lead to a significant reduction in aquaporin activity[53,54]. The key enzymes in the ROS scavenging system are SOD and POD, where SOD catalyzes the disproportionation of O_2^-, while POD metabolizes H_2O_2[33]. For plants to adapt to hypoxic stress, it is critical to maintain high antioxidant enzyme activity[28]. It has been shown that increasing nitrogen supply rates can reduce H_2O_2 and O_2^- accumulation in maize leaves, reducing high-water-stress-induced oxidative membrane damage[33]. Waterlogging causes an increase in ROS production in plants, which affects cell membrane stability through lipid peroxidation[33]. Furthermore, the MDA content was lower in N3 and N4 under HW than at other nitrogen application rates, implying that increasing nitrogen fertilizer supply rates could reduce lipid peroxidation in maize leaves caused by high irrigation.

Antioxidant enzyme activity regulation is an innate plant response to prevent oxidative stress caused by a variety of external biotic and abiotic stress factors[55,56]. In the present study, a significant treatment effect on the antioxidant defenses in maize leaves was detected. Results show that in HW irrigation, there were lower SOD, POD, and APX activities at each N treatment compared to LW (Figure 5). The maximum SOD, POD, and APX contents were observed in N3 under LW irrigation at R3, whereas the minimum activities were observed in N0 and N4 under HW. The optimum higher N3 and LW irrigation increased the enzymatic activity contact with improved plant growth, root development, and yield, according to the correlations between the parameters investigated (Figure 7). According to previously published findings, waterlogging damages the membrane of maize[55]. Furthermore, increasing N fertilizer application in winter rape can improve plant enzyme activities in leaves[28]. Similarly,[57] reported that antioxidant enzyme activities (SOD and POD) were increased at higher N rates. Ahmad et al.[31] reported that N application significantly increased antioxidant enzyme activities and reduced leaf senescence in maize leaves up to the optimum level of N, and that a further increase in N had an adverse effect on antioxidant enzyme activities. The application of N has been shown to have a significant positive effect on SOD and POD synthesis[58]. Furthermore, under high irrigation water stress, appropriate nitrogen fertilizer application can help induce the expression of related antioxidant enzyme genes, resulting in increased antioxidant enzyme activity[59]. These findings suggest that increasing nitrogen supply rates can reduce H_2O_2 and O_2^- accumulation in maize leaves, reducing waterlogging-induced oxidative membrane damage.

Soil enzymes are commonly utilized as markers of soil quality because of their relationship to soil biology, their accessibility, and their quick reaction to alterations in soil management[60]. Excessive amounts of nitrogen can cause the buildup of harmful compounds such as ammonia, which harms plants and hamstrings microbial growth, as well as lower soil pH, which inhibits enzyme activity[61,62]. Similarly, a recent study reported that high doses of mineral nitrogen fertilizer can cause a considerable decrease in enzyme activity[62]. Compared to a control, the application of medium and high doses of N into the soil significantly increased urease activity by 42.9% and 23.6%, respectively[62]. In the current study, LW irrigation, together with appropriate N fertilizer (300 kg N/ha) input, significantly increased soil nutrient availability and provided a suitable environment for soil microorganisms, which resulted in higher enzyme activities[60] (Figure 6). Under LW irrigation, all enzymatic activities were enhanced by N3 treatment, although N3 is statistically similar to N4 under HW irrigation, which is consistent with Zhou et al.[60] and Pathan et al.[63]. Acid phosphatase, acid invertase, β-glucoside, catalase, cellulase, and urease enzymatic activity were markedly affected by irrigation and nitrogen application, and all treatments showed great variability in their enzyme activities. This is mostly because of the usage of urea fertilizer, which supplies a substrate for the urease reaction[64]. Interestingly, the activity of cellulase showed slight variations with irrigation, and this may be because the substrate-binding portion of the enzyme developed resistance to fertilizer anions[5].

5　Conclusions

Low irrigation water (LW) resulted in higher plant height, leaf area, root and shoot dry matter, SOD, POD, APX in leaves, kernels per row, kernels per ear, root length, root volume, root diameter and kernel yield of hybrid maize plants. Related to yield, antioxidant enzymes, soil enzymes, and root parameters also improved with increased N application rates up to 300 kg N/ha under LW irrigation. However, MDA, proline, H_2O_2 and O_2^- contents in leaves showed the opposite response to N application rates. Similarly, SOD, POD, and APX activities were associated with a relatively higher N content in the soil. Thus, increasing application rate of N (up to 300 kg N/ha) improved leaf-physiological characteristics and consequently produced a considerable maize yield under LW irrigation. This knowledge can be applied to need-based N applications, reducing potential N loss and non-point pollution. More research is needed to determine the impact of N fertilizer rates and irrigation on N uptake, kernel quality, and the leachate of N to groundwater.

Acknowledgments

This study was financially supported by the National Natural Science Foundation of China (31760354), and the Natural Science Foundation of Guangxi Province (2019GXNSFAA185028). The authors extend their appreciation to the researchers supporting project number

(RSP2022R483), King Saud University, Riyadh, Saudi Arabia. The authors express their special gratitude to Xun-Bo Zhou and Ahmad Khan for strong support revising the final version of the manuscript; we are also thankful to all the funding sources and especially to Guangxi University for the financial assistance.

References

[1] Chi YX, Gao F, Muhammad I, et al. Effect of water conditions and nitrogen application on maize growth, carbon accumulation and metabolism of maize plant in subtropical regions. *Archiv. Agron. Soil Sci.* 2022, https://doi.org/10.1080/03650340.2022.2026931.

[2] Khan A, Zahir Afridi M, Airf M, et al. A sustainable approach toward maize production: effectiveness of farm yard manure and urea N. *Ann. Biol. Sci.* 2017, 5: 7-13.

[3] Hou P, Gao Q, Xie R, et al. Grain yields in relation to N requirement: Optimizing nitrogen management for spring maize grown in China. *Field Crops Res.* 2012, 129: 1-6.

[4] Wang GY, Hu YX, Liu YX, et al. Effects of supplement irrigation and nitrogen application levels on soil carbon-nitrogen content and yield of one-year double cropping maize in subtropical region. *Water.* 2021, 13: 1180.

[5] Jia X, Shao L, Liu P, et al. Effect of different nitrogen and irrigation treatments on yield and nitrate leaching of summer maize (*Zea mays* L.) under lysimeter conditions. *Agric. Water Manag.* 2014, 137: 92-103.

[6] Muhammad I, Khan F, Khan A, et al. Soil fertility in response to urea and farmyard manure incorporation under different tillage systems in Peshawar, Pakistan. *Int. J. Agric. Biol.* 2018, 20: 1539-1547.

[7] Zhou XB, Yang L, Wang GY, et al. Effect of deficit irrigation scheduling and planting pattern on leaf water status and radiation use efficiency of winter wheat. *J. Agron. Crop. Sci.* 2021, 207: 437-449.

[8] Mansouri-Far C, Sanavy SAMM, Saberali SF. Maize yield response to deficit irrigation during low-sensitive growth stages and nitrogen rate under semi-arid climatic conditions. *Agric. Water Manag.* 2010, 97: 12-22.

[9] Soler C, Hoogenboom G, Sentelhas P, et al. Impact of water stress on maize grown off-season in a subtropical environment. *J. Agron. Crop Sci.* 2007, 193: 247-261.

[10] Meng Y, Liu X, Gu W, et al. Effects of a chemical plant growth regulator and planting density on the leaf senescence and yield of spring maize in northeast china. *Appl. Ecol. Environ. Res.* 2020, 18: 3297-3311.

[11] Pandey R, Maranville J, Admou A. Deficit irrigation and nitrogen effects on maize in a Sahelian environment: I. Grain yield and yield components. *Agric. Water Manag.* 2000, 46: 1-13.

[12] Moser SB, Feil B, Jampatong S, et al. Effects of pre-anthesis drought, nitrogen fertilizer rate, and variety on grain yield, yield components, and harvest index of tropical maize. *Agric. Water Manag.* 2006, 81: 41-58.

[13] Subedi K, Ma B. Nitrogen uptake and partitioning in stay-green and leafy maize hybrids. *Crop Sci.* 2005: 45740-45747.

[14] Binder DL, Sander DH, Walters DT. Maize response to time of nitrogen application as affected by level of nitrogen deficiency. *Agron. J.* 2000, 92: 1228-1236.

[15] Gheysari M, Mirlatifi SM, Homaee M, et al. Nitrate leaching in a silage maize field under different irrigation and nitrogen fertilizer rates. *Agric. Water Manag.* 2009, 96: 946-954.

[16] Gholamhoseini M, AghaAlikhani M, Sanavy SM, et al. Interactions of irrigation, weed and nitrogen on corn yield, nitrogen use efficiency and nitrate leaching. *Agric. Water Manag.* 2013, 126: 9-18.

[17] Nyfeler D, Huguenin-Elie O, Suter M, et al. Strong mixture effects among four species in fertilized agricultural grassland led to persistent and consistent transgressive overyielding. *J. Appl. Ecol.* 2009, 46: 683-691.

[18] Kenkel P, Fitzwater B. Causes of Fertilizer Price Volatility. Oklahoma State University. Oklahoma Cooperative Extension Service. AGEC-261. Available online: http://articles.extension.org/pages/72692/causes-of-fertilizer-price-volatility. 2009.

[19] Brown B, Hart J, Horneck D, et al. Nutrient management for field corn silage and grain in the Inland Pacific Northwest. *University of Idaho*. 2010, 9.

[20] Hu H, Ning T, Li Z, et al. Coupling effects of urea types and subsoiling on nitrogen-water use and yield of different varieties of maize in northern China. *Field Crops Res.* 2013, 142: 85-94.

[21] Berenguer P, Santiveri F, Boixadera J, et al. Nitrogen fertilisation of irrigated maize under Mediterranean conditions. *Eur. J. Agron.* 2009, 30: 163-171.

[22] Zhou XB, Wang GY, Yang L, et al. Double-double row planting mode at deficit irrigation regime increases winter wheat yield and water use efficiency in North China Plain. *Agronomy.* 2020, 10: 1315.

[23] Akmal M, Janssens M. Productivity and light use efficiency of perennial ryegrass with contrasting water and nitrogen supplies. *Field Crops Res.* 2004, 88: 143-155.

[24] Nilahyane A, Islam MA, O Mesbah A, et al. Evaluation of silage corn yield gap: An approach for sustainable production in the semi-arid region of USA. *Sustainability.* 2018, 10: 2523.

[25] Huang J, Xu CC, Ridoutt BG, et al. Nitrogen and phosphorus losses and eutrophication potential associated with fertilizer application to cropland in China. *J. Cleaner Product.* 2017, 159: 171-179.

[26] Wang X, Fan J, Xing Y, et al. The effects of mulch and nitrogen fertilizer on the soil environment of crop plants. *Adv. Agron.* 2019, 153: 121-173.

[27] Mack UD, Feger KH, Gong Y, et al. Soil water balance and nitrate leaching in winter wheat-summer maize double-cropping systems with different irrigation and N fertilization in the North China Plain. *J. Plant Nutr. Soil Sci.* 2005, 168: 454-460.

[28] Chen H, Chen S, Zheng S, et al. Regulation effects of adding nitrogen on physiologi-

cal properties and yield of rapeseed after waterlogging during seedling. *Soil.* 2017, 49: 519-526.

[29] Chen ZK, Tao XP, Khan A, et al. Biomass accumulation, photosynthetic traits and root development of cotton as affected by irrigation and nitrogen - fertilization. *Front. Plant Sci.* 2018, 9: 00173.

[30] Su W, Kamran M, Xie J, et al. Shoot and root traits of summer maize hybrid varieties with higher grain yields and higher nitrogen use efficiency at low nitrogen application rates. *Peer J.* 2019, 7: e7294.

[31] Ahmad I, Ahmad S, Kamran M, et al. Uniconazole and nitrogen fertilization trigger photosynthesis and chlorophyll fluorescence, and delay leaf senescence in maize at a high population density. *Photosynthetica.* 2021, 59: 192-202.

[32] Liu H, Song FB, Liu SQ, et al. Physiological response of maize and soybean to partial root-zone drying irrigation under N fertilization levels. *Emir. J. Food Agr.* 2018, 30: 364-371.

[33] Tian G, Qi D, Zhu J, et al. Effects of nitrogen fertilizer rates and waterlogging on leaf physiological characteristics and grain yield of maize. *Archiv. Agron. Soil Sci.* 2021, 67: 863-875.

[34] Lamptey S, Li L, Xie J, et al. Photosynthetic response of maize to nitrogen fertilization in the semiarid western loess plateau of China. *Crop Sci.* 2017, 57: 2739-2752.

[35] Yu P, White PJ, Hochholdinger F, et al. Phenotypic plasticity of the maize root system in response to heterogeneous nitrogen availability. *Planta.* 2014, 240: 667-678.

[36] Lynch JP. Steep, cheap and deep: An ideotype to optimize water and N acquisition by maize root systems. *Ann. Botany.* 2013, 112: 347-357.

[37] Su W, Ahmad S, Ahmad, et al. Nitrogen fertilization affects maize grain yield through regulating nitrogen uptake, radiation and water use efficiency, photosynthesis and root distribution. *Peer J.* 2020, 8: e10291.

[38] Saengwilai P, Nord EA, Chimungu JG, et al. Root cortical aerenchyma enhances nitrogen acquisition from low - nitrogen soils in maize. *Plant Physiol.* 2014, 166: 726-735.

[39] Mu X, Chen F, Wu Q, et al. Genetic improvement of root growth increases maize yield via enhanced post-silking nitrogen uptake. *Eur. J. Agron.* 2015, 63: 55-61.

[40] Wang X, Yang W, Chen G, et al. Effects of spraying uniconazole on leaf senescence and yield of maize at late growth stage. *J. Maize Sci.* 2009, 17: 86-88.

[41] Yong CW, Wan RG, Le FY, et al. Physiological mechanisms of delaying leaf senescence in maize treated with compound mixtures of DCPTA and CCC. *J. Northeast Agric. Univ.* 2015, 22: 1-15.

[42] Ahmad I, Kamran M, Su W, et al. Application of uniconazole improves photosynthetic efficiency of maize by enhancing the antioxidant defense mechanism and delaying leaf senescence in semiarid regions. *J. Plant Growth Regul.* 2019, 38:

855-869.

[43] Ye YX, Wen ZR, Huan Y, et al. Effects of post-silking water deficit on the leaf photosynthesis and senescence of waxy maize. *J. Integr. Agric.* 2020, 19: 2216-2228.

[44] He P, Osaki M, Takebe M, et al. Endogenous hormones and expression of senescence-related genes in different senescent types of maize. *J. Exp. Bot.* 2005, 56: 1117-1128.

[45] Cairns JE, Sonder K, Zaidi P, et al. Maize production in a changing climate: Impacts, adaptation, and mitigation strategies. *Adv. Agron.* 2012, 114: 1-58.

[46] Ahmad S, Wang GY, Muhammad I, et al. Interactive effects of melatonin and nitrogen improve drought tolerance of maize seedlings by regulating growth and physiochemical attributes. *Antioxidants.* 2022, 11: 359.

[47] Markelz RC, Strellner RS, Leakey AD. Impairment of C_4 photosynthesis by drought is exacerbated by limiting nitrogen and ameliorated by elevated [CO_2] in maize. *J. Exp. Botany.* 2011, 62: 3235-3246.

[48] Zong YZ, Shangguan ZP. Nitrogen deficiency limited the improvement of photosynthesis in maize by elevated CO_2 under drought. *J. Integrative Agric.* 2014, 13: 73-81.

[49] Perdomo JA, Capó-Bauçà S, Carmo-Silva E, et al. Rubisco and rubisco activase play an important role in the biochemical limitations of photosynthesis in rice, wheat, and maize under high temperature and water deficit. *Front. Plant Sci.* 2017, 8: 490.

[50] Zhang YJ, Zhang X, Chen CJ, et al. Effects of fungicides JS399-19, azoxystrobin, tebuconazloe, and carbendazim on the physiological and biochemical indices and grain yield of winter wheat. *Pest. Biochem. Physiol.* 2010, 98: 151-157.

[51] Khan MN, Siddiqui MH, Mohammad F, et al. Salinity induced changes in growth, enzyme activities, photosynthesis, proline accumulation and yield in linseed genotypes. *World J. Agric. Sci.* 2007, 3: 685-695.

[52] Siddiqui MH, Mohammad F, Khan MN, et al. Nitrogen in relation to photosynthetic capacity and accumulation of osmoprotectant and nutrients in Brassica genotypes grown under salt stress. *Agric. Sci. China.* 2010, 9: 671-680.

[53] Ding L, Gao C, Li Y, et al. The enhanced drought tolerance of rice plants under ammonium is related to aquaporin (AQP). *Plant Sci.* 2015, 234: 14-21.

[54] Qiao Y, Ren J, Yin L, et al. Exogenous melatonin alleviates PEG-induced short-term water deficiency in maize by increasing hydraulic conductance. *BMC Plant Biol.* 2020, 20: 1-14.

[55] Zhang Y, Yu XX, Zhang WJ, et al. Interactions between endophytes and plants: beneficial effect of endophytes to ameliorate biotic and abiotic stresses in plants. *J. Plant Biol.* 2019, 62: 1-13.

[56] Muhammad I, Yang L, Ahmad S, et al. Melatonin application alleviates stress-induced photosynthetic inhibition and oxidative damage by regulating antioxidant defense system of maize: a meta-analysis. *Antioxidants.* 2022, 11: 512.

[57] Gup W, Chen B, Liu R, et al. Effects of nitrogen application rate on cotton leaf antioxidant enzyme activities and endogenous hormone contents under short-term waterlogging at flowering and boll-forming stage. *Yingyong Shengtai Xuebao*. 2010, 21: 53-60.

[58] Zhang LX, Li SX. Effects of nitrogen, potassium and glycinebetaine on the lipid peroxidation and protective enzyme activities in water-stressed summer maize. *Acta Agron. Sin.* 2007, 33: 482-490. (in Chinese with English abstract)

[59] Özçubukçu S, Ergün N, Ilhan E. Waterlogging and nitric oxide induce gene expression and increase antioxidant enzyme activity in wheat (*Triticum aestivum* L.). *Acta Biol. Hung.* 2014, 65: 47-60.

[60] Zhou SM, Zhang M, Zhang KK, et al. Effects of reduced nitrogen and suitable soil moisture on wheat (*Triticum aestivum* L.) rhizosphere soil microbiological, biochemical properties and yield in the Huanghuai Plain, China. *J. Integr. Agric.* 2020, 19: 234-250.

[61] Muhammad I, Yang L, Ahmad S, et al. Irrigation and nitrogen fertilization alter soil bacterial communities, soil enzyme activities, and nutrient availability in maize crop. *Front. Microbiol.* 2022, 3: 105.

[62] Sawicka B, Krochmal-Marczak B, Pszczółkowski P, et al. Effect of differentiated nitrogen fertilization on the enzymatic activity of the soil for sweet potato (*Ipomoea batatas* L. [Lam.]) cultivation. *Agronomy*. 2020, 10: 1970.

[63] Pathan SI, Ceccherini MT, Pietramellara G, et al. Enzyme activity and microbial community structure in the rhizosphere of two maize lines differing in N use efficiency. *Plant Soil.* 2015, 387: 413-424.

[64] Xing S, Chen C, Zhou B, et al. Soil soluble organic nitrogen and active microbial characteristics under adjacent coniferous and broadleaf plantation forests. *J. Soils Sedim.* 2010, 10: 748-757.

[65] Gloser V, Zwieniecki MA, Orians CM, et al. Dynamic changes in root hydraulic properties in response to nitrate availability. *J. Exp. Bot.* 2007, 58: 2409-2415.

[66] MacAdam JW, Nelson CJ, Sharp RE. Peroxidase activity in the leaf elongation zone of tall fescue: I. Spatial distribution of ionically bound peroxidase activity in genotypes differing in length of the elongation zone. *Plant Physiol.* 1992, 99: 872-878.

[67] Nakano Y, Asada K. Hydrogen peroxide is scavenged by ascorbate-specific peroxidase in spinach chloroplasts. *Plant Cell Physiol.* 1981, 22: 867-880.

[68] Arnon DI. Copper enzymes in isolated chloroplasts. Polyphenoloxidase in Beta vulgaris. *Plant Physiol.* 1949, 24: 1.

[69] Schneider K, Schlegel H. Production of superoxide radicals by soluble hydrogenase from Alcaligenes eutrophus H16. *Biochem. J.* 1981, 193: 99-107.

[70] Ohto MA, Onai K, Furukawa Y, et al. Effects of sugar on vegetative development and floral transition in Arabidopsis. *Plant Physiol.* 2001, 127: 252-261.

[71] Weisany W, Sohrabi Y, Heidari G, et al. Changes in antioxidant enzymes activity

and plant performance by salinity stress and zinc application in soybean (*Glycine max* L.). *Plant Omics*. 2012, 5: 60.

[72] Bates LS, Waldren RP, Teare I. Rapid determination of free proline for water-stress studies. *Plant Soil*. 1973, 39: 205-207.

Effect of Previous Crop Nitrogen Application on Yield of Following Maize Under Different Planting Patterns

Y. X. Zhao, X. M. Mao, J. H. Huang, D. H. Jiang and X. B. Zhou

(Agricultural College of Guangxi University, Nanning 530004, China)

Abstract: Nitrogen application could affect the crop growth and yield, and then affect the water use efficiency (WUE). This study aimed to determine the effects of previous winter wheat (*Triticum aestivum* L.) nitrogen application and following summer maize (*Zea mays* L.) planting pattern on the yield and WUE of maize in the North China Plain. The experiments consisted of the winter wheat 112.5 (N1) and 225.0 (N2) kg/ha nitrogen application, and summer maize flat planting (FP) and ridge tillage planting (RTP) treatments arranged in a split-plot design (4 m × 4 m) with three replications in 2014 and 2015. Results showed that planting pattern and previous crop nitrogen had significant effect on yield components and WUE. Compared with FP, the RTP increased leaf relative water content, soil water content, soil water storage, WUE, and yield by 1.9%, 2.8%, 2.0%, 3.8%, and 7.1%, respectively. The yield, harvest index, stem diameter, and ear diameter of N2 treatment were 13.0%, 11.9%, 5.5%, and 2.3% higher than those of N1, respectively. Nitrogen of the previous winter wheat and the RTP pattern improved the water status and yield component of summer maize. It may be concluded previous crop nitrogen and RTP pattern can improve population structure, increase the yield and WUE of summer maize and thus is a promising method for farmers in North China.

Keywords:

1 Introduction

Winter wheat and summer maize are the main rotation crops in the North China Plain, with farmers often ignoring the previous crop residue fertilizer. The nitrogen application influences the growth and yield of maize (Javeed et al., 2013); however, the application of excess nitrogen can effect on environment, particularly for soil and groundwater (Shi et al., 2016). The previous crop residue nitrogen is becoming more and more important problem. Some studies have shown that planting pattern influences conserving water, fertilizer use efficiency, water use efficiency (WUE), and grain yield (Mao et al., 2017). Gheysari et al. (2009) provide a effective method to control NO_3^- leaching out of the root zone by a proper combination of irrigation and fertilizer management.

Author for correspondence: whyzxb@gmail.com

The level of NO_3-N leaching may be minimized during agricultural practices by combining irrigation and fertilizer management (Jia et al., 2014).

The double cropping system may significantly increase nitrogen use efficiency (Hartmann et al., 2015). However, soil and ground water are affected by excess application of nitrogen in the North China Plain, with approximately 70% of the applied nitrogen accumulated in 0-500 cm soil layer (Li et al., 2016). Cultivation problems were encountered with excessive nitrogen fertilizer application in the summer maize (Jin et al., 2012). Appropriate management was substantial to maximize nitrogen efficiency (Zhang et al., 2016).

Planting patterns and nitrogen application rate significantly affected the yield component factors (Vos et al., 2005; Wang et al., 2015). Zhang et al. (2007) showed that the yield and the WUE of furrow-irrigated raised bed-planting and mulched ridge and furrow planting were higher than that of conventional flat planting. In the wheat-maize crop rotation system, the irrigation combination with straw mulch or straw-mulched furrows practices increased crop yield and WUE (Ma et al., 2017). The objective of this study was to determine the effects of previous crop nitrogen level and later crop planting patterns on WUE and yield of the summer maize.

2 Materials and Methods

The study was conducted at the Agronomy Experimental Station of Shandong Agricultural University, Taian, Shandong Province, China (36°09′ N, 117°09′ E) during 2014 and 2015. The field test conditions were sandy soil and the average nutrients of the experimental field (at a depth of 0 cm to 20 cm) were tested. The soil physical and chemical properties of the experimental field are shown in Table 1.

Table 1 The soil physical and chemical properties of the experimental field

pH	Total N (mg/kg)	A-P (mg/kg)	A-K (mg/kg)	Soil bulk density (g/cm^3)	Soil organic matter (g/kg)	Field capacity (V%)
6.9	123.2	40.6	124.5	1.50	18.9	38.6

The region has a warm and semi-humid continental monsoon-type climate. At the experimental field, the precipitation averages of the 2014 and 2015 growth seasons were 372.5 mm and 282.6 mm, respectively, and the annual temperature mean was 25.0℃ (1971-2015). Approximately 70% to 80% of the precipitation occurred from July to September (Table 2).

Table 2 The rainfall and temperature of the experimental field in 1971-2015

Months	Jan.	Feb.	Mar.	Apr.	May	June	July	Aug.	Sept.	Oct.	Nov.	Dec.
Rainfall (mm)	5.4	9.9	15.8	30.0	52.3	85.3	209.3	147.4	69.6	33.7	19.3	8.0
Temperature (℃)	-1.7	1.1	7.2	14.3	19.8	24.7	26.3	25.2	20.5	14.4	6.6	0.1

The experiment was a split-plot design with summer maize planting patterns (main plot): flat pattern (FP) and ridge tillage pattern (RTP), previous winter wheat nitrogen levels (split plot): 112.5 kg/ha (N1) and 225.0 kg/ha (N2) (Figure 1). Each plot was 60 cm row spacing, 4 m × 4 m in size, three replications, and the field plots were superimposed on the same position each year. The planting density of the previous winter wheat (var. Jimai 22) was 200×10^4/ha on 9 October 2013 and 8 October 2014, and was harvested on 5 June 2014 and 7 June 2015. Irrigation amount was 200 mm during winter wheat growth season. After harvesting winter wheat, the following summer maize (var. Zhengdan 958) was planted with no tillage with a planting density of 62,500/ha on 15 June 2014 and 15 June 2015, and was harvested on 27 September 2014 and 30 September 2015. The summer maize growing period did not undergo fertilization and irrigation, and used hand weeding in managing weeds.

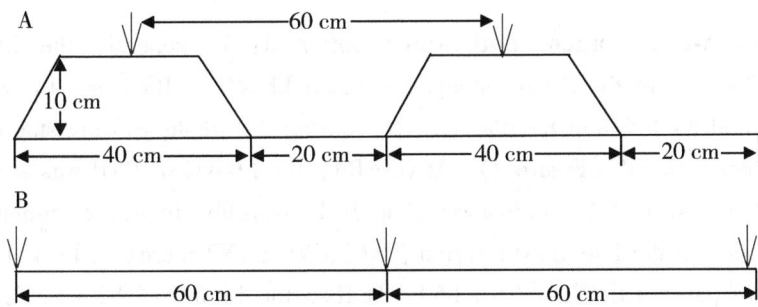

Figure 1 The summer maize ridge tillage planting (A) and flat planting (B) pattern

Leaf relative water content (LRWC) was monitored at V6 (6 leaves fully emerged), R0 (silking), R2 (blister), R3 (milking), and R4 (dough) (Zadoks et al., 1974). Five ear leaves were measured for each treatment.

$$LRWC = (FW-DW)/(SFW-DW) \times 100\% \tag{1}$$

where FW is the fresh weight, DW is dry weight, and SFW is the saturated fresh weight (Galmés et al., 2007).

The water potential (Ψw) of the leaves was sampled and measured with Psypro Water Potential System (Wescor, Inc., Logan, UT) at VE (emergence), V6, R0, R2, R3, R4, and R5 (dent). Before the test, the measured leaves were wiped clean and dried. Samples were obtained using a 0.6 cm diameter round puncher, placed into the sample room, waiting approximately 15 min before reading.

The soil water content (SWC, v/v) was monitored at VE, V6, R0, R2, R3, R4, and R5, determined at 0 cm to 120 cm depths by neutron moisture meter (CNC503B, Super Energy Nuclear Technology, Ltd., Beijing, China). All levels were 10 cm, with a total of 12 levels.

$$S \text{ (mm)} = \sum (\Delta\theta_i \times Z_i) \tag{2}$$

where S is the soil water storage (SWS) (mm), $\Delta\theta_i$ is the volumetric water content of a certain level of soil, Z_i is the depth of the soil layer (mm).

The evapotranspiration (ETa) was computed using climate data obtained from the Taian Agrometeorological Experimental Station with the following equations:

$$ETa = \Delta W + I + P \tag{3}$$

where ETa is the total amount of seasonal evapotranspiration (mm), ΔW is the change in the stored soil water (mm, Mao et al., 2017), I is the irrigation amount (mm), P is the rainfall (mm). Basing on the observations for the summer maize growing seasons, the researchers found that the surface run-off was negligible.

$$WUE = Y/Eta \tag{4}$$

where Y is the grain yield (kg/ha).

The test data were analyzed with SAS9.2, and SigmaPlot10.0 (SPSS Inc., Chicago, IL) was used for drawing.

3 Results

Leaf relative water content and water potential: In general, the LRWC decreased gradually from R2 to R4 in the 2 year study. The mean LRWC of RTP and FP were 87.0% and 84.5% (2014) and 84.8% and 83.6% (2015) during the whole growing stages, respectively, and RTP was higher than FP (Figure 2). At R3-R4, the LRWC of RTP was significantly higher than that of FP ($P<0.05$), indicating that RTP possibly improved summer maize plant physiological function in the late growth period. The LRWC of N2 increased by 1.1% (2014) and 0.8% (2015) compared with N1 ($P>0.05$). At R4, the LRWC of N2 was significantly higher than that of N1 ($P<0.05$). Thus, nitrogen application to winter wheat was beneficial to improve the LRWC of summer maize.

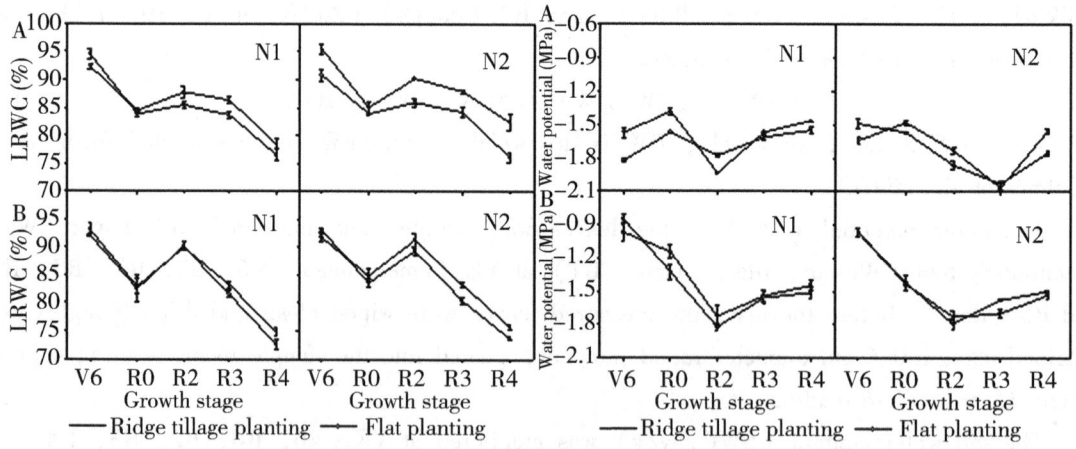

Figure 2 Effects of planting pattern and previous crop nitrogen on leaf relative water content (LRWC) and water potential in 2014 (A) and 2015 (B). The bars are the SE

The Ψw decreased rapidly at R2, but at R3 for N2 in 2014. The Ψw averages of N1 and N2 were -1.62 MPa and -1.72 MPa in 2014, and -1.39 MPa and -1.46 MPa in 2015, respectively, with a significant difference of $P<0.05$. At R0 and R4, the Ψw of RTP was significantly higher than that of FP ($P<0.05$) from a 2 year experimental study, except for N2 in 2015. The Ψw of RTP and FP were -1.64 MPa and -1.70 MPa in 2014, respectively. The Ψw of RTP was 27% higher than that of FP in 2015, and showed that the RTP enhanced Ψw of summer maize. The mean values of V6, R0, R2, R3, and R4 water potentials for 2 years were -1.28 MPa, -1.46 MPa, -1.79 MPa, -1.71 MPa, -1.54 MPa, respectively, and Ψw gradually decreased and then increased during the growth period.

Soil water content and soil water storage: The SWC in 0 cm to 40 cm increased with the increase of depth. The SWC of 40 cm to 80 cm soil layer was greater than that of 0 cm to 40 cm, and exhibited a small fluctuation scope. The SWC presented an irregular Z-shaped curve. The average SWC of the RTP and FP were different (Figure 3). The SWC of RTP in 0 cm to 40 and 0 cm to 120 cm were increased by 3.1% and 0.4% (2014) and 3.9% and 3.2% (2015) compared with FP, respectively, and a significant difference was found between their planting patterns ($P < 0.05$). According to the 2-year results, the SWC of the N1 and N2 treatments was basically the same as in 0 cm to 40 cm and 0 cm to 120 cm, showing that the crop nitrogen had no significant effect on SWC. The average SWC of RTP in 0 cm to 40 cm and 0 cm to 120 cm was 3.6% and 2.0% higher than that of FP, respectively.

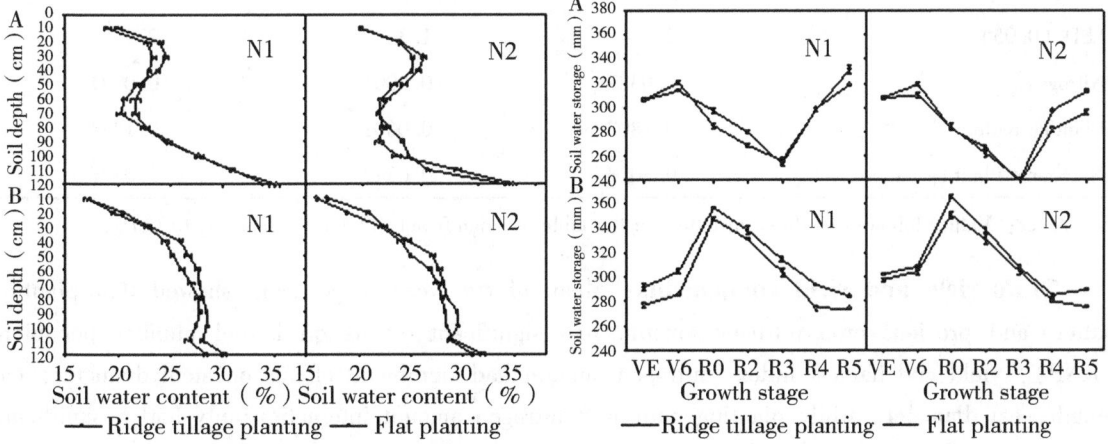

Figure 3 Effects of planting pattern and previous crop nitrogen on soil water content and soil water storage in 2014 (A) and 2015 (B). The bars are the SE

During the 2015 growing seasons, the ranking of SWS average was RTP>FP, and the values were 312.8 mm and 302.3 mm, respectively. The SWS of N1 and N2 were 295.5 mm and 286.3 mm in 2014, and 309.9 mm and 305.5 mm in 2015, respectively. For SWS, no significant difference between RTP and FP was found in 2014 ($P>0.05$), whereas a significant difference was found in 2015 ($P<0.05$). The SWS of RTP was 2.0% higher than that of FP, and was similar between N1 and N2 treatment in the 2-year study. This indicated that previous wheat nitrogen had no significant effect on SWS of summer maize, and ridge tillage was effective to collect rainfall and reduce evaporation.

Water use efficiency: The previous crop nitrogen amount affected the WUE of summer maize (Table 3), whereas planting pattern and planting pattern × nitrogen amount did not. In the 2-year study, the WUE of N2 treatment was significantly higher than that of N1 treatment ($P<0.05$). The WUE of RTP was 2.7% (2014) and 4.8% (2015) higher than that of FP ($P>0.05$). Test results showed that for 2 years, the WUE of N2 and RTP were significantly higher than those of N1 and FP ($P<0.05$) during the growing season.

Table 3 Effects of planting patterns and previous crop nitrogen amount on water use efficiency (WUE) of summer maize

Treatments	2014 WUE [kg/(ha·mm)]	2015 WUE [kg/(ha·mm)]	Mean WUE [kg/(ha·mm)]
Nitrogen (kg/ha)			
N1 (112.5)	25.0b	32.4b	28.7b
N2 (225.0)	28.4a	36.0a	32.2a
LSD (0.05)	1.7	2.0	2.7
Planting pattern			
Ridge tillage planting	27.0a	35.7a	31.3a
Flat planting	26.3a	32.8b	29.5a
LSD (0.05)	2.3	1.3	2.9
Nitrogen	0.0387	0.0001	0.0721
Planting pattern	0.1803	0.0001	0.1371
Nitrogen× Planting pattern	0.8073	0.0842	0.7476

Notes: Values followed by different letters in the table are significant difference according to $LSD_{0.05}$.

Grain yield and yield components: Mean of two years experiment showed that planting pattern and previous crop nitrogen amount had significant effects on kernel number per plant (KNP), yield and harvest index, nitrogen amount had significant effects on stem diameter, ear length, ear diameter, while planting pattern × nitrogen amount interaction only had a significant effect on ear length, row number, and kernel number (Table 4). The yield, harvest index, stem diameter, and ear diameter of N2 were 13%, 11.9%, 5.5%, and 2.3% significant higher than those of N1, respectively ($P<0.05$). The KNP, yield, and harvest index of RTP were 5.9%, 7.1%, and 5.4% significantly higher than those of FP, respectively ($P<0.05$). The results showed that yield and harvest index of summer maize were increased with increasing KNP under RTP.

Table 4 The effects of planting pattern (PP) and nitrogen on the yield components and yield

Treatments	Plant height (cm)	Stem diameter (cm)	Ear length (cm)	Ear diameter (cm)	Row number (/ear)	Kernel number (/plant)	Kernel weight (mg)	Yield (kg/ha)	Harvest index
Nitrogen (kg/ha)									
N1 (112.5)	217a	15.7b	15.6b	5.0b	14.4a	480b	391a	9 238b	0.48b
N2 (225.0)	216a	16.5a	16.7a	5.2a	14.4a	524a	397a	10 438a	0.53a
LSD	14	1.0	0.7	0.1	0.6	14	24	908	0.09

(continued)

Treatments	Plant height (cm)	Stem diameter (cm)	Ear length (cm)	Ear diameter (cm)	Row number (/ear)	Kernel number (/plant)	Kernel weight (mg)	Yield (kg/ha)	Harvest index
PP									
Ridge tillage planting	218a	16.1a	16.2a	5.1a	14.5a	517a	392a	10 177a	0.52a
Flat planting	215b	16.0a	16.1a	5.1a	14.3a	488b	396a	9 498b	0.49b
LSD	2	1.3	1.3	0.2	0.6	14	26	869	0.18
Nitrogen	0.4458	0.0070	0.0001	0.0212	0.9098	0.0017	0.7819	0.0001	0.0001
PP	0.0005	0.6892	0.5449	0.2550	0.7817	0.0150	0.8330	0.0001	0.0001
Nitrogen × PP	0.0983	0.0717	0.0133	0.6961	0.0033	0.0039	0.7988	0.1296	0.1296

Notes: Mean of 2014 and 2015; the different letters in the table are significant difference according to $LSD_{0.05}$.

Water status relations with yield: The SWC, LRWC, Ψw, WUE, and yield correlation analysis of summer maize for 2 years are shown in Table 5. A negative correlation between LRWC, WUE, yield, and SWC was found. A positive correlation was found between SWC and Ψw; LRWC showed a positive correlation with WUE and yield. The Ψw had an extremely significant negative correlation with WUE ($P<0.01$) and a negative correlation with yield ($P<0.05$). The WUE exhibited a significantly positive correlation with respect to yield ($P<0.01$).

Table 5 The correlation analysis between yield and soil water content (SWC), leaf relative water content (LRWC), water potential (Ψw) and water use efficiency (WUE)

Variable	SWC	LRWC	Ψw	WUE	Yield
SWC	1.000	-0.293	0.782	-0.604	-0.581
LRWC		1.000	-0.738	0.681	0.742
Ψw			1.000	-0.932**	-0.914*
WUE				1.000	0.946**
Yield					1.000

Notes: *, ** Correlation is significant at the 0.05 and 0.01 level, respectively.

4 Discussion

In this study, the increase of nitrogen amount in winter wheat was beneficial to improve LRWC of summer maize under RTP, which probably be attributed to promoted maize growth because the previous crop residue nitrogen, and had higher LRWC under ridge tillage (Tao et al., 2013). During both years, the LRWC mean of RTP was higher than FP; it showed that the RTP could enhance the ability of crop to resist drought stress. The WUE of N2 and RTP were significantly higher than those of N1 and FP. The RTP changed the surface shape, prevented runoff, and increased

WUE. The results were similar to the previous finding that planting pattern affects crop yield (Tao et al., 2013).

The SWC determined crop water status (Jongdee et al., 2002). Previous research reported that ridge tillage can increase root growth and improve soil moisture, thereby enhancing the plant's ability to absorb water, and energetically adjust the osmotic balance (Wang et al., 2012). In our study, the SWC average of RTP in 0 cm to 40 cm and 0 cm to 120 cm was higher than that of FP, indicating that RTP improved soil moisture status and was beneficial to root growth of summer maize. These results were consistent with previous studies (Serme et al., 2015; Gu et al., 2016).

Grain number per plant is the key factor to increase the yield, and N application could improve yield and kernel quality (Ortega et al., 2016). The RTP increased maize yield and harvest index, with a yield increase of 30% compared with FP (Hassan et al., 2005). Stem diameter, ear height, plant height, panicle length, ear width, rows per ear, and other production-related indicators affected yield formation (Tsimba et al., 2013). Tao et al. (2013) also found that the filling rate, prolonged metaphase, increased grain number per spike, the 1000-grain weight, and yield of maize were improved by ridge tillage. Two year results showed that planting pattern and previous crop nitrogen amount had significant effects on KNP, yield, harvest index, stem diameter, ear length, and ear diameter. The increase of the amount of nitrogen could increase KNP, stem diameter, ear diameter, and ear length of summer maize.

The nitrogen of previous crop winter wheat could increase SWC, SWS, LRWC, Ψw, promoting the increase in stem diameter, ear diameter, ear length, and grain number per plant of the succeeding crop summer maize. The RTP pattern was beneficial to the increase of grain yield and WUE by improving water status and yield component of summer maize. For summer maize production, combining the previous crop nitrogen with ridge tillage is a promising method in North China.

Acknowledgments

The research was sponsored by the National Natural Science Foundation of China (31760354), Guangxi Natural Science Foundation (2017GXNSFAA198036). We would like to thanks Zhang Zhen for the work contribution.

References

Galmés J, Flexas J, Savéand R, et al., 2007. Water relations and stomatal characteristics of Mediterranean plants with different growth forms and leaf habits: responses to water stress and recovery. Plant Soil. 290: 139-155.

Gheysari M, Mirlatifi SM, Homaee M, et al., 2009. Nitrate leaching in a silage maize field under different irrigation and nitrogen fertilizer rates. Agric. Water Manage. 96: 946-954.

Gu XB, Li YN, and Du YD, 2016. Continuous ridges with film mulching improve soil water content, root growth, seed yield and water use efficiency of winter oilseed rape. Ind. Crop Prod. 85: 139-148.

Hartmann TE, Yue SC, Schulz R, et al., 2015. Yield and N use efficiency of a maize-wheat cropping system as affected by different fertilizer management strategies in a farmer's field of the North China Plain. Field Crops Res. 174: 30-39.

Hassan I, Hussain Z, and Akbar G, 2005. Effect of permanent raised beds on water productivity for irrigated maize-wheat cropping system. Roth, C. H., Fischer, R. A. and Meisner, C. A. editor. Evaluation and performance of permanent raised bed cropping systems in Asia, Australia and Mexico. Aust. Centre Int. Agric. Res. Proc. 121: 59-65.

Javeed HMR, and Zamir MSI, 2013. Influence of tillage practices and poultry manure on grain physical properties and yield attributes of spring maize (*Zea mays* L.). Pak. J. Agr. Sci. 50: 177-183.

Jia XC, Shao LJ, Liu P, et al., 2014. Effect of different nitrogen and irrigation treatments on yield and nitrate leaching of summer maize (*Zea mays* L.) under lysimeter conditions. Agric. Water Manage. 137: 92-103.

Jin LB, Cui HY, Li B, et al., 2012. Effects of integrated agronomic management practices on yield and nitrogen efficiency of summer maize in North China. Field Crops Res. 134: 30-35.

Jongdee B, Fukai S, and Cooper M, 2002. Leaf water potential and osmotic adjustment as physiological traits to improve drought tolerance in rice. Field Crops Res. 76: 153-163.

Li Y, Liu H, Huang G, et al., 2016. Nitrate nitrogen accumulation and leaching pattern at a winter wheat: summer maize cropping field in the North China Plain. Environ. Earth Sci. 75: 118.

Ma C, Liu X, Bian C, et al., 2017. Straw mulching can realize soil/plants carbon sequestration and yield increasing of summer maize in north china. Rom. Agric. Res. 34: 129-136.

Mao XM, Zhong WW, Wang XY, et al., 2017. Effects of precision planting patterns and irrigation on winter wheat yields and water productivity. J. Agric. Sci. Cambridge. 155: 1394-1406.

Ortega AL, Torres NR, Carrillo GV, et al., 2016. Environment and nitrogen influence on rainfed maize yield and quality. Crop Sci. 56: 1257-1264.

Serme I, Ouattara K, Logah V, et al., 2015. Impact of tillage and fertility management options on selected soil physical properties and sorghum yield. Int. J. Biol. Chem. Sci. 9: 1154-1170.

Shi D, LI YH, Zhang JW, et al., 2016. Increased plant density and reduced N rate lead to more grain yield and higher resource utilization in summer maize. *J. Integr. Agric.* 15: 2515-2528.

Tao ZQ, Sui P, Chen YQ, et al., 2013. Subsoiling and Ridge Tillage Alleviate the High Temperature Stress in Spring Maize in the North China Plain. J. Integr. Agric. 12: 2179-2188.

Tsimba R, Edmeades GO, Millner JP, et al., 2013. The effect of planting date on maize grain yields and yield components. Field Crops Res. 150: 135-144.

Vos J, Van Der Putten PEL, Birch CJ, et al., 2005. Effect of nitrogen supply on leaf appearance, leaf growth, leaf nitrogen economy and photosynthetic capacity in maize (*Zea mays* L.). Field Crops Res. 93: 64-73.

Wang XB, Wu HJ, Dai K, et al., 2012. Tillage and crop residue effects on rainfed wheat and maize production in northern China. Field Crops Res. 132: 106-116.

Wang XY, Zhou XB, and Chen YH, et al., 2015. Planting pattern effects on soil water and yield of summer maize. Maydica 60: M18, 1-6.

Zadoks JC, Chang TT, and Konzak CF, et al., 1974. A decimal code for the growth stages of cereals. Weed Res. 14: 415-421.

Zhang JY, Sun JS, Duan AW, et al., 2007. Effects of different planting patterns on water use and yield performance of winter wheat in the Huang-Huai-Hai plain of China. Agric. Water Manage. 92: 41-47.

Zhang YL, Li CH, Wang YW, et al., 2016. Maize yield and soil fertility with combined use of compost and inorganic fertilizers on a calcareous soil on the North China Plain. Soil Till. Res. 155: 85-94.

Impact of the mixture verses solo residue management and climatic conditions on soil microbial biomass carbon to nitrogen ratio: A systematic review

I. Muhammad[a], J. Wang[b], A. Khan[c], S. Ahmad[a], L. Yang[a], I. Ali[a], M. Zeeshan[a], S. Ullah[a], S. Fahad[b], S. Ali[d] and X. B. Zhou[a]

([a] Guangxi Colleges and Universities Key Laboratory of Crop Cultivation and Tillage, Agricultural College, Guangxi University, Nanning 530004, China;
[b] Shaanxi Key Laboratory of Earth Surface System and Environmental Carrying Capacity, College of Urban and Environmental Science, Northwest University, Xi'an 710127, China;
[c] Department of Agronomy, The University of Agriculture, Peshawar, Pakistan;
[d] Department of Soil and Environmental Sciences, The University of Agriculture, Peshawar, Pakistan)

Abstract: Cover crops (CCs) have been increasingly cultivated to boost soil quality, crop yield and minimize environmental degradation compared with no cover crops (NCCs). There is no consensus of CCs under different climatic conditions on soil microbial biomass carbon (SMBC), soil microbial biomass nitrogen (SMBN), and soil microbial biomass carbon and nitrogen ratio (SMBC/SMBN) are yet documented. Thus, a global meta-analysis of 40 currently available literature was carried out to elucidate the effect of CCs on SMBC and SMBN, and its ratio for cash and cover cropping systems was conducted. Our findings demonstrated that CCs increased SMBC, SMBN, and SMBC/SMBN ratios by 39%, 51%, and 20%, respectively, as compared to NCCs. The categorical meta-analyses showed that the mixture of legume and non-legume CCs decreased the SMBC, SMBN, and SMBC/SMBN ratios relative to the sole legume or non-legume CCs. Non-legume CCs enhanced the SMBC, SMBN, and SMBC/SMBN ratio compared to legume CCs. When CCs residues were incorporated into the soil or surface mulched, the SMBC and SMBN increased compared to the removal of residues. The effect of CCs on the SMBN and SMBC/SMBN ratio was higher in medium-textured soils compared to coarser or fine-textured soils, but coarser-textured soils have a higher SMBC. The effect of CCs on SMBN and SMBC/SMBN ratio was prominent on medium-textured soils having soil organic carbon (SOC) in the range of 10-20 mg/g, pH>6.5, and total nitrogen (TN) in the range of 1%-2%. It was concluded that CCs enhanced SMBC, SMBN, and its ratio compared to NCCs. The response, however, varied depending on the

---

Corresponding author E-mail address: xunbozhou@gmail.com

soil properties and climatic region. Cover crops can boost the biological soil's health by increasing the microbial population's abundance compared to NCCs.

Keywords: Soil organic carbon; Soil microbial biomass; Köppen climate; Meta-regressions; Cover crops

1 Introduction

Cover crops (CCs) of agroecosystems have provided multiple ecological and environmental benefits, including enhancement of soil fertility, carbon (C) sequestration, leaching reduction, erosion control, and pest and disease suppression (Alfonso Gomez et al., 2018; Alliaume et al., 2014; Berlanas et al., 2018; Isik et al., 2009). Cover crops have been grown worldwide in recent decades to provide a variety of ecosystem services, particularly improved soil fertility, increased soil aggregation, soil organic matter (SOM) and weed control, and reduced nutrient losses through leaching and soil erosion when compared to no cover crop. (Jackson 2000; Wawan et al., 2019). Likewise, CCs accelerated nutrient cycling, biological nitrogen fixation (BNF), biological diversity, weed control, and crop yields (Carrera et al., 2007; Muhammad et al., 2019; Reddy 2003). Despite the fact that cover crops can improve soil health and environmental quality by increasing C and N cycling, soil fertility and sustainability (Muhammad et al., 2019). The mechanisms of improving soil sustainability by CCs practices include low usage of synthetic fertilizer, improving soil microbial diversity, and SOM (Hubbard et al., 2013; Nevins et al., 2018; Sanz-Cobena et al., 2014). Thus, multiple agronomic and environmental benefits (such as increasing crop production, decreasing water pollution and improving soil fertility) can be achieved by establishing a highly productive CCs community (de Carvalho et al., 2013; Destain et al., 2010).

Previous researchers have documented the impact of the CCs on SOM in temperate zones, Mediterranean and semiarid annual agroecosystems (Austin et al., 2017; Cardenas et al., 2012; Peregrina et al., 2010; Zhou et al., 2016). Cover crop perform a variable function in the main crop due to varied growth rate, quantity of biomass produced, uniformity of the soil cover, and C/N ratio. Numerous studies demonstrate that legume CCs increase soil quality and so provide more favorable circumstances for the establishment, development, and producing of major crops, while also playing an important role in weed infestation reduction (Amossé et al., 2013; Somenahally et al., 2018). However, in tile-drained soils, the crop rotation (corn-soybean) reduced the N leaching in the nitrate form (Dinnes et al., 2002). Similarly, in Iowa, climate oat and rye CCs in maize-soybean rotation decreased nitrate losses by 70% during three years (Logsdon et al., 2002), who reported that CCs mitigated some of the adverse effects of disruption using residue as a surface mulch and helped soil moisture preservation (Hall et al., 2010), prevented weed germination and growth, minimized soil erosion (Alfonso Gomez et al., 2018; Amossé et al., 2013), and reduced soil temperature (Bagley et al., 2012; Li et al., 2013; Unger et al., 1997). The addition of non-legume CCs such as oats tends to reduce mineralization and increase immobilization, thus reducing the N availability for the following crop. The multi-purpose nature of legume CCs is shown in their ability to reduce soil compaction and deterioration, improve structural and hydraulic characteristics, increase organic matter content and soil microorganism activity, and nitrogen content through symbiotic N_2 fixation (Kokalis-Burelle et al., 2017; Muhammad et al., 2019).

Cover crop cultivation provides physical protection to the soil by reducing the impact of rainfall (DuPont et al., 2009; Sainju et al., 2003) and can increase soil structure and aggregation (Tang et al., 2017). Thus, it improves soil microbial activities (Araujo et al., 2019). The addition of CCs improves soil organic carbon (SOC), specifically the labile component of SOC, which is considered a key driver for improving soil functions and the sustainability of agricultural ecosystems (Chalise et al., 2019; Zhou et al., 2012; Wang et al., 2021). Soil microbes easily access labile organic carbon and serve as an essential short-term storage of nutrients for crop growth (Haynes 2005). Though the effects of CCs on microbial activities and soil fertility are well documented in the literature. However, information on the impact of CCs species, such as legumes, non-legumes, oilseed crops, and their mixtures on the microbial biomass carbon to nitrogen ratio is limited. Therefore, the current global meta-analysis was undertaken to reveal the impact of different CCs species types and climatic factors on the SMBC, SMBN, SMBC/SMBN ratio. The aim of the studies is to discover the relationship of climatic factors, Köppen climatic conditions and CCs species on SMBC, SMBN, and SMBC/SMBN ratio, in various soil and climates across the globe. We hypothesized that 1) cover cropping has an overall positive effect on SMBC, SMBN, and SMBC/SMBN ratios when compared to no cover cropping, 2) this effect will vary depending on cover crop species, residues quality and management, and 3) the effect of cover crop on microbial biomass (C and N) will vary depending on soil texture and climatic zone of different regions.

2 Materials and method

2.1 Data collection

To quantify the effect of cover crops (CCs) on soil microbial biomass carbon (SMBC), nitrogen (SMBN), and the SMBC/SMBN ratio, a total of 40 studies across the globe were collected from the available literature of the Web of Science (www.sciencedirect.com) for this meta-analysis (Figure 1). Keywords, such as soil microbes, microbial biomass, microbial biomass carbon, microbial biomass nitrogen, microorganisms, and cover crops, were used for the search. The climatic conditions of various regions selected for the studies are shown in Figure 1. Soil texture, precipitation, temperature, latitude, and longitude of the selected studies were also recorded (Table S1).

The following criteria were used for publication selection in the current meta-analysis:

i. Only field studies were included for the data collection on cover crops with no cover crops in the same experimental area.

ii. Oilseed and grass cover crops were treated as non-legumes.

iii. Those studies included at least three replications and a standard deviation (SD) or standard error (SE).

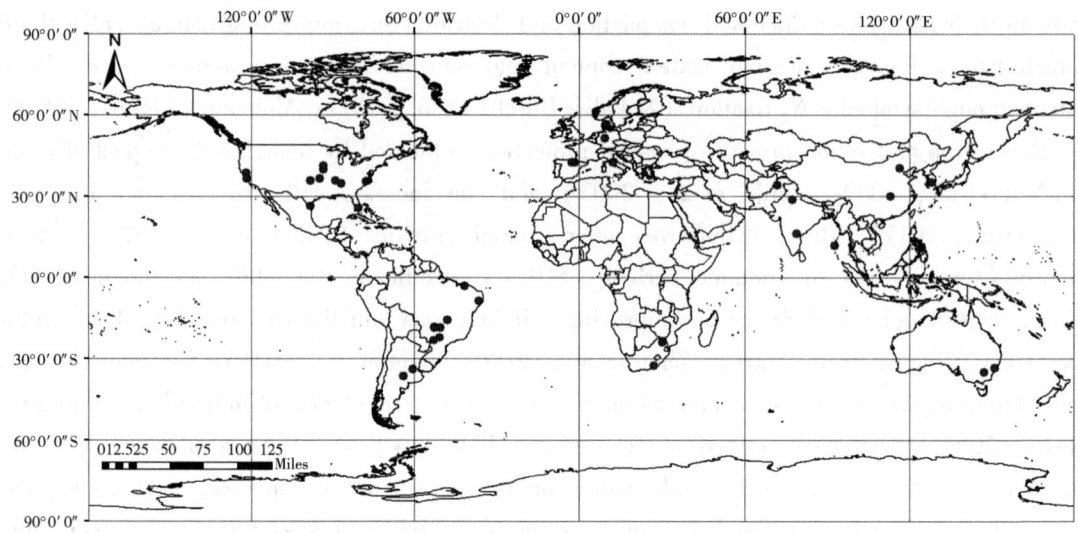

Figure 1 Origin of the study sites around the world included in the current meta-analysis

 iv. Studies with no proper control (no cover crop) were discarded.
 v. The following equation was used to calculate the SD from standard error.

$$SD = SE \times \sqrt{n} \tag{1}$$

Where n is the number of replications.

For data collection, the following criteria were used:

 i. Data was taken from the recommended fertilizer if the experiment had more than one level of fertilization.

 ii. Average data was used for the studies if different fertilizer sources and types were used.

 iii. Data was collected from publications only comparing cover crops with no cover crops on the soil, SMBC + SMBN.

 iv. Different cover crop types and rotations were used as individual comparisons.

Data for SMBC, SMBN, and SMBC/SMBN ratio were extracted from the literature. The data from the tables was taken directly. For data extraction if the data was presented in figures, the graph digitizer software 'GetData 2.26' was used. When the publication didn't show climatic data, then the latitude and longitude were used to identify the study areas' climatic zones through a high-resolution Google map (Peel et al., 2007b).

2.2 Variance estimators and weighting functions

Reporting the full results of variance estimators and weighting functions is one of the difficulties in carrying out a systematic review (Wiebe et al., 2006). In the current meta-analysis, the effect size and related inferences were estimated using the weighting method (Hungate et al., 2009), based on the inverse of pooled variance (Van Groenigen et al., 2011). Studies where variances were not reported and could lead to bias were excluded from the analysis (Wiebe et al., 2006). In studies where no variances were reported, the coefficient of variation (CV) was calculated using the following equation.

$$CV = \frac{SD}{M} \times 100 \tag{2}$$

Where SD and M are standard deviation and mean, respectively. When there was a lack of SD, it was estimated as:

$$SD_{missing} = \frac{CV \times M}{100} \tag{3}$$

Where CV and M are the coefficient of variance and means, respectively.

2.3 Data analysis

The effect of the CCs on microbial variables was measured using the response ratio (RR), which is the normal log of the treatments with and without CCs (Hedges et al., 1999) as follows:

$$RR = Ln\left(\frac{Soil\ microbial\ biomass\ from\ cover\ crop\ treatment}{Soil\ microbial\ biomass\ from\ no\ cover\ crop\ treatment}\right)$$

$$RR = Ln\left(\frac{X_{CCs}}{X_{NCCs}}\right) = Ln(X_{CCs}) - Ln(X_{NCCs}) \tag{4}$$

X_{CCs} and X_{NCCs} are arithmetic means of the microbial biomass (SMBC, SMBN, and ratio) in CCs and NCCs. In the numerator and denominator, the natural log ratio confirms that changes are similarly affected. If the distributions of X_{CCs} and X_{NCCs} are normal and X_{NCCs} are likely to be positive, the RR would be normally distributed with a mean equal to the true log response ratio (Gurevitch 1993). Besides, the variance of RR (VLn RR) for each sample was determined using the following equation (Hedges et al., 1999).

$$V_{LnRR} = \frac{S_{CCs}^2}{n_{CCs} X_{CCs}^2} + \frac{S_{NCCs}^2}{n_{NCCs} X_{NCCs}^2} \tag{5}$$

Where S_{CCs} and S_{NCCs} are the standard deviations, n_{CCs} and n_{NCCs} represent replications of the study, and X_{CCs} and X_{NCCs} are the means of CCs and NCCs treatments, respectively.

The random effect model with a 95% CI using reciprocal variance (V) and weight (W) was calculated as below for each RR (Borenstein et al., 2011):

$$W = \frac{1}{V} \tag{6}$$

The overall mean response ratio (RR_{E++}) for individual CC treatment was calculated through the following formula:

$$RR_{E++} = \frac{\sum_{i=1}^{n} \sum_{j=1}^{m} W_{ij} RR_{ij}}{\sum_{i=1}^{n} \sum_{j=1}^{m} W_{ij}} \tag{7}$$

Where "n" is the number of treatments and "m" is the comparisons of the SMBC, SMBN, and the SMBC/SMBN ratio. The SE of RR_{E++} was calculated as follows:

$$SE(RR_{E++}) = \sqrt{\frac{1}{\sum_{i=1}^{n} \sum_{j=1}^{m} W_{ij}}} \tag{8}$$

The mean effect size was estimated using the random model MetaWin 2.1 (Sinaure Associate Inc., Sunderland, USA) at a 95% confidence interval (CI) to evaluate the effect of the CCs on microbial variables (Muhammad et al., 2019). When the 95% CI did not overlap the vertical zero line, the impact of the CCs was deemed significant.

The significance of the meta-analysis was assessed based on the 95% CI. The test treatment group was considered to show a significant increase or decrease relative to the control group ($P<0.05$) if the 95% CI did not overlap with the zero lines.

In this study, the meta-analysis's heterogeneity was estimated using the weighted sum of squares (Q) (Borenstein et al., 2011; Hedges et al., 1999).

$$Q = \sum_{i=1}^{k} W_i (In R_i)^2 - \frac{(\sum_{i=1}^{k} W_i InR_i)^2}{\sum_{i=1}^{k} W_i} \tag{9}$$

In addition, we used the Origin software (version 2018) to chart the kernel density estimates (a smoothed version of the histogram) for SMBC, SMBN, and SMBC/SMBN ratio.

3 Results

3.1 Publication bias and data heterogeneity

Globally, CCs significantly increase SMBC, SMBN, and SMBC/SMBN by 39.3%, 51.3%, and 19.7%, respectively, when averaged across all cover crop types (Figure 2a, 2b, and 2c). Our meta-analysis of 581 pairwise comparisons from 40 published studies indicated that there was significant residual heterogeneity in the global SMBC (Q_b = 304.2485; $P<0.001$; n = 330), SMBN (Q_b = 94.4141; $P<0.001$; n = 149) and SMBC/SMBN ratios (Q_b = 97.9771; $P<0.001$; n = 102). We attempted to explain the heterogeneity by considering different factors (Figure 3a, 3b, and 3c). Among the ten moderators (CC types, CC management, plant category, soil texture, soil pH, SOC, TN, Köppen climate, precipitation, and temperature), soil texture, soil pH, and SOC had no significant effects on changes in SMBC in cover cropping systems ($P>0.05$; Table 1). However, all moderators had significant effects on the SMBN and the SMBC/SMBN ratio ($P>0.01$) except CCs types and mean annual precipitation, which was non-significant for the SMBN.

Furthermore, there were significant differences in SMBC among CC types (Q_b = 15.77; $P<0.001$), CC management (Q_b = 12.68; $P<0.05$), plant category (Q_b = 68.09; $P<0.001$), TN (Q_b = 24.35; $P<0.001$), Köppen climate (Q_b = 20.04; $P<0.001$), precipitation (Q_b = 24.43; $P<0.01$) and temperature (Q_b = 26.72; $P<0.001$), but soil texture, soil pH and SOC did not show significant differences in SMBC ($P>0.05$; Table 1). The CC management, plant category, soil texture, soil pH, SOC, TN, Köppen climate, and temperature significantly influenced the SMBN in cover crops (Table 1) and explained 10.81% ($P<0.01$), 6.13% ($P<0.05$), 30.47% ($P<0.001$), 6.90% ($P<0.05$), 20.95% ($P<0.001$), 10.68% ($P<0.01$), 21.50% ($P<0.001$), and 19.93% ($P<0.001$) of the response, respectively. Significant ($P<0001$) differences in SMBC/SMBN ratio among CC types (Q_b = 16.99), plant category (Q_b = 20.15), soil texture (Q_b = 36.66), SOC (Q_b = 29.667), Köppen climate

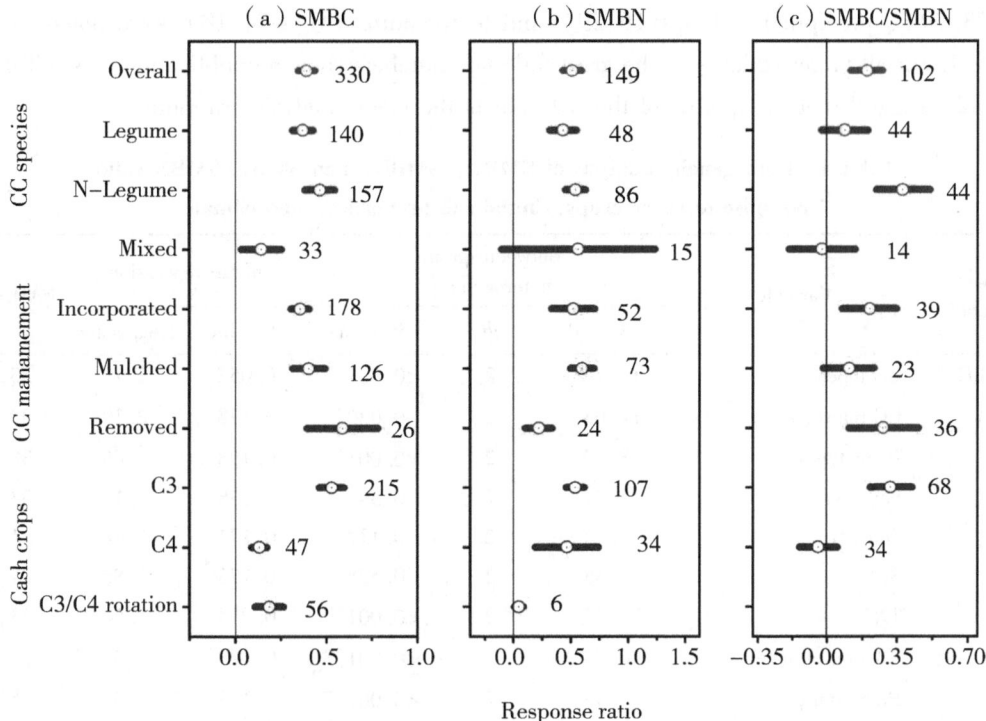

Figure 2 Mean response ratio of cover crops (CCs) compared to no cover crops (NCCs) and 95% confidence intervals (horizontal box) for soil microbial biomass carbon (SMBC) and soil microbial biomass nitrogen (SMBN) and SMBC/SMBN ratios affected by cover crop species, cover crop management and cash crop types

Notes: The vertical line (RR=0) indicates no difference between cover cropping and no cover cropping systems. Numbers following the box indicate the number of observations for comparison.

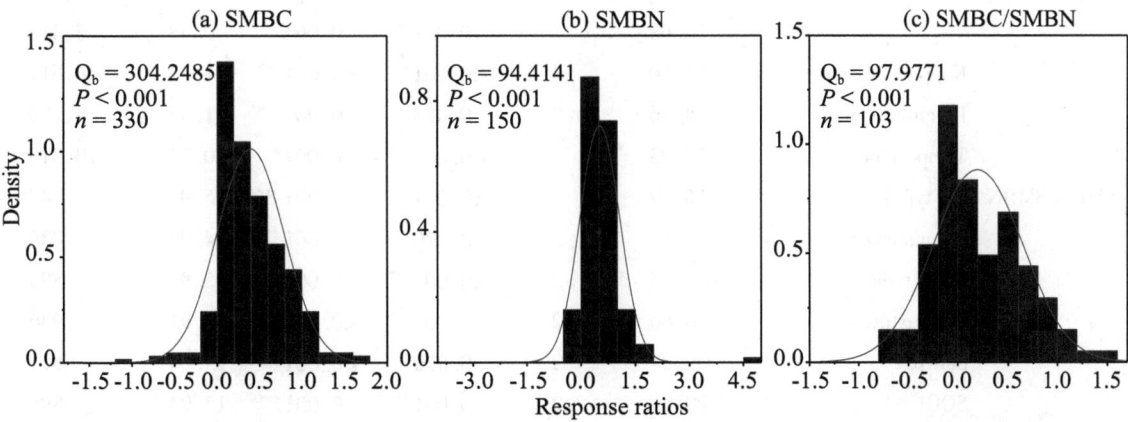

Figure 3 Kernel density estimates (smoothed version of the histogram) for soil microbial biomass carbon (SMBC; a), soil microbial biomass nitrogen (SMBN; b), and soil microbial biomass carbon and nitrogen (c)

Notes: The red curves were fitted by the Gaussian function using the software originPro 2019b. Q_b is the between-group heterogeneity result for all observations in the meta-analysis.

($Q_b = 73.79$), precipitation ($Q_b = 16.29$) and temperature ($Q_b = 74.18$) were noted as shown in Table 1. For all target variables, the great fail-safe numbers show no publication bias (Table 1). Our results suggest that the quality of the data meets the meta-analysis criterion.

Table 1 Homogeneity analysis of SMBC, SMBN, and SMBC/SMBN ratio response to cover crops, initial soil properties, and climate

Response variable	Variable	Between-group heterogeneity			Meta-regression		FAIL-SAFE
		Q_b	df	P-value	P-value	Regression	
SMBC	CC types	15.77	2	<0.001***	0.052	3.79	33,529
	CC management	12.68	2	0.030*	0.138	2.19	34,802
	Plant types	68.09	2	<0.001***	0.198	1.65	38,228
	Soil texture	1.94	2	0.347	0.138	2.19	33,303
	Soil pH	3.74	2	0.137	0.171	1.88	31,915
	SOC	1.38	2	0.525	0.173	1.85	27,507
	TN	24.35	2	<0.001***	0.214	1.54	33,127
	Köppen climate	20.04	3	<0.001***	0.123	2.37	32,317
	Precipitation	24.43	2	<0.001***	0.123	2.37	35,570
	Temperature	26.72	2	<0.001***	0.123	2.37	33,810
SMBN	CC types	1.81	2	0.404	<0.001***	11.45	8,201
	CC management	10.51	2	0.005**	<0.001***	11.45	8,404
	Plant types	6.13	2	0.046*	0.001**	10.70	8,497
	Soil texture	30.47	2	<0.001***	<0.001***	11.45	11,459
	Soil pH	6.90	2	0.031*	0.021*	5.31	6,814
	SOC	20.95	2	<0.001***	0.002**	9.48	7,163
	TN	10.68	2	0.001**	0.002**	9.48	6,312
	Köppen climate	21.50	3	<0.001***	<0.001***	11.45	8,611
	Precipitation	4.86	2	0.088	<0.001***	11.45	8,199
	Temperature	19.93	2	<0.001***	0.001**	10.71	10,319
SMBC/SMBN	CC types	16.99	2	<0.001***	<0.001***	15.91	847
	CC management	10.32	2	0.020*	<0.001***	14.93	836
	Plant types	20.15	2	<0.001***	<0.001***	15.91	898
	Soil texture	36.66	2	<0.001***	<0.001***	15.91	936
	Soil pH	9.37	2	0.020*	<0.001***	15.04	811
	SOC	29.67	2	<0.001***	<0.001***	15.04	896
	TN	6.20	2	0.040*	<0.001***	15.04	808
	Köppen climate	73.79	3	<0.001***	<0.001***	15.91	1,225
	Precipitation	16.29	2	<0.001***	<0.001***	15.91	858
	Temperature	74.18	2	<0.001***	<0.001***	15.91	1,229

Notes: The total sum of squares (Q_t), between-group sum of square (Q_b), P-value (P), number of observations (n), and heterogeneity (I^2).

3.2 Meta-regressions

According to the meta-regression results, all moderators did not show significant differences in the change of SMBC ($P>0.05$; Table 1). The SMBN was significantly affected by CCs types, CCs management, soil texture, Köppen climate, and precipitation by 11.45% ($P<0.001$), plant category, SOC, TN, and temperature by 10.70%, 9.48%, 9.48%, and 10.71%, respectively ($P<0.01$), and soil pH by 5.31% ($P<0.05$) as shown in Table 1. Similarly, CC types, plant category, soil texture, Köppen climate, precipitation, and temperature significantly affected the SMBC/SMBN ratio by 15.91%, CC management by 14.93%, while soil pH, SOC, and TN by 15.04% ($P<0.01$; Table 1).

3.3 Microbial biomass response to cover cropping

The study's objective was to investigate whether significant differences in SMBC, SMBN, and its ratio under cover crop types, management, and cash crops exist on a global scale. The investigated target variables (SMBC, SMBN, and SMBC/SMBN ratio) were significantly increased in CCs types, CCs management, and cash crops. The overall cover crop enhanced SMBC by 39%, SMBN by 51%, and SMBC/SMBN ratios by 19.7% compared to NCCs (Figure 2). The categorical meta-analysis revealed that among CCs types, non-legume CCs had a 25% and 223% higher SMBC than that of legume and mixed CCs, respectively. While the legume CCs significantly increased SMBC by 158% compared to mixed CCs. Our results highlighted that the non-legume and legume CCs increase in SMBC was considerably higher than the mixed cover crop. Non-legume CCs exhibited higher mean values than legume CCs. It is important to point out that, even if non-legume CCs had the highest mean value, this increment was not significantly greater than the legume CCs (Figure 2). Similarly, the results showed that legume and non-legumes CCs significantly increased SMBN compared to NCCs, while no-statistically significant differences were found among CCs species. The SMBC/SMBN ratios were significantly higher for non-legume CCs than other treatment groups. Legume and mixed CCs have no effect on the SMBC/SMBN ratio.

Cover crops management, i.e., incorporation, surface mulching, and removing, influenced the soil microbial biomass and soil fertility. The CCshad significantly increased SMBC, SMBN, and SMBC/SMBN ratio than NCCs plots. Soil microbial biomass carbon was 36%, 40%, and 59% higher for incorporated, mulched, and removed CCs than NCCs, respectively (Figure 3). Average across CCs, the residue of CCs removal had improved the SMBC than surface mulched or incorporated CCs. However, this increment was non-significant. Soil microbial biomass carbon and SMBC/SMBN ratio were non-significantly different among CCs management practices. However, the CCs management practices significantly affected the SMBN, and mulching had increased the SMBN by 171% than removed CCs (Figure 3).

The RR of CCs for SMBC, SMBN, and SMBC/SMBN ratio varied across the C_3 and C_4 plants categories (Figure 3). Under a cover cropping system, plants with the C_3 cycle have a positive effect on the SMBC, SMBN, and the SMBC/SMBN ratio. Crops with a C_3 cycle increased SMBC by 301% and 184% compared to C_4 and C_3/C_4 cycles, respectively. The mean value of the SMBC is higher for C_3/C_4 cycle rotations than for C_4 cycle rotations, but the differences are not statistically significant. Similarly, the C_3 cycle resulted in a higher SMBN than the C_3/C_4 rotation cycle, however not statistically different from the C_4 cycle plants. Our results showed that the C_3 plant significantly increases the SMBC/SMBN ratio compared to the C_4 cycle. The meta-regression results showed that legume CCs or CCs residue removal had significantly increased the RRs of SMBC by 5.01% ($P<0.05$), and 23.82% ($P<0.001$), respectively (Table S2). Furthermore, mulched CC and C_3-plants have significantly increased the SMBN by 83.50% and 22.59% ($P<0.001$), respectively, whereas non-legume CC, mulched, removed and C_3-plants have significantly increased the SMBC/SMBN ratio by 18.80% ($P<0.001$), 8.29% ($P<0.05$), 26.48 ($P<0.001$) and 4.32% ($P<0.05$), respectively (Table S2).

3.4 Microbial biomass response to different climatic regions

Cover crop management in different climatic regions results in distinct responses to soil microbial attributes and different soil fertility. Average across Köppen climate classification, the continental climatic regions had significantly higher values for SMBC and SMBC/SMBN ratios than topical, and temperate climatic regions, however, have no effects on SMBN (Figure 4). Continental climatic regions had higher SMBC than tropical and temperate by 119% and 82%, respectively. The SMBC/SMBN ratio was significantly increased by 63.6% under CCs by continental climatic regions compared to NCC. However, no effects of temperate and tropical climatic regions on the SMBC/SMBN ratio were noted. The CCs in temperate climatic regions had a higher SMBC than in tropical and dry climatic regions; however, the increase was not significantly different.

The changes occurred in SMBC, SMBN, and SMBC/SMBN ratio in response to the mean annual air temperature under CCs was higher at low mean air temperatures ($<15°C$) than at higher temperatures ($>15°C$). Results showed that mean values for SMBC were greater for high temperatures ($>20°C$) than moderate ($15-20°C$). The in-contrast SMBN and SMBC/SMBN ratios were higher at moderate than at high temperatures. Lower mean air temperature resulted in 103% and 65% higher SMBC than moderate temperature and high temperature, respectively. Similarly, the effect of lower temperatures on SMBN was significantly higher than that of high temperatures, but statistically similar to that of moderate temperatures, implying that SMBN at low temperatures was increased by 174% when compared to high temperatures (Figure 4). Our meta-analysis showed that the SMBC/SMBN ratio was not different between moderate and high temperatures, where the bars overlap the zero line. The results of the meta-regression showed that a dry climate (4.6%; P 0.05), precipitation >1,500 mm (28.2%; $P<0.001$), and precipitation of 1,000–1,500 mm (13.9%; $P<0.001$) had a positive effect on the SMBC (Table S3). Among all moderators, the tropical (4.80%; $P<0.05$), temperate (27.24%; $P<0.001$), precipitation <1,000 mm (13.38%; $P<0.001$), temperature >20°C (7.11%; $P<0.01$) and temperature 15–20°C (7.50%; $P<0.01$) had significant effects on SMBN. Whereas, precipitation <1,000 mm and 1,000–1,500 mm affected SMBC/SMBN ratios by (15.94%; $P<0.001$) and (9.43%; $P<0.01$), respectively (Table S3).

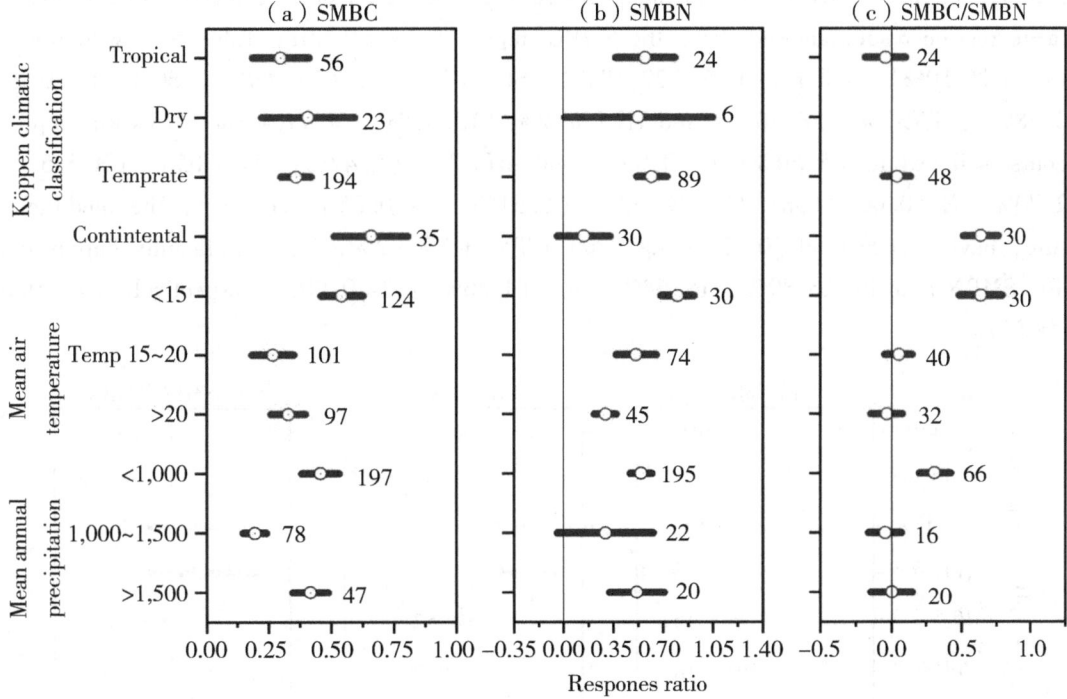

Figure 4 Mean response ratio of cover crops (CCs) compared to no cover crops (NCCs) and 95% confidence intervals (horizontal box) for soil microbial biomass carbon (SMBC) and soil microbial biomass nitrogen (SMBN) and SMBC/SMBN ratios affected by Köppen climate, mean air temperature and mean annual precipitation

Notes: The vertical line (RR = 0) indicates no difference between cover cropping and no cover cropping systems. Numbers following the box indicate the number of observations for comparison.

3.5 Microbial biomass response to soil properties

The effect of CCs on SMBC was greater in courser-textured soils compared to medium and fine-textured soils, although these findings were not statistically significant. According to the results, SMBC had a higher mean value in courser-texture soil, with organic carbon of 10-20 mg/g and soil pH>7.5 (Figure 5). The SMBC significantly increased in soil having STN of 1-2 mg/g than STN>2 mg/g. Our findings revealed that SMBN was significantly higher in medium (435%) and courser-texture soil (243%) than fine; however, medium-texture soil did not differ statistically from courser-texture soil (Figure 5). These results indicate that CCs enhanced SMBN compared to fallow in medium and courser soils. Soils with pH 6.5-7.5 and pH<6.5, SOC of 10-20 mg/g, and STN of 1-2 mg/g improved the SMBN and thus the RR of the SMBN.

The ratio of SMBC/SMBN significantly increased with medium soil texture than both courser and fine soil texture under the CCs system. The results showed that the SMBC/SMBN ratio was 50% higher in medium soil texture under the CCs system than the NSSc system. The SMBC/SMBN ratio was not significantly different between courser and fine soil texture which overlapped the zero line, but higher with medium soil texture. The high RR of SMBC/SMBN ratio was noted in soils having pH of 6.5-7.5 or<6.5, SOC of 10-20 mg/g or<10 mg/g and STN of 1-2 mg/g (Figure 5). The meta-regression results showed that, the SMBC improved ($P<0.001$; Table S4) with fine soil texture (41.10%), soil pH<6.5 (22.18%), SOC>20 g/kg (31.25%), SOC 10-20 g/kg (20.38%), TN>2% (28.20%) and TN 1%-2% (13.85%). Whereas the SMBN was improved in course soil texture (5.50%; $P<0.05$), soil pH>7.5 (5.47%; $P<0.05$), pH 6.5-7.5 (22.70%; $P<0.001$) and TN 1%-2% (5.51%; $P<0.05$). However, the medium soil texture, having an SOC of 10-20 g/kg, and a TN of 1%-2%, had significantly improved the SMBC/SMBN ratio by 26.89%, 18.68%, and 13.86% ($P<0.001$) respectively ($P<0.01$; Table S4).

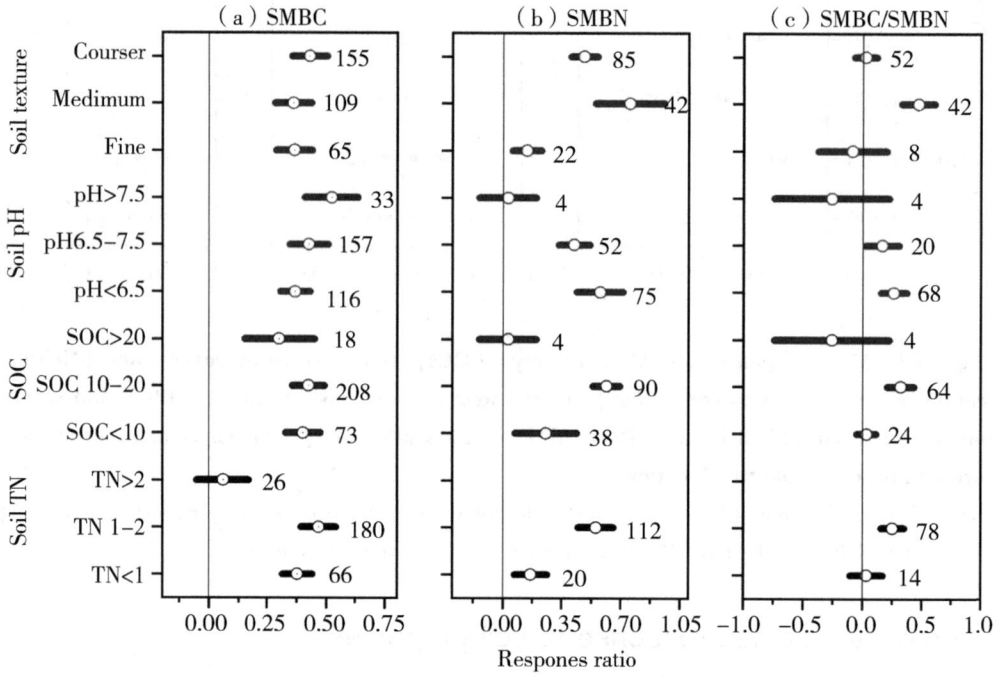

Figure 5 Mean response ratio of cover crops (CCs) compared to no cover crops (NCCs) and 95% confidence intervals (horizontal box) for soil microbial biomass carbon (SMBC) and soil microbial biomass nitrogen (SMBN) and SMBC/SMBN ratios affected by soil texture, soil pH, soil SOC, and soil TN

Notes: The vertical line (RR=0) indicates no difference between cover cropping and no cover cropping systems. Numbers following the box indicate the number of observations for comparison.

4 Discussion

Cover cropping is considered one of the most important agronomic practices for improving soil fertility and sustainability. The improvement in soil fertility due to CCs occurs mainly by two mechanisms: a) direct nutrition addition via CCs residue return (Frasier et al., 2016; Hubbard et al., 2013) and b) improved microbial activity, resulting in improved nutrient cycling and availability (Nevins et al., 2018; Sanz-Cobena et al., 2014). Our meta-results showed that CCs enhanced

SMBC, SMBN, and SMBC/SMBN ratio compared to NCCs (Figure 1a, 1b, 1c). The increase in microbial indices could be related to the first mechanism, i.e., enhanced C and N inputs from CCs residue (Hubbard et al., 2013). The beneficial effects of CCs on microbial biomass have also been attributed to increased soil C and N inputs from CCs residue (Frasier et al., 2016). The varying impact of legume, non-legumes and mixed CCs on SMBC and SMBN is due to the CCs residue return quality (C/N ratio) and amount (Coppens et al., 2006). The greater content of C content or lower N content in non-legume might have increased C input in CCs residue returns, which increased MBC with non-legumes relative to legume and mixed cover crops (Bradford et al., 2013). In contrast, greater N input increased SMBN with legumes than non-legumes and mixed cover crops (Mayer et al., 2003). It is worth mentioning that legume cover crops resulted in higher SMBN but were not substantially different from non-legume or mixed cover crops. The high variability in SMBN is related to the inconsistency of the C/N ratio for mixed cover crops. Several researchers (Mbuthia et al., 2015; Muhammad et al., 2019; Sainju et al., 2007; Sainju et al., 2003) found that non-legume CCs increase SMBC and soil respiration, while legume CCs have increased SMBN relative to no CCs. According to our findings, CCs with a higher C than N content are likely to increase the SMBC/SMBN ratio. The SMBC/SMBN ratio of non-legume CCs was higher than that of legume CCs. The lower SMBC and SMBN with mixed CCs than legume and non-legume CCs might be due to greater nutrient cycling, which might have caused lower final microbial indices. Non-legume cover crops, on average, produce more above-and below-ground biomass, enhance carbon supply, and have a higher C : N ratio than legume CCs (Muhammad et al., 2019).

Cover crop residue management affects soil microbial biomass (C and N) by changing residue interaction with soil microorganisms. The increases in SMBN with CCs were probably due to increased C and N inputs from above and below-ground cover crop residue, which increased substrate availability for soil microorganisms, boosting their growth and biomass (Schmidt et al., 2019). The residue removal practice had a higher SMBN than residue mulched or incorporated practices (Figure 2b). However, the residue mulched or incorporated practices had a higher SMBC and SMBC/SMBN ratio (Figure 2a, and Figure 2c). The SMBN improved with CCs surface placement, perhaps due to decreased soil temperature and increased water content as a result of surface mulching (Karuku et al., 2014). It is well documented that CCs residue incorporation stimulates soil microbial growth (Nevins et al., 2018), enhances soil quality (fertility and structure), and hence crop yields (Brozovic et al., 2018).

Higher quality C_3 residue (e.g., lower lignin/N) should favor bacteria and support a higher abundance of root-feeding nematodes (Dornbush et al., 2008). Furthermore, new studies of ^{13}C-labeled lignin monomers have revealed that certain lignin may decompose faster than bulk SOC (Dornbush et al., 2008). Basche et al. (2014) revealed that the effect of cover crops on emissions and microbial activity varied depending on cover crop species, biomass yield, residue C/N ratio, lignin content, and residue management strategy. The change in C_3-C_4 vegetation influences the SMBC, SMBN, and SMBC/SMBN ratio, and their C assimilation is generated from the relevant plant type (Figure 2). For soils that displayed a C_3 plant, the signature was 1.6‰, higher ^{13}C-enriched SMBC than total C but ^{13}C-depleted soils and enriched C_4 signature had 1.1‰

lower ^{13}C-enriched SMBC (Dijkstra et al., 2006). Similarly, increases in SMBC for ^{13}C enriched were 1.7%–5.6% higher relative to total soil C (Dijkstra et al., 2006; Liang et al., 2002; Potthoff et al., 2003).

The SMBC in courser soils was greater than fine and medium soil textures (Figure 5a) due to CCs. The higher SMBC in CCs than NCCs could be associated with improved activities of microbes in courser-textured soil. However, the RRs for SMBN and SMBC/SMBN ratio were higher with medium soil texture than course/fine soil texture soil (Figure 3b, and 3c). Soil can be considered ideal for microbial development if it has 50% porosity and 50% solid, and pore spaceis filled by water and air equally (Moore and Bradley 2018). Anaerobic conditions limit microbial growth and activity in fine-textured soil (Drury et al., 1991). Similarly, the C and N substrate availability is reduced with increased soil organic matter mineralization and leaching losses in coarse-textured soil (Peregrina 2016; Peregrina et al., 2014; Sakamoto et al., 2010). The pH had affected the microbial biomass carbon but have no effects on SMBN. It is not surprising that the SMBN and SMBC/SMBN ratio are higher with CCs in soils with a pH of 6.5–7.5 or>6.5 pH because the microorganisms thrive better in a neutral soil pH. Both acidic and alkaline soils limit soil microorganisms' growth (Fierer and Jackson 2006; Yang et al., 2017). Increased C substrate increased SMBC, SMBN, and SMBC/SMBN ratio due to CCs of soils having 10–20 mg/g SOC. Soil total nitrogen of 1%–2% had significantly higher SMBN, SMBC, and SMBC/SMBN ratios than that of total N<1%. The increases in microbial indices could be associated with higher plant below-ground activities, the release of root exudate, and improved soil moisture availability (He et al., 2013a), which caused better rhizosphere and microbial indices. Previous reports suggested that when the C/N ratio of the soil was below 30 : 1, microorganisms had limited access to C (He et al., 2013b; Kaye and Hart 1997) during a shift from N limiting to C limiting.

A categorical meta-analysis using Ko'ppen climatic zones, i.e., dry, continental, temperate, and tropical, was used to determine climatic conditions' potential effect on microbial indices. No-significant differences in SMBC and SMBN were reported in the dry, temperate, and tropical climatic zones. In contrast, the continental climate zone significantly increased SMBC and SMBC/SMBN while decreasing SMBN (Figure 4a, 4b, and 4c). It is well known that low moisture content inhibits the rapid mineralization of SOC and freshly added organic matter and thus results in lower SMBC due to moisture sensitivity in the dry climatic zone (Khan et al., 2018, Peregrina 2016). The tropical climatic zone, also known as the super thermal climate, has an average annual temperature of 18℃ and minimum monthly rainfall of 60 mm (Lori et al., 2017, Peel et al., 2007a), thus would have lower microbial indices. The Continental climate zone, also known as the microthermal climate, is characterized by average temperatures above 10℃ and below-3℃ during the hottest and coldest months (Lori et al., 2017; Peel et al., 2007a). Thus, the CCs might have regulated the temperature and caused increases in SMBC and SMBC/SMBN. The variations in soil microbial indices between CCs and NCCs have been shown to be climate-dependent to some degree, particularly soil temperature and moisture (Khan et al., 2018; Pena-Pena and Irmler 2016; Singh et al., 2017). It has been demonstrated that CCs improve soil physical and chemical properties such as soil organic matter content (Bechara et al., 2018; Bukovsky-

Reyes et al., 2019), soil microbial community abundance and structure (Bacq-Labreuil et al., 2019), aggregate stability, and soil organic matter content (de Souza et al., 2016), which could account for improvements in SMBC, SMBN, and SMBC/SMBN ratio.

The effect of CCs on SMBC, SMBN and SMBC/SMBN was higher with a mean air temperature of < 15 ℃ compared to other treatments in the group. Microbial biomass (carbon, nitrogen, and ratio) was significantly increased with a mean air temperature of < 15 ℃ and a mean annual precipitation of < 1,000 mm (Figure 5a, 5b, and 5c). Increased C substrate availability accelerated soil microorganisms due to greater SOC content and optimum temperature (Patkowska et al., 2016). Microorganisms thrive better in regions with optimum air temperatures (10–15 ℃) and abundant precipitation (Pietikäinen et al., 2005). Heat stress destroys microorganisms at high air temperatures > 30 ℃ (Pietikäinen et al., 2005), whereas microbial activity is decreased at low air temperatures < 10 ℃ (Alster et al., 2016). Similarly, mostly anaerobic microorganisms thrive in flooded regions, while microorganisms remain dormant in dry regions (Elfstrand et al., 2007). Increased C substrate availability due to higher SOC content and optimal temperature increased microbial activity and biomass (Kong and Six, 2012; Maul and Drinkwater, 2010; Muhammad et al., 2018). Cover crops enhanced the soil microbial communities and abundance due to higher mycorrhizal abundance, microbial biomass, and phosphatase activity compared to fallow (Hallama et al., 2019).

5 Conclusions

This study aimed to provide a comprehensive quantitative synthesis of the impact of CCs on soil microbial biomass using a meta-analysis. Cover crops had 39%, 51%, and 20% higher SMBC, SMBN, and SMBC/SMBN ratios than NCCs across the globe. Non-legume CCs resulted in higher SMBC and SMBC/SMBN ratios than mixed CCs in the C_3 planting system. All CCs management practices increased the SMBC, SMBN, and SMBC/SMBN ratio significantly; however, CCs mulching had no effect on the SMBC/SMBN ratio. The SMBC and SMBC/SMBN ratios induced by CCs increased in continental climatic regions with mean air temperature < 15 ℃ and annual precipitation < 1,000 mm. The effect of CCs on SMBN and SMBC/SMBN ratios was higher in medium-textured soil, < 6.5 soil pH, 10–20 mg/g SOC and 1%–2% of TN. Increased C and N inputs from CCs increased SMBC and SMBC/SMBN ratios. The SMBC, SMBN, and SMBC/SMBN ratio were positively influenced by cover crop types (i.e., legumes, non-legumes, and mixed) and management practices.

Acknowledgments

This study was supported by the National Natural Science Foundation of China (31760354) and the Natural Science Foundation of Guangxi (2019GXNSFAA185028).

References

Alfonso Gomez J, Campos M, Guzman G, et al., 2018. Soil erosion control, plant diversity, and arthropod communities under heterogeneous cover crops in an olive orchard. Environmental Science and Pollution Research. 25: 977-989.

Alliaume F, Rossing WAH, Tittonell P, et al., 2014. Reduced tillage and cover crops improve water capture and reduce erosion of fine textured soils in raised bed tomato systems. Agriculture Ecosystems & Environment. 183: 127-137.

Alster CJ, Koyama A, Johnson NG, et al., 2016. Temperature sensitivity of soil microbial communities: An application of macromolecular rate theory to microbial respiration. Journal of Geophysical Research-Biogeosciences. 121: 1420-1433.

Amossé C, Jeuffroy M-H, Celette F, et al., 2013. Relay-intercropped forage legumes help to control weeds in organic grain production. European Journal of Agronomy. 49: 158-167.

Araujo FS, Barroso JR, Freitas LdO, et al., 2019. Chemical attributes and microbial activity of soil cultivated with cassava under different cover crops. Revista Brasileira De Engenharia Agricola E Ambiental. 23: 614-619.

Austin EE, Wickings K, McDaniel MD, et al., 2017. Cover crop root contributions to soil carbon in a no-till corn bioenergy cropping system. Global Change Biology Bioenergy. 9: 1252-1263.

Bacq-Labreuil A, Crawford J, Mooney SJ, et al., 2019. Cover crop species have contrasting influence upon soil structural genesis and microbial community phenotype. Scientific Reports. 9.

Bagley JE, Desai AR, Dirmeyer PA, et al., 2012. Effects of land cover change on moisture availability and potential crop yield in the world's breadbaskets. Environmental Research Letters. 7.

Basche AD, Miguez FE, Kaspar TC, et al., 2014. Do cover crops increase or decrease nitrous oxide emissions? A meta-analysis. Journal of Soil and Water Conservation. 69: 471-482. doi: 10.2489/jswc.69.6.471.

Bechara E, Papafilippaki A, Doupis G, et al., 2018. Nutrient dynamics, soil properties and microbiological aspects in an irrigated olive orchard managed with five different management systems involving soil tillage, cover crops and compost. Journal of Water and Climate Change. 9: 736-747.

Berlanas C, Andres-Sodupe M, Lopez-Manzanares B, et al., 2018. Effect of white mustard cover crop residue, soil chemical fumigation and Trichoderma spp. root treatment on black-foot disease control in grapevine. Pest Management Science. 74: 2864-2873.

Borenstein M, Hedges LV, Higgins JP, et al., 2011. Introduction to meta-analysis. John Wiley & Sons.

Bradford MA, Keiser AD, Davies CA, et al., 2013. Empirical evidence that soil carbon formation from plant inputs is positively related to microbial growth. Biogeochemistry. 113: 271-281.

Brozovic B, Jug D, Jug I, et al., 2018. Influence of Winter Cover Crops Incorporation on Weed Infestation in Popcorn Maize (Zea mays everta Sturt.) Organic Production. Agriculturae Conspectus Scientificus. 83: 77-81.

Bukovsky-Reyes S, Isaac ME, Blesh J, 2019. Effects of intercropping and soil properties on root functional traits of cover crops. Agriculture Ecosystems & Environment. 285.

Cardenas M, Castro J, Campos M, 2012. Short-term response of soil spiders to cover-crop removal in an organic olive orchard in a Mediterranean setting. Journal of Insect Science. 12.

Carrera LM, Buyer JS, Vinyard B, et al., 2007. Effects of cover crops, compost, and manure amendments on soil microbial community structure in tomato production systems. Applied Soil Ecology. 37: 247-255.

Chalise KS, Singh S, Wegner BR, et al., 2019. Cover crops and returning residue impact on soil organic carbon, bulk density, penetration resistance, water retention, infiltration, and soybean yield. Agronomy Journal. 111: 99-108.

Coppens F, Garnier P, De Gryze S, et al., 2006. Soil moisture, carbon and nitrogen dynamics following incorporation and surface application of labelled crop residues in soil columns. European journal of soil science. 57: 894-905.

de Carvalho WP, de Carvalho GJ, Abbade Neto DdO, et al., 2013. Agronomic performance of Cover crops used as ground cover mulching in the fallow period. Pesquisa Agropecuaria Brasileira. 48: 157-166.

de Souza GP, de Figueiredo CC, Gomes de Sousa DM, 2016. Soil organic matter as affected by management systems, phosphate fertilization, and cover crops. Pesquisa Agropecuaria Brasileira. 51: 1668-1676.

Destain J-P, Reuter V, Goffart J-P, 2010. Autumn cover crops and green manures: environment protection and agronomic interest. Biotechnologie Agronomie Societe Et Environnement. 14: 73-78.

Dijkstra P, Ishizu A, Doucett R, et al., 2006. 13C and 15N natural abundance of the soil microbial biomass. J Soil Biology Biochemistry. 38: 3257-3266.

Dornbush M, Cambardella C, Ingham E, et al., 2008. A comparison of soil food webs beneath C3-and C4-dominated grasslands. Biology & Fertility of Soils. 45: 73-81.

Drury C, McKenney D, Findlay W, 1991. Relationships between denitrification, microbial biomass and indigenous soil properties. Soil Biology and Biochemistry. 23: 751-755.

DuPont ST, Ferris H, Van Horn M, 2009. Effects of cover crop quality and quantity on nematode-based soil food webs and nutrient cycling. Applied Soil Ecology. 41: 157-167.

Elfstrand S, Hedlund K, Martensson A, 2007. Soil enzyme activities, microbial community composition and function after 47 years of continuous green manuring. Applied Soil Ecology. 35: 610-621.

Fierer N, Jackson RB, 2006. The diversity and biogeography of soil bacterial communities. Proceedings of the National Academy of Sciences. 103: 626-631.

Frasier I, Noellemeyer E, Figuerola E, et al., 2016. High quality residues from cover crops favor changes in microbial community and enhance C and N sequestration. Global Ecology and Conservation. 6: 242-256.

Gurevitch J, 1993. Meta-analysis: combining the results of independent experiments. Design and analysis of ecological experiments.

Hallama M, Pekrun C, Lambers H, et al., 2019. Hidden miners-the roles of cover crops and soil microorganisms in phosphorus cycling through agroecosystems. Plant and soil. 434: 7-45.

Hall H, Li Y, Comerford N, et al., 2010. Cover crops alter phosphorus soil fractions and organic matter accumulation in a Peruvian cacao agroforestry system. Agroforestry systems. 80: 447-455.

Haynes R, 2005. Labile organic matter fractions as centralcomponents of the quality of agricultural soils: anoverview. Advances in agronomy. 85: 221-268.

Hedges LV, Gurevitch J, Curtis PS, 1999. The meta-analysis of response ratios in experimental ecology. Ecology. 80: 1150-1156.

He Y, Qi Y, Dong Y, et al., 2013a. Effects of nitrogen fertilization on soil microbial biomass and community functional diversity in temperate grassland in Inner Mongolia, China. J Clean-Soil, Air, Water. 41: 1216-1221.

He Y, Qi Y, Dong Y, et al., 2013b. Effects of nitrogen fertilization on soil microbial biomass and community functional diversity in temperate grassland in Inner Mongolia, China. Clean-Soil, Air, Water. 41: 1216-1221.

Hubbard RK, Strickland TC, Phatak S, 2013. Effects of cover crop systems on soil physical properties and carbon/nitrogen relationships in the coastal plain of southeastern USA. Soil and Tillage Research. 126: 276-283.

Hungate BA, Van GROENIGEN KJ, Six J, et al., 2009. Assessing the effect of elevated carbon dioxide on soil carbon: a comparison of four meta-analyses. Global Change Biology. 15: 2020-2034.

Isik D, Kaya E, Ngouajio M, et al., 2009. Weed suppression in organic pepper (Capsicum annuum L.) with winter cover crops. Crop Protection. 28: 356-363.

Jackson LE, 2000. Fates and losses of nitrogen from a nitrogen-15-labeled cover crop in an intensively managed vegetable system. Soil Science Society of America Journal. 64: 1404-1412.

Karuku GN, Gachene C, Karanja N, et al., 2014. Effect of different cover crop residue management practices on soil moisture content under a tomato crop (Lycopersicon esculentum). Tropical and Subtropical Agroecosystems. 17: 509-523.

Kaye JP, Hart SC, 1997. Competition for nitrogen between plants and soil microorganisms. Trends in Ecology Evolution. 12: 139-143.

Khan MI, Hwang HY, Kim GW, et al., 2018. Microbial responses to temperature sensitivity of soil respiration in a dry fallow cover cropping and submerged rice mono-cropping system. Applied Soil Ecology. 128: 98-108.

Kokalis-Burelle N, McSorley R, Wang K-H, et al., 2017. Rhizosphere microorganisms affected by soil solarization and cover cropping in Capsicum annuum and Phaseolus lunatus agroecosystems. Applied Soil Ecology. 119: 64-71.

Kong AY, Six J, 2012. Microbial community assimilation of cover crop rhizodeposition within soil microenvironments in alternative and conventional cropping systems. Plant and soil. 356: 315-330.

Liang B, Wang X, Ma B, 2002. Maize root-induced change in soil organic carbon pools. J Soil Science Society of America Journal. 66: 845-847.

Li R, Hou X, Jia Z, et al., 2013. Effects on soil temperature, moisture, and maize yield of cultivation with ridge and furrow mulching in the rainfed area of the Loess Plateau, China. Agricultural Water Management. 116: 101-109.

Logsdon SD, Kaspar TC, Meek DW, et al., 2002. Nitrate leaching as influenced by cover crops in large soil monoliths. Agronomy Journal. 94: 807-814.

Lori M, Symnaczik S, Mäder P, et al., 2017. Organic farming enhances soil microbial abundance and activity—A meta-analysis and meta-regression. PLoS One. 12: e0180442.

Maul J, Drinkwater L, 2010. Short-term plant species impact on microbial community structure in soils with long-term agricultural history. Plant and Soil. 330: 369-382.

Mayer J, Buegger F, Jensen ES, et al., 2003. Residual nitrogen contribution from grain legumes to succeeding wheat and rape and related microbial process. Plant and Soil. 255: 541-554.

Mbuthia LW, Acosta-Martinez V, DeBruyn J, et al., 2015. Long term tillage, cover crop, and fertilization effects on microbial community structure, activity: Implications for soil quality. Soil Biology & Biochemistry. 89: 24-34.

Moore K, Bradley LK, 2018. North Carolina Extension gardener handbook. NC State Extension, College of Agriculture and Life Sciences, NC State ….

Muhammad I, Khan F, Khan A, et al., 2018. Soil fertility in response to urea and farmyard manure incorporation under different tillage systems in Peshawar, Pakistan. International Journal of Agriculture and Biology. 20: 1539-1547.

Muhammad I, Sainju UM, Zhao F, et al., 2019. Regulation of soil CO_2 and N_2O emissions by cover crops: A meta-analysis. Soil & Tillage Research. 192: 103-112.

Nevins CJ, Nakatsu C, Armstrong S, 2018. Characterization of microbial community response to cover crop residue decomposition. Soil Biology and Biochemistry. 127: 39-49.

Patkowska E, Błażewicz-Woźniak M, Konopiński M, et al., 2016. The effect of cover crops on the fungal and bacterial communities in the soil under carrot cultivation. Plant, Soil and Environment. 62: 237-242.

Peel MC, Finlayson BL, McMahon TA, 2007a. Updated world map of the Köppen-Geiger climate classification.

Peel MC, Finlayson BL, McMahon TA, 2007b. Updated world map of the Köppen-Geiger climate classification. Hydrology and earth system sciences discussions. 4: 439-473.

Pena-Pena K, Irmler U, 2016. Moisture seasonality, soil fauna, litter quality and land use as drivers of decomposition in Cerrado soils in SE-Mato Grosso, Brazil. Applied Soil Ecology. 107: 124-133.

Peregrina F, 2016. Surface Soil Properties Influence Carbon Oxide Pulses After Precipitation Events in a Semiarid Vineyard Under Conventional Tillage and Cover Crops. Pedosphere. 26: 499-509.

Peregrina F, Larrieta C, Ibanez S, et al., 2010. Labile organic matter, aggregates, and stratification ratios in a semiarid vineyard with cover crops. Soil Science Society of America Journal. 74: 2120-2130.

Peregrina F, Pilar Pérez-Álvarez E, García-Escudero E, 2014. Soil microbiological properties and its stratification ratios for soil quality assessment under different cover crop management systems in a semiarid vineyard. Journal of Plant Nutrition and Soil Science. 177: 548-559.

Pietikäinen J, Pettersson M, Bååth E, 2005. Comparison of temperature effects on soil respiration and bacterial and fungal growth rates. FEMS microbiology ecology. 52: 49-58.

Potthoff M, Loftfield N, Buegger F, et al., 2003. The determination of δ13C in soil microbial biomass using fumigation-extraction. J Soil Biology Biochemistry. 35: 947-954.

Reddy KN, 2003. Impact of rye cover crop and herbicides on weeds, yield, and net return in narrow-row transgenic and conventional soybean (Glycine max). Weed Technology. 17: 28-35.

Sainju UM, Singh BP, Whitehead WF, et al., 2007. Accumulation and crop uptake of soil mineral nitrogen as influenced by tillage, cover crops, and nitrogen fertilization. Agronomy Journal. 99: 682-691.

Sainju UM, Whitehead WF, Singh BR, 2003. Cover crops and nitrogen fertilization effects on soil aggregation and carbon and nitrogen pools. Canadian Journal of Soil Science. 83: 155-165.

Sakamoto T, Wardlow BD, Gitelson AA, et al., 2010. A two-step filtering approach for detecting maize and soybean phenology with time-series MODIS data. Remote Sensing of Environment. 114: 2146-2159.

Schmidt R, Mitchell J, Scow K, 2019. Cover cropping and no-till increase diversity and symbiotroph: saprotroph ratios of soil fungal communities. Soil Biology and Biochemistry. 129: 99-109.

Singh RN, Praharaj CS, Kumar R, et al., 2017. Influence of rice (Oryza sativa) habit groups and moisture conservation practices on soil physical and microbial properties in rice plus lathyrus relay cropping system under rice fallows in Eastern Plateau of India. Indian Journal of Agricultural Sciences. 87: 1633-1639.

Somenahally A, DuPont JI, Brady J, et al., 2018. Microbial communities in soil profile are more responsive to legacy effects of wheat-cover crop rotations than tillage systems. Soil Biology and Biochemistry. 123: 126-135.

Tang H, Xiao X, Tang W, et al., 2017. Returning Winter Cover Crop Residue Influences Soil Aggregation and Humic Substances under Double-cropped Rice Fields. Revista Brasileira De Ciencia Do Solo. 41.

Unger PW, Schomberg HH, Dao TH, et al., 1997. Tillage and crop residue management practices for sustainable dryland farming systems.

Van Groenigen KJ, Osenberg CW, Hungate BA, 2011. Increased soil emissions of potent greenhouse gases under increased atmospheric CO_2. Nature. 475: 214.

Wang GY, Hu YX, Liu YX, et al., 2021. Effects of supplement irrigation and nitrogen application levels on soil carbon-nitrogen content and yield of one-year double cropping maize in subtropical region. Water. 13: 1180.

Wawan, Dini IR, Hapsoh, et al., 2019. The effect of legume cover crop Mucuna bracteata on soil physical properties, runoff and erosion in three slopes of immature oil palm plantation, International Conference on Sustainable Agriculture for Rural Development 2018. IOP Conference Series-Earth and Environmental Science.

Wiebe N, Vandermeer B, Platt RW, et al., 2006. A systematic review identifies a lack of standardization in methods for handling missing variance data. Journal of clinical epidemiology. 59: 342-353.

Yang Y, Li X, Liu J, et al., 2017. Bacterial diversity as affected by application of manure in red soils of subtropical China. Biology and Fertility of Soils. 53: 639-649.

Zhou X, Chen C, Lu S, et al., 2012. The short-term cover crops increase soil labile organic carbon in southeastern Australia. Biology and Fertility of Soils. 48: 239-244.

Zhou X, Wu H, Li G, et al., 2016. Short-term contributions of cover crop surface residue return to soil carbon and nitrogen contents in temperate Australia. Environmental Science and Pollution Research. 23: 23175-23183.

Table S1　Information of cover crop and cash crop, location, climate, and soil texture for each experimental site used for meta-analysis in this study

No	Author	Site	Country	Köppen climate	Climatic zone	Köppen-Geiger	Temperature	Precipitation	Soil texture	Latitude	Longitude	Cover crop	Cash crop
1	Chavarría et al., (2016)	Pergamino, buenos aires	Argentina	Temperate climate	Humid subtropical	Cfa	16.5	971	Sandy clay loam	33°51′S	60°40′W	Oat/radish	Soybean-maiz
2	Khan et al., (2020)	Jinju	Korea	Temperate climate	Humid subtropical	Cwa	13.1	1,499	Clay loam	35°10′47.935″N	128°6′27.436″E	Vetch	Rice
3	Sánchez-Moreno et al., (2006)	Davis, CA	USA	Temperate climate	Hot-summer mediterranean climate	Csa	18	394	Silty clay loam	38°32′41.665″N	121°44′25.861″W	Oat	Tomato
4	Burger et al., (2005)	Davis, CA	USA	Temperate climate	Hot-summer mediterranean climate	Csa	18	394	Silt loam	38°32′41.665″N	121°44′25.861″W	Hairy vetch, winter pea	Tomato-maize
5	Minoshima et al., (2007)	Davis, CA	USA	Temperate climate	Hot-summer mediterranean climate	Csa	18	394	Silty clay loam	38°32′41.665″N	121°44′25.861″W	Oat-hay crops	Tomato
6	Rankoth et al., (2019)	Chariton, MS	USA	Continental climate	Hot-summer humid continental climate	Dfa	11.7	1,026	Silty clay loam	39°50′N	92°72′W	Winter barley, winter cereal rye	Maize-soybean
7	Araújo et al., (2019)	Teresina, Piauí	Brazil	Tropical Climate	Tropical savanna	Aw	27.9	1,004	Fine soil	3°5′S	41°46′W	Jack bean	Cassava
8	Cruz et al., (2019)	Brasília, Distrito Federal	Brazil	Tropical climate	Tropical savanna	Aw	21	1,749	Sandy soil	15°44′07.59″S	47°52′56.75″W	Mixed	Macadamia trees
9	Cagnini et al., (2018)	Cidade Gaúcha, Paraná state,	Brazil	Temperate climate	Humid subtropical	Cfa	20	1,340	Loamy fine sand	23°20′S	52°54′W	Millet	Sugarcane
10	Njamwe et al., (2018)	Lenye, ZIS	South Africa	Temperate climate	Humid subtropical	Cfa	23.2	590	Clay loam	32°45′S	27°04′E	Wheat/Oat	Maize
11	Singhe et al., (2018)	New Delhi	India	Dry Climate	Hot semi-arid	Bsh	24	642	Sandy loam	28°36′51″N	77°12′8.1″E	Pearl millet with Leucaena residue	Chickpea (Cicer arietinum)

(continued)

No	Author	Site	Country	Köppen climate	Climatic zone	Köppen-Geiger	Temperature	Precipitation	Soil texture	Latitude	Longitude	Cover crop	Cash crop
12	Somenahally et al., (2018)	El Reno, OK	USA	Temperate climate	Humid subtropical	Cfa	15	847	Silty clay loam	35°31'55"N	97°57'18" W	Cowpea	Winter-wheat
13	Tenelli et al., (2019)	Quirinópolis, GO	Brazil	Tropical climate	Tropical savanna climate	Aw	24.4	1,378	Clay	18°32' S	50°26' W	Sunn hemp	Sugarcane
14	Zibilske and Makus (2009)	Texas	USA	Dry climate	Hot semi-arid	Bsh	23	676	Sandy clay loam	26° 9' N	97° 57' W	Black oat	Cotton/maize
15	Tian et al., (2011)	Changping, Beijing	China	Continental climate	Humid continental climate	Dwa	12.1	610	Silty loam	40° 13' 14.376" N	116° 13' 52.334" E	Sweet corn residue	Cucumber
16	Mbuthia et al., (2015)	Jackson, TN	USA	Temperate climate	Humid subtropical	Cfa	16.5	1,375	Silty loam	35° 36' 52.142" N	88° 48' 49.91" W	Hairy vetch	Cotton
17	Linsler et al., (2016)	Kassel, Hesse	Germany	Temperate climate	Oceanic climate	Cfb	10	676	Silty clay loam	51° 18' 45.76" N	9° 28' 47.086" E	Oil radish	Rapesead-wheat
18	Zhu et al., (2012)	Huarong, Hunan	China	Temperate Climate	Humid subtropical	Cfa	17	1,330	Clay loamy	29° 52'N	112° 55' E	Ryegrass	Rice
19	Virto et al., (2011)	Olite, Navarra	Spain	Temperate climate	Oceanic climate	Cfb	15	525	Loamy	42° 28' 53.13" N	1° 39' 3.517" W	Grass cover	Vineyards
20	Marinari et al., (2015)	Viterbo, Rome	Italy	Temperate climate	Hot-summer Mediterranean	Csa	18	760	Sandy loam	45° 25'N	12° 04' E	Hairy vetch	Tomato
21	Frasier et al., (2016)	Anguil, La Pampa	Argentina	Temperate climate	Humid subtropical	Cfa	15.5	700	Sandy loam	36° 36'37.95" S	63° 58'48.2" W	Vetch rye	Sorghum
22	Abdollahi and Munkholm (2014)	Foulum	Denmark	Continental climate	Warm, humid continental	Dfb	7.3	626	Sandy loam	56° 30'N	9°35'E	Fodder radish	Spring barley
23	Peregrina (2016)	La Rioja	Spain	Temperate climate	Oceanic climate	Cfb	18	659	Loamy	42° 26' 3418" N	2° 26' 5307" W	Bromus catharticus	Vineyards

(continued)

No	Author	Site	Country	Köppen climate	Climatic zone	Köppen-Geiger	Temperature	Precipitation	Soil texture	Latitude	Longitude	Cover crop	Cash crop
24	Steenwerth and Belina (2008)	Monterey, CA	USA	Temperate climate	Dry – summer maritime sub-alpine climate	Csc	13.5	469	Sandy loam	36° 18′ 49.032″ N	121° 21′ 14.987″ W	Trios	Vineyard
25	Collins et al., (2006)	Benton County, WA	USA	Dry climate	Cold semi-arid	Bsk	12	204	Silt loam	36° 22′ 21″ N	94° 12′ 22″ W	White Mustard	Rice–Maize
26	De Medeiros et al., (2019)	Garanhuns, Pernambuco	Brazil	Tropical climate	Tropical savanna	Aw	24.5	1,000	Sandy soil	08° 48′ 34, 2″ S	36° 24′ 29, 3″ W	Arachis pintoi	Maize
27	Dinesh (2004)	Andaman, Islands	India	Tropical climate	Tropical monsoon climate	Am	30.1	3,100	Sandy clay loam	10° 30′ N	92° 14′ W	Pueraria phaseoloides	Coconut
28	Dinesh et al., (2009)	Andaman, Islands	India	Tropical climate	Tropical monsoon climate	Am	30.1	3,100	Sandy clay loam	11° 41′ N	92° 39′ W	Pueraria phaseoloides	Coconut
29	Diniz et al., (2019)	Chapadão do Sul, MS	Brazil	Tropical climate	Tropical savanna	Aw	22.7	1,598	Silty soil	18°41′33″S	52°40′45″W	Brachiaria	Soybean
30	Liang et al. (2014)	Kinston, NS	USA	Temperate climate	Humid subtropical	Cfa	16	1,287	Sandy loam	35° 16′ 42.611″ N	77° 36′ 55.267″ W	Crimson clover	Cotton
31	Motta et al., (2007)	Alabama	USA	Temperate climate	Humid subtropical	Cfa	18.3	1,500	Silt loam soil	34° 41′ 30″ N	86° 53′ 25″ W	Wheat/soybean	Cotton
32	Shah et al, (2010)	Tarnab Peshawar	Pakistan	Dry climate	Hot semi-arid	Bsh	22.7	384	Loamy	34° 04′ N	71° 40′ E	Pigeonpea	Rice–wheat
33	Venkateswarlu et al., (2007)	Andhra Pradesh	India	Tropical climate	Tropical savanna	Aw	25.2	738	Loamy	16° 6′ N	78° 40′ E	Horsegram	Sorghum/sunflower
34	Wang et al. (2007)	Homestead, Florida	USA	Tropical climate	Tropical savanna	Aw	23.6	1,534	Loamy	25° 28′ N	80° 29′ W	Sunn hemp	Tomato
35	Wells et al., (2000)	Sydney, SW	Australia	Temperate climate	Humid subtropical	Cfa	17.6	1,309	Sandy loam	33°23′ S	151°21′ E	Field pea, cowpea, subterranean clover, Oat, white lupin	Tomato, spinach, cucumber, broccoli,

(continued)

No	Author	Site	Country	Köppen climate	Climatic zone	Köppen-Geiger	Temperature	Precipitation	Soil texture	Latitude	Longitude	Cover crop	Cash crop
36	Zhou et al., (2016)	New South Wales,	Australia	Temperate climate	Humid sub-tropical	Cfa	21.85	389	Sandy clay loam	35°05′ S	147°20′ E	Mixture	Winter wheat
37	Steenwerth and Belina (2008)	Monterey, CA	USA	Temperate climate	Hot-summer	Csa	18.2	432	Sandy loam	36°18′49.032″ N	121°21′14.987″ W	Rye	Vineyard
38	Njaimwe et al., (2018)	Burnshill, ZIS	South Africa	Temperate climate	Humid sub-tropical	Cfa	23.2	590	Sandy clay loam	32°45′ S	27°04′ E	Wheat/Oat	Maize
39	Tenelli et al., (2019)	Quata, SP	Brazil	Temperate climate	Humid sub-tropical	Cfa	23.7	1,391	Sand-loam	22°14′	50°42′	Sunn hemp	Sugarcane
40	Tenelli et al., (2019)	Chapadãodo Ce'u, GO	Brazil	Tropical climate	Tropical savanna climate	Aw	22.5	1,654	Clay	18°25′	52°33′	Sunn hemp	Sugarcane

Table S2 Subgroup homogeneity analysis for the moderator of soil microbial biomass carbon (SMBC), soil microbial biomass nitrogen (SMBN), and SMBC/SMBN ratio in response to cover crops (CCs) types CC management, and plant types

Moderator			Intercept	SE	Slope	SE	Regression (Q_m)	P-value	Residual (Q_e)	P-value
SMBC										
CC types		Legume	0.62	0.11	-0.06	0.03	5.01	0.025*	122.58	0.838
		N-legume	0.51	0.06	-0.01	0.01	0.91	0.341	109.11	0.993
		Mixed	-0.21	0.26	0.11	0.08	2.11	0.147	37.19	0.205
CC management		Incorporated	0.48	0.05	-0.01	0.01	0.55	0.458	103.97	0.800
		Mulched	-0.33	0.92	0.16	0.29	0.29	0.591	15.58	0.997
		Removed	1.80	0.29	-0.39	0.08	23.82	<0.001***	56.07	0.291
Plant Category		C3-plants	0.58	0.05	-0.01	0.01	1.57	0.211	118.92	1.000
		C4-plants	0.24	0.09	-0.03	0.02	1.49	0.222	48.98	0.317
		C3 and C4	0.21	0.40	-0.01	0.13	0.00	0.948	67.68	0.100
SMBN										
CC types		Legume	0.59	0.23	-0.04	0.06	0.48	0.490	51.37	0.271
		N-legume	1.25	0.16	-0.21	0.05	21.09	0.000	124.96	0.003**
		Mixed	1.98	2.22	-0.42	0.65	0.41	0.520	12.55	0.483
CC management		Incorporated	1.00	0.16	-0.15	0.04	14.45	0.480	57.22	0.425
		Mulched	1.56	0.11	-0.26	0.03	83.50	<0.001***	130.724	<0.001***
		Removed	1.31	0.46	-0.20	0.13	2.12	0.519	42.16	0.429
Plant category		C3-plants	1.12	0.13	-0.17	0.04	22.59	<0.001***	163.15	0.001**
		C4-plants	0.63	1.11	-0.05	0.34	0.02	0.886	33.28	0.405
		C3 and C4	0.10	0.07	-0.01	0.01	0.46	0.496	1.24	0.872
SMBC/SMBN ratio										
CC types		Legume	0.11	0.16	-0.01	0.03	0.03	0.873	48.34	0.232
		N-legume	1.36	0.23	-0.27	0.06	18.80	<0.001***	46.52	0.292
		Mixed	-0.11	0.56	0.02	0.16	0.02	0.878	12.01	0.445
CC management		Incorporated	0.45	0.21	-0.04	0.04	0.91	0.681	31.00	0.354
		Mulched	2.29	0.74	-0.67	0.23	8.29	0.022*	19.89	0.050
		Removed	2.75	0.48	-0.69	0.13	26.48	<0.001***	33.25	0.504
Plant category		C3-plants	0.62	0.15	0.04	-0.08	4.32	0.038*	76.29	0.181
		C4-plants	-0.20	0.19	0.04	0.04	0.74	0.391	35.20	0.319
		C3 and C4	—	—	—	—	—	—	—	—

Table S3 Subgroup homogeneity analysis for the moderator of soil microbial biomass carbon (SMBC), soil microbial biomass nitrogen (SMBN), and SMBC/SMBN ratio in response to climate, precipitation (mm), and air temperature (℃)

Moderator			Intercept	SE	Slope	SE	Regression (Q_m)	P-value	Residual (Q_e)	P-value
SMBC										
Climate		Tropical	0.06	0.22	0.052	0.05	1.31	0.251	46.21	0.765
		Dry	1.41	0.47	-0.31	0.14	4.56	0.032*	18.43	0.621
		Temperate	0.39	0.05	-0.01	0.01	1.08	0.298	140.85	0.997
		Continental	1.58	0.75	-0.28	0.23	1.51	0.219	29.74	0.630
Precipitation (mm)		>1,500	1.53	0.20	-0.20	0.04	28.20	<0.001***	25.56	0.487
		1,000–1,500	0.57	0.10	-0.09	0.02	13.88	<0.001***	143.55	0.001**
		<1,000	0.49	0.06	-0.01	0.01	0.48	0.489	117.74	1.000
Temperature (℃)		>20	0.28	0.13	0.01	0.03	0.12	0.724	78.10	0.896
		15–20	0.28	0.06	0.00	0.01	0.19	0.666	54.36	1.000
		<15	1.71	0.61	-0.38	0.20	3.70	0.054	115.47	0.649
SMBN										
Climate		Tropical	1.99	0.65	-0.37	0.17	4.80	0.028*	22.18	0.448
		Dry	0.93	0.44	-0.11	0.12	0.83	0.361	4.00	0.406
		Temperate	1.58	0.19	-0.27	0.05	27.24	<0.001***	127.35	0.003**
		Continental	0.03	0.19	0.03	0.06	0.29	0.589	28.00	0.464
Precipitation (mm)		>1,500	1.84	1.30	-0.43	0.42	1.05	0.304	31.54	0.025*
		1,000–1,500	0.80	0.59	-0.13	0.15	0.78	0.235	31.91	0.018*
		<1,000	1.08	0.15	-0.15	0.04	13.38	<0.001***	129.90	0.038*
Temperature (℃)		>20	0.59	0.12	-0.08	0.03	7.11	0.008**	107.81	<0.001***
		15–20	1.50	0.37	-0.29	0.11	7.50	0.006**	80.07	0.241
		<15	0.73	0.41	0.02	0.12	0.03	0.853	26.90	0.524
SMBC/SMBN ratio										
Climate		Tropical	-2.39	1.25	0.59	0.31	3.51	0.060	22.50	0.430
		Dry	—	—	—	—	—	—	—	—
		Temperate	0.21	0.16	-0.04	0.04	1.20	0.272	54.55	0.181
		Continental	1.26	0.69	-0.19	0.22	0.82	0.364	27.27	0.503
Precipitation (mm)		>1,500	-0.30	0.19	0.06	0.04	2.85	0.091	19.08	0.387
		1,000–1,500	-0.76	0.24	0.13	0.04	9.43	0.002**	14.57	0.985
		<1,000	1.77	0.37	-0.44	0.11	15.94	<0.001***	73.04	0.205
Temperature (℃)		>20	-0.29	0.14	0.06	0.03	3.82	0.051	31.51	0.391
		15–20	0.32	0.19	-0.06	0.04	2.18	0.140	46.88	0.153
		<15	1.26	0.69	-0.20	0.22	0.82	0.364	27.28	0.503

Table S4 Subgroup homogeneity analysis for the moderator of soil microbial biomass carbon (SMBC), soil microbial biomass nitrogen (SMBN), and SMBC/SMBN ratio in response to soil texture, soil pH, initial soil organic carbon (SOC) and initial total nitrogen (TN)

Moderator		Intercept	SE	Slope	SE	Regression (Q_m)	P-value	Residual (Q_e)	P-value
SMBC									
Soil texture	Course	0.66	0.14	-0.06	0.04	2.77	0.096	75.54	1.000
	Medium	0.37	0.05	0.00	0.01	0.17	0.682	127.14	0.090
	Fine	1.04	0.11	-0.18	0.03	41.10	<0.001***	143.82	<0.001***
Soil pH	pH>7.5	0.46	0.25	0.02	0.07	0.07	0.791	34.52	0.303
	pH 6.5-7.5	0.45	0.06	-0.01	0.01	0.38	0.537	88.16	1.000
	pH<6.5	1.00	0.14	0.17	0.17	22.18	<0.001***	151.86	0.010*
SOC (mg/kg)	SOC>20	-1.38	0.31	0.52	0.09	31.25	<0.001***	14.80	0.540
	SOC 10-20	0.46	0.05	-0.01	0.01	0.94	0.333	143.04	1.000
	SOC<10	0.92	0.12	-0.13	0.03	20.38	<0.001***	145.48	<0.001***
TN (%)	TN>2	1.53	0.20	-0.20	0.04	28.20	<0.001***	25.56	0.487
	TN 1-2	0.57	0.10	-0.09	0.02	13.85	<0.001***	142.77	0.001**
	TN<1	0.49	0.06	-0.01	0.01	0.49	0.484	113.28	1.000
SMBN									
Soil texture	Course	1.20	0.31	-0.22	0.10	5.50	0.019*	119.72	0.005**
	Medium	0.46	0.74	0.09	0.22	0.16	0.690	48.37	0.171
	Fine	0.34	0.17	-0.04	0.03	1.47	0.226	30.34	0.064
Soil pH	pH>7.5	-0.26	-0.26	0.06	0.03	5.47	0.019*	1.79	0.409
	pH 6.5-7.5	1.08	0.15	-0.18	0.04	22.70	<0.001***	95.57	<0.001***
	pH<6.5	-0.11	0.54	0.21	0.16	1.67	0.197	77.14	0.348
SOC (mg/kg)	SOC>20	-0.26	0.14	0.06	0.03	5.47	0.019	1.79	0.409
	SOC 10-20	1.01	0.29	-0.12	0.09	1.95	0.162	96.03	0.262
	SOC<10	0.61	0.31	-0.09	0.08	1.42	0.234	65.49	0.002**
TN (%)	TN>2	—	—	—	—	—	—	—	—
	TN 1-2	1.14	0.26	-0.17	0.07	5.51	0.019*	125.32	0.151
	TN<1	0.27	0.15	-0.02	0.03	0.54	0.463	19.41	0.367
SMBC/SMBN ratio									
Soil texture	Course	-0.03	0.14	0.01	0.03	0.14	0.709	56.41	0.248
	Medium	1.29	0.17	-0.21	0.04	26.89	<0.001***	45.10	0.267
	Fine	0.23	0.92	-0.09	0.27	0.12	0.728	6.34	0.386
	pH>7.5	-2.01	1.10	0.47	0.29	2.55	0.110	2.00	0.368
Soil pH	pH 6.5-7.5	0.13	0.97	0.01	0.32	0.00	0.970	26.75	0.084
	pH<6.5	0.71	0.15	-0.11	0.03	9.43	0.002**	70.31	0.335
	SOC>20	-2.01	1.10	0.47	0.29	2.55	0.110	2.00	0.368
SOC (mg/g)	SOC 10-20	1.90	0.37	-0.47	0.11	18.68	<0.001***	68.10	0.278
	SOC<10	-0.21	0.12	0.04	0.02	4.80	0.029	26.16	0.245
	TN>2	—	—	—	—	—	—	—	—
TN (%)	TN 1-2	0.90	0.18	-0.17	0.05	13.86	<0.001***	90.51	0.122
	TN<1	-0.20	0.14	0.05	0.02	3.56	0.059	14.27	0.284

Effects of Spatial Distribution on Photosynthesis and Yield of Summer Maize

X. M. Mao, P. J. Shen, Y. X. Zhao and X. B. Zhou

(Guangxi Colleges and Universities Key Laboratory of Crop Cultivation and Farming System, Agricultural College of Guangxi University, Nanning 530004, China)

Abstract: The experiment was aimed to determinate reasonable group structure for maintaining stable and higher grain yield by adjusting row spacing. High leaf area index, chlorophyll content index, photosynthetic rate (Pn) and radiation use efficiency of row spacing 50 cm was observed in this study. Pn of row spacing 40 cm and 80 cm were lower than those of other treatments. Three-year average values of the daily increase in dry matter of row spacing 40 cm, 50 cm, 60 cm, 70 cm, and 80 cm were 199 kg/(ha · d), 198 kg/(ha · d), 182 kg/(ha · d), 185 kg/(ha · d), and 184 kg/(ha · d), respectively. Grain yield of row spacing 50 cm was significantly different as compared to row spacing of only 80 cm. Therefore, row spacing 80 cm was minimal spatial structure whereas 50 cm spatial structure found to be optimal compared to rest of the row spacings that positively affect summer maize grain yield under rain-fed condition.

Keywords: Zea mays; Leaf area index; Photosynthetic rate; Radiation use efficiency

1 Introduction

Maize (*Zea mays* L.) is one of the most important cereal crops and plays an important role in expanding overall grain production capacity, especially in the North China Plain, which is main producing areas in China. However, lack of surface water sources has led to long-term and massive exploitation of groundwater resources for development of irrigated agriculture in the region (Sun et al., 2010), which has caused the water level to fall and created several environmental problems (Wang et al., 2016). Therefore, water-saving agricultural practices system is necessary. Cultivation practices could affect maize population significantly and break these restrictions on yield (Guan et al., 2014). Row spacing determines the spatial distribution of the plants, which affects canopy structure, light interception and radiation use efficiency (RUE) (Mattera et al., 2013); row spacing is expected to be an economical method for enhancing grain yield by utilizing increased radiation capture (Caviglia and Andrade, 2010).

Different spatial arrangements can affect resource competition relationships (Brant et al., 2009), and the reasonable row spacing is necessary to improve the relation between group and individual plants (Norsworthy and Shipe, 2005). The bilinear response of dry matter (DM) accumu-

---

Author for correspondence: whyzxb@gmail.com.

lation to plant spatial distribution was determined by RUE (Mattera et al., 2013). Cropping systems are also proposed as a better method to enhance crop yields (Caviglia and Andrade, 2010). A leaf area index (LAI) of 3.5-4.0 in early reproduction is necessary to increasing crop yield (Liu et al., 2016). Zarate-Valdez et al. (2012) predicted that LAI and chlorophyll content should be determined in early stages to enhance distribution and utilization of crop resources. Therefore, the hypothesis was that reduced row spacing generates reasonable spatial arrangements that minimize interspecific competition, and favorable utilization of crop resources likely maintain higher grain yield of summer maize under rain-fed condition in the North China Plain.

2 Materials and Methods

The experiment was carried out at the Agronomy Experiment Stations of Shandong Agricultural University (36°09′ N, 117°09′ E) at Tai'an, China from 2011 to 2013. The soil was silt loam (from surface 0 cm to 20 cm depth; pH 6.9) with the following average contents: soil organic matter 16.3 mg/g, total nitrogen 1.3 mg/g, available P 35 mg/kg, and available K 95 mg/kg. The long-term (from 1971 to 2010) annual average rainfall and temperature were 693.5 mm and 13.1℃, respectively. Total solar radiation data in 2011, 2012, and 2013 summer maize growing season were 4,747 MJ/m^2, 4,951 MJ/m^2, and 4,980 MJ/m^2, respectively. The weather data were collected from the Tai'an Agrometerological Experimental Station.

The experiment consisted of five row spacings as 40 cm (RS40), 50 cm (RS50), 60 cm (RS60), 70 cm (RS70), 80 cm (RS80) under the same plant density (62,500/ha). The experiments had three replications with a randomized plot design. Summer maize (cv. Luyu14) was planted in plots (4 m × 4 m) on June 18, 2011, June 17, 2012, and June 19, 2013 and harvested on September 24, 2011, October 2, 2012, and October 2, 2013, respectively. The experimental plot was applied with 202.5 g diammonium phosphate, 202.5 g urea, and 152.1 g kalium chloratum before sowing. Summer maize is a rain-fed crop thus the experiment conducted without irrigation during growth period.

The photosynthetic rate (Pn), LAI, chlorophyll content index (CCI), and DM were measured at V6, R0, R2, R3, R4, and R5 growth stages following (Wang et al., 2015). Plants were dried in an oven at 105℃ for 30 min followed by at 80℃ until reach to a constant weight to determine DM.

The Pn was measured using a LI-6400XT (LI-COR Inc., Lincoln, USA) with an artificial light source [1,400 μmol/(m^2 · s)]; CCI was measured using chlorophyll content meter 200 (Optic-Sciences Inc., Tyngsboro, USA). LAI was measured with the following formulae:

LAI = Leaf length × leaf width × 0.75 (Amanullah et al., 2007).

Where leaf length is the distance between the leaf pillow and leaf tip, and leaf width is the widest part of the leaf. Use of radiation efficiency was calculated as:

RUE = ΔW × H/ΣS × 100% (Zhang et al., 2016), where ΔW is the aboveground biomass of sample (g/m^2), H is the product value of heat (summer maize kernel is 16.5 KJ/g; both stem and leaf are 14.4 KJ/g), it is same for all treatments, ΣS is the total radiation of the unit area

(MJ/m^2).

Ten plants with similar growth vigour were harvested as samples by using a sickle in each plot for measuring per-plant kernel number (KNP), kernel weight (KW). A total of 2 m^2 summer maize was harvested to measure grain yield and harvest index (HI).

The data were statistically analyzed by SAS 9.2 software. All graphs were drawn using Sigma Plot 10.0 (SPSS Inc., Chicago, IL). Effects were considered significant with the least significant difference (LSD) test at $P \leq 0.05$.

3 Results and Discussion

The three-year experiment showed Pn more or less decreasing tendency with growth stage development (Figure 1). A significant linear regression was also noted between Pn and growth stage. The linear correlation equation was y [Pn, $\mu mol\ CO_2/(m^2 \cdot s)$] $= -6.201x$ (growth stage) $+ 49.944$, $R^2 = 0.803$ (2011, $P \leq 0.001$); $y = -2.441x + 38.088$, $R^2 = 0.544$ (2012, $P \leq 0.001$); and $y = -4.393x + 42.95$, $R^2 = 0.624$ (2013, $P \leq 0.001$). At V6 to R4, the three-year average values of row spacing for 40 cm, 50 cm, 60 cm, 70 cm, 80 cm were 29.9 $\mu mol\ CO_2/(m^2 \cdot s)$, 31.0 $\mu mol\ CO_2/(m^2 \cdot s)$, 32.1 $\mu mol\ CO_2/(m^2 \cdot s)$, 31.3 $\mu mol\ CO_2/(m^2 \cdot s)$, and 28.8 $\mu mol\ CO_2/(m^2 \cdot s)$, respectively. At V6 and R3, no significant difference was noted among the Pn of the treatments ($P > 0.05$). At R0 and R2, the Pn of RS80 was significantly lower than that of the other row spacing; at R4, the Pn of RS50, RS60, RS70 was higher than that of RS40. The photosynthesis in narrow spacings (40 cm and 50 cm) was better than those of wide spacing (60 cm, 70 cm and 80 cm), suggesting reasonable canopy closure and plants distribution in narrow spacing was an indicator of more favorable growing condition (Caviglia and Andrade, 2010).

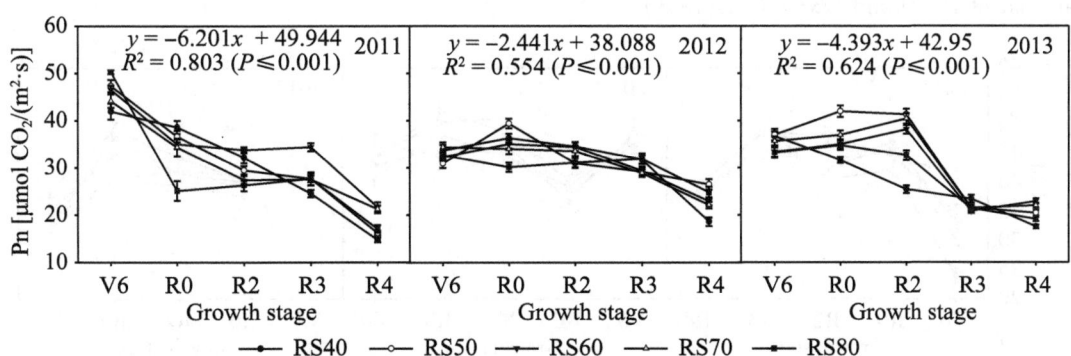

Figure 1 Net photosynthetic rate (Pn) of different row spacings and growth stages
The bars are the SE.

Different treatments showed similar trend of inverted-U shaped ('∩') curve in LAI during the growing season, especially in 2012 (Figure 2). Three regression equations in 2011, 2012, and 2013 were: y (LAI) $= -0.1731x^2 + 0.9669x$ (growing stage) $+ 3.5087$, $R^2 = 0.6931$ ($P \leq 0.001$); $y = -0.450x^2 + 3.055x - 0.281$, $R^2 = 0.935$ ($P \leq 0.001$); and $y = -0.167x^2 +$

$0.800x + 3.082$, $R^2 = 0.877$ ($P \leqslant 0.001$), respectively. In V6 to R4, the LAIs of row spacing 40 cm, 50 cm, 60 cm, 70 cm, 80 cm were listed as follows: 4.49, 4.38, 4.60, 4.44, and 4.62 in 2011; 3.76, 4.06, 3.99, 3.95, and 3.92 in 2012; and 3.52, 3.83, 3.62, 3.54, and 3.68 in 2013, respectively. Han et al. (2016) reported that seedling phase to physiological maturity, LAI were negative correlated with row spacing for winter wheat under deficit irrigation.

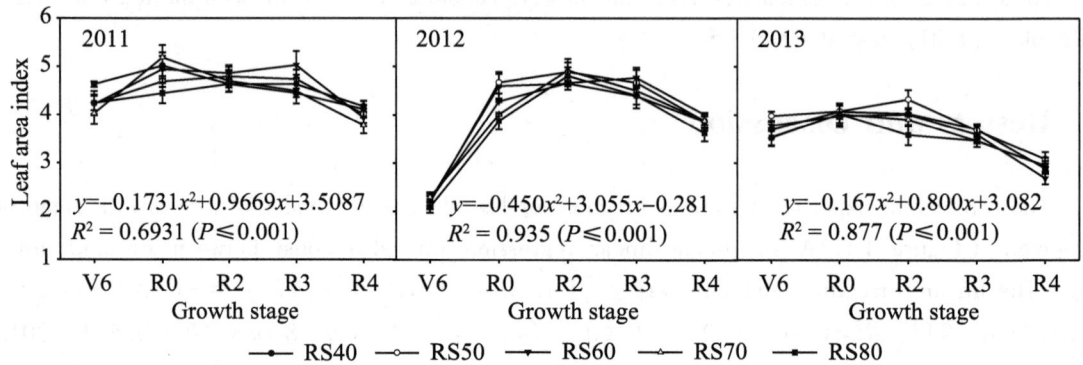

Figure 2　Leaf area index of different row spacings and growth stages
The bars are the SE.

In 2011, the CCI of RS40 at V6 was significantly higher than that of other treatments in contrast, the CCI of RS80 was $P<0.05$. The CCIs of all treatments were relatively low because of low amount of solar radiation. A relatively high CCI in R0 to R4 was attributed to high amount of precipitation. At R2, the CCI of RS50 was the highest in all treatments. At V6 to R4, the CCIs of row spacing 40 cm, 50 cm, 60 cm, 70 cm, 80 cm were 43.5, 43.3, 41.5, 40.7 and 41.5, respectively. The CCI of RS40 11.1% was higher than that of RS80. At V6, the CCI of 2012 was higher than that of 2011 and 2013. Whereas the CCI mean of RS40, RS50, RS60 was 7.4% higher than that of RS70 and RS80 (Figure 3).

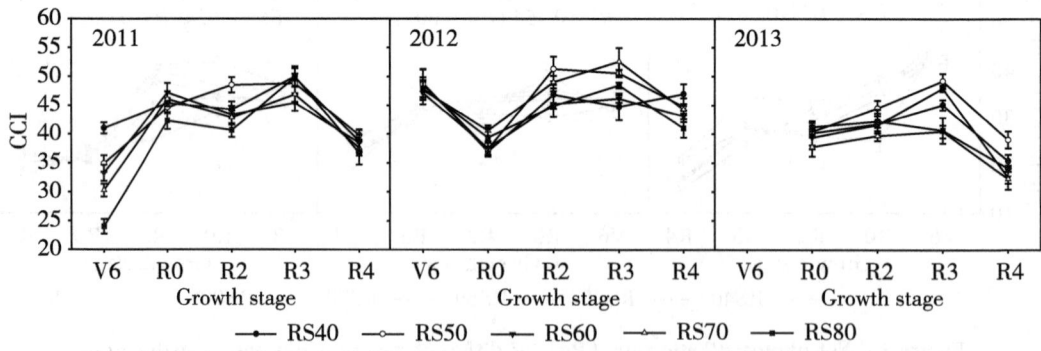

Figure 3　Chlorophyll content index (CCI) of different row spacings and growth stages
The bars are the SE.

The three-year CCI average of row spacing 40 cm, 50 cm, 60 cm, 70 cm, 80 cm were 43.0, 44.4, 41.9, 41.7, and 41.0, respectively. The pattern of CCIs observed as: RS50 > RS40 > RS60 > RS70 > RS80. The CCI of RS80 was the lowest, but CCI of RS50 was the highest among all treatments. The discrepancy may contribute to optimum spatial distribution and improve light distribution. CCI and LAI for RS40 and RS50, resulted in high intercept and capture of solar radiation in rain-fed condition. Photosynthesis was significantly affected by plant spatial distribution (Hamzei and Soltani, 2012).

DM was positively correlated with growing stage (Figure 4). The linear equation was y (DM, kg/ha) $= 3,896.1 x$ (growth stage) $+ 853.88$, $R^2 = 0.902$ (2011, $P \leqslant 0.001$); $y = 5,230.9 x - 3,580.3$, $R^2 = 0.989$ (2012, $P \leqslant 0.001$), $y = 3,664.5 x + 509.01$, $R^2 = 0.906$ (2013, $P \leqslant 0.001$), respectively. In 2011, the DM of RS60 was lower than that of other treatments at R0; the DM of RS40 increased daily by 13.9% higher than that of RS50. In 2012, DM slightly differed among the treatments; the average DM values of RS40 and RS50 were 10.3% higher than those of other treatments. In 2013, the average DM was relatively high. The three-year average values of daily increase in DM of row spacing 40, 50, 60, 70, 80 cm were 199, 198, 182, 185, and 184 kg/(ha · d). Narrow spacings improved spatial distribution and increased Pn, hence, crop plants produced high DM (Gonias and Oosterhuis, 2011).

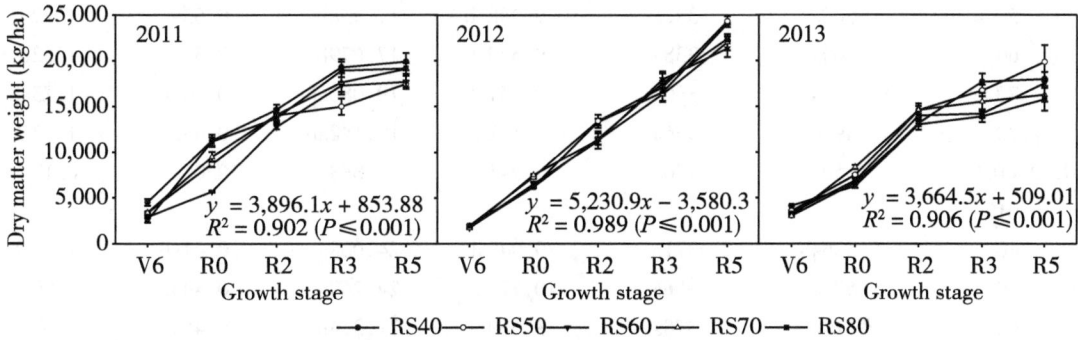

Figure 4 Dry matter weight of different row spacings and growth stages
The bars are the SE.

The RUE of crops are determined by environmental factors, such as canopy structures, LAI, radiation regimes (diffuse or direct), temperatures, water contents, and cropping systems (Brodrick et al., 2013). In 2011, row spacing did not significantly affect KNP and KW ($P > 0.05$). The yield and RUE of RS40 were the highest among all of the treatments. The HI of RS80 was also significantly lower than that of RS50 and RS60 ($P \leqslant 0.05$). In 2012, the KNP of RS70 and RS80 was significantly lower than that of RS50 ($P \leqslant 0.05$). The yield, DM, and RUE of RS60 were significantly lower than that of RS50 ($P \leqslant 0.05$).

The average of the three-year experiments showed that row spacing did not significantly affect KW. In addition, the KNP of RS40 and RS50 were significantly higher than that of RS80. The order of yield from high to low was recorded as RS50, RS40, RS60, RS70, and RS80, respectively. The DMs of RS40 were significantly higher than those of RS60 and RS70 ($P \leqslant 0.05$). The HI of RS60 was 7.1% higher than that of RS80 (Table 1). Although DM accumulation and RUE of RS40 were higher than those of RS50, uniform distribution was not advisable to maintain high Pn in the late growth stage. Yield also slightly decreased. Therefore, RS50 is an optimal pattern for summer maize cultivation. In all treatments the RUE of RS40 was the highest in 2011; whereas RUE of RS50 was relatively higher in 2012 and 2013. Three years study demonstrated that there was no difference in row spacings regarding KW, KNP was significantly higher in RS50 than that in RS80, correspondingly, plants in RS50 had the highest grain yield, while that in RS80 had the lowest. These indicate that RS50 had significantly increased grain yield when compared to other row spacings and positively related with the increase of KNP. This finding shows similarity to the result of mungbean crop reported by Rachaputi et al. (2015).

Table 1 Row spacing effects on summer maize per-plant kernel number (KNP), kernel weight (KW), yield, dry matter (DM), harvest index (HI) and radiation use efficiency (RUE) in 2011-2013

Row spacing (cm)	KNP	KW (mg)	Yield (kg/ha)	DM (kg/ha)	HI	RUE (%)
2011						
40	538a	348a	10,415a	19,878a	0.53a	1.81a
50	506a	332a	9,288ab	17,458b	0.53a	1.59b
60	500a	338a	9,569ab	17,679b	0.54a	1.62b
70	510a	330a	9,551ab	19,080ab	0.50ab	1.73ab
80	496a	336a	9,152b	19,142ab	0.48b	1.73ab
LSD (0.05)	50	20	941	1,888	0.05	0.17
2012						
40	536ab	371a	9,948ab	24,008a	0.41ab	1.87ab
50	572a	356a	10,779a	24,269a	0.44ab	1.90a
60	529ab	378a	9,660b	21,314b	0.45a	1.67b
70	515bc	366a	9,035b	21,992ab	0.41ab	1.71ab
80	475c	366a	9,011b	22,367ab	0.40b	1.74ab
LSD (0.05)	40	39	993	2,668	0.05	0.20
2013						
40	552ab	314b	8,352b	18,003ab	0.46b	1.37ab
50	572a	351a	9,443a	19,853a	0.48ab	1.51a
60	519b	332ab	8,693ab	17,536ab	0.50ab	1.34ab
70	539ab	313b	8,238b	16,278b	0.51a	1.25b
80	535ab	313b	8,099b	15,778b	0.51a	1.21b
LSD (0.05)	39	33	1,586	3,509	0.04	0.26
Mean						
40	543a	344a	9,572ab	20,629a	0.47bc	1.68a
50	550a	347a	9,837a	20,503ab	0.48ab	1.67a
60	515ab	349a	9,307bc	18,843c	0.50a	1.54bc
70	521ab	336a	8,956cd	18,995c	0.47bc	1.56b
80	502b	338a	8,754d	19,076bc	0.46c	1.56b
LSD (0.05)	39	17	520	1,445	0.02	0.12

Notes: Values in a column with different letters are significantly different ($P \leq 0.05$).

The values of each variable decreased with the increase of row spacing and significantly negatively correlated with yield and LAI ($P<0.05$). Yield was highly significantly and positively correlated with LAI ($P<0.01$). Pn increased with other variables and there was no significant differences between them. CCI significantly and positively correlated with LAI ($P<0.05$) (Table 2). The change in row spacing would have changed of the canopy structure, and as spacing reduced, leaves may had been exposed to lower irradiance resulting higher LAI, and thus showed increase RUE (Mattera et al., 2013). High KNP in RS50 and its positive association with RUE, suggested that different row spacings changed the responses of RUE to KNP, and grain yield.

Table 2 Correlation matrix of row spacing-related variables, grain yield, and photosynthetic rate (Pn), chlorophyll content index (CCI) and leaf area index (LAI) in summer maize grown in 2011-2013

Variable	Row spacing	Yield	Pn	CCI	LAI
Row spacing	1.0000	-0.9020*	-0.2212	-0.8004	-0.8893*
Yield		1.0000	0.3217	0.9556**	0.9580**
Pn			1.0000	0.2460	0.5606
CCI				1.0000	0.8787*
LAI					1.0000

Notes: * Presented at $P<0.05$; ** presented at $P<0.01$.

This study showed that plant spatial distribution had significant effect on Pn, CCI, LAI, HI, grain yield, yield components and RUE. The results suggested that RS50 is an optimal spatial structure that positively improves CCI, LAI and RUE of summer maize and consequently is relatively high KNP and plant grain yield in rain-fed condition.

Acknowledgements

This work was supported by the National Natural Science Foundation of China (31760354), and Guangxi Natural Science Foundation (2017GXNSFAA198036, 2015GXNSFAA139049). The authors are specially thankful to Zhong Wen Wen, Wang Xing Ya, Zhang Zhen for their valuable contributions.

References

Amanullah, Hassan MJ, Nawab K and Ali A, 2007. Response of specific leaf area (SLA), leaf area index (LAI) and leaf area ratio (LAR) of maize (*Zea mays* L.) to plant density, rate and timing of nitrogen application. World Appli. Sci. J. 2: 235-243.

Brant V, Neckář K, Pivec J, et al., 2009. Competition of some summer catch crops and volunteer cereals in the areas with limited precipitation. Plant Soil Environ. 55: 17-24.

Brodrick R, Bange MP, Milroy SP, et al., 2013. Physiological determinants of high yielding ultra-narrow row cotton: canopy development and radiation use efficiency. Field Crops Res. 148: 86-94.

Caviglia OP and Andrade FH, 2010. Sustainable Intensification of agriculture in the Argentinean Pampas: capture and use efficiency of environmental resources. Plant Biotechnol. J. 3: 1-8.

Gonias DM and Oosterhuis ACB, 2011. Light interception and radiation use efficiency of okra and normal leaf cotton isoclines. Environ. Exp. Bot. 72: 217-222.

Guan D, Al-Kaisi MM, Zhang Y, et al., 2014. Tillage practices affect biomass and grain yield through regulating root growth: root-bleeding sap and nutrients uptake in summer maize. Field Crops Res. 157: 89-97.

Hamzei J and Soltani J, 2012. Deficit irrigation of rapeseed for water-saving: effects on biomass accumulation, light interception and radiation use efficiency under different N rates. Agr. Ecosyst. Environ. 155: 153-160.

Han YY, Wang XY and Zhou XB, 2016. Precision planting patterns effect on growth, photosynthetic characteristics and yield of winter wheat under deficit irrigation. Int. J. Agric. Biol. 18: 741-746.

Liu JQ, Li MD and Zhou XB, 2016. Row spacing effects on radiation distribution, leaf water statues and yield of summer maize. J. Anim. Plant Sci. 26: 697-705.

Mattera J, Romero LA, Cuatrín AL, et al., 2013. Yield components, light interception and radiation use efficiency of Lucerne (*Medicago sativa* L.) in response to row spacing. Eur. J. Agron. 45: 87-95.

Norsworthy JK and Shipe ER, 2005. Effect of row spacing and soybean genotype on mainstem and branch yield. Agron. J. 97: 919-923.

Rachaputi RCN, Yashvir C, Col D, et al., 2015. Physiological basis of yield variation in response to row spacing and plant density of mungbean grown in subtropical environments. Field Crops Res. 183: 14-22.

Sun H, Shen Y, Yu Q, et al., 2010. Effect of precipitation change on water balance and WUE of the winter wheat-summer maize rotation in the North China Plain. Agr. Water Manage. 97: 1139-1145.

Wang GY, Zhou XB and Chen YH, 2016. Planting pattern and irrigation effects on water status of winter wheat. J. Agric. Sci. Cambridge154: 1362-1377.

Wang XY, Zhou XB and Chen YH, 2015. Planting pattern effects on soil water and yield of summer maize. Maydica 60: 1-6.

Zarate-Valdez JL, Whiting ML, Lampinenc BD, et al., 2012. Prediction of leaf area index in almonds by vegetation indexes. Comput. Electron. Agr. 85: 24-32.

Zhang Z, Zhou XB and Chen YH, 2016. Effects of irrigation and precision planting patterns on photosynthetic product of wheat. Agron. J. 108: 1-7.

Photosynthetic Characteristics of Summer Maize under Different Planting Patterns and the Responses to Nitrogen Application of Previous Crop

X. B. Zhou, P. J. Shen, Y. X. Zhao, D. H. Jiang, J. H. Huang

(Guangxi Colleges and Universities Key Laboratory of Crop Cultivation and Farming System, Agricultural College of Guangxi University, Nanning 530004, China)

Abstract: Maize (*Zea mays* L.) is one of the most important grain crops in the North China Plain. Management practices affect the photosynthetic characteristics and the production of summer maize. This 2 year (2014–2015) study examined the effects of different planting patterns and the application of nitrogen to previous winter wheat (*Triticum aestivum* L.) on the photosynthetic characteristics, yield and radiation use efficiency (RUE) of summer maize. Field experiments used a two-factor split-plot design with three replicates at Taian, Shandong Province, China (36°09′ N, 117°09′ E). The experiments involved two planting patterns (ridge planting, RP; and uniform row planting, UR) and two nitrogen application levels of previous winter wheat (N1, 112.50 kg/ha; N2, 225.00 kg/ha). The results indicated that the application of nitrogen on previous crop and ridge planting of the following crop had significant effects on the photosynthetic characteristics and yields of summer maize. Compared with UR, this study found that RP increased the chlorophyll content index (CCI), leaf area index (LAI), net photosynthetic rate (Pn), dry matter (DM), yield and grain RUE by 4.1%, 6.3%, 5.2%, 6.4%, 8.9% and 9.4%, respectively. The CCI, LAI, Pn, yield, and grain RUE of N2 were 9.7%, 3.3%, 3.7%, 10.0% and 10.1% higher than those of N1, respectively. RP combined with the application of nitrogen on previous crop of winter wheat could increase the CCI, LAI, Pn, DM, ultimately increasing the grain yield and RUE of the following summer's maize. It was concluded that previous crop nitrogen application and RP pattern treatment resulted in optimal cropping conditions for the North China plain.

Keywords: *Zea mays* L.; Ridge planting; Nitrogen level; Photosynthetic rate; Leaf area index; Yield

1 Introduction

Maize (*Zea mays* L.) has become an important pillar of agricultural development in China

---

Corresponding author's E-mail: xunbozhou@gmail.com

and is the third primary crop. Currently, there are many kinds of planting patterns that are used to attain high grain yields in maize production. The maize planting patterns used in China include flat planting, ridge planting, and furrow planting (Wu et al., 2017). Planting pattern affects the population structure of crops, in addition to physiological characteristics, such as light utilization (Zhang et al., 2017). This study hypothesizes that ridge planting is superior to conventional flat planting, which has been widely promoted and has been significantly improved in many ways, including conserving water, improving fertilizer use efficiency, and increasing grain yield (Majeed et al., 2015). Zhang et al. (2012) and Yao (2015) indicate that ridge planting can affect enzyme activities and microbial functional groups, increase photosynthetic parameters, gas exchange rates, root absorption, and reduce water consumption. In addition, the canopy structure, net photosynthetic rate, stomatal conductance, transpiration rate, and the relative content of chlorophyll in maize leaves is also increased under ridge planting (Tao et al., 2013).

Some studies have shown that the application of nitrogen fertilizers positively influencesphotosynthetic characteristics and maize grain yield (Vos et al., 2005; Javeed et al., 2013). The application of nitrogen fertilizer can result in considerable increases in maize yields and profitability (Jama et al., 2017); however, the application of excess nitrogen can lead to environmental problems, particularly for soil and groundwater quality (Li et al., 2016). The key to addressing this problem is to apply nitrogen rationally, taking into account the nitrogen applied to the previous crop. The maize yield of tillage systems that combine fertilizer applications is significantly higher than that of conventional systems (Tueche and Hauser, 2011). Little attention has been given to how planting patterns combined with nitrogen application of previous crops affect summer maize. The objective of this study was to investigate the combined effects of ridge planting and the application of nitrogen on previous winter wheat (*Triticum aestivum*) on the photosynthetic characteristics and grain yield of summer maize.

2 Materials and Methods

A field experiment was conducted at the Agronomy Experimental Station of Shandong Agricultural University, Taian, Shandong Province, China (36°09′ N, 117°09′ E). The winter wheat-summer maize rotation is an important cropping system in the North China Plain. In the experimental area, winter wheat (var. Jimai 22) was cultivated before the planting of summer maize (var. Zhengdan 958) in 2014 and 2015. Summer maize was sowed on June 15, 2014 and June 15, 2015, respectively, 10 days after winter wheat harvested. The planting density was 62,500 plants/ha and harvested on September 27, 2014 and September 30, 2015. This study utilized a two-factor split-plot design, with two planting patterns (ridge planting, RP; and uniform row planting, UR) for the main plot, and two nitrogen application levels on previous winter wheat (N1, 112.5 kg/ha; N2, 225.0 kg/ha) for the sub-plot (Figure 1). The summer maize field experiment had three replications, and no nitrogen was applied for field plots. Each plot was 4 m × 4 m in size.

The study site characterizes the main summer maize growth area of North China with a warm-

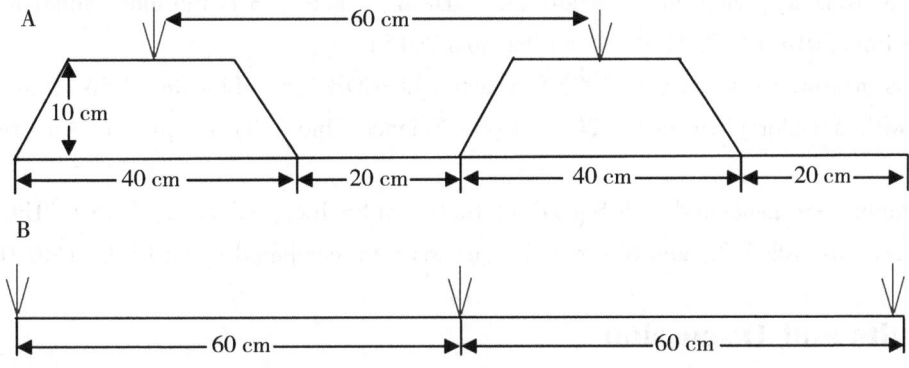

Figure 1　The summer maize ridge planting (A) and uniform row planting (B)

temperate monsoon climate. During the summer maize growing season, precipitation values were 372.5 mm and 282.6 mm for 2014 and 2015, respectively (Table 1), and the annual temperature mean was 25.0℃ (1971–2015). The physical and chemical properties of the experimental site soil layer at a depth from 0 cm to 20 cm were tested before planting the maize (Table 2).

Table 1　Monthly rainfall (mm) in the 2014 and 2015 summer maize growing season

	June (Sowing)	July	August	September (Harvest)	Total
2014	65.0	135.2	35.3	137.0	372.5
2015	74.7	74.2	120.7	13.0	282.6

Table 2　The soil physical and chemical properties of the experimental field

pH	Total N (mg/kg)	Available P (mg/kg)	Available K (mg/kg)	Soil bulk density (g/cm^3)	Soil organic matter (g/kg)	Field capacity (V%)
6.9	123.2	40.6	124.5	1.50	18.9	38.6

For vegetative growth stages: V6 (sixth leaf), reproductive development stages: R0 (silking stage), R2 (blister stage), R3 (milk stage), and R4 (dough stage), the chlorophyll content index (CCI) and the net photosynthetic rate (Pn) of the ear leaf, the dry matter (DM) and the leaf area index (LAI) were measured (RITCHIE et al., 1986).

The leaf, stem, and ear samples were separated and placed in a drying oven at 105℃ for 30 min and then dried at 80℃ until they reached a constant weight. The following equations were used to calculate the leaf area and radiation use efficiency (RUE):

$$\text{Leaf area} = \text{leaf length} \times \text{leaf width} \times 0.83 \tag{1}$$

where leaf length is the distance between the leaf pillow and the leaf tip, and leaf width is the widest part of the leaf.

$$\text{RUE}(\%) = \Delta W \times H / \sum S \times 100\% \tag{2}$$

In this equation, ΔW is grain yield (T/ha) of summer maize, H is the rate of product heat (grain is 16.5 MJ/kg; stem and leaf are 14.4 MJ/kg), and $\sum S$ is the total radiation (1.95 × 10^7 MJ/ha from 2014 and 2.21 × 10^7 MJ/ha from 2015).

Pn was measured using a LI-6400XT system (LI-COR Inc., Lincoln, USA), and CCI was measured with a Chlorophyll meter 200 (Opti-Sciences Inc., Tyngsboro, USA) from 09:00 to 11:00.

All graphs were generated with SigmaPlot 10.0 (SPSS Inc., Chicago, USA). The trail data were analyzed via SAS 9.2, and the treatment means were compared using LSD ($P<0.05$).

3 Results and Discussion

3.1 Chlorophyll content index (CCI)

Figure 2 shows an increasing initial CCI trend, followed by a decreasing trend, and it indicates the turning CCI points of different treatments of R0 and R2 for 2014 and 2015, respectively. Compared with UR, the CCI of RP increased by 3.6% (2014) and 4.6% (2015). The CCI of summer maize with RP and UR decreased by 38.8% and 39.3% from R0 to R4 (2014) and 28.4% and 31.6% from R2 to R4 (2015), respectively. The CCI averages of RP and UR were 40.7 in 2014 and 39.1 in 2015, respectively, and the CCI of RP increased by 4.1% compared to that of UR. This finding illustrated that RP was beneficial to the improvement of CCI. According to the 2 year results, the average CCI of N1 and N2 was 37.4 and 42.4 (2014) and 38.0 and 41.7 (2015) during each growing season, respectively. The average CCI values of N2 were significantly higher than those of N1.

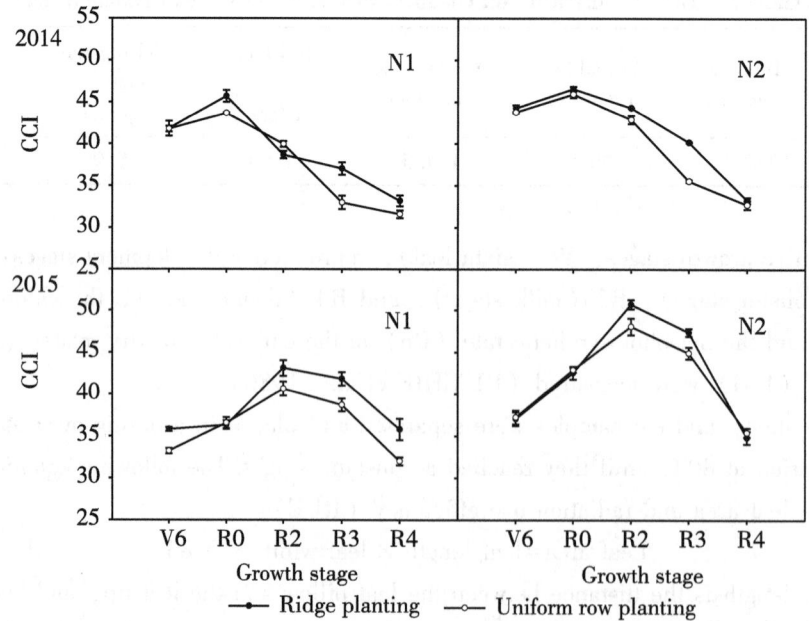

Figure 2 Effects of planting pattern and nitrogen on chlorophyll content index (CCI) of summer maize. N1 (112.5 kg/ha) and N2 (225.0 kg/ha) were nitrogen content; bars were SE

3.2 Leaf area index (LAI)

With the growth of summer maize, the LAI of RP and UR reached the highest point at R2. The average LAI values of RP and UR were 5.35 and 3.32, respectively, and the LAI of RP was 6.3% higher than that of UR over 2 years (Figure 3). The LAI descent ranges of RP and UR were 45.3% and 56.6% from R2 to R4, indicating that RP could significantly improve the LAI of summer maize and depress the decrease of late growth stages. In two-year study, the ranking of the LAI average was N2>N1, with specific values of 3.60 and 3.25, respectively. For N1 and N2 treatments, the maximum change range of LAI variation was 5.7% and 6.9%, respectively. The LAI of RP was 3.3% higher than that of UR under N2. This showed that the RP was beneficial to increasing LAI values for summer maize.

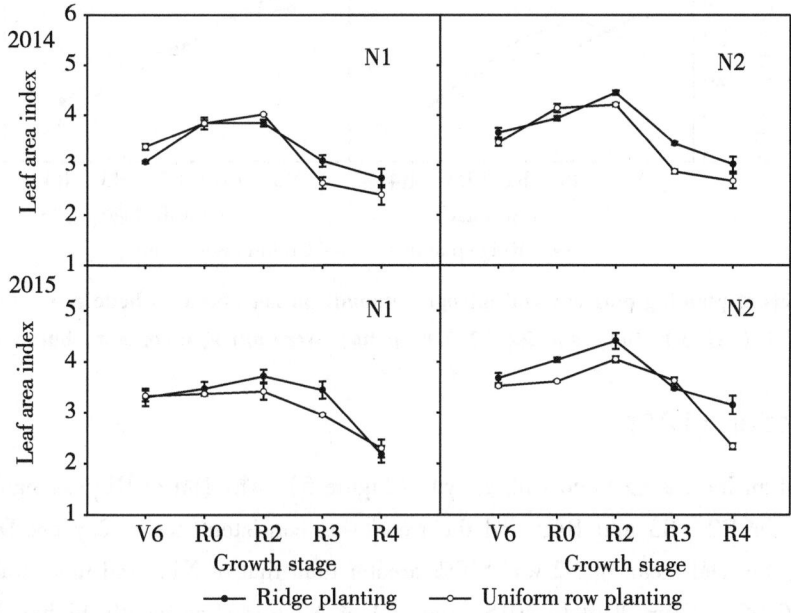

Figure 3 Effects of planting pattern and nitrogen amount on leaf area index of summer maize. N1 (112.5 kg/ha) and N2 (225.0 kg/ha) were nitrogen content; bars were SE

3.3 Photosynthetic rate (Pn)

The different planting patterns and nitrogen fertilization influenced the Pn of the summer maize in 2014 and 2015 (Figure 4). The Pn mean decreased gradually with growth for two years, and the Pn values for V6, R0, R2, R3 and R4 were 41.7 $\mu mol\ CO_2/(m^2 \cdot s)$, 39.5 $\mu mol\ CO_2/(m^2 \cdot s)$, 34.0 $\mu mol\ CO_2/(m^2 \cdot s)$, 28.8 $\mu mol\ CO_2/(m^2 \cdot s)$ and 15.1 $\mu mol\ CO_2/(m^2 \cdot s)$, respectively. The curve displayed that Pn decreased slowly at the early stages (V6-R2), and decreased rapidly in the late growth stages (R2-R4). Compared with UR, the Pn mean of the RP treatment from 2014 to 2015 increased by 5.2%. Pn values across the whole growth period were characterized by a N2>N1 relationship, with specific Pn mean values for N1 and N2 of 32.5 $\mu mol\ CO_2/(m^2 \cdot s)$ (2014) and 35.2 $\mu mol\ CO_2/(m^2 \cdot s)$ (2015), respectively. The Pn values of

N1 and N2 at R4 decreased by 66.0% and 55.7% compared with R2, respectively. These results show that N2 could maintain a relatively high Pn and help to increase the yield of summer maize. The Pn of N2 was 3.7% higher than that of N1.

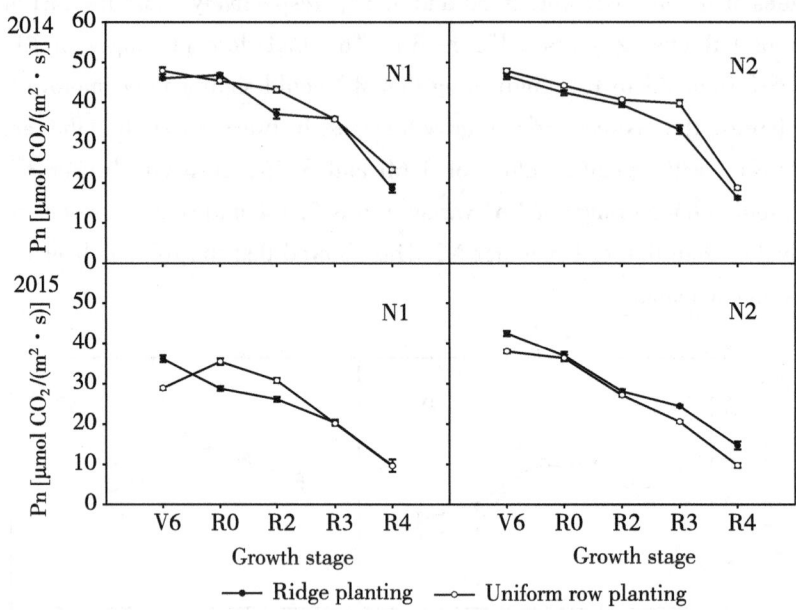

Figure 4 Effects of planting patterns and nitrogen amount on net photosynthetic rate (Pn) of summer maize. N1 (112.5 kg/ha) and N2 (225.0 kg/ha) were nitrogen content; bars were SE

3.4 Dry matter (DM)

DM showed an increasing trend with growth (Figure 5). The DM of RP was significantly higher than that of UR for R2, R3 and R4, and the trend was consistent across 2 years. During the 2015 growing season, the DM mean of N2 was 5.8% greater than that of N1, and there was no significant difference ($P>0.05$). From 2014 to 2015, the DM of N2 was significantly higher than that of N1 ($P<0.05$). DM means for RP and UR were 9,810 kg/ha and 9,322 kg/ha (2014), and 8,444 and 7,603 kg/ha (2015), respectively. The DM of RP was significant higher than that of UR.

3.5 Yield and radiation use efficiency (RUE)

In 2014 growth season, biomass yield and biomass RUE of UR (N1) and RP (N2) were significantly higher than that of RP (N1), grain yield and grain RUE of RP (N2) were significantly higher than that of UR (N1) ($P<0.05$); in 2015 growth season, biomass yield and biomass RUE were no significantly between different treatments ($P>0.05$), grain yield and grain RUE of RP (N2) were significantly higher than those of other treatments ($P<0.05$). Under N1, means of tow growth seasons of biomass yield, grain yield, biomass RUE and grain RUE were 1.76 T/ha, 0.95 T/ha, 1.31%, 0.76% (RP) and 1.89 T/ha, 0.85 T/ha, 1.40%, 0.68% (UR); under N2, means of tow growth seasons of biomass yield, grain yield, biomass RUE and grain RUE were 1.88 T/ha, 1.02 T/ha, 1.41%, 0.82% (RP) and 1.84 T/ha, 0.96 T/ha,

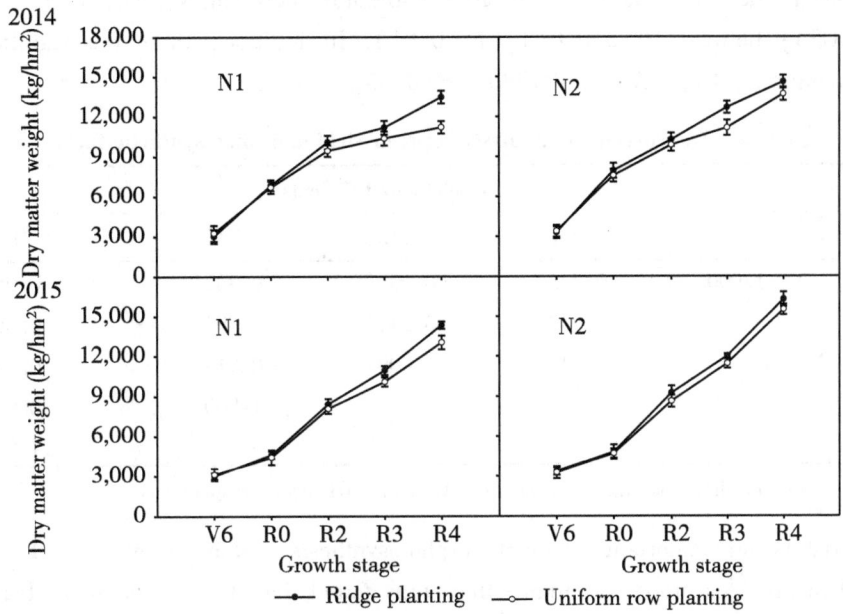

Figure 5 Effects of planting patterns and nitrogen amount on dry matter weight of summer maize. N1 (112.5 kg/ha) and N2 (225.0 kg/ha) were nitrogen content; bars were SE

1.38%, 0.76% (UR) (Table 3). These results indicated that nitrogen fertilization of previous crop, yield of the following maize were increased under the same planting pattern; yield of RP was 11.7% (N1) and 6.3% (N2) higher than those of UR. The results showed that biomass yield and RUE there were significantly interaction effect between application nitrogen in previous crop and following summer maize planting pattern ($P<0.05$).

Table 3 Effects of planting patterns (PP) and nitrogen amount on yield and radiation use efficiency (RUE) of summer maize

Treatments		Biomass yield (T/ha)		Grain yield (T/ha)		Biomass RUE (%)		Grain RUE (%)	
		2014	2015	2014	2015	2014	2015	2014	2015
N1	RP	1.63b	1.88a	0.98ab	0.91b	1.31b	1.31a	0.83ab	0.68b
	UR	1.85a	1.93a	0.84b	0.86b	1.46a	1.34a	0.71b	0.64b
N2	RP	1.87a	1.89a	1.01a	1.03a	1.49a	1.33a	0.86a	0.77a
	UR	1.76ab	1.92a	0.98ab	0.93b	1.41ab	1.34a	0.83ab	0.69b
P value	N	0.7195	0.8125	0.3507	0.1640	0.0940	0.6561	0.1381	0.0042
	PP	0.8101	0.1209	0.3482	0.2085	0.3775	0.3342	0.1372	0.0134
	N×PP	0.0205	0.5990	0.2060	0.2837	0.0241	0.4693	0.5261	0.2837

Notes: N1, 112.5 kg/ha; N2, 225.0 kg/ha; RP, ridge planting; UR, uniform row planting. Values followed by a different small letter within a column are significantly different at 5% probability level.

3.6 The correlation analysis between yield and photosynthetic factors

The N, P_n, CCI, LAI and yield correlation analysis of summer maize for the 2 years results is shown in Table 4. There was significant positive correlation between N, Pn, CCI, and LAI with yield ($P<0.05$); between Pn and CCI ($P<0.01$). In addition, there was significant positive correlation among N, CCI, LAI, and yield ($P<0.05$).

Table 4 The correlation analysis between yield and photosynthetic factors

Variable	Person correlation coefficients				
	N	Pn	CCI	LAI	Yield
N	1.000	0.926**	0.990**	0.882*	0.988**
Pn		1.000	0.941**	0.791	0.897*
CCI			1.000	0.913*	0.982**
LAI				1.000	0.933**
Yield					1.000

Notes: *, ** correlation is significant at the 0.05 and 0.01 level, respectively.

Chlorophyll is an important pigment forphotosynthesis (Peng et al., 2011). When the nitrogen level in previous crops was high, this study found that the CCI of maize leaves was also high, which indicated that CCI was strongly influenced by previous crop nitrogen levels. Other studies observed similar results (Clevers and Kooistra, 2012; Schlemmer et al., 2013). Leaves are the material carrier of intercepted light energy, and leaf size in maize is responsive to nitrogen supply and had a close relationship with photosynthetic rate. LAI is also one of the main physiological determinants of crop yield (Hamzei and Soltani, 2012). The results of this study indicated that the relationship for the LAI and Pn averages was N2>N1 and that N2 had advantage for increasing LAI, maintaining a relatively high Pn and increasing the yield of summer maize. In general, the increase of previous crop nitrogen content combined with using RP pattern for the following maize crops could enhance CCI and increase both Pn and grain yield.

RP has been widely used for wheat, maize, rice, and soybean crops, with varying increases of yield (Song et al., 2013). Ren et al. (2016) illustrates that RP is conducive to improved photosynthetic characteristics, contribute to photo assimilate accumulation and enhance maize productivity (Jiang et al., 2016). Hassan et al. (2005) observe that there are increases of 30% in the grain yield of RP compared to flat planting pattern. This study demonstrated that previous winter wheat nitrogen levels significantly affected photosynthetic factors and yield of the following summer maize crop. The yields of different nitrogen contents can be significantly different owing to their differential impacts on growth and yield (Hamzei and Soltani, 2012). The photosynthetic characteristics, grain RUE and yields of RP were higher than those of UR.

4 Conclusions

Previous crop nitrogen levels and following crop ridge planting pattern had a significant effect on the photosynthetic characteristics and yield of summer maize. The use of RP pattern, combined with the application of nitrogen to previous winter wheat, could increase CCI, LAI, Pn, and DM, ultimately increasing the grain yield and RUE of the following summer maize crops. This study demonstrated that the previous crop N2 (225.0 kg/ha) combined with following crop ridge planting pattern could be utilized for summer maize in the North China Plain.

Acknowledgements

The research was sponsored by the National Natural Science Foundation of China (31760354), Guangxi Natural Science Foundation (2017GXNSFAA198036). We would like to thanks Zhang Zhen for the work contribution.

References

CLEVERS JGPW, KOOISTRA L, 2012. Using hyperspectral remote sensing data for retrieving canopy chlorophyll and nitrogen content. Journal of Selected Topics inApplied Earth Observations and Remote Sensing, v. 5, p. 574-583. https://doi.org/10.1109/JSTARS.2011.2176468.

HAMZEI J, SOLTANI J, 2012. Deficit irrigation of rapeseed for water-saving: Effects on biomass accumulation, light interception and radiation use efficiency under different N rates. Agriculture Ecosystems and Environment, v. 155, p. 153-160. https://doi.org/10.1016/j.agee.2012.04.003.

HASSAN I, HUSSAIN Z, AKBAR G, 2005. Effect of permanent raised beds on water productivity for irrigated maize-wheat cropping system. ROTH, E. C.; FISCHER, R. A.; MEISNER, C. A. Eds. Evaluation and Performance of Permanent Raised Bed Cropping Systems in Asia, Australia and Mexico. p. 59-65.

JAMA B, KIMANI D, HARAWA R, et al., 2017. Maize yield response, nitrogen use efficiency and financial returns to fertilizer on smallholder farms in southern Africa. Food Security, v. 9, p. 1-17. https://doi.org/10.1007/s12571-017-0674-2.

JAVEED HMR, ZAMIR MSI, 2013. Influence of tillage practices and poultry manure on grain physical properties and yield attributes of spring maize (*Zea mays* L.). Pakistan Journal of Agricultural Sciences, v. 50, p. 177-183.

JIANG R, LI X, ZHOU M, et al., 2016. Plastic film mulching on soil water and maize (*Zea mays* L.) yield in a ridge cultivation system on Loess Plateau of China. Soil Science and Plant Nutrition, v. 62, p. 1-15. https://doi.org/10.1080/00380768.2015.1104642.

LI Y, LIU H, HUANG G, et al., 2016. Nitrate nitrogen accumulation and leaching pattern at a winter wheat: summer maize cropping field in the North China Plain. Environmental Earth Sciences, v. 75, p. 118. https://doi.org/10.1007/s12665-015-4867-8.

MAJEED A, MUHMOOD A, NIAZ A, et al., 2015. Bed planting of wheat (*Triticum aestivum* L.) improves nitrogen use efficiency and grain yield compared to flat planting. The Crop Journal, v. 3, p. 118-124. https://doi.org/10.1016/j.cj.2015.01.003.

PENG Y, GITELSON AA, KEYDAN G, et al., 2011. Remote estimation of gross primary production in maize and support for a new paradigm based on total crop chlorophyll content. Remote Sensing of Environment, 115, p. 978 – 989. https://doi.org/10.1016/j.rse.2010.12.001.

REN BZ, DONG ST, LIU P, et al., 2016. Ridge tillage improves plant growth and grain yield of waterlogged summer maize. Agricultural Water Management, v.177, p.392 – 399. https://doi.org/10.1016/j.agwat.2016.08.033.

RITCHIE SW, HANWAY JJ, BENSON GO, 1986. How a Corn Plant Develops. Special report no.48, Iowa State University, Ames, Iowa.

SCHLEMMER M, GITELSON A, SCHEPERS J, et al., 2013. Remote estimation of nitrogen and chlorophyll contents in maize at leaf and canopy levels. International Journal of Applied Earth Observation and Geoinformation, v.25, p.47 – 54. https://doi.org/10.1016/j.jag.2013.04.003.

SONG ZW, GUO JR, ZHANG ZP, et al., 2013. Impacts of planting systems on soil moisture, soil temperature and corn yield in rainfed area of Northeast China. European Journal of Agronomy, v.50, p.66−74. https://doi.org/10.1016/j.eja.2013.05.008.

TAO ZQ, SUI P, CHEN YQ, et al., 2013. Subsoiling and ridge tillage alleviate the high temperature stress in spring maize in the North China Plain. Journal of Integrative Agriculture, v.12, p.2179−2188. https://doi.org/10.1016/S2095-3119(13)60347.

TUECHE JR, HAUSER S, 2011. Maize (*Zea mays* L.) yield and soil physical properties as affected by the previous plantain cropping systems, tillage and nitrogen application. Soil and Tillage Research, v.115−116, p.88−93. https://doi.org/10.1016/j.still.2011.07.004.

VOS J, VAN DER PUTTEN PEL, BIRCH CJ, 2005. Effect of nitrogen supply on leaf appearance, leaf growth, leaf nitrogen economy and photosynthetic capacity in maize (*Zea mays* L.). Field Crops Research, v.93, p.64 – 73. https://doi.org/10.1016/j.fcr.2004.09.013.

WU Y, HUANG FY, JIA ZK, et al., 2017. Response of soil water, temperature, and maize (*Zea may* L.) production to different plastic film mulching patterns in semi−arid areas of northwest China. Soil and Tillage Research, v.166, p.113−121. https://doi.org/10.1016/j.still.2016.10.012.

YAO YZ, 2015. Effects of ridge tillage on photosynthesis and root characters of rice. Chilean-Journal of Agricultural Research, v.75, p.35 – 41. https://doi.org/10.4067/S0718-58392015000100005.

ZHANG XL, MA L, GILLIAMC FS, et al., 2012. Effects of raised−bed planting for enhanced summer maize yield on rhizosphere soil microbial functional groups and enzyme activity in Henan Province, China. Field Crops Research, v.130, p.28 – 37. https://doi.org/10.1016/j.fcr.2012.02.008.

ZHANG Z, MAO XM, ZHONG WW, et al., 2017. Photosynthetic production of wheat under precision planting patterns in northern China. Bioscience Journal, v.33, p.1−9. https://doi.org/10.14393/BJ-v33n1a2017-30328.

Planting Pattern Effects on Soil Water and Yield of Summer Maize

X. Y. Wang[1,2], X. B. Zhou[1,2], Y. H. Chen[2]

([1] Agricultural College of Guangxi University, Nanning 530004, China;
[2] State Key Laboratory of Crop Biology, Shandong Agricultural University, Tai'an 271018, China)

Abstract: Productivity and water use efficiency are important problems in sustainable agriculture, especially in high-demand water resource crops such as maize (*Zea mays* L.). The aims of this research were to study plant and row spacing in maize, evaluating soil water content (SWC), yield and water use efficiency (WUE). A 3 year field experiment (2011-2013) was carried out in the north of China. The summer maize experiment consisted of five types of row spacing under the same planting density. The results showed that the SWC in 90-120 cm was higher than 0-30 cm, and soil water storage was a significant regression with advancing growth stage. A negative correlation was observed among yield, WUE and row spacing. The average yield of RS50 and RS40 was by 9.6% higher than that of RS70 and RS80, and the WUE of the RS40 and RS50 were significantly higher than RS60, RS70 and RS80. The study also indicated that increased productivity and WUE of rainfed summer maize can be reached via row spacing reduction and plant spacing widening under same planting density, and RS50 is regarded as the best planting system selection for the plains of Northern China.

Keywords: *Zea mays*; Rainfed; Row spacing; Soil water content; Water use efficiency

1 Introduction

Plants compete among themselves for some resources. The main competition factors can be identified as light, temperature, water, nutrients and weed (Brant et al., 2009). One plant was sufficiently close to another to influence its soil or atmospheric environment and thereby decrease its rate of growth (De Bruin and Pedersen, 2008). Many planting patterns and agricultural practices have been used to make full use of resources, and by adjusting the row spacing can promote crop growth and improve efficiency of resource use. Different row spacings changed the local environment of individual plant. In Alabama, USA, the sorghum grain yield of row spacing 45 cm was significantly higher than that of row spacing 60 cm and 90 cm when the seeding rate was 20 grain per meter (Bishnoi et al., 1990). Under the same summer soybean plant population density, yields of narrow row spacing were significantly higher than that of wide row spacing (Zhou et al., 2015).

Global demand for agricultural products is expected to double in the coming decades (Godfray

Corresponding E-mail address: whyzxb@gmail.com

et al., 2010). Maize is one of the staple food crops, and China is currently the world's second largest maize producer (Meng et al., 2006). Summer maize in north China is not irrigated during the growing season, and water supply is an important factor to the yield. Zhou et al. (2010) indicated that enhanced productivity and water use efficiency of rainfed summer soybean can be achieved via row spacing reduction under same planting density. For rainfed crops, relatively uniform row spacing would made reasonable absorb moisture inter-plant, minimize unproductive consumption caused by the soil evapotranspiration (Debaeke and Aboudrare, 2004). Water loss, due to evapotranspiration, was also significantly greater in the row position than in the interrow position (Timlin et al., 2001).

Many previous researches have focused on the water use efficiency (WUE) under the condition of water restrict, but only a few have studied the effects of crop row spacing on yield and WUE (Bowers et al., 2000). The aim of this study was to explore the effects of row spacing on soil evaporation, water-consumption characteristics, grain yield and WUE for rainfed summer maize in the North China Plain.

2 Materials and Methods

2.1 Experimental design and weather data collection

The study was conducted at Agronomy Experimental Station, Shandong Agricultural University, which located in Tai'an, China (36°09′N, 117°09′E). The soil type in this region was a silt loam with the average soil organic matter of 16.3 g/kg, N 1.3 g/kg, P 35 mg/kg, K 95 mg/kg, and pH of 6.9. The long-term yearly average (1971-2010) rainfall was 693.5 mm, and the average temperature was 13.1℃. Data on monthly rainfall through the year are shown in Figure 1. Precipitation during the summer maize growing season was 572.5 mm in 2011, 337.1 mm in 2012, and 461.8 mm in 2013.

Experiments were established in 2011, 2012, 2013 and consisted of 5 planting patterns under the same planting density (6.25×10^4 plant/ha); row spacing (RS, cm) × plant spacing (cm) was 40 cm × 40 cm (RS40), 50 cm × 32 cm (RS50), 60 cm × 27 cm (RS60), 70 cm × 23 cm (RS70), 80 cm × 20 cm (RS80). Treatments were randomized plot design and replicated three times. Non-irrigated summer maize (cv. Luyu14) was hand planted on June 18, 2011, June 17, 2012, June 19, 2013, and harvested on September 24, 2011, October 2, 2012, and October 2, 2013. The experiment plot area was 4 m × 4 m. The growth stage of VE、V6、R0、R2、R3、R4、R5 were measured in this experiment (Ritchie et al., 1996).

Soil water content (SWC, v/v) was measured every 10 d using a neutron moisture meter (CNC503B, Super Energy Nuclear Technology, Ltd., Beijing, China) throughout the summer maize growing season at 10 cm intervals in the 0-120 cm soil profile.

2.2 Computation and statistical analyses

The evapotranspiration (ET) was calculated using the following equations (Zhang et al.,

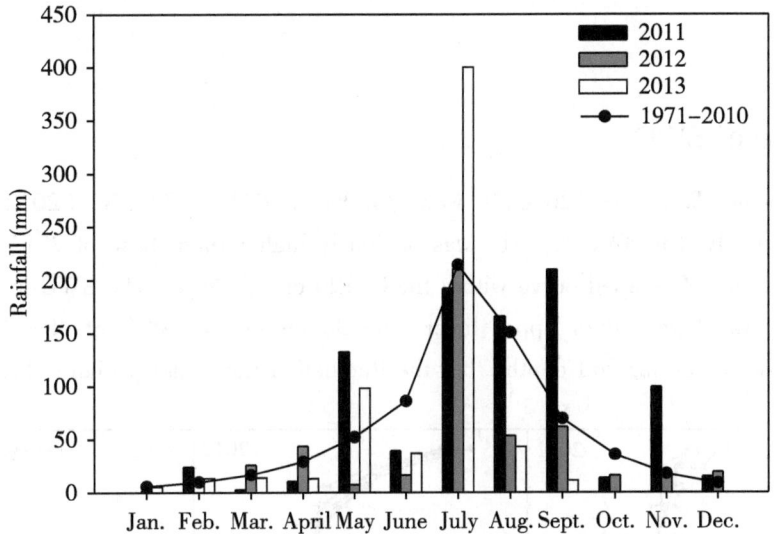

Figure 1　The monthly rainfall average in 1971–2010 and monthly rainfall in 2011–2013

2011):

$$ET = \Delta W + R - SI - Q$$

Where: ΔW is change of soil water stored (SWS, mm), R is rainfall (mm), SI is deep percolation (mm), Q is surface run-off (mm). SI was estimated using the approach proposed by Gong and Li (1995).

$$SI = \Delta W - FK$$

Where: FK is field capacity.

$$\Delta W = \sum (\Delta \emptyset_i \times Z_i)$$

Where: $\Delta \emptyset_i$ is change in soil volumetric water content (m^3/m^3) and Z_i is depth of the soil layer (mm).

$$Q = (R - 0.2S)^2 / (R + 0.8S)$$

Where: S is potential maximum retention after runoff begins (mm) (Bosznay, 1989).

$$S = (25,400/CN) - 254$$

Where: CN is runoff curve number.

The WUE formula is as follows (Neal et al., 2011):

$$WUE = Y/ET$$

Where: Y is grain yield (kg/ha) of summer maize, ET is total seasonal evapotranspiration.

All graphs were prepared from means and drawn using SigmaPlot 10.0 (SPSS Inc., Chicago, IL). All data were analyzed using SAS 9.2 (SAS Institute Inc., Cary, USA). Multiple comparisons were conducted for significant effects with the least significant difference test at $P = 0.05$.

3 Results

3.1 Change of SWC

The means of SWC (0–120 cm) were 34.0% (2011), 27.3% (2012), and 29.8% (2013), respectively. The SWC of 2011 was obviously higher than those of 2012 and 2013, and presented an irregular Z-shaped curve within the 0–120 cm soil layer. The SWC in the deeper layer (90–120 cm) was higher than upper layer (0–30 cm); the SWC of 40–80 cm soil layer increased with the increasing soil depth, but the fluctuation was small (Figure 2).

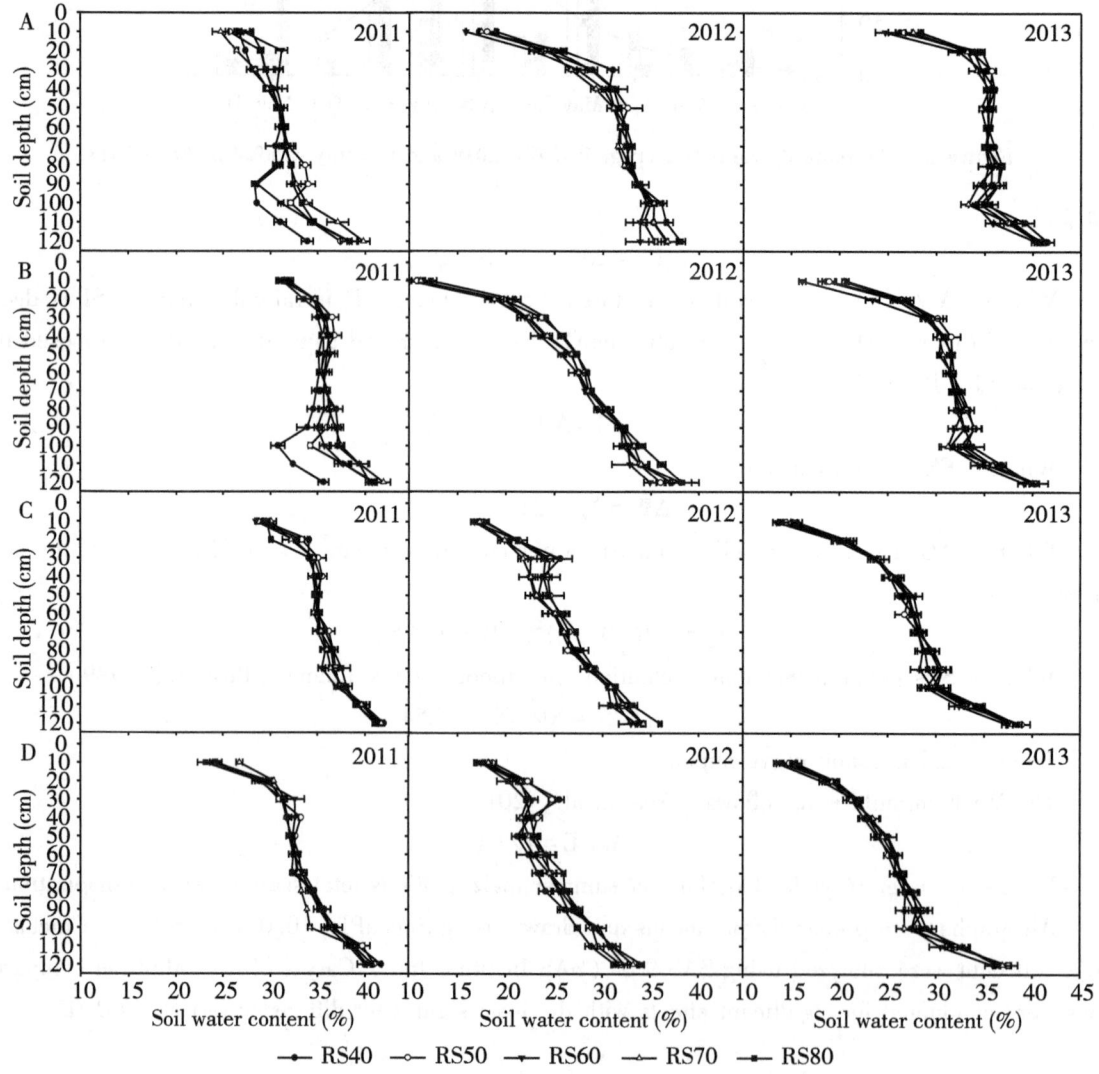

Figure 2 Soil water content at 0–120 cm layer in 2011–2013
Notes: A, B, C, D are V6, R0, R2, R3 respectively; the bars are the SE.

The SWC had a large difference in the different growth stages. At V6, the fluctuation of SWC in 2011 was smaller than those of 2012 and 2013 at 0-120 cm layer, and the SWC in 0-30 cm layer was 28.1% (2011), 23.5% (2012), and 31.7% (2013), respectively. At R0, the SWC in 0-30 cm layer was highest in 2011 and the value was 33.9%, but it was lowest in 2012 and the value was 17.8%. The SWC of 2012 was lower than those of other two years in 0-120 cm layer. At R2, the SWC in 0-120 cm layer were 35.6% (2011), 26.1% (2012), 27.3% (2013), and were lower than R0. The SWC of R3 was 6.3% (2011), 4.3% (2011) and 6.5% (2011) lower than those of R2, respectively. At R2 and R3, the SWC of 2013 was lower than that of 2012 in 0-30 cm layer.

The SWC changed with different RSs. At V6 and R0, the SWC in 90-120 cm layer of RS40 was lower than those of other RS treatments in 2011, which maybe attributed to more rainfall in the early growth stage, and was growing quickly and had a higher consumption under RS40. The SWC averages of RS40, RS50, RS60, RS70, RS80 were 33.3%, 34.3%, 34.1%, 34.2%, 34.2% (2011), 27.2%, 27.5%, 26.7%, 27.1%, 28.0% (2012), 30.0%, 29.9%, 29.4%, 29.6%, 30.0% (2013), respectively; the means of three years were 30.1%, 30.6%, 30.0%, 30.3%, 30.7%, respectively

3.2 Change of SWS

In 0-120 cm, the SWS of 2011, 2012, 2013 were 393.9, 304.1, and 334.4 mm, and it may be relate to rainfall; the SWS of VE, V6, R0, R2, R3, R5 were 286.2, 388.3, 376.5, 356.4, 335.8, and 321.7 mm, respectively; relative low SWS values were observed at VE and R5. In the middle and late summer maize growth seasons (R0-R5) maintained relative high SWS values attribute to more rainfall. A similar trend was observed in 2012 and 2013. The SWS reached to a peak at V6 and decreased gradually which related to less rainfall and intense water consumption (Figure 3).

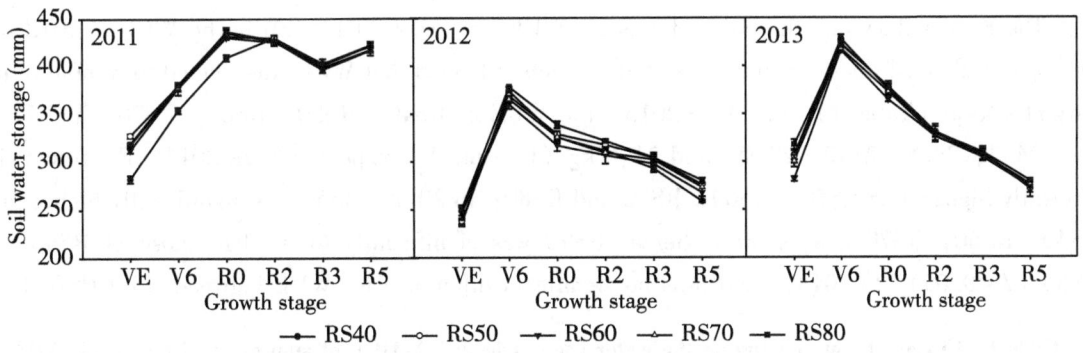

Figure 3　Effects of row spacing on soil water storage

Notes: The bars are the SE.

A correlation analysis showed that there was a significant regression trend between SWS and GS, and the equation can be denoted as y (SWS, mm) $= -11.682x^2$ (GS) $+ 72.971x + 241.03$, with an $R^2 = 0.4798$ ($P = 0.0090$). The means of RS40, RS50, RS60, RS70, RS80 (2011-2013) were 339.8 mm, 346.7 mm, 340.5 mm, 344.4 mm, and 349.2 mm, respectively. There was no significant regression between RS and SWS, the SWS of RS40 was 2.7% lower than that of RS80, and RS50 and RS80 were higher than those of the other RSs.

3.3 ET, yield and WUE

ET versus grain yield was plotted for all treatment conditions (Figure 4). The ET in 2011, 2012, and 2013 were 465.0, 310.1, and 489.7 mm, respectively; the ET of 2012 was lower than those of 2011 and 2013. In 2011, the results indicated that yields were increased with increased ET; yield and ET were significantly positive correlated, and the correlation coefficient (R^2) was 0.4113 ($P=0.0100$). In 2012 and 2013, yield and ET were not significantly correlated, and the R^2 were 0.1490 ($P=0.2706$) and 0.0707 ($P=0.3381$). Those results showed that rainfed summer maize was different from irrigated winter wheat due to rainfall, light intensity, temperature and other environmental factors, and high water consumption may not promoting yield. The yields of RS40 (2011) and RS50 (2012) were higher than those of other RS treatments, but ETs were low (Figure 4).

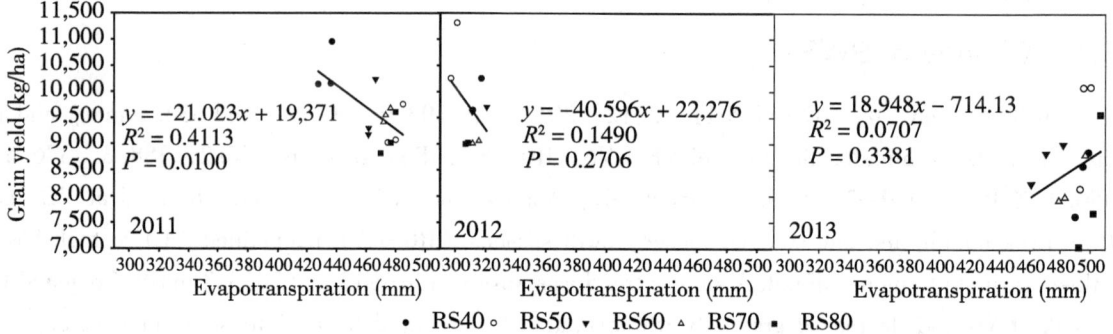

Figure 4 Regression of evapotranspiration vs. grain yield for the rainfed summer maize in 2011-2013

The results show that the order of average WUE was 2012>2011>2013. The WUE of 2012 was the highest although least amount of rainfall, which indicated that WUE was related to rainfall of the current season (Table 1). In 2011-2013, the WUE of RS40, RS50, RS60, RS70, and RS80 were 24.2, 24.8, 23.2, 22.0, and 21.6 kg/(ha·mm), respectively. In 2011, RS40 was significantly higher than RS50, RS60, RS70 and RS80; in 2012, RS50 was significantly higher than RS40, RS60, RS70 and RS80, whereas RS70 was significantly lower than those of RS40 and RS50 ($P<0.05$). In 2013, there was no significant difference for WUE between RS ($P>0.05$).

Table 1 Effects of row spacing on the water use efficiency (WUE) of summer maize in 2011-2013

Row spacing (cm)	2011		2012		2013	
	Yield (kg/ha)	WUE [kg/(ha·mm)]	Yield (kg/ha)	WUE [kg/(ha·mm)]	Yield (kg/ha)	WUE [kg/(ha·mm)]
40	10,415a*	24.0a	9,948ab	31.6b	8,352b	16.9
50	9,288ab	19.4b	10,779a	36.0a	9,443a	19.0
60	9,569ab	20.7b	9,660b	30.6bc	8,693ab	18.4

(continued)

Row spacing (cm)	2011		2012		2013	
	Yield (kg/ha)	WUE [kg/(ha·mm)]	Yield (kg/ha)	WUE [kg/(ha·mm)]	Yield (kg/ha)	WUE [kg/(ha·mm)]
70	9,551ab	20.2b	9,035b	28.8c	8,238b	16.9
80	9,152b	19.3b	9,011b	29.3bc	8,099b	16.2
LSD (0.05)	941	2.04	993	2.78	1,586	2.95

Notes: * Values followed by the same letter in a column are not significantly different according to $LSD_{0.05}$.

The means of three years results showed that the order of yield was RS50 > RS40 > RS60 > RS70 > RS80, and the average yield of RS50 and RS40 was 9.6% higher than that of RS70 and RS80. In 2011, the grain yield of RS40 was significantly higher than that of RS80; in 2012, RS50 was significantly higher than those of RS60, RS70, RS80; in 2013, RS50 was significantly higher than those of RS40, RS70, RS80 ($P<0.05$).

3.4 Soil water relations with yield

The study over 3 years showed that a significant negative correlation was observed between RS and yield, and the correlation coefficient (r) was -0.9020 ($P<0.05$); a significant positive correlation was observed between ET and SWC, and the r was 0.9017 ($P<0.05$). The result indicated that the increased SWC would improve crop transpiration and soil evaporation, increased ET. There was a positive correlation between ET, SWC and RS, and the r was 0.7169 and 0.5067, respectively; the ET and SWC increased with the increasing of RS. A negative correlation was observed between ET and yield, and high ET did not increase the yield of summer maize. The results indicated that the natural rainfall was not consistent with crop water demand, and relative low water resource utilization was observed under rainfed condition (Table 2).

Table 2 Correlation coefficients between row spacing (RS), yield, evapotranspiration (ET), and soil water content (SWC) of summer maize in 2011-2013

	RS	Yield	ET	SWC
RS	1.0000	-0.9020*	0.7169	0.5067
Yield		1.0000	-0.4392	-0.2621
ET			1.0000	0.9017*
SWC				1.0000

Notes: * r values presented at $P<0.05$.

4 Discussion

The SWC of summer maize was greatly influenced by rainfall. For SWC, 2011 was evidently higher than that of 2012 and 2013, and the deeper layer (90-120 cm) was higher compared to upper layer (0-30 cm). The density and depth of root penetration are greatly affect by the soil profile water status and the factor can also limit crops full use of available soil water (Angadi and Entz, 2002; Zuo et al., 2006). An upward capillary flux and hydraulic gradient would appear in the deeper soil layers of the crop root zone (Bandyopadhyay et al., 2005; Li et al., 2010).

The changes of the SWC curve of different growth stages had related reports (Wang et al., 2014). In 0–30 cm soil layer, the high SWC at V6 in 2013 growing season may attribute to 399.8 mm of rainfall in July, and the high SWC at R2 and R3 in 2012 growing season might have been affected with 115.0 mm of rainfall in August and September. In the three years, the SWC average of RS50 and RS80 was higher than that of other treatments, this result indicate that changes of row spacing of summer maize effected extracting water in soil. There was a descending trend with the advance of the GS, which may be relative to less rainfall in September and more water consumption in the middle and late periods of GS.

In 2011, the overall yield trend indicated that yields were increased with increased ET. The result is similar to previous findings (Schneider and Howell, 1997; Huang et al., 2004). But significant correlation was not found in 2012 and 2013, which time and amount of rainfall was difficult to completely consistent with crop water requirement under the rainfed condition.

The WUE of RS40 and RS50 were significantly higher than RS60, RS70 and RS80, which attributed to greater early-season light interception for narrow row spacing and accelerated crop growth. Relative narrow row was the important factor to increase light interception when the key period of yield formation, and this was the crucial factor to make a high yield (Andrade et al., 2002). It was negative correlation between WUE and RS and positive correlation between WUE and yield, which were alike to the research of soybean study (Ethredge et al., 1989).

5 Conclusion

The study 3 years has shown that high yields and WUE of summer maize can be gained by reducing row spacing under the same planting density in the plains of northern China. The conclusion of the study was that RS50 may be an optimum planting pattern to improve WUE and yield of summer maize under rainfed conditions.

Acknowledgements

This research was supported by the National High Technology Research and Development Program of China (2013AA102903). The authors especially thank Huang Juan, Feng Zhibo, Wang Guoyun and Han Yuanyuan for his valuable contributions.

References

Andrade FH, Calvino P, Cirilo A, et al., 2002. Yield responses to narrow rows depend on increased radiation interception. Agron J. 94: 975–980.

Angadi SV, Entz MH, 2002. Root system and water use patterns of different height sunflower cultivars. Agron J. 94: 136–145.

Bandyopadhyay PK, Mallick S, Rana SK, 2005. Water balance and crop coefficients of summer-grown peanut (Arachis hypogaea L.) in a humid tropical region of India. Irrig Sci. 23: 161-169.

Bishnoi UR, Mays DA, Fabasso MT, 1990. Response of no-till and conventionally planted grain sorghum to weed control method and row spacing. Plant Soil. 129: 117-120.

Bosznay M, 1989. Generalization of SCS curve number method. J Irrig Drain Eng. 155: 139-144.

Bowers GR, Rabb JL, Ashlock LO, et al., 2000. Row spacing in the early soybean production system. Agron J. 92: 524-531.

Brant V, Neckář K, Pivec J, et al., 2009. Competition of some summer catch crops and volunteer cereals in the areas with limited precipitation. Plant Soil Environ. 55: 17-24.

Debaeke P, Aboudrare A, 2004. Adaptation of crop management to water-limited environments. Eur J Agron. 21: 433-446.

De Bruin JL, Pedersen P, 2008. Effect of row spacing and seeding rate on soybean yield. Agron J. 100: 704-710.

Ethredge WJ, Ashley DA, Woodruff JM, 1989. Row spacing and plant population on yield components of soybean. Agron J. 81: 947-951.

Godfray HCJ, Beddington JR, Crute IR, et al., 2010. The challenge of feeding 9 billion people Science. Food Secur. 327: 812-818.

Gong YS, Li BG, 1995. Using field water balance model to estimate the percolation of soil water. Adv Water Sci. 6: 16-21 (In Chinese).

Huang M., Gallichand J. and Zhong L., 2004. Water-yield relationships and optimal water management for winter wheat in the Loess Plateau of China. Irrig Sci. 23: 47-54.

Li F, Wei C, Zhang F, et al., 2010. Water-use efficiency and physiological responses of maize under partial root-zone irrigation. Agric Water Manage. 97: 1156-1164.

Meng E, Hu R, Shi X, et al., 2006. Maize in China: production systems, constraints, and research priorities. CIMMYT, Mexico.

Neal JS, Fulkerson WJ, Sutton BG, 2011. Differences in water-use efficiency among perennial forages used by the dairy industry under optimum and deficit irrigation. Irrig Sci. 29: 213-232.

Ritchie SW, Hanway JJ, Benson GO, 1996. How a corn plant develops. 1996. Spec. Rep. 48. Rev. ed. Iowa State Univ. Coop. Ext. Serv., Ames.

Schneider, A. D. and Howell T. A., 1997. Methods, amounts, and timing of sprinkler irrigation for winter wheat. Trans ASAE. 40: 137-142.

Timlin D, Pachepsky Y, Reddy VR, 2001. Soil water dynamics in row and interrow positions in soybean (Glycine max L.). Plant Soil. 237: 25-35.

Wang GY, Han YY, Zhou XB, et al., 2014. Planting pattern and irrigation effects on water-use efficiency of winter wheat. Crop Sci. 54: 1166-1174.

Zhang Y, Kang S, Ward EJ, et al., 2011. Evapotranspiration components determined by sap flow and microlysimetry techniques of a vineyard in northwest China: Dynamics and influential factors. Agric Water Manage. 98: 1207-1214.

Zhou XB, Chen YH, Ouyang Z, 2015. Spacing between rows: effects on water-use efficiency of double-cropped wheat and soybean. J Agric Sci. 153: 90-101.

Zhou XB, Yang GM, Sun SJ, et al., 2010. Plant and row spacing effects on soil water and yield of rainfed summer soybean in the northern China. Plant Soil Environ. 56: 1-7.

Zuo Q, Shi J, Li Y, et al., 2006. Root length density and water uptake distributions of winter wheat under sub-irrigation. Plant Soil. 285: 45-55.

Regulation of soil microbial community structure and biomass to mitigate soil greenhouse gasses emission

I. Muhammad[a,c], J. Z. Lv[b], J. Wang[c], S. Ahmad[a], F. Saqib[a], S. Ali[d], and X. B. Zhou[a]

([a] Guangxi Colleges and Universities Key Laboratory of Crop Cultivation and Tillage, Agricultural College, Guangxi University, Nanning 530004, China; [b] Maize Research Institute of Guangxi Academy of Agricultural Sciences, Nanning 530007, China; [c] Shaanxi Key Laboratory of Earth Surface System and Environmental Carrying Capacity, College of Urban and Environmental Science, Northwest University, Xi'an 710127, China; [d] Department of Soil and Environment Science, University of Agriculture, Peshawar, Pakistan)

Abstract: Sustainable reduction of fertilization with technology acquisition for improving soil quality and realizing green food production is a major strategic demand for global agricultural production. Introducing legume (LCCs) and/or non-legume cover crops (NLCCs) during the fallow period before planting main crops such as wheat and corn increases surface coverage, retains soil moisture content, and absorbs excess mineral nutrients, thus reducing pollution. In addition, the cover crops (CCs) supplements the soil nutrients upon decomposition and play a green manure effect. Compared to the traditional bare-land, the introduction of CCs systems has multiple ecological benefits, such as improving soil structure, promoting nutrient cycling, improving soil fertility and microbial activity, controlling soil erosion, and inhibiting weed growth, pests, and diseases. The residual decomposition process of cultivated crops after being pressed into the soil will directly change the soil carbon (C) and nitrogen (N) cycle and greenhouse gas emissions (GHGs), and thus affect the soil microbial activities. This key ecological process determines the realization of various ecological and environmental benefits of the cultivated system. Understanding the mechanism of these ecological environmental benefits provides a scientific basis for the restoration and promotion of cultivated crops in dry farming areas of the World. These findings provide an important contribution for the understanding of the mutual interrelationships and the research in this area, as well as increasing the use of CCs in the soil for better soil fertility, GHGs mitigation, and improving soil microbial community structure. This literature review studies the effects of crop biomass and quality on soil GHGs emissions, microbial biomass, and community structure of the crop cultivation system, aiming to clarify crop cultivation in theory.

Keywords: Cover crops; Soil microbial community structure; Greenhouse gas emission; De-

---

These authors contributed equally
Corresponding E-mail address: xunbozhou@gmail.com (Z. X. B); wangj@nwu.edu.cn (W. J)

composition; Cover crop management practices

1 Introduction

Cover crops (CCs) within agroecosystems impart ecological and environmental benefits, like enhancement of soil fertility, C sequestration, leaching reduction, erosion control, and pest and disease suppression (Alfonso Gomez et al., 2018; Berlanas et al., 2018). Cover crops also increase nutrient cycling and biological N fixation, soil organic matter (SOM), biological diversity (e.g., microbes, insects, and birds), weed control, and crop yields (Muhammad et al., 2019), decreasing drainage, increasing infiltration, and maintaining soil nutrients (Wawan et al., 2019). In addition, CCs provide a friendly agronomic environment with suppression of weeds and thus decreasing the dependency for the herbicides uses (Gavazzi et al., 2010; de Barros et al., 2013). Dhima et al. (2006) reported that winter CCs, such as cereal rye (*Secale cereale* L.) and barley (*Hordeum vulgare* L.), could release inhibitory substances known as allelochemicals that can affect the initial growth of grass weeds like barnyard grass.

Previous researchers studied the impact of CCs on SOM in temperate zones and some ephemeral and long-term pools in the Mediterranean and semi-arid annual agroecosystems (Zhou et al., 2016; Austin et al., 2017). Meisinger et al. (1991), reviewed past studies and demonstrated that CCs minimized 20% to 80% of nitrate losses through leaching, whereas NLCCs are more effective than leguminous CCs. They found that winter CCs (small grains) could reduce the nitrate load through leaching and nitrate concentrations by 64% and 50%, respectively. Potential nitrate N leaching in the drainage was minimized by proper crop rotation using CCs (Dinnes et al., 2002). Logsdon et al. (2002) demonstrated that oat and rye CCs significantly decreased nitrate N by 70% in maize-soybean rotation in three simulated years.

Cover crop residues mitigate the negative effects of soil disruption as a result of improving SOM, soil moisture, preventing the germination and emergence of weed seeds, and defending against erosion (Teasdale, 1996; Hall et al., 2010; Alfonso Gomez et al., 2018). Residues mulching maintain the soil moisture by reducing the soil temperature (Unger et al., 1997; Bagley et al., 2012; Li et al., 2013). Cover crops can also affect crop yields through changes in N dynamics in the soil. The addition of NLCCs such as oats tends to reduce mineralization and increase immobilization, lowering the inorganic N availability for the following crop (Haramoto and Brainard, 2012). However, farmers prefer cold-tolerant and productive cereal CCs in the Upper Midwest of the USA and seldom experiment with LCCs (Snapp and Borden, 2005). Leguminous CCs is a rich source of soil N and decompose faster than NLCCs, which results in higher N_2O emissions, however, NLCCs have a higher C : N ratio and are a rich source of soil organic carbon (SOC), which have higher CO_2 emissions (Muhammad et al., 2019).

The introduction of CCs into agricultural soil is an important management practice (Figure 1). As shown in Figure 1, the CCs mulching and incorporation increase soil fertility, soil microbial growth, and hence SOM and residues decomposition. The schematic diagram showed that the incorporated residues decomposed faster than mulching, which released more GHGs and rapid availability of nutrients to plants. It has been used extensively to boost SOM and consequently increase cash crop productivity (Sainju and Singh, 2008). Cover crops cultivation not only provides physical protection to the soil by reducing the impact of rainfall but can also improve soil structure aggregation and microorganisms (Araujo et al., 2019; Tang et al., 2017). Soil organic carbon (SOC) is the critical component of SOM, soil functions, and agricultural ecosystems sustainability (Chalise et al., 2019; Muhammad et al., 2018). Labile organic C is the most active part of SOC and can be used in the short – term experiments as an indicator for assessing the soil quality (Ghimire et al., 2017; Sharma et al., 2018). Furthermore, soil microbes easily access labile organic C and serve as essential nutrients for crop growth (Haynes, 2005). The current review aims 1) to examine the effect of CCs types and amount on soil microbial biomass, community abundance and structure, and soil GHGs emissions; 2) to understand the decomposition pattern of CCs residue based on C : N ratio and its impact on microbial dynamics and GHGs emissions. Excessive fertilization of crops leads to watercourses pollution, thus adding interest to the area of research in green manure practices and its management. Even through a considerable amount of work has been done regarding mineralization, yet the whole concept of residues decomposition, into different frictions under different CCs types which need further detailed exploration. Therefore there is a need to identify impact of residue management on soil property, soil microbes and GHGs emission from various residues types and management.

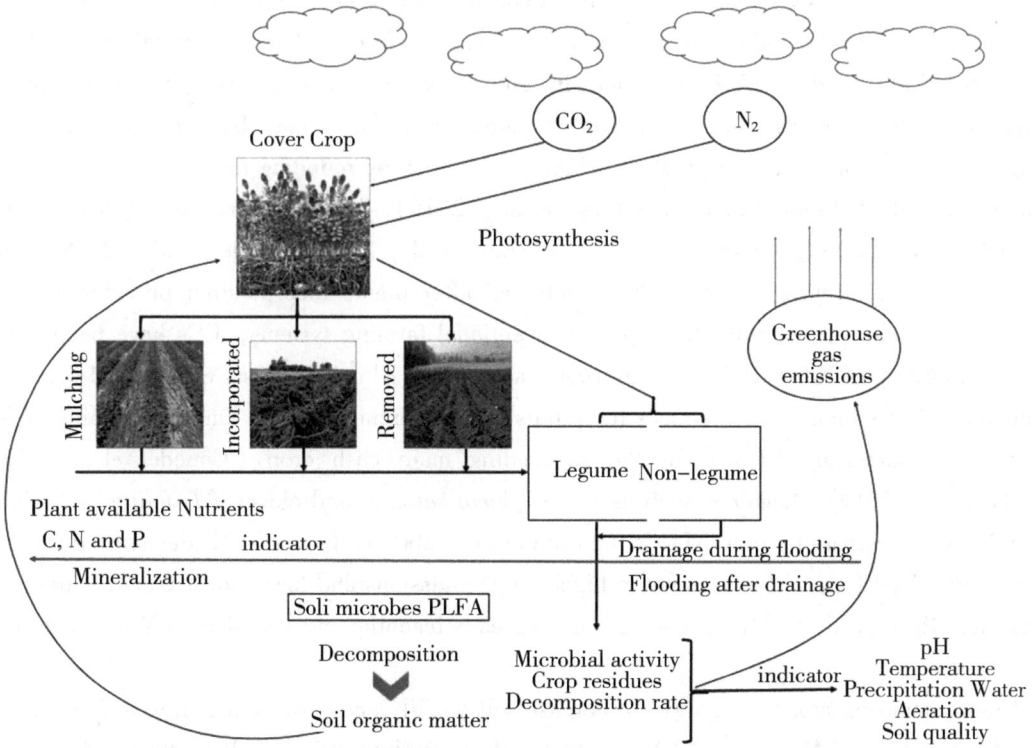

Figure 1 Schematic diagram of cover crops growing, termination methods and their relationship with soil microbes and greenhouse gas emissions

2 Manipulating Cover Crop Types to Influence Soil Microbial Biomass, Community, and GHGs Emissions

Cover crops have been increasingly grown for improving soil health and crop production and minimizing environmental impact compared to NCCs. It is making a tremendous contribution to the supply of food. Cover crop growing and its residue management is an essential cultural practice for improving productivity on a sustainable basis. This study is much relevant to the national development and socio-economic stability of the world. This research will lead to utilizing the organic waste in a better and less harmful manner for crop improvement and soil fertility.

2.1 Legume Cover Crops

Legume CCs have been used to increase SOM and N concentration (Maltais-Landry and Crews, 2019), the amount of N fixed by legumes, dependent upon legume species and environmental conditions (Liebman et al., 2018), and hence increases soil N_2O emissions (Peyrard et al., 2016). It has been estimated that some LCCs can fix 115 kg of N/(ha · a) from atmospheric N (N_2) (Peyrard et al., 2016). Kornecki et al. (2016) reported that crimson clover (*Trifolium inrcanatum* L.) increased yield by 30% when compared to NCCs plots. However, (Reddy, 2001) reported that LCCs had reduced the soybean yield when compared to NCCs plots. Reddy (2003) observed a 50% reduction in grass weeds, such as barnyard grass, broadleaf signal grass, brown top millet, and a 55% reduction in entire leaf morning-glory (*Ipomoea purpurea*) emergence when using crimson clover. The N taken up by CCs may be subsequently available through mineralization after incorporation (Figure 2), thereby reducing the commercial N fertilizer requirement of the subsequent crop (Ackroyd et al., 2019). The incorporation of LCCs had higher N_2O emissions than NLCCs and mixed CCs (Peyrard et al., 2016; Kandel et al., 2018). In low input and organic farming systems, the N released after plants incorporation provides a valuable source of N for the following arable crop. In conventional farming systems, CCs have been found to retain up to 60 kg of N/ha during the growing seasons (de Almeida Acosta et al., 2014). The introduction of LCCs into the soil reduces the inputs of commercial N, thus limiting leaching of N and acts as green manuring (GM) for the succeeding main cash crop (Couedel et al., 2018a; Abdalla et al., 2019). Legumes such as vetch (*Vicia sativa*) and clover (*Trifolium sp*) CCs has higher N fixation capability than NLCCs (Sainju et al., 2007). Legume CCs decompose faster than NLCCs and mixed CCs, which results in higher N_2O emissions and lower soil CO_2 emissions (Gonsiorkiewicz Rigon et al., 2018), and thus decreases N leaching and emissions (Muhammad et al., 2019).

Green manuring crops are grown to increase soil fertility and provide a source of N for the subsequent crops. Since GM improves SOM content and especially nutrition value, thus GM is often incorporated in an early immature stage before the cash crop grows (Venkateswarlu et al., 2007; Hwang et al., 2015; Madsen et al., 2016). Legumes are commonly grown as GM due to their high-quality residues (lower C:N ratio) and fixed biological N_2 from the atmosphere, which

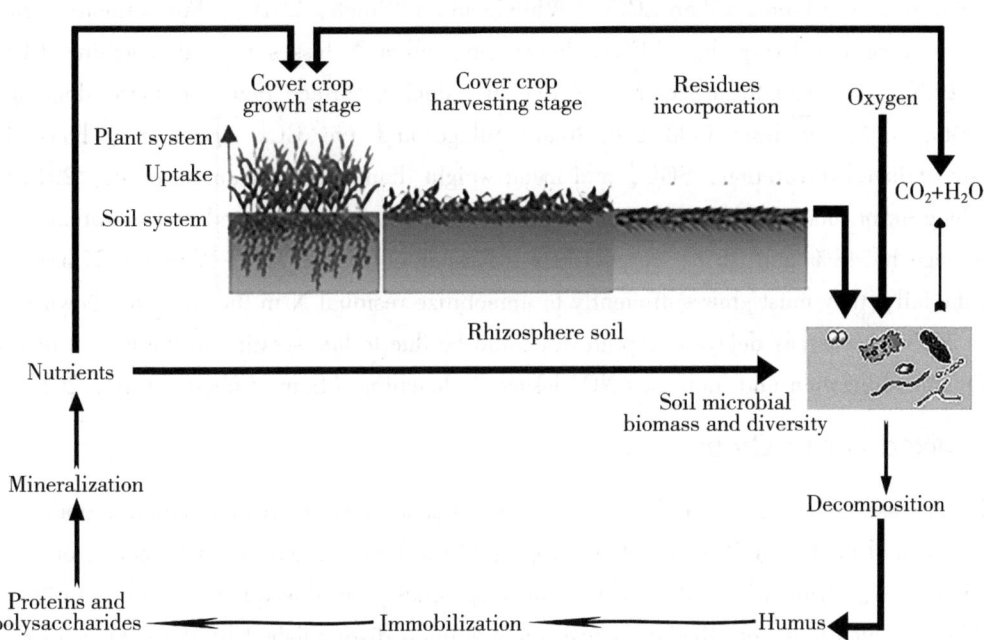

Figure 2 Schematic relationship of residues incorporation, decomposition, nutrient immobilization, and mineralization through soil microbes

leads to decreased N_2O emissions by 55% (Sanz-Cobena et al., 2014a), and decreased N_2O emissions by 86% in barley crop (Sanz-Cobena et al., 2017). Although CCs during the growth period may reduce gaseous losses by N uptake. However, their incorporation may result in increased N_2O production during nitrification and/or denitrification, release inorganic N in decomposition, and create anaerobic zones in the soil (Baggs et al., 2000; Couedel et al., 2018b). These emissions are generally higher where high N residues, such as LCCs are incorporated (Peyrard et al., 2016). The importance of such gaseous losses, in relation to a crop recovery and leaching losses, needs to be quantified to improve N-use efficiency in organic systems.

2.2 Non-Legume Cover Crops

The capacity of non-legume CCs is to minimize soil erosion (wind & water) and water runoff, increase soil aeration, available water holding, SOM and nutrient cycling (Wawan et al., 2019), reduce NO_3^--N losses in fallow soils and provide more N for subsequent cash crops (Sainju and Singh, 2008). The impact of NLCCs on N dynamics is not fully understood, particularly in vegetable fields (White et al., 2020). The mineralization of NLCCs by soil microbes can take place on a long fallow period, and readily available mineral N lost as leachate or emissions from fallow (Rodrigues Torres et al., 2008). It was found that CCs can provide 20% to 55% of the recovered N for succeeding crops (Malpassi et al., 2000) beside these losses. However, to reduce the dependency of nitrogenous fertilizers without compromising yield, the mineralization of crop residues N in soil should be synchronized with the N demand of the main crop (Weinert et al., 2002). Cover crop types, growth period, precipitation, and temperature affect N accumulation, N use efficiency, and soil fertility of subsequent cash crops (Toom et al., 2019). Non-legume CCs such as oats and rye produced greater biomass than LCCs (Whitehead and Singh, 2010). Furthermore, grasses use residual N more effectively than LCCs, hence preventing N losses through leaching (Campiglia et al., 2009). Winter CCs are more effective in reducing surface flow and increasing the evapo-transpiration of water from field soil. Rotary tillage and rye CCs significantly boosted fungal substrate-induced respiration, SOC, and mean weight diameter (Nakamoto et al., 2012). Similarly, long-term rotations of maize soybean, rye, and oat CCs decreased concentrations of NO_3^- in tile drainage by 48% and 26%, respectively (Kaspar et al., 2012). Winter CCs are effective, but in the fall, they must grow sufficiently to immobilize residual N in the soil, as shown by (Mays et al., 2003), whereas delayed crop growth could be due to late seeding in the fall, which decreases NO_3^- immobilization and increases NO_3^- losses by leaching (Sanz-Cobena et al., 2012).

2.3 Mixed Cover Crops

To increase the utilization of CCs, growers need specific regional information to understand how the biomass and quality of CCs can affect crop yields and reduce emissions. Selection of CCs types, tillage practices, termination date, residue decomposition, residue quality, and quantity is highly desirable in a such situation. Residue consistency is often distinguished by the content of C and N, lignification, C : N ratios, and the content of polyphenols (Muhammad et al., 2019; Wang et al., 2021; Liu et al., 2021). If the C : N of crop residues is low, it is generally considered to be high-quality crops and nutrients will be released, which affects crop yield strongly and vice versa (Marahatta et al., 2012). A strategy for increasing the quality of crop residues (C : N ratio) and minimizing N immobilization is needed. It has been reported that a mixture of NLCCs and LCCs monocultures is the better choice to improve soil fertility, crop production, and minimized environmental contaminations (Odhiambo and Bomke, 2000). Mixed CCs provide another strategy to mitigate environmental problems because they have a relatively high C : N ratio compared with LCCs and consequently reduce N_2O emissions (Aita and Giacomini, 2003; Schmeer et al., 2014).

3 Cover Crop Uses and Benefits for Soil Microbial Biomass and Community Improvement

Cover crop cultivation and its residues management practices are the main factors that improve soil water holding capacity, soil microbial abundance and structure, and weed suppressions. Cover crops incorporation, mulching, and removal from the field after harvesting have a critical impact on soil microorganisms. The influence of CCs types, residues management, and restudies quality on soil microbes and soil moisture content.

3.1 Soil Microbes

Decades of intensive farming have reduced SOM content, thus plummeting soil fertility and arable land biodiversity (Gardiano et al., 2013). Subsequently, important services for soil ecosystems like nutrient cycling, water management, C storage, and functional biodiversity have been impaired. Microbial communities are vital for improving soil structure conservation and act as main decomposers of fresh organic material and drive biogeochemical nutrients transformation (Pina, 2019). The impacts of management practices on microbial populations are well known, at least regarding the increase in bacterial abundance and enzymatic activity (Muhammad et al., 2021a). Soil with sweet corn residue removed, incorporated, or garland chrysanthemum had 5.0%, 5.4%, and 6.2% higher microbial populations and 22%, 32%, and 26% higher fruit yield, respectively, than control soil (Tian et al., 2011). Similarly, the perennial CCs increased the N mineralization rate and MBC by 37% and 41%, respectively, compared to the NCCs (Pandey and Begum, 2010). In organic farming, huge amounts of C are usually incorporated into the soil, replacing mineral fertilizers ultimately increase the SOM content (Lal, 2009). In a recent study, the SOM content was increased in organic farming as compared to non-organic farming (Cagnini et al., 2019). The introduction of CCs increased the quantity of SOC and improved SOM, microbial biomass carbon, and microbial community structure (Finney et al., 2017). These modifications are essentially based on the characteristics of CCs chemistry and the biotic interactions between plant and soil. Leguminous CCs can fix more atmospheric N due to rhizobia increasing the mineralization and N pool of soil (Schroth et al., 2001). Similarly, previous studies have shown that different CCs species had strong correlations with soil microbial biomass, suggesting that milk vetch had the highest microbial biomass N (15.4 mg/kg) followed by ryegrass (11.3 mg/kg), while the lowest 6.1 mg/kg was observed for the NCCs (Zhu et al., 2012).

Growing CCs create a conducive environment for microbial growth and activities, and upon the decay of CCs, the fungi are attacking residues first and followed by bacteria (Hodge et al., 2001). Cover crops had a positive impact on soil microbial abundance, microbial activity, plant metabolism affecting soil respiration, and plant mineral nutrition (Setyawan et al., 2011), depending on the weather, plants species, and the season of the year. Grasses as CCs played an important role in soil management in citrus orchards and showed that grass roots release stimulating compounds for arbuscular mycorrhizal fungi, which is beneficial for plant growth (Lekberg and Koide, 2005). Legumes cover crops are more effective at fixing N, which is necessary for protein synthesis and plants growth. According to our estimation, the LCCs increases AMF, fungi, bacteria, MBC, MBN, and total PLFA by 17.37%, 19.74%, 74.28%, 47.19%, 33.1%, and 34.4% compared to NCCs, respectively (Figure 3), however fungi to bacteria ratio (F : B) and MBC : MBN ratio decreased by 0.3%, and 5.63%. These results are in line with the finding of Cavalca et al., (2013), who reported that the CCs increase SOM content and microbial activity. The cultivation of CCs adds both the above-ground and below-ground biomass to soil (Wang et al., 2010), which raised the soil C and N stocks (Sainju et al., 2006; Buechi et al., 2015). Growing CCs are encouraged to optimize the productive use of N for a subsequent cash crop and increase productivity due to decreased nutrient losses through leaching (Valkama et al., 2016). Furthermore, reducing dependency on mineral fertilizers, improved water holding capacity, suppressing insect pests and weeds (Dorn et al., 2015; Brooks et al., 2018; Maltais-Landry and Crews, 2019).

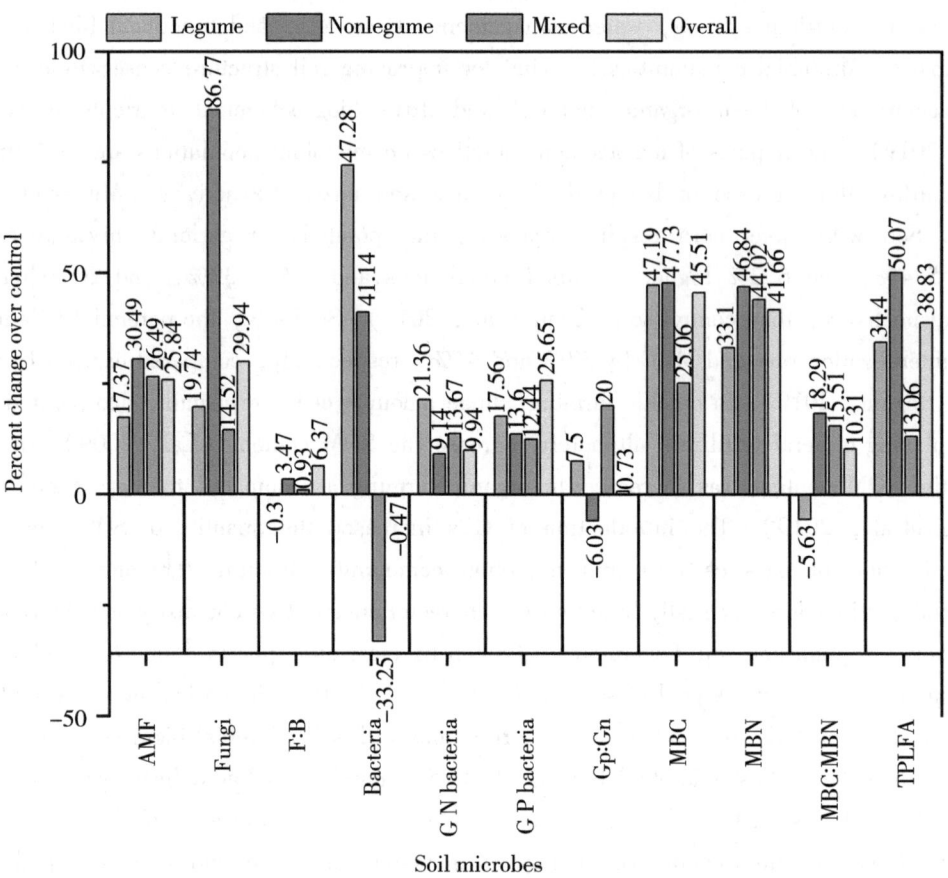

Figure 3 Percent changes of AMF, fungi, F: B, bacteria, gram positive bacteria (G P bacteria), gram negative bacteria (G N bacteria), gram positive: gram negative bacteria (Gp: Gn), microbial biomass carbon (MBC), microbial biomass nitrogen (MBN), MBC: MBN and total phospholipid fatty acid analysis (PLFA) in cover crop (CCs) treatments over control (Ncc)

3.2 Microbial Communities

The introduction of CCs into the agricultural system improves soil and environmental quality through increasing soil microbial population and reducing the application of chemical fertilization (Mitchell et al., 2017). Blanco-Canqui and Jasa (2019) testified that CCs enhanced the soil chemical, physical, and biological properties (Blanco-Canqui and Jasa, 2019). Cover crop utilization reduces soil erosion by covering the soil and improves soil quality by cycling nutrients, SOM, and MBC (Kerri and Belina, 2008). Cover crops significantly impact microbial communities size, operation, and structure by increasing soil C inputs (Xi et al., 2010). Soil microorganisms play a significant role in soil feeding, development, and restoration (CalderÓN et al., 2016; Zhou et al., 2017). According to Peregrina et al. (2014), soil microbial population characteristics are strongly linked to microbial biodiversity, soil and plants quality, and ecosystem sustainability. Researchers argued that nutrient cycling and C conservation are driven by soil microbial communities and vegetation species diversity (Carney and Matson, 2006). Soil microorganisms release enzymes that facilitate the breakdown of complex components in organic materials, correlate enzymes with soil organic C and N substances, which is an indicators of microbial community due to changes in management systems (Tian et al., 2010).

Cover crops significantly improve SOM and boost soil microbial communities (Bacq-Labreuil et al., 2019). In addition, CCs used as GM could boost soil AMF, bacteria, microbial biomass, and total PLFA compared to NCCs (Figure 3). Higo et al. (2018b) concluded that LCCs in rotation (corn-legume) form a symbiosis relationship with particular mycorrhizal fungi and bacterial groups. The continuous crop rotations are collectively referred to as legacy effects, such as beneficial legacy impacts enhancing competent AMF richness, AMF spore density, AMF root colonization, and microbial diversity, eventually improving soil health and agricultural production (Higo et al., 2019). Nonetheless, some of these legacy effects may not be necessary, such as replacing capable AMF with non-host crops and escalating the potential for nutrient immobilization and mineralization in subsurface soil in winter-wheat (Somenahally et al., 2018).

3.3 Soil Bacteria

The maintenance of biological health is important for restoring deteriorated soil because the living components of the soil are necessary for ecosystem functions and utilities (Lehman et al., 2015). Cover crops rotation and minimal tillage are approaches for enhancing the sequestration of organic compounds in agroecosystems that seem to be the most significant ecosystem services and significantly impact the soil biota. It was reported that the first principal component (PC1) distinguished the vetch treatments from the NCCs and wheat treatments, accounting for 23.4% of the total variability (Mbuthia et al., 2015). Gram-positive bacteria (i17 : 0, i16 : 0, a15 : 0, a17 : 0), Gram-negative bacteria (cy19 : 0ω8c, 16.1ω7c), and actinomycetes (10Me16 : 0, 10Me17 : 0, 10Me18 : 0) were found in greater abundance in communities under the vetch CCs treatment. Communities with NCCs and wheat treatments, on the other hand, were linked to the mycorrhizae fungi fatty acid methyl ester biomarker (18 : 1ω9c) and the saprophytic fungi biomarker (18 : 2ω6c; Mbuthia et al., 2015). Cover crops and biological fertilizers are critical aspects of soil quality and fertility in organic management systems. Researchers discovered that CCs prevents soil C, boosts SOM, reduce nutrient leaching, and LCCs fixes N biologically (Snapp and Borden, 2005). Cover crops influence below ground soil functioning through soil microbial communities, such as nutrient cycling and availability, decomposition and transformation of crop residues, and disease suppression (Garbeva et al., 2004; Van Der Heijden et al., 2008). Cover crops generally increase overall microbial activity, nutrient cycling, and microbial diversity (Lori et al., 2017). Similarly, different compositions of CCs cause various changes in soil microbial populations, such as microbial infection, gram-positive bacteria, gram-negative bacteria, and bacterial/fungal ratios (Wortman et al., 2013; Buyer et al., 2017). It was reported that calopo (*Calopogonium mucunoides*) and callisia (*Callisia repens*) residues have 30% and 25% higher AMF than NCCs, respectively. However, mixed CCs significantly increased all the soil microbial biomass and community structure but decreased the soil bacteria by 33.25% (Figure 3). A previous study reported that root C content is more critical than residues C in maintaining stable C (Kong and Six, 2012) suggesting that the effect of litter quality on the sequestration of C is crucial

(Mueller et al., 2017). Frasier et al. (2016) demonstrated that high quality litter and more effective soil biota would improve SOC stability and ultimately increase sits storage.

Cover crops, the most commonly grown vegetation between cash crops, serve as organic matter modifications to the agroecosystems and increase SOC (Sharma et al., 2018). Studies have confirmed that CCs increased SOM and nutrient use efficiency in agroecosystems by reducing nitrate N losses through leaching and drainage, thereby increasing soil bacteria (Rice and Gowda, 2011; Alahmad et al., 2019). According to the microbial community structure estimated by PLFA profiles, the total bacterial and Gram-positive bacteria were significantly higher in CCs than in NCCs treatments. The PCoA analysis showed that the first two principal components, PC1, and PC2, accounted for 13.5% and 56.4% of the total bacterial variability. According to the PCoA, bacterial communities were clearly clustered according to their utilization of CCs (Figure 4), suggesting that the PC2 clearly separated the bacterial communities of CCs and NCCs plots. Furthermore, the bacterial community was not significantly affected by CCs with N fertilization in 10-30 cm soil depth (Alahmad et al., 2019); however, it reduced wind and water erosion (Moreira Rovedder and Foletto Eltz, 2008). The production of CCs improved soil quality and soil biota (Morales Salmeron et al., 2019). Cover crop residues increased the amount of labile C in

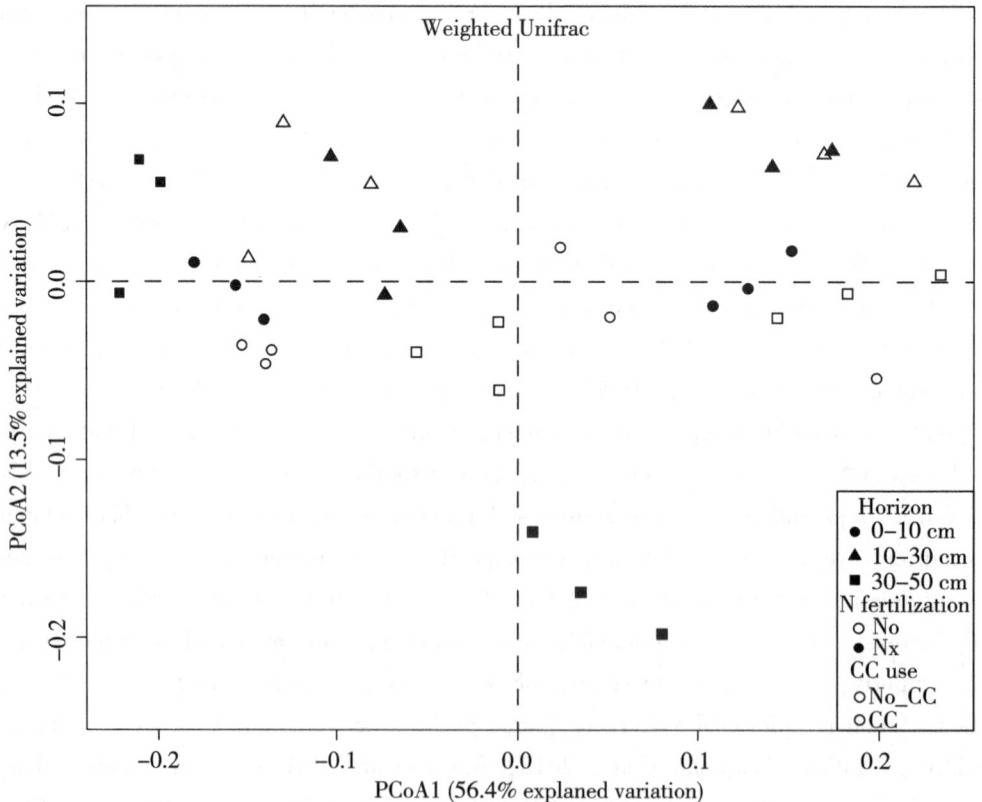

Figure 4 Principal coordinates analysis (PCoA) of microbial communities based on weighted UniFrac distances across soil horizons and among the different experimental treatments (Nx: conventional N fertilization; N0: no N fertilization; CC: presence of cover crops; No-CC: bare soil conditions). [copied with permission from Ref (Alahmad et al., 2019), copyright (2018) John Wiley & Sons, Inc.]

the agroecosystem, especially in the spring season after harvest, and the primary consumer of this labile C is the microbial community (Fernandez et al., 2016). These results show that glucose is a vital source of energy in microbial metabolism (Mukumbareza et al., 2016).

It was observed that cluster 5 (Figure 5) was associated with NCCs, dominated by Proteobacteria and characterized by *Blastocatella fastidiosa* and *Sphingomonas starnbergensisi*. The CCs were linked to Cluster 6, which included species from all major and minor phyla and was as described by *Aciditerrimonas ferrireducens* and *Dehalogenimonas alkenigignensi* (Figure 5). In CCs, the community was dominated by Actinobacteria, with *Oscillochloris trichoides* and *Streptomyces grisemus* as characteristic species and containing many species from "other phyla" (Alahmad et al., 2019). The community in NCCs was characterized by species from Cluster 4, which included Proteobacteria, Actinobacteria, and Firmicutes, with *Actinocatenispora rupis* and *Streptomyces chartreusis* as

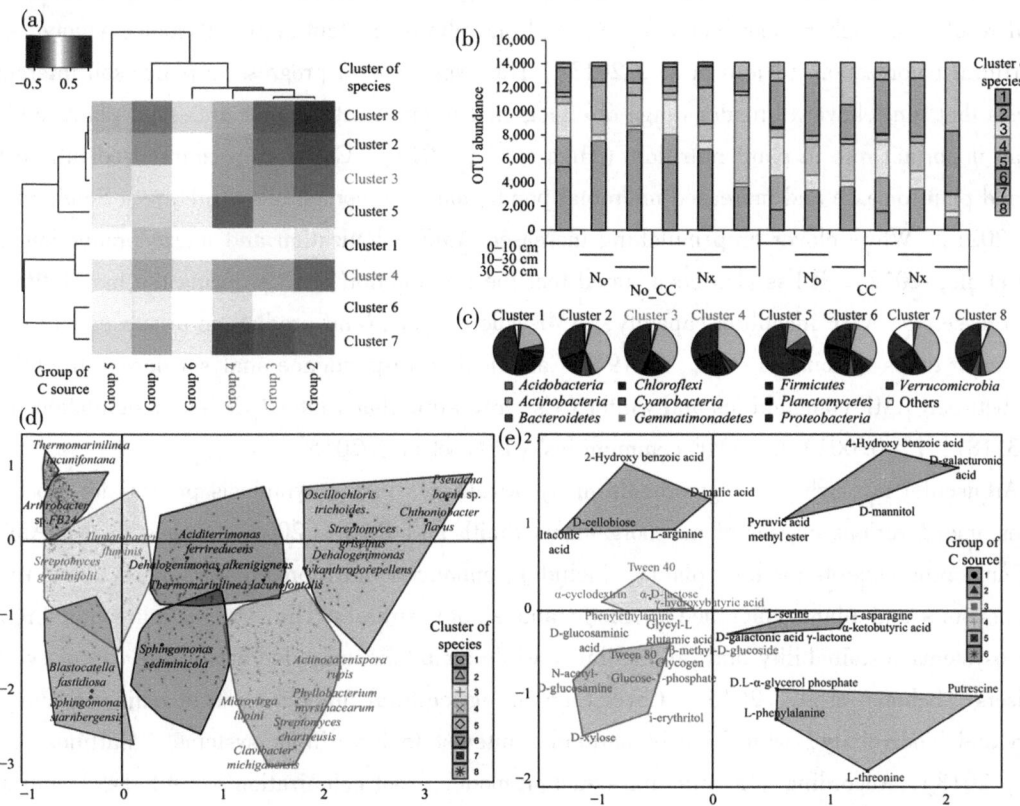

Figure 5　Results of the hierarchical clustering of bacterial species composition and used C-sources

Notes: (a) Heat-map of C-source groups and species clusters based on their values in the cross-table of the BGCoIA.

(b) Relative importance of the clusters across soil depths and among treatments (Nx: conventional N fertilization; N0: no N fertilization; CC: presence of cover crops; No_CC: bare soil conditions).

(c) Distribution of the main bacterial phyla among clusters.

(d) Projection of the clusters and their constitutive species in the diagram defined by the first two BGCoIA axes. Only the name of the most characteristic species for each cluster is reported.

(e) Projection of the groups and their constitutive C sources in the diagram defined by the first two BGCoIA axes. [copied with permission from Ref (Alahmad et al., 2019), copyright (2018) John Wiley & Sons, Inc.].

representative species. Nitrogen fertilization resulted in functional divergence regardless of CCs, but this was supported by compositional divergence only in the presence of NCCs (Figure 5).

3.4 Soil Fungi

Kingdom Fungi include a morphologically diverse group of species extending from single lad yeast to macro fungi, forming networks in soil over many meters. Fungi have attracted attention as major crop pathogens in cultivated agriculture. Nevertheless, they also play a key role in nutrient cycling via dead organic matter catabolism and mycorrhizal symbionts (Chavarria et al., 2016). In cultivated and grassy soils, AMF such as Glomeromycota is the primary mycorrhizal symbionts. Mycorrhizal colonization increased by 35%, 29.4%, and 20.9%, with hairy vetch, mixed CCs, and Indian mustard in maize crops, respectively. This suggests that releasing isothiocyanates in soil resulted in higher shoot biomass, N, and phosphorus content across all maize genotypes with mycorrhizal colonization (Njeru et al., 2013). However, recent progress in plant-soil interactions suggests that fungi have a broader range of effects that interact with higher grassland plants and thus play an important role in plant nutrition (Higo et al., 2017). Cover crops minimized nitrate leaching and plant disease and increased microbial populations and community structure (Singh and Kumar, 2021). White clover crop mulching increased AMF colonization and maize production (Deguchi et al., 2012). It has also been stated that the introduction of CCs during the bare fallow season increases the AMF inoculum capacity, AMF colonization, and production of subsequently cultivated major crops (Hontoria et al., 2019). Canonical correspondence analysis showed the relationships between AMF communities and winter CCs, revealing that winter CCs had a significant impact ($F=3.187$, $P=0.001$) on AMF communities (Higo et al., 2015).

Arbuscular mycorrhizal fungi provide many advantages in the symbiosis process for most plant species among various classes of microorganisms. Smith and Read (2008) suggested that AMF offered numerous benefits for host plants, including enhanced nutrition uptake, particularly in poor nutrients soils and plants tolerance to biotic and abiotic stresses. The AMF significantly improves agroecosystems sustainability and productibility while simultaneously decreasing the use of synthetic fertilizers (Lehman et al., 2015). Cover crops in agricultural practices permit native inoculum recovery and biodiversity, which are of particular interest to low-input systems (Martinez-Garcia et al., 2018). According to a structural equation model, root colonization ($\lambda=1.149$) and maize phosphorus uptake ($\lambda=1.185$) had immediate strong positive effects on crop performance, whereas AMF diversity ($\lambda=0.395$) had intermediate positive effects (Higo et al., 2018a). A recent study has shown that replacing fallow with CCs during intercropping time improves AMF root colonization in the succeeding cash crop (Cagnini et al., 2019).

Cover crop with negligible or zero tillage facilitate mycelium and make colonization faster (Brito et al., 2012). Likewise, CCs improve nutrient reusability, regulating weed growth, reducing erosion, subduing soil disease, and decreasing nutrient leakage (Briar et al., 2011). However, a limited number of literature studies are published to elaborate CCs effect on the AMF community (Bowles et al., 2017). Previous studies suggested that winter CCs influence and change the structure of the AMF population in the subsequent cash crops (Higo et al., 2018a; Morimoto et al., 2018). However, instead of long-term experiments, researchers mostly studied short- to mid-term experiments in the past. Therefore, a literature review on how CCs types and management practices influence the operation and microbial communities is required. Blanco-Canqui and Jasa (2019) reported that the AMF population could be significantly affected by various environmental factors, such as soil properties, nutrient availability, and air temperature. Arbuscular mycorrhizal fungi communities in legume CCs plots are differed from those in grass and herb plots (Figure 6a) and explain 16.1% of the variation in AMF communities. Likewise, archaeal communities in

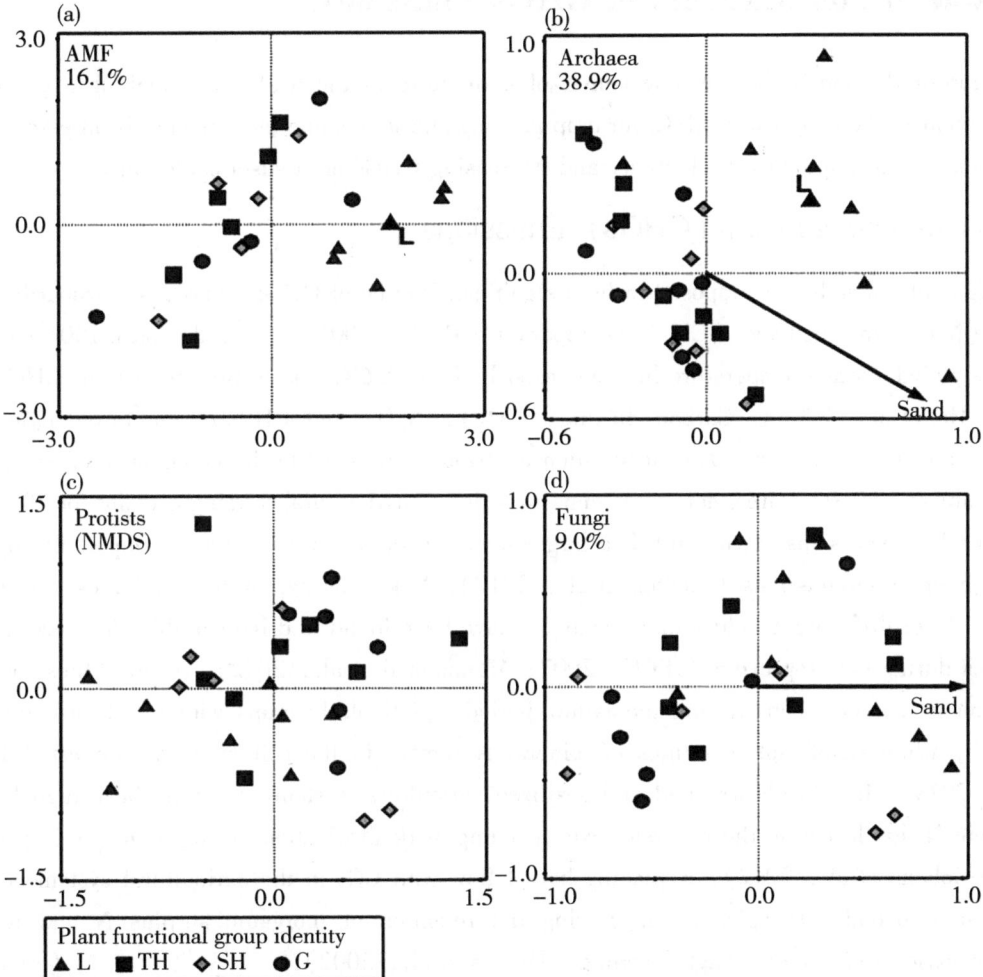

Figure 6 Distance-based redundancy analysis (db-RDA) plot showing the relationship of abiotic soil factors and plant functional group identities to community composition of AMF (a), archaea (b), and fungi (d). Community composition of protists (c) could not be explained by any of the factors measured; therefore, a nonmetric multidimensional scaling (NMDS) of the community is shown instead. The plant functional groups tested were grasses (G; green), legumes (L; red), small herbs (SH; yellow) and tall herbs (TH; blue). The ordination is based on Bray-Curtis distance. With forward selection, factors were chosen that significantly ($P_{adj}<0.05$) contributed to the model. In each window, the percentage of explained variation is shown. [Adapted from Ref (Dassen et al., 2017) under the terms of the Creative Commons Attribution License 4.0]

legume plots differed from grass and herb plots (Figure 6b), and the plant functional group accounted for 14.0% of the total variation. Regarding protists community composition, plant functional group had no significant explanatory power (Figure 6c). Abiotic properties appeared important for fungi and archaea, as shown in Figure 6d. The abiotic properties seemed important for fungal and archaea communities, which explained 9.0% of fungal community composition and 24.9% of archaeal community composition (Dassen et al., 2017).

4 Role of Soil Microbes in GHGs Emissions

Crop production is the focusing center of agriculture in the world. It is making a tremendous contribution to the supply of food. Cover crop growing and its residue management is an essential cultural practice for improving productivity and decreasing GHGs on a sustainable basis.

4.1 Greenhouse Gas (GHGs) Emissions

Agricultural soils are supposed to be a significant source of GHGs emissions, primarily nitrous oxide (N_2O) and methane (CH_4) as reported (IPCC, 2007). Over the past 100 years, the average global surface temperature has increased by 0.3–0.6℃ due to the increase in GHGs emissions (He et al., 2009). Carbon dioxide (CO_2), N_2O, and CH_4 are the three main GHGs gases, and the increase in CO_2 concentration contributes the most to the greenhouse effect (Ozturk and Acaravci, 2010). Greenhouse gas emissions due to agricultural farming account for 10%–12% of total GHGs emissions. Agriculture is a large-scale human activity and has an important impact on global greenhouse emissions (Shelton et al., 2017). Numerous agricultural activities (e.g. irrigation, N fertilization, residues management) have been found to drive variables for these gases emissions during the crop cycle (IPCC, 2007; Muhammad et al., 2022). Some of these activities often impact gas emissions during the fallow period, particularly crops with low N use efficiency (NUE), while significant quantities of mineral N remain in the soil after the harvest (Sanchez et al., 2019). Intensively irrigated and fertilized agricultural systems generally have high levels of inorganic N that lost in to the ecosystem via leaching or denitrification during fallow or crop periods (Sanz-Cobena et al., 2012). Replacing bare fallow with CCs in the agricultural system is one of the most agronomic strategies for increasing the retention of inorganic surplus N and reducing nutrient losses (nitrate) through leaching (Dinnes et al., 2002).

In Mediterranean climate zones, winter CCs are used as catch crops (Gabriel and Quemada, 2011). However, due to CCs, the N and water requirements changes and C pools may affect the processes leading to GHGs emissions. Cover crop root depth, crop N, water requirements, rhizosphere nutrient release, and climate change adaption are strategic factors for mitigation GHGs (Gabriel and Quemada, 2011). In this context, cereals crops generally reduce soil N content in the early growth stages due to higher N uptake and reduced losses through drainage water (Thorup-Kristensen et al., 2003; Moeller et al., 2008). Whereas cruciferous plants have a greater capacity to absorb N in deeper rooting systems at later growth stages and are more easily extracted from the deeper areas (Thorup-Kristensen et al., 2003). Legumes were assessed as catch crops, and it was shown that 50% to 60% of the N in legume tissue mainly comes from N absorption (Pappa et al., 2011). The soil alterations that affect GHGs gases during the CCs information and its relation with GHGs gases are limited (Bayer et al., 2016).

After harvesting, CCs biomass is typically used as GM to reuse the N reserved in residues biomass and contribute to sustainable long-term soil fertility (Chirinda et al., 2010). The quality of CCs are mainly depends on the C : N ratio, crop lignin content, and the N fixing capability, which affect the dynamics of C and N (Sofi et al., 2018). Incorporating crop residues into fertile soils will increase N_2O emissions from cultivated soils (Bair et al., 2008), which depends largely on the residual biomass composition. Suggesting that residues with low C : N ratios generally increase N_2O emissions (Huang et al., 2016). Temporary N fixation often reduces N_2O loss, and thus delaying N availability for the crops as a result of CCs (Marahatta et al., 2012). In crop rotation, the N use efficiency of the main crops (cereals) is affected by GM of previous CCs and other N sources such as fertilizer. The joint effects of residues and synthetic N sources will also affect N_2O emissions. Mineral N in a fertilized plot with greater C : N ratio residues promotes and increases N_2O and CO_2 emissions compared to sole N fertilizer because soluble residual organic C can be used as energy for denitrification (Ghimire et al., 2017; Ozturk, 2017). In terms of CH_4 emissions, the amount of mineral N in the soil may influence the oxidation of this compound (Sanz-Cobena et al., 2014b). The higher emissions might be due to the mineral N pool increasing through mineralization of N-rich plant material from legumes used as GM. Specific GHGs emissions enable us to recommend specific crop management methods in irrigated corn systems to mitigate these losses (Rosolem et al., 2004; Sogbedji et al., 2006). At present, there are many studies on the effects of CCs types and management practices on soil GHGs emissions, but limited studies are available on CCs biomass rates and residue C : N ratios. The characteristics of GHGs emissions after CCs harvesting and whether the management of the residues (e.g. mulch, incorporation, and removal) are unclear. Thus, this review documented the effect of residue quality and quantity on different GHGs gases under different climatic reigns.

4.2 Carbon Dioxide (CO_2) Emissions

Greenhouse gas emissions from agricultural soils are affected by a variety of factors. After CCs introduction to the soil, the C and N cycle, pH, and biomass (microorganisms, roots, etc.) change, which affects GHGs emissions in agriculture soil. Behnke and Villamil, (2019) reported that CCs introductions to the soil increased GHGs emissions, which affected the initial C sequestration. In contrast, CCs mulching effectively reduced the GHGs emissions (Tribouillois et al., 2018). Cover crops may accelerate soil CO_2 efflux compared to conventional farming by increasing SOC and microbiological properties at the 0 to 2.5 cm soil surface (Peregrina, 2016). Cover crop mulching increases the total porosity of the soil, and increases CO_2 content of soil solution, which is conducive to the diffusion of CO_2 (Mondal and Lenka, 2012). The range of CO_2 emissions were 158 mg/kg to 1,884 mg/kg and 154 mg/kg to 1,613 mg/kg for surface-applied, and incorporated residues, respectively, and the highest emissions were found under the clover crop while the lowest for the fallow (Linsler et al., 2016).

To use CCs, farmers select species that fulfill two functions: to keep soil protection and fertility in balance. Xiao et al. (2013) reported that CCs increased SOC content and changed it to a water-soluble form in the soil, thereby affecting the CO_2 emission mode. Compared to NCC, the winter LCCs in the growing season had significantly higher CO_2 emissions than in the rainy season. The range between NCCs [96.92 mg CO_2/($m^2 \cdot h$) to 259.05 mg CO_2/($m^2 \cdot h$)] and winter LCCs [106.28 mg CO_2/($m^2 \cdot h$) to 353.01 mg CO_2/($m^2 \cdot h$)] are shown in Table 1. Common vetch as a GM resulted in lowering soil C:N and higher CO_2 emissions [range of CO_2 8.09 mg CO_2/($m^2 \cdot h$) to 10.58 g CO_2/($m^2 \cdot h$)], enzyme activity, and N release. Conversely, the GM of rye grass had a slightly lower CO_2 emission than vetch CCs (range of CO_2 emissions; 6.63 mg CO_2/($m^2 \cdot h$) to 8.19 g CO_2/($m^2 \cdot h$)) and early soil N immobilization (Table 1). In contrast, NLCCs decreased CO_2 emissions. Winter LCCs treatment had the greatest effect on soil CO_2 and N_2O emissions under furrow irrigation (Kallenbach et al., 2010). Zhou et al. (2019) suggested that CCs provides a C source for microorganisms, which promotes microbial growth and metabolism, thereby increasing CO_2 emissions. A researcher demonstrated that CCs incorporation increases the content of water-stable aggregates in surface soil, which positively correlated with CO_2 emissions (Uliarte et al., 2013). In addition, CCs mulching can indirectly affect soil CO_2 emissions by changing conditions such as preventing soil water losses and maintaining soil temperature. Kokalis-Burelle et al. (2017) reported that mulching of sun hemp as a CCs improves the living environment of soil microorganisms, increases microbial activity, and may lead to increased CO_2 emissions. Researchers found that CO_2 emissions from bare land were much higher than those of CCs treatment, mainly because mulching reduced surface soil temperature and prevented the soil from emitting CO_2 into the atmosphere (Li et al., 2013; Liang et al., 2017).

Table 1 Role of different cover crop types (legume, non-legume and mixed) in soil carbon dioxide emissions (CO_2)

Author	Soil-types	Vegetation	Cover crop	CC type	Range of CO_2 emissions		Units
					Control	Cover crop	
Kallenbach et al., (2010)	Coarse-loamy	Tomato	Winter legume	Legume	259.05	307.66	mg CO_2/($m^2 \cdot h$)
Kallenbach et al., (2010)	Coarse-loamy	Tomato	Winter legume	Legume	215.36	353.01	mg CO_2/($m^2 \cdot h$)
Kallenbach et al., (2010)	Coarse-loamy	Tomato	Winter legume	Legume	96.92	106.29	mg CO_2/($m^2 \cdot h$)
Kallenbach et al., (2010)	Coarse-loamy	Tomato	Winter legume	Legume	106.07	162.22	mg CO_2/($m^2 \cdot h$)
Mancinelli et al., (2013)	Sandy loam	Sweet pepper	Rye grass	Non legume	5.17 to 6.53	6.63 to 8.19	g CO_2/($m^2 \cdot h$)
Mancinelli et al., (2013)	Sandy loam	Sweet pepper	Common vetch	Legume	5.17 to 6.53	8.09 to 10.58	g CO_2/($m^2 \cdot h$)

(continued)

Author	Soil-types	Vegetation	Cover crop	CC type	Range of CO$_2$ emissions Control	Range of CO$_2$ emissions Cover crop	Units
Uliarte et al., (2013)	Clay/silt	Vineyards	*Festuca arundinacea*	Non legume	0.28 to 0.58	-1.4738	g CO$_2$/(m^2·h)
Sanz-Cobena et al., (2014)	Silty clay loam	Maize	Barley	Non legume	2.71 to 9.01	5.16 to 16.54	kg CO$_2$-C/(ha·d)
Sanz-Cobena et al., (2014)	Silty clay loam	Maize	Rape	Non legume	2.71 to 9.01	3.97 to 12.28	kg CO$_2$-C/(ha·d)
Sanz-Cobena et al., (2014)	Silty clay loam	Maize	Vetch	Legume	2.71 to 9.01	2.98 to 16.06	kg CO$_2$-C/(ha·d)
Bavin et al., (2009)	Silt loam	Soybean	Rye	Non legume	2.64 to 4.06	3.23 to 3.68	μmol (m^2·s)
Negassa et al., (2015)	Fine loamy	Maize/soybean	Rye	Non legume	73.03 to 136.18	87.17 to 151.73	mg CO$_2$/(m^2·h)
Hwang et al., (2017)	Clay loam	Rice	Barley	Non legume	72.62 to 172.81	188.33 to 219.04	mg CO$_2$/(m^2·h)
Hwang et al., (2017)	Clay loam	Rice	Hairy Vetch	Legume	72.62 to 172.81	173.32 to 274.31	mg CO$_2$/(m^2·h)
Hwang et al., (2017)	Clay loam	Rice	Barley+Vetch	Mixed	72.62 to 172.81	200.43 to 297.24	mg CO$_2$/(m^2·h)
Abdalla et al., (2014b)	Sandy loam	Spring barley	Mustered	Non legume	24.51	26.09	mg CO$_2$-C/(m^2·h)
Xavier et al., (2013)	Sandy loam	Dwarf cashew	Butterfly-pea and pigeon pea	Legume	7.47 to 5.54	7.47 to 8.57	mg CO$_2$-C/(kg soil·d)
Peregrina (2016)	Fine-loamy	Vitis vinifera L. Vineyard	Resident vegetation	Non legume	0.231 to 0.231	0.41 to 0.45	μmol CO$_2$/(m^2·S)
Steenwerth and Belina (2008)	Coarse-loamy	Vineyard	Rye	Non legume	2.5 to 13.41	6.47 to 16.2	μg CO$_2$-C/(m^2·S)
Steenwerth and Belina (2008)	Coarse-loamy	Vineyard	Trios	Non legume	2.50 to 13.41	4.37 to 22.92	μg CO$_2$-C/(m^2·S)
Barrios-Masias et al., (2011)	Silt loam	Tomato	Winter mustard	Non legume	52.17 to 151.63	71.73 to 228.26	mg CO$_2$-C/(m^2·h)
Rosecrance et al., (2000)	Coarse-loamy	Maize	Rye	Non legume	1.37 to 59.32	4.99 to 544.36	mg/core
Rosecrance et al., (2000)	Coarse-loamy	Maize	Vetch	Legume	1.37 to 59.32	9.99 to 558.60	mg/core
Rosecrance et al., (2000)	Coarse-loamy	Maize	Rye + Vetch	Mixed	1.37 to 59.32	3.84 to 552.71	mg/core
Bhattacharyya et al., (2012)	Sandy clay loam	Rice	*Sesbania aculeata*	Legume	14.80 to 35.25	14.81 to 68.06	mg/(m^2·h)

(continued)

Author	Soil-types	Vegetation	Cover crop	CC type	Range of CO_2 emissions		Units
					Control	Cover crop	
Liebig et al., (2010)	Silt loam	Wheat	Rye	Non legume	5.52 to 45.71	8.23 to 49.80	mg CO_2-C/($m^2 \cdot h$)
Baggs et al., (2003)	Silt loam soil	Maize	Rye	Non legume	194 to 490	132 to 937	kg CO_2-C/ha
Baggs et al., (2003)	Silt loam soil	Maize	Bean	Legume	194 to 490	314 to 901	kg CO_2-C/ha

4.3 Nitrous Oxide (N_2O) Emissions

Cover crop will minimize N losses from agricultural practices by minimizing both nitrate leaching and ammonia plus nitrous oxide transport to the atmosphere (Lacey and Armstrong, 2015). Nitrous oxide is the major contributor to global warming from the agricultural farming system (IPCC, 2007), which is released in soils primarily through two coupled microbial processes; nitrification in aerobic and denitrification in anaerobic environments (Bowen et al., 2018). The frequency and strength of these procedures are strongly influenced by the availability of soil mineral N, soluble C, water, oxygen, temperature, pH, and soil texture (Li et al., 2019). As changes in farming techniques directly affect substrate quality and the environmental conditions in the soil, they are expected to have an effect on N_2O emissions (Abdalla et al., 2012). Steenwerth and Belina, (2008) demonstrated that legumes CCs might lead to greater N_2O emissions, and much of this increase could be due to the decomposition of crop residues. The global impact of legume CCs during crop rotation on N_2O emissions tends to be mostly neutral, although it may differ in relation to the quality (C : N ratios) of crop residues and management techniques (Muhammad et al., 2019). The seasonal N_2O flux in barley cultivation did not differ from that in the NCCs (3.6 kg/ha). However, LCCs (hairy vetch and/or barley + hairy vetch mixture) treatments increased seasonal N_2O flux by 1.8 times than the barley treatment alone (Table 2).

Table 2 Role of different cover crop types (legume, non-legume and mixed) in nitrous oxide emissions (N_2O)

Author	Soil-types	Vegetation	Cover crop	Cc type	Range of N_2O emissions		Units
					Control	Cover crop	
Kallenbach et al., (2010)	Coarse-loamy	Tomato	Winter legume	Legume	21.46	55.46	μg N_2O/($m^2 \cdot h$)
Kallenbach et al., (2010)	Coarse-loamy	Tomato	Winter legume	Legume	26.05	82.76	μg N_2O/($m^2 \cdot h$)
Kallenbach et al., (2010)	Coarse-loamy	Tomato	Winter legume	Legume	45.98	148.66	μg N_2O/($m^2 \cdot h$)

(continued)

Author	Soil-types	Vegetation	Cover crop	Cc type	Range of N$_2$O emissions		Units
					Control	Cover crop	
Kallenbach et al., (2010)	Coarse-loamy	Tomato	Winter legume	Legume	103.45	183.14	μg N$_2$O/(m^2·h)
Garland et al., (2011)	Silty clay	Grape vineyard	Mix legumes	Legume	1.71 to 1.90	0.60 to 1.06	kg N$_2$O-N/(ha·d)
Mitchell et al., (2013)	Clarion loam series	Maize-soybean	Rye	Non Legume	1.52 to 5.06	1.03 to 5.14	kg N$_2$O-N/(ha·d)
Sanz-Cobena et al., (2014)	Silty clay loam	Maize	Barley	Non legume	-0.26	0.02	g N$_2$O-N/(ha·d)
Sanz-Cobena et al., (2014)	Silty clay loam	Maize	Rape	Non legume	-0.26	0.72	g N$_2$O-N/(ha·d)
Sanz-Cobena et al., (2014)	Silty clay loam	Maize	Vetch	Legume	-0.26	0.03	g N$_2$O-N/(ha·d)
Sanz-Cobena et al., (2014)	Silty clay loam	Maize	Barley	Non legume	0.31 to 0.50	0.93 to 3.09	g N$_2$O-N/(ha·d)
Sanz-Cobena et al., (2014)	Silty clay loam	Maize	Rape	Non legume	0.31 to 0.50	0.34 to 1.35	g N$_2$O-N/(ha·d)
Sanz-Cobena et al., (2014)	Silty clay loam	Maize	Vetch	Legume	0.31 to 0.50	1.98 to 3.17	g N$_2$O-N/(ha·d)
Pramanik et al., (2014)	Silt loam	Rice	Barley+ hairy vetch	Mixed	0.41	0.32 to 0.37	mg/(m^2·h)
Steenwerth and Belina (2008)	Loam series	Rice	Rye	Non legume	1.51	2.33	g N$_2$O-N/(ha·d)
Steenwerth and Belina (2008)	Loam series	Rice	Trios	Legume	1.51	1.98	g N$_2$O-N/(ha·d)
Jarecki et al., (2009)	Silty clay loam	Maize-soybean	Rye	Non legume	2.10 to 8.87	3.04 to 6.37	kg N$_2$O-N/ha
Bavin et al., (2009)	Silt loam	Soybean	Winter rye	Non legume	0.162 to 1.79	0.27 to 1.17	nmol/(m^2·s)
Abdalla et al., (2014b)	Sandy loam	Spring barley	Mustard	Non legume	41.21 to 56.92	51.45 to 70.79	g N$_2$O-N/(ha·d)
Sarkodie-Addo et al., (2003)	Silt loam	Maize	Rye	Non legume	8.17 to 13.97	7.83 to 19.30	g N$_2$O-N/(ha·d)
Sarkodie-Addo et al., (2003)	Silt loam	Maize	Winter wheat	Non legume	7.51 to 22.11	6.21 to 36.96	g N$_2$O-N/(ha·d)
Negassa et al., (2015)	Fine loamy	Maize	Rye	Non legume	0.176 to 0.202	0.17 to 0.20	mg/(m^2·h)
Hwang et al., (2017)	Clay loam	Rice	Barley	Non legume	0.07 to 0.1083	0.05 to 0.07	mg/(m^2·h)

(continued)

Author	Soil-types	Vegetation	Cover crop	Cc type	Range of N_2O emissions		Units
					Control	Cover crop	
Hwang et al., (2017)	Clay loam	Rice	Hairy vetch	Legume	0.07 to 0.1083	0.11 to 0.12	mg/($m^2 \cdot h$)
Hwang et al., (2017)	Clay loam	Rice	Barley + hairy vetch	Mixed	0.07 to 0.1083	0.10 to 0.14	mg/($m^2 \cdot h$)
Abdalla et al., (2014a)	Sandy loam	Spring barley	Mustard	Non legume	44.20	58.04	g N_2O-N/(ha $\cdot$ d)
Petersen et al., (2011)	Loamy sandy soil	Spring barley	Fodder radish	Non legume	44.44 to 58.60	62.59 to 97.28	g N_2O-N/(ha $\cdot$ d)
Zhao et al., (2015)	Anthrosols	Rice	Rice/fava bean	Legume	0.51 to 3.69	0.20 to 1.28	μg N_2O/($m^2 \cdot h$)
Zhao et al., (2015)	Anthrosols	Rice	Rice/milk vetch	Legume	0.51 to 3.69	0.27 to 1.18	μg N_2O/($m^2 \cdot h$)
Guardia et al., (2016)	Silty clay loam	Maize	Barley	Non legume	0.03 to 0.81	0.07 to 0.50	mg N_2O-N $m^{-2} d^{-1}$
Guardia et al., (2016)	Silty clay loam	Maize	Vetch	Legume	0.03 to 0.81	0.04 to 1.00	mg N_2O-N $m^{-2} d^{-1}$
Parkin and Kaspar (2006)	Fine loamy	Maize	Rye	Non legume	39.85 to 59.06	33.03 to 65.03	g N_2O-N/(ha $\cdot$ d)
Parkin and Kaspar (2006)	Fine loamy	Soybean	Rye	Non legume	13.86 to 28.67	15.57 to 27.71	g N_2O-N/(ha $\cdot$ d)
Parkin et al., (2006)	Clay loam	Soybean	Rye	Non legume	0.02 to 0.37	0.04 to 0.23	g N/m^2
Barrios-Masias et al., (2011)	Silt loam	Tomato	Oat	Non legume	0.01 to 0.24	0.01 to 0.32	mg N_2O-N/($m^2 \cdot h$)
Rosecrance et al., (2000)	Coarse-loamy	Maize	Cereal rye	Non legume	0.08 to 2.04	0.10 to 0.72	ng/(g $\cdot$ d)
Rosecrance et al., (2000)	Coarse-loamy	Maize	Hairy vetch	Legume	0.08 to 2.04	0.59 to 10.71	ng/(g $\cdot$ d)
Rosecrance et al., (2000)	Coarse-loamy	Maize	Hairy vetch + cereal rye	Mixed	0.08 to 2.04	0.155 to 1.49	ng/(g $\cdot$ d)
Rosecrance et al., (2000)	Silt loam	Maize	Cereal rye	Non legume	0.29 to 4.89	2.59 to 42.65	ng/(g $\cdot$ d)
Rosecrance et al., (2000)	Silt loam	Maize	Hairy vetch	Legume	0.29 to 4.89	3.70 to 33.70	ng/(g $\cdot$ d)
Bhattacharyya et al., (2012)	Sandy clay loam	Rice	*Sesbania aculeata*	Legume	3.57 to 7.63	4.61 to 25.30	μg/($m^2 \cdot h$)
Liebig et al., (2010)	Silt loam	Wheat	Rye	Non legume	-0.25 to 18.62	-8.26 to 24.32	μg N_2O-N/($m^2 \cdot h$)

(continued)

Author	Soil-types	Vegetation	Cover crop	Cc type	Range of N_2O emissions		Units
					Control	Cover crop	
Xiong et al., (2002)	Hydragric anthrosols	Early rice	Vetch	Legume	9.90 to 21.00	11.10 to 167	$\mu g\ N/(m^2 \cdot h)$
Baggs et al., (2003)	Silt loam soil	Maize	Rye	Non legume	104 to 312	158 to 3,542	$g\ N_2O\text{-}N/ha$
Baggs et al., (2003)	Silt loam soil	Maize	Bean	legume	104 to 312	113 to 2,581	$g\ N_2O\text{-}N/ha$
Omonode et al., (2009)	Silty clay loam	Maize	Soybean	legume	0.02 to 651.81	0.71 to 558.64	$g\ N_2O\text{-}N/(ha \cdot d)$
Dietzel et al., (2011)	Howard gravelly loam	Maize	Rye	Non legume	16.44 to 94.68	7.69 to 93.12	$ng\ N_2O\text{-}N/(cm^2 \cdot h)$
Fronning et al., (2008)	Sandy loam	Maize	Rye	Non legume	-0.44 to 46.56	0.08 to 31.01	$g\ N_2O\text{-}N/(ha \cdot d)$

Cover crops are most commonly used as catch crops to mitigate nitrates leaching during fall and winter periods (Plaza-Bonilla et al., 2015). Legume CCs alone or in combination with NLCCs as a GM provide additional N for the subsequent crop (Tribouillois et al., 2015). When compared to unfertilized plots, cover crops had higher mean daily N_2O emissions but lower yearly N_2O emissions when compared to fertilized cropping systems. Cover crop treatments had a 2 to 4-fold higher potential for nitrification, mineralization, and denitrification than conventional cultivation. Thus, CCs improved the soil capacity to support higher MBN, potential N mineralization, and microbiological functions of nitrification and denitrification. Kerri and Belina (2008) found that total soil C content was 40% - 50% higher in soils under five consecutive years of annual CCs than continuously cultivated soil. Cover crops also influence the quality of soil water by rising transpiration rate than bare soil. In general, existing studies show little effect of CCs (LCCs or NLCCs) on N_2O emissions, particularly when results are integrated on the basis of residues quality and quantity (Muhammad et al., 2019). According to a meta-analysis, the effect of CCs on N_2O emissions is mostly influenced by the residues C : N ratio, climatic condition, and residues management practices (Muhammad et al., 2019). Similarly, another meta-analysis demonstrated that the introduction of CCs residues into soil often leads to a short-term increase in N_2O emissions, especially for LCCs (Basche et al., 2014).

Higher precipitations have a deleterious effect on N_2O emissions from CCs fields. Cover crop residue incorporation through tillage has a significant impact on soil structure, soil water dynamics, soil nutrients, and organic residues, which may have an impact on crop production (Gregorich et al., 2008; Negassa et al., 2015). Soil N_2O is mainly produced by soil microorganisms through nitrification and denitrification. Similarly, farmland cultivation measures also affect the soil N_2O emission process by affecting soil temperature, humidity, and nutrient status (Sanz-Cobena et al., 2012). Previous study reported that CCs mulching significantly reduces N_2O emissions (Fiorini et al. , 2020), whereas some researchers have found that soil N_2O emissions significantly increased with CCs incorporation (Snapp and Borden, 2005; Tang et al. , 2015). Compared to saturated moisture, higher soil moisture content (45% - 75%) had significantly higher N_2O emissions. Nitrous oxide emissions increase with higher soil water content and gradually decrease after reaching saturated water content (Sheppard et al. , 2013).

4.4 Methane (CH_4) Emissions

Cold-resistant legumes like Chinese milk vetch and hairy vetch and NLCCs like rye and barley could be the only option for winter CCs in temperate countries with cold and dry weather during winter seasons. Vetch is considered the most common GM in rice fields due to its high N-fixing capacity to adapt to harsh winter conditions and better growth in wet paddy soil. Hwang et al. (2017) demonstrated that hairy vetch and barley mixtures as GM are favored in rice paddy soil because of their higher biomass productivity than sole vetch or barley crops and stronger resistance to winter drought (Ahmad et al. , 2022). The addition of NLCCs (barley or rye) would not efficiently increase rice production due to the high C : N ratio and slow mineralization process of crop residues. The combination of barley and hairy vetch therefore reduce the C : N ratio below 20, which support the mineralization of organic substrates in soil (Marton et al. , 2014). During rice cultivation, CCs biomass application as GM increased soil C balance by 39%-142% over NCCs and increased the seasonal net global warming potential by 3.2-5.7 times due to significantly higher CH_4 emissions under flooded soil conditions.

As shown in Table 3, the CH_4 emission ranges from 73.52-83.08 mg $CH_4/(m^2 \cdot h)$ in milk vetch CCs which is significantly higher than control [0.74-26 mg $CH_4/(m^2 \cdot h)$], and rye CCs [1.11-80.14 mg $CH_4/(m^2 \cdot h)$] in rice paddy soil (Sang et al. , 2012). Most lowlands are under entirely irrigated or rain-fed conditions, and their cultivation serves as an important source of CH_4 emissions (Zschornack et al. , 2016). The average global CH_4 emissions from rice paddy fields are reported to be around 11% of the overall anthropogenic CH_4 emissions (Yan et al. , 2009). It was stated previously that the production of rice, which was 473 million tons in 1990, needs to be increased by 39.43% in 2020 to meet the world population food demand and that anthropogenic CH_4 emissions have been raised by 40%-50% (Lee et al. , 2010). Since CH_4 is primarily produced under strictly anaerobic conditions, it might be because of the decay of organic matter through archaeal methanogens (Chandio et al. , 2020). Similarly, the introduction of organic materials such as GM into flooded rice fields will promote CH_4 emissions by methanogens with readily available C. Agricultural Mediterranean soils produce significant CH_4 emissions through methanogens in flooded crops (e. g. rice), which represent 6% of the total agricultural production (Sanz-Cobena et al. , 2017). Methane emissions mainly depend on residue incorporation, and the CH_4 emissions increase from rice paddy soil by 100-500 kg $CH_4/(ha \cdot a)$ when the straw is added at the rate of 0-7t/ha (Sanchis et al. , 2012). Improved crop production approaches are likely to clash with the extenuation of CH_4 emissions (Yan et al. , 2009).

Table 3 Role of different cover crop types (legume, non-legume and mixed) in soil methane emissions (CH$_4$)

Author	Soil-types	Vegetation	Cover Crop	CC type	Range of CH$_4$ emissions		Units
					Control	Cover crop	
Sang et al., (2012)	Fine silty	Rice	Rye	Non legume	0.74 to 25	1.11 to 80.14	mg CH$_4$/(m^2·h)
Sang et al., (2012)	Fine silty	Rice	Milk vetch	Legume	0.74 to 26	73.52 to 83.08	mg CH$_4$/(m^2·h)
Sanz-Cobena et al., (2014)	Silty clay loam	Maize	Barley	Non legume	0.08 to 5.05	0.084 to 5.96	g CH$_4$-C/(ha·d)
Bavin et al., (2009)	Silt loam	Soybean	Winter rye	Non legume	0.05 to 0.27	0.03 to 0.25	nmol/(m^2·s)
Changhoon et al., (2010)	Fine silty mixed mesic,	Rice	Chinese milk vetch	Legume	8.99 to 85.39	5.99 to 316.11	mg/(m^2·h)
Haque et al., (2013)	Silt loam	Rice	Barley+Vetch	Mixed	3.75 to 13.13	7.15 to 180.17	mg/(m^2·h)
Peyrard et al., (2016)	Calcareous clay	Wheat – sunflower	Barley	Non legume	0.09 to 1.04	0.51 to 1.55	mg/(m^2·h)
Peyrard et al., (2016)	Calcareous clay	Wheat – sunflower	Hairy Vetch	Legume	0.09 to 1.04	0.16 to 3.17	mg/(m^2·h)
Peyrard et al., (2016)	Calcareous clay	Wheat – sunflower	Barley+Vetch	Mixed	0.09 to 1.04	0.21 to 1.96	mg/(m^2·h)
Peyrard et al., (2016)	Calcareous clay	Wheat – sunflower	Barley	Non legume	2.01 to 19.15	1.00 to 172.37	mg/(m^2·h)
Peyrard et al., (2016)	Calcareous clay	Wheat – sunflower	Hairy Vetch	Legume	2.01 to 19.15	1.00 to 50.40	mg/(m^2·h)
Peyrard et al., (2016)	Calcareous clay	Wheat – sunflower	Barley+Vetch	Mixed	2.01 to 19.15	2.01 to 186.49	mg/(m^2·h)
Hwang et al., (2017)	Clay loam	Rice	Barley	Non legume	-1.60 to 25	18 to 170	mg/(m^2·h)
Hwang et al., (2017)	Clay loam	Rice	Hairy Vetch	Legume	-1.60 to 25	-1 to 130	mg/(m^2·h)
Hwang et al., (2017)	Clay loam	Rice	Barley+Vetch	Mixed	-1.60 to 25	1 to 190	mg/(m^2·h)
Abdalla et al., (2014b)	Sandy loam	Spring barley	Mustard	Non legume	0.05 to 0.50	0.01 to 0.47	g CH$_4$-C/(ha·d)
Bhattacharyya et al., (2012)	Sandy clay loam	Rice	Sesbania aculeata	Legume	1.92 to 3.02	2.81 to 8.64	mg/(m^2·h)
Fronning et al., (2008)	Sandy loam	Maize	Rye	Non legume	1.20 to 1.50	1.10 to 50	g CH$_4$-C/(ha·d)

5 Residues C : N RAtios and GHGs EMission

Cover crop residue C : N ratios have a direct impact on soil GHGs (CO_2, N_2O, and CH_4) emissions. Results showed that residues with a high C : N ratio significantly increase CO_2 and CH_4 fluxes while decreasing N_2O flux. These results are in line with the finding of earlier researchers (Muhammad et al., 2019), who documented that CO_2 emissions are positively and N_2O emissions are negatively affected by increasing the residues C : N ratios. Huang et al. (2002) reported that seasonal N_2O emission from wheat-cultivated soil was negatively correlated with increasing soil C : N ratio. The N_2O/NO_3^- ratio and N_2O emission rate increased with decreasing C : N ratio in organic amendments in a well-aerated soil (Melling et al., 2007). Reddy and Crohn (2014) demonstrated that the rate of N_2O production is partially controlled by C susceptible through the mineralization process. Compared with wheat monoculture, wheat-chickpea crop rotation showed a C sequestration rate of 0.53 Mg C/(ha · a) during 20 years (López-Bellido et al., 2010) which could be related to the fluxes of CO_2. Huang et al. (2004) found that the root decay content was strongly negatively correlated with the increasing root residue C : N ratio. Similarly, Khalil et al. (2002) reported that N_2O production was increased by decreasing the C : N ratio of different organic materials. In addition, Kato et al. (2011) found a strongly negative correlation between the mean annual N_2O emission and C : N ratio of tropical rain forest soils. Residues with lower C : N decomposed more rapidly, might provide a greater opportunity for producing more dissolved organic C, hence resulting in higher N_2O emissions.

5.1 Low C : N Ratio Residues

Cover crops develop resistance against pests, weeds, and potential environmental degradation, improve soil quality, and thus increase the production of subsequent cash crops (Neely et al., 2018). Leguminous CCs release N into the soil and thus enhance crop yield (Holman et al., 2018). Lentil (*Lens culinaris Medik*), field pea (*Pisum sativum* L.), and faba bean (*Vicia faba* L.) were used as LCCs for potential N fixation in rotation with cash crops. The biennial clover (*Melilotus officinalis* L.) has advantages over annual legumes because of its low seed costs, a strong competitor to weeds, and high productivity in biomass production and N fixation in semi-arid environments (Ravenel et al., 2015). Intercropping LCCs with winter cereals may have fixed atmospheric N_2, but after winter cereal harvest, they may show maximum growth and N fixation. Cover crops such as alfalfa (*Medicago sativa* L.) and red clover (*Trifolium pratense* L.) had higher production of biomass and added N to winter wheat crops due to their faster decomposition rate (Chen et al., 2006). The research revealed that alfalfa and red clover as relay crops and lentil and chickling vetch (*Lathyrus sativus* L.) as double crops were productive legumes in a cereal cropping system in winter (Martens and Entz, 2001). It has been shown that alfalfa and red clover can fix significant amounts of N without adversely affecting the winter wheat yield in moderate to high rainfall regions (Martens and Entz, 2001), while the incorporation of these crops into the soil increases soil health due to mineralization through soil microorganisms (Li et al., 2019).

Similarly, alfalfa and red clover showed potential for N fixation in moderate to high rainfall areas without adversely affecting the yield of winter wheat (Martens et al., 2001).

5.2 High C : N Ratio Residue

The CCs are selected based on N-fixing capacity, biomass quality (C : N ratios) and quantities in many agricultural systems. However, there are insufficient evidence to suggest that crops increase soil fertility, crop yield, and soil microbial populations (Blesh, 2018). Non-legume CCs are also the best choice for spring forage because of their regrowth behavior during the growth stage as compared to small grain cereals (White and Weil, 2010). Cover crops during the following seasons decrease the soil water and wind erosion due to covering of soil surface (Kaspar et al., 2001). The rye crop biomass as CCs help to sustain SOM (Kaspar et al., 2006). Researchers proposed that CCs decrease soil bulk density and increase water retention potential, soil microbial processes, and porosity of soils (Villamil et al., 2006). However, other researchers reported that CCs have either improved or have no effect on soil properties (Mupambwa and Wakindiki, 2012; Mupambwa, 20012). Soil response to agronomic methods is also influenced by soil and environmental conditions (Bescansa et al., 2006) in addition to CCs management. The introduction of rye CCs conserved soil resources and had environmental benefits immediately and long-term economic benefits during the cropping system (Igos et al., 2016).

5.3 Optimum C : N Ratio Residues or Mixed Residues

Mixed CCs are an important management method used extensively to increase SOM and subsequently increase cash crop productivity (Sainju and Singh, 2008). Growing CCs can provide physical protection to the soil by reducing the impact of rainfall and also improve soil structure and aggregation (Tang et al., 2017), and similarly improving soil microorganisms (Araujo et al., 2019). Morales Salmeron et al. (2019) found that soil structure and fertility are improved by the presence of legumes and legume-grain mixtures in row crop systems. In addition, the residues of the main crop rotated with a potato crop are supposed to affect the accumulation of soil N and C over time. Planting CCs mixtures (multi-species) could be a viable solution to enhance the ecological stability, microbial diversity, and resilience of CCs communities, which contribute to higher and consistent productivity. The production benefits of multi-species plant communities include the potential for increased efficiency in resource use and crop yields (Tosti et al., 2012).

6 Termination Method to Mitigate GHGs Emissions

The goal of cultivation and the availability of machinery are the two main factors that mitigate the GHGs emission. The CCs are terminated through incorporation, surface mulch, and/or removal, as well as using herbicide for killing for improving the productivity of the following cash crop. The influence of CCs variety, termination timing, and termination method on mulch, weed cover, and soil nitrate in organic systems play a vital role in GHGs emissions.

6.1 Residues Incorporation

A traditional termination method involves using a fixed machine-driven disk to NCCs and CCs plot (Figure 7a) to incorporate residues into the soil (Jani et al., 2016; Rojas et al., 2018). Termination of CCs through disking results in rapid residue decomposition and release of nutrients due to close contact with soil microorganisms. As a result, microbial decomposition has been facilitated by better residue soil interaction and higher levels of soil oxygen. Cover crop incorporating through disking improved the total bacteria population while decreasing or stabilizing the populations of fungi and actinomycetes (Elfstrand et al., 2007). The incorporation of barley and rape CCs residue increased the soil respiration by 21% and 28%, respectively. The CH_4 emissions were decreased with CCs incorporation with mean values of -0.12 and -0.10 kg CH_4-C/ha for plots with and without CCs, respectively (Table 3; Sanz-Cobena et al., 2014a). Fungi abundance a reduction in the disked soils could be due to high lignin materials, increased soil disturbance, and reduced soil moisture (Elfstrand et al., 2007). Incorporating CCs residue into the soil increased CO_2 and N_2O emissions compared to residue placed on the soil surface because of increased contact with soil microorganisms.

Figure 7 Examples of cover cropping termination methods (Incorporation, mulching and removing), and the concept of cover crop residues biomass and soil microbes interaction

Tillage practices expose more residue to microorganisms, which enhances aeration and microbial activity, resulting in higher CO_2 and N_2O emissions (Alluvione et al., 2010). Several studies have found that incorporating CCs residues into the soil increases N_2O emissions when compared to surface placement (Xiong et al., 2002; Basche et al., 2014; Muhammad et al., 2021b). Further research is needed to examine the impact of disking on CCs residues breakdown, retention of soil water content, OM, dynamics of N releases, and subsequent cash crops. Cover crops with a high C : N ratio, such as rye, stimulated more CH_4 emissions than CCs with a low C : N ratio per grain yield, but LCCs vetch was more effective in improving the rice growth and yield (Table 3). To improve soil properties and rice productivity while limiting CH_4 release, it may be more beneficial to use a low C : N ratio, such as milk vetch, as a CCs than a high C : N ratio, such as rye (Sang et al., 2012).

6.2 Residues Mulching

Using a rotary or flail mower to terminate CCs, cuts the residue into slices and shreds the residue on the soil surface (Rose et al., 2016). In early fall, the soil microbial biomass was 150 mg C/kg soil, which is 50% higher than in late spring (100 mg C/kg soil; Zibilske and Makus, 2009). Mowing rye at the vegetative growth stage resulted in regrowth, which eventually resulted in water and nutrient competition with subsequent cash crops (Kornecki et al., 2009). The termination of CCs with rotating mowers can be difficult due to the irregular size and dispersal of plant remains. Compared to the arable weed community present in NCCs plots, LCCs increased total PLFA concentration by 5.37 nmol/g and 10.20 nmol/g in the fall and spring (Finney et al., 2017). Hairy vetch was positively associated with non-AMF (*Vicia villosa* L.). The associations between CCs and microbial groups were found in monocultures and in multispecies CCs mixtures (Finney et al., 2017). In comparison to CCs incorporation, CCs termination with a flail mower may be important in controlling soil nutrient transformations and SOM maintenance in subtropical climates, which regulate soil temperature and moisture at more suitable levels for microbial activity and residue decomposition (Zibilske and Makus, 2009; Khan et al., 2021). Moreover, another study found that flail-mowing is difficult to achieve uniform mulching, allowing weeds to emerge via thin openings (Teasdale et al., 2007; Wegner et al., 2018).

Mechanical mowing of rye, mustard, and hairy vetch mixture improved N mineralization. Compared to the NCCs treatment, white mustard (*Sinapis alba* L.), Lacy phacelia (*Phacelia tanacetifolia* Benth. LP), and Hairy vetch (*Vicia villosa* Roth) mulching increased microbial biomass C by 38%, 80%, and 44%, respectively. These increases could be due to the increase in soil C and N by 19% and 44% in white mustard, 6% and 2% in Lacy phacelia, and 10% and 13% in Hairy vetch, respectively, when compared to NCCs (Marinari et al., 2015). Several studies have found that incorporating CCs residues increases N_2O emissions when compared to surface placement (Basche et al., 2014; Muhammad et al., 2019). The CCs residue removal had no effect on N_2O emissions when compared to NCCs but resulted in higher CO_2 emissions because of CCs roots in the soil that were not removed. Only the above-ground biomass of CCs was removed from the soil for animal forage in the residue removal treatment.

Timing is the most crucial factor when operating a roller-crimper for CCs termination. Termination of grass CCs should occur once flowering has begun (Figure 7b), while LCCs should be terminated after pods formation. The CCs will not be terminated successfully; if it is performed too late (McMechan et al., 2021; Werle et al., 2021). The use of a roller-crimper is one of the best ways to suppress weeds with CCs residue (Mirsky et al., 2009). To avoid competing with cash crops for water and nutrients, effective termination of CCs with a roller-crimper is critical (Bavougian et al., 2019). It has been demonstrated that crimping LCCs with roller crimped produce 10 kg N/ha to 217 kg N/ha reliant on CCs species and killing times (Parr et al., 2014), and soil inorganic N peaks after 4 to 6 weeks of roller crimping (Parr et al., 2014). Further study is required to discover how mulching and removing the CCs termination method affects soil C, N release, and soil microbial diversity in agricultural ecosystems.

6.3 Residue Removal with Herbicide Application

Non-selective herbicides are commonly used for terminating the CCs because they are effective at all growth phases (Clark et al., 2007). These have low application costs and the ability to terminate CCs on a large scale within a short period of time (Hay et al., 2019; Farooq et al., 2022). As the role of herbicides in non-target species has increased, along with their negative impact on these species, there is an increased desire to learn about their impact on plant nutrition and nutrient cycling (Cornelius and Bradley, 2017). Previous studies have demonstrated that glyphosate has no impact on soil microorganisms diversity and activity (Weaver et al., 2007; Lane et al., 2012), although, after one week of glyphosate treatment, a significant drop in total microbial biomass in soybean rhizosphere soil was detected (Lane et al., 2012). Selection of herbicide with residual soil activity when planning to use CCs again in the fall must be done carefully in order to avoid impacts on CCs species establishment. Glyphosate should be applied when temperatures reach 55° F during the day and 40° F during the night. Applying glyphosate before the boot stage will help to improve effective cereal rye. Improved spray coverage will increase the efficacy of contact herbicides such as paraquat (Gramoxone) and glufosinate (Liberty, Cheetah, and Scout; Werle et al., 2021). During the spring thaw until the end of rye, CO_2 emissions were higher in green fallow than in chemical fallow ($P=0.0071$). The atmospheric CH_4 uptake was the dominant exchange process, and it was significantly ($P=0.0124$) higher under chemical fallow [2.7 g CH_4-C/(ha·d)] than under green fallow [1.5 g CH_4-C/(ha·d)]. Cumulative CO_2, CH_4, and N_2O emissions did not differ between the chemical and green-fallow phases ($P=0.1293$, 0.2629, and 0.9979, respectively) during the 19-month period (Liebig et al., 2010).

7 Conclusions

The effect of CCs types, biomass, and residue C∶N ratios on soil GHGs emissions, soil microbial biomass, and community was investigated in this study. Compared to NCCs, the CCs increased microbial biomass and community abundance due to additional organic matter input. A higher fungi/bacteria ratio with CCs suggests that CCs have a greater impact on fungi than bacteria. The LCCs had lower actinomycete levels but a higher MBC/MBN ratio than the NLCCs. The benefits of CCs on soil microbial biomass were reduced when mixed LCCs and NLCCs were used instead of LCCs or NLCCs alone. Carbon dioxide and N_2O emissions vary according to CCs species, biomass residue quality and quantity, and method of residue placement in the soil. When compared to NCCs, CCs increased CO_2 emissions. Legume CCs emitted more N_2O than NLCCs or mixed CCs. Increased CCs biomass resulted in higher CO_2 emissions but lower N_2O emissions. The increases in N_2O emissions can cause changes in global warming potential, thus affecting the C sinks (soil organic C input via plants) and losses (CO_2 emissions/mineralization). The LCC and NLCCs combined application as CCs reduced the GHGs emissions and improved soil health and crop yields. Although CCs increases CO_2 emissions compared to NCCs, they have positive effects on soil and C sequestration and other known soil health and environmental quality parameters. Further studies are needed to clarify the effect of CCc biomass rates and residues quality (C : N ratios) on soil microbial community structure and abundance and their influence on soil GHGs emissions.

Author Contributions

Conceptualization, I. M., X. B. Z., and W. J.; Methodology, I. M. and S. A.; Investigation, I. M., S. F. and S. A.; Resources, X. B. Z.,; Data curation, I. M.,; Writing original draft preparation, I. M.; Writing review and editing, X. B. Z and W. J., S, A; Supervision, X. B. Z., All authors have read and agreed to the published version of the manuscript.

Funding

This study was supported by the National Natural Science Foundation of China (31760354) and the Natural Science Foundation of Guangxi (2019GXNSFAA185028); CAS: "Light of West China" Program for introduced talent in the west, The National Natural Science Foundation of China (Grant No. 31570440, 31270484); the Key international scientific and Technological cooperation and exchange project of Shaanxi Province China (2020KWZ-010).

References

Abdalla M, Hastings A, Cheng K, et al., 2019. A critical review of the impacts of cover crops on nitrogen leaching, net greenhouse gas balance and crop productivity. *Global Change Biology*. 25: 2530-2543.

Abdalla M, Rueangritsarakul K, Jones, M, et al., 2012. How effective is reduced tillage-cover crop management in reducing n2o fluxes from arable crop soils? *Water Air and Soil Pollution*. 223: 5155-5174.

Ackroyd VJ, Cavigelli MA, Spargo JT, et al., 2019. Legume cover crops reduce poultry litter application requirements in organic systems. *Agronomy Journal*. 111: 2361-2369.

Ahmad S, Wang GY, Muhammad I, et al., 2022. Interactive Effects of Melatonin and Nitrogen Improve Drought Tolerance of Maize Seedlings by Regulating Growth and Physiochemical Attributes. *Antioxidants*. 11: 359.

Aita C, and Giacomini SJ, 2003. Crop residue decomposition and nitrogen release in single and mixed cover crops. *Revista Brasileira De Ciencia Do Solo*. 27: 601-612.

Alahmad A, Decocq G, Spicher F, et al., 2019. Cover crops in arable lands increase functional complementarity and redundancy of bacterial communities. *Journal of Applied Ecology*. 56: 651-664.

Alfonso Gomez J, Campos M, Guzman G, et al., 2018. Soil erosion control, plant diversity, and arthropod communities under heterogeneous cover crops in an olive orchard. *Environmental Science and Pollution Research*. 25: 977-989.

Alluvione F, Bertora C, Zavattaro L, et al., 2010. Nitrous oxide and carbon dioxide emissions following green manure and compost fertilization in corn. *Soil Science Society of A-*

merica Journal. 74: 384-395.

Araujo FS, Barroso JR, Freitas LDO, et al., 2019. Chemical attributes and microbial activity of soil cultivated with cassava under different cover crops. *Revista Brasileira De Engenharia Agricola E Ambiental.* 23: 614-619.

Austin EE, Wickings K, Mcdaniel MD, et al., 2017. Cover crop root contributions to soil carbon in a no-till corn bioenergy cropping system. *Global Change Biology Bioenergy.* 9: 1252-1263.

Bacq-Labreuil A, Crawford J, Mooney SJ, et al., 2019. Cover crop species have contrasting influence upon soil structural genesis and microbial community phenotype. *Scientific Reports.* 9: 7473.

Baggs EM, Watson CA, and Rees RM, 2000. The fate of nitrogen from incorporated cover crop and green manure residues. *Nutrient Cycling in Agroecosystems.* 56: 153-163.

Bagley JE, Desai AR, Dirmeyer PA, et al., 2012. Effects of land cover change on moisture availability and potential crop yield in the world's breadbaskets. *Environmental Research Letters.* 7: 014009.

Bair KE, Davenport JR, and Stevens RG, 2008. Release of available nitrogen after incorporation of a legume cover crop in concord grape. *Hortscience.* 43: 875-880.

Basche AD, Miguez FE, Kaspar TC, et al., 2014. Do cover crops increase or decrease nitrous oxide emissions? A meta-analysis. *Journal of Soil and Water Conservation.* 69: 471-482.

Bavougian CM, Sarno E, Knezevic S, et al., 2019. Cover crop species and termination method effects on organic maize and soybean. *Biological Agriculture & Horticulture.* 35: 1-20.

Bayer C, Gomes J, Zanatta JA, et al., 2016. Mitigating greenhouse gas emissions from a subtropical Ultisol by using long-term no-tillage in combination with legume cover crops. *Soil & Tillage Research.* 161: 86-94.

Behnke GD, and Villamil MB, 2019. Cover crop rotations affect greenhouse gas emissions and crop production in Illinois, USA. *Field Crops Research.* 241: 107580.

Berlanas C, Andres-Sodupe M, Lopez-Manzanares B, et al., 2018. Effect of white mustard cover crop residue, soil chemical fumigation and Trichoderma spp. root treatment on black-foot disease control in grapevine. *Pest Management Science.* 74: 2864-2873.

Bescansa P, Imaz M, Virto I, et al., 2006. Soil water retention as affected by tillage and residue management in semiarid Spain. *Soil and Tillage Research.* 87: 19-27.

Blanco-Canqui H, and Jasa PJ, 2019. Do grass and legume cover crops improve soil properties in the long term? *Soil Science Society of America Journal.* 83: 1181-1187.

Blesh J, 2018. Functional traits in cover crop mixtures: Biological nitrogen fixation and multifunctionality. *Journal of Applied Ecology.* 55: 38-48.

Bowen H, Maul JE, Poffenbarger H, et al., 2018. Spatial patterns of microbial denitrification genes change in response to poultry litter placement and cover crop species in an agricultural soil. *Biology and Fertility of Soils.* 54: 769-781.

Bowles TM, Jackson LE, Loeher M, et al., 2017. Ecological intensification and arbuscular mycorrhizas: a meta-analysis of tillage and cover crop effects. *Journal of Applied Ecology*. 54: 1785-1793.

Briar SS, Fonte SJ, Park I, et al., 2011. The distribution of nematodes and soil microbial communities across soil aggregate fractions and farm management systems. *Soil Biology and Biochemistry*. 43: 905-914.

Brito I, Goss M, and De Carvalho M, 2012. Effect of tillage and crop on arbuscular mycorrhiza colonization of winter wheat and triticale under Mediterranean conditions. *Soil use and management*. 28: 202-208.

Brooks JP, Tewolde H, Adeli A, et al., 2018. Effects of subsurface banding and broadcast of poultry litter and cover crop on soil microbial populations. *Journal of Environmental Quality*. 47: 427-435.

Buechi L, Gebhard CA, Liebisch F, et al., 2015. Accumulation of biologically fixed nitrogen by legumes cultivated as cover crops in Switzerland. *Plant and Soil*. 393: 163-175.

Buyer JS, Baligar VC, He Z, et al., 2017. Soil microbial communities under cacao agroforestry and cover crop systems in Peru. *Applied Soil Ecology*. 120: 273-280.

Cagnini CZ, Garcia DM, Silva NDS, et al., 2019. Cover crop and deep tillage on sandstone soil structure and microbial biomass. *Archives of Agronomy and Soil Science*. 65: 980-993.

Calderón FJ, Nielsen D, Acosta-Martínez V, et al., 2016. Cover Crop and Irrigation Effects on Soil Microbial Communities and Enzymes in Semiarid Agroecosystems of the Central Great Plains of North America. *Pedosphere*. 26: 192-205.

Campiglia E, Paolini R, Colla G, et al., 2009. The effects of cover cropping on yield and weed control of potato in a transitional system. *Field Crops Research*. 112: 16-23.

Carney KM, and Matson PA, 2006. The influence of tropical plant diversity and composition on soil microbial communities. *Microbial Ecology*. 52: 226-238.

Cavalca L, Corsini A, Zaccheo P, et al., 2013. Microbial transformations of arsenic: perspectives for biological removal of arsenic from water. *Future Microbiology*. 8: 753-768.

Chalise KS, Singh S, Wegner BR, et al., 2019. Cover crops and returning residue impact on soil organic carbon, bulk density, penetration resistance, water retention, infiltration, and soybean yield. *Agronomy Journal*. 111: 99-108.

Chandio AA, Magsi H, and Ozturk I, 2020. Examining the effects of climate change on rice production: case study of Pakistan. *Environmental Science and Pollution Research*. 27: 7812-7822.

Chavarria DN, Verdenelli RA, Serri DL, et al., 2016. Effect of cover crops on microbial community structure and related enzyme activities and macronutrient availability. *European Journal of Soil Biology*. 76: 74-82.

Chen S, Wyse DL, Johnson GA, et al., 2006. Effect of cover crops alfalfa, red clover, perennial ryegrass, and rye on soybean cyst nematode population and soybean and corn yields in Minnesota. *Journal of Nematology*. 38: 267.

Chirinda N, Olesen JE, Porter JR, et al., 2010. Soil properties, crop production and greenhouse gas emissions from organic and inorganic fertilizer-based arable cropping systems. *Ag-

riculture, ecosystems & environment. 139: 584-594.

Clark AJ, Meisinger JJ, Decker AM, et al., 2007. Effects of a grass-selective herbicide in a vetch-rye cover crop system on corn grain yield and soil moisture. *Agronomy Journal*. 99: 43-48.

Cornelius CD, and Bradley KW, 2017. Herbicide Programs for the Termination of Various Cover Crop Species. *Weed Technology*. 31: 514-522.

Couedel A, Alletto L, Kirkegaard J, et al., 2018a. Crucifer glucosinolate production in legume-crucifer cover crop mixtures. *European Journal of Agronomy*. 96: 22-33.

Couedel A, Alletto L, Tribouillois H, et al., 2018b. Cover crop crucifer-legume mixtures provide effective nitrate catch crop and nitrogen green manure ecosystem services. *Agriculture Ecosystems & Environment*. 254: 50-59.

Dassen S, Cortois R, Martens H, et al., 2017. Differential responses of soil bacteria, fungi, archaea and protists to plant species richness and plant functional group identity. *Molecular Ecology*. 26: 4085-4098.

De Almeida Acosta JA, Carneiro Amado TJ, Da Silva LS, et al., 2014. Fitomass decomposition and nitrogen release of cover crops in function of the level of residue input to soil under no-tillage system. *Ciencia Rural*. 44: 801-809.

De Barros DL, Oliveira Gomide PH, and De Carvalho GJ, 2013. Cover crops and their effects on succession culture. *Bioscience Journal*. 29: 308-318.

Deguchi S, Uozumi S, Touno E, et al., 2012. Arbuscular mycorrhizal colonization increases phosphorus uptake and growth of corn in a white clover living mulch system. *Soil science and plant nutrition*. 58: 169-172.

Dhima KV, Vasilakoglou IB, Eleftherohorinos IG, et al., 2006. Allelopathic potential of winter cereal cover crop mulches on grass weed suppression and sugarbeet development. *Crop Science*. 46: 1682-1691.

Dinnes DL, Karlen DL, Jaynes DB, et al., 2002. Nitrogen management strategies to reduce nitrate leaching in tile-drained Midwestern soils. *Agronomy journal*. 94: 153-171.

Dorn B, Jossi W, and Van Der Heijden MGA, 2015. Weed suppression by cover crops: comparative on-farm experiments under integrated and organic conservation tillage. *Weed Research*. 55: 586-597.

Elfstrand S, BåTh B, and MåRtensson A, 2007. Influence of various forms of green manure amendment on soil microbial community composition, enzyme activity and nutrient levels in leek. *Applied Soil Ecology*. 36: 70-82.

Farooq S, Wu H, Nie J, et al., 2022. Application, advancement and green aspects of magnetic molecularly imprinted polymers in pesticide residue detection. *Science of The Total Environment*. 804: 150293.

Fernandez AL, Sheaffer CC, Wyse DL, et al., 2016. Structure of bacterial communities in soil following cover crop and organic fertilizer incorporation. *Applied Microbiology and Biotechnology*. 100: 9331-9341.

Finney DM, Buyer JS, and Kaye JP, 2017. Living cover crops have immediate impacts on soil microbial community structure and function. *Journal of Soil and Water Conservation*. 72: 361-373.

Fiorini A, Maris SC, Abalos D, et al., 2020. Combining no-till with rye (*Secale cereale* L.) cover crop mitigates nitrous oxide emissions without decreasing yield. *Soil & Tillage Research*: 196.

Frasier I, Noellemeyer E, Figuerola E, et al., 2016. High quality residues from cover crops favor changes in microbial community and enhance C and N sequestration. *Global Ecology and Conservation*. 6: 242-256.

Gabriel JL, and Quemada M, 2011. Replacing bare fallow with cover crops in a maize cropping system: Yield, N uptake and fertiliser fate. *European Journal of Agronomy*. 34: 133-143.

Garbeva PV, Van Veen J, and Van Elsas J, 2004. Microbial diversity in soil: selection of microbial populations by plant and soil type and implications for disease suppressiveness. *Annu. Rev. Phytopathol*. 42: 243-270.

Gardiano CG, Krzyzanowski AA, Abib Saab OJ, et al., 2013. Population reduction of the reniform nematode with the incorporation of soil cover crops in greenhouse. *Nematropica*. 43: 138-142.

Gavazzi C, Schulz M, Marocco A, et al., 2010. Sustainable weed control by allelochemicals from rye cover crops: from the greenhouse to field evidence. *Allelopathy Journal*. 25: 259-273.

Ghimire B, Ghimire R, Vanleeuwen D, et al., 2017. Cover Crop Residue Amount and Quality Effects on Soil Organic Carbon Mineralization. *Sustainability*. 9: 2316.

Gonsiorkiewicz Rigon JP, Calonego JC, Rosolem CA, et al., 2018. Cover crop rotations in no-till system: short-term CO_2 emissions and soybean yield. *Scientia Agricola*. 75: 18-26.

Gregorich E, Rochette P, St-Georges P, et al., 2008. Tillage effects on N_2O emission from soils under corn and soybeans in Eastern Canada. *Canadian Journal of Soil Science*. 88: 153-161.

Hall H, Li Y, Comerford N, et al., 2010. Cover crops alter phosphorus soil fractions and organic matter accumulation in a Peruvian cacao agroforestry system. *Agroforestry systems*. 80: 447-455.

Haramoto ER, and Brainard DC, 2012. Strip tillage and oat cover crops increase soil moisture and influence N mineralization patterns in cabbage. *HortScience*. 47: 1596-1602.

Hay MM, Dille JA, and Peterson DE, 2019. Integrated pigweed (Amaranthus spp.) management in glufosinate-resistant soybean with a cover crop, narrow row widths, row-crop cultivation, and herbicide program. *Weed Technology*. 33: 710-719.

Haynes R, 2005. Labile organic matter fractions as centralcomponents of the quality of agricultural soils: anoverview. *Advances in agronomy*. 85: 221-268.

He J, Wang Q, Li H, et al., 2009. Effect of alternative tillage and residue cover on yield and water use efficiency in annual double cropping system in North China Plain. *Soil & Tillage Research*. 104: 198-205.

Higo M, Isobe K, Yamaguchi M, et al., 2015. Impact of a soil sampling strategy on the spatial distribution and diversity of arbuscular mycorrhizal communities at a small scale in two winter cover crop rotational systems. *Annals of microbiology.* 65: 985-993.

Higo M, Sasaki R, Unoki T, et al., 2017. Influence of cover crop residue management on the indigenous arbuscular mycorrhizal fungi, corn growth and yield. *Canadian Journal of Plant Pathology.* 39: 560-560.

Higo M, Takahashi Y, Gunji K, et al., 2018a. How are arbuscular mycorrhizal associations related to maize growth performance during short-term cover crop rotation? *Journal of the Science of Food and Agriculture.* 98: 1388-1396.

Higo M, Tatewaki Y, Gunji K, et al., 2019. Cover cropping can be a stronger determinant than host crop identity for arbuscular mycorrhizal fungal communities colonizing maize and soybean. *Peerj.* 7.

Hodge A, Campbell CD, and Fitter AH, 2001. An arbuscular mycorrhizal fungus accelerates decomposition and acquires nitrogen directly from organic material. *Nature.* 413: 297-299.

Holman JD, Arnet K, Dille J, et al., 2018. Can Cover or Forage Crops Replace Fallow in the Semiarid Central Great Plains? *Crop Science.* 58: 932-944.

Hontoria C, Garcia-Gonzalez I, Quemada M, et al., 2019. The cover crop determines the AMF community composition in soil and in roots of maize after a ten-year continuous crop rotation. *Sci Total Environ.* 660: 913-922.

Huang S, Tang J, Liao P, et al., 2016. Interaction of winter legume manure covering (*Astragalus sinicus* L.) and atraw retention on yield and soil properties in a double rice cropping system. *Acta Agriculturae Universitatis Jiangxiensis.* 38: 215-222.

Huang Y, Jiao Y, Zong L, et al., 2002. N_2O emission from wheat cultivated soils as influenced by soil physicochemical properties. *Acta Scientiae Circumstantiae.* 22: 598-602.

Huang Y, Zou J, Zheng X, et al., 2004. Nitrous oxide emissions as influenced by amendment of plant residues with different C:N ratios. *Soil Biology and Biochemistry.* 36: 973-981.

Hwang HY, Kim GW, Kim SY, et al., 2017. Effect of cover cropping on the net global warming potential of rice paddy soil. *Geoderma.* 292: 49-58.

Hwang HY, Kim GW, Lee YB, et al., 2015. Improvement of the value of green manure via mixed hairy vetch and barley cultivation in temperate paddy soil. *Field Crops Research.* 183: 138-146.

Igos E, Golkowska K, Koster D, et al., 2016. Using rye as cover crop for bioenergy production: An environmental and economic assessment. *Biomass & Bioenergy.* 95: 116-123.

Ipcc CC, 2007. The physical science basis. Contribution of working group I to the fourth assessment report of the Intergovernmental Panel on Climate Change. *Cambridge University Press, Cambridge, United Kingdom and New York, NY, USA.* 996: 2007.

Jani AD, Grossman J, Smyth TJ, et al., 2016. Winter legume cover-crop root decomposition and N release dynamics under disking and roller-crimping termination approaches. *Renewable Agriculture and Food Systems.* 31: 214-229.

Kallenbach CM, Rolston DE, and Horwath WR, 2010. Cover cropping affects soil N_2O and

CO_2 emissions differently depending on type of irrigation. *Agriculture, Ecosystems & Environment*. 137: 251-260.

Kandel TP, Gowda PH, Somenahally A, et al., 2018. Nitrous oxide emissions as influenced by legume cover crops and nitrogen fertilization. *Nutrient Cycling in Agroecosystems*. 112: 119-131.

Kaspar TC, Jaynes DB, Parkin TB, et al., 2012. Effectiveness of oat and rye cover crops in reducing nitrate losses in drainage water. *Agricultural Water Management*. 110: 25-33.

Kaspar TC, Parkin TB, Jaynes DB, et al., 2006. Examining changes in soil organic carbon with oat and rye cover crops using terrain covariates. *Soil Science Society of America Journal*. 70: 1168-1177.

Kaspar TC, Radke JK, and Laflen JM, 2001. Small grain cover crops and wheel traffic effects on infiltration, runoff, and erosion. *Journal of Soil and Water Conservation*. 56: 160-164.

Kato T, Hirota M, Tang Y, et al., 2011. Spatial variability of CH_4 and N_2O fluxes in alpine ecosystems on the Qinghai-Tibetan Plateau. *Atmospheric Environment*. 45: 5632-5639.

Kerri S, and Belina KM, 2008. Cover crops and cultivation: Impacts on soil N dynamics and microbiological function in a Mediterranean vineyard agroecosystem. *Applied Soil Ecology*. 40: 370-380.

Khan I, Lei H, Shah AA, et al., 2021. Climate change impact assessment, flood management, and mitigation strategies in Pakistan for sustainable future. *Environmental Science and Pollution Research*. 28: 29720-29731.

Kokalis-Burelle N, Mcsorley R, Wang KH, et al., 2017. Rhizosphere microorganisms affected by soil solarization and cover cropping in Capsicum annuum and Phaseolus lunatus agroecosystems. *Applied Soil Ecology*. 119: 64-71.

Kong AYY, and Six J, 2012. Microbial community assimilation of cover crop rhizodeposition within soil microenvironments in alternative and conventional cropping systems. *Plant and Soil*. 356: 315-330.

Kornecki TS, Price AJ, Raper RL, et al., 2009. New roller crimper concepts for mechanical termination of cover crops in conservation agriculture. *Renewable Agriculture and Food Systems*. 24: 165-173.

Kornecki TS, Prior SA, and Torbert HA, 2016. Effects of a custom cover crop residue manager in a no-till cotton system. *Applied Engineering in Agriculture*. 32: 333-340.

Lacey C, and Armstrong S, 2015. The efficacy of winter cover crops to stabilize soil inorganic nitrogen after fall-applied anhydrous ammonia. *Journal of Environmental Quality*. 44: 442-448.

Lal R, 2009. Soil quality impacts of residue removal for bioethanol production. *Soil and tillage research*. 102: 233-241.

Lane M, Lorenz N, Saxena J, et al., 2012. Microbial activity, community structure and potassium dynamics in rhizosphere soil of soybean plants treated with glyphosate. *Pedobiologia*. 55: 153-159.

Lee CH, Do Park K, Jung KY, et al., 2010. Effect of Chinese milk vetch (*Astragalus* sinicus L.) as a green manure on rice productivity and methane emission in paddy soil. *Agriculture, ecosystems & environment*. 138: 343-347.

Lehman RM, Cambardella CA, Stott DE, et al., 2015. Understanding and enhancing soil biological health: the solution for reversing soil degradation. *Sustainability*. 7: 988-1027.

Lekberg Y, and Koide R, 2005. Is plant performance limited by abundance of arbuscular mycorrhizal fungi? A meta-analysis of studies published between 1988 and 2003. *New Phytologist*. 168: 189-204.

Liang H, Hu K, Qin W, et al., 2017. Modelling the effect of mulching on soil heat transfer, water movement and crop growth for ground cover rice production system. *Field Crops Research*. 201: 97-107.

Liebig M, Tanaka D, and Gross J, 2010. Fallow effects on soil carbon and greenhouse gas flux in central North Dakota. *Soil Science Society of America Journal*. 74: 358-365.

Liebman AM, Grossman J, Brown M, et al., 2018. Legume cover crops and tillage impact nitrogen dynamics in organic corn production. *Agronomy Journal*. 110: 1046-1057.

Li F, Sorensen P, Li X, et al., 2019. Carbon and nitrogen mineralization differ between incorporated shoots and roots of legume versus non-legume based cover crops. *Plant and Soil*. 446: 243-257.

Linsler D, Kaiser M, Andruschkewitsch R, et al., 2016. Effects of cover crop growth and decomposition on the distribution of aggregate size fractions and soil microbial carbon dynamics. *Soil Use and Management*. 32: 192-199.

Li R, Hou X, Jia Z, et al., 2013. Effects on soil temperature, moisture, and maize yield of cultivation with ridge and furrow mulching in the rainfed area of the Loess Plateau, China. *Agricultural Water Management*. 116: 101-109.

Liu Y, Pan Y, Yang L, et al., 2021. Stover return and nitrogen application affect soil organic carbon and nitrogen in a double-season maize field. *Plant Biology*. 24: 387-395.

Logsdon SD, Kaspar TC, Meek DW, et al., 2002. Nitrate leaching as influenced by cover crops in large soil monoliths. *Agronomy Journal*. 94: 807-814.

Lori M, Symnaczik S, Mäder P, et al., 2017. Organic farming enhances soil microbial abundance and activity—A meta-analysis and meta-regression. *PloS one*: 12.

López-Bellido RJ, Fontán JM, López-Bellido FJ, et al., 2010. Carbon sequestration by tillage, rotation, and nitrogen fertilization in a Mediterranean Vertisol. *Agronomy Journal*. 102: 310-318.

Madsen H, Talgre L, Eremeev V, et al., 2016. Do green manures as winter cover crops impact the weediness and crop yield in an organic crop rotation? *Biological Agriculture & Horticulture*. 32: 182-191.

Malpassi R, Kaspar T, Parkin T, et al., 2000. Oat and rye root decomposition effects on nitrogen mineralization. *Soil Science Society of America Journal*. 64: 208-215.

Maltais-Landry G, and Crews TE, 2019. Hybrid systems combining manures and cover crops can substantially reduce nitrogen fertilizer use. *Agroecology and Sustainable Food Systems*.

Marahatta SP, Wang KH, and Sipes BS, 2012. DOES INTEGRATION OF HIGH AND LOW

C : N RATIO COVER CROPS BENEFIT SOIL HEALTH MANAGEMENT? *Journal of Nematology*. 44: 476-477.

Marinari S, Mancinelli R, Brunetti P, et al., 2015. Soil quality, microbial functions and tomato yield under cover crop mulching in the Mediterranean environment. *Soil and Tillage Research*. 145: 20-28.

Martens JRT, and Entz MH, 2001. Availability of late-season heat and water resources for relay and double cropping with winter wheat in prairie Canada. 81: 273-276.

Martens JRT, Hoeppner JW, and Entz MH, 2001. Legume cover crops with winter cereals in southern Manitoba: Establishment, productivity, and microclimate effects. *Agronomy Journal*. 93: 1086-1096.

Martinez-Garcia LB, Korthals G, Brussaard L, et al., 2018. Organic management and cover crop species steer soil microbial community structure and functionality along with soil organic matter properties. *Agriculture Ecosystems & Environment*. 263: 7-17.

Marton JM, Fennessy MS, and Craft CB, 2014. USDA conservation practices increase carbon storage and water quality improvement functions: an example from Ohio. *Restoration ecology*. 22: 117-124.

Mays DA, Sistani KR, and Malik RK, 2003. Use of winter annual cover crops to reduce soil nitrate levels. *Journal of Sustainable Agriculture*. 21: 5-19.

Mbuthia LW, Acosta-Martínez V, Debruyn J, et al., 2015. Long term tillage, cover crop, and fertilization effects on microbial community structure, activity: Implications for soil quality. *Soil Biology and Biochemistry*. 89: 24-34.

Mcmechan AJ, Hodgson EW, Varenhorst AJ, et al., 2021. Soybean Gall Midge (Diptera: Cecidomyiidae), a New Species Causing Injury to Soybean in the United States. *Journal of Integrated Pest Management*. 12: 8.

Meisinger J, Hargrove W, Mikkelsen R, et al., 1991. Effects of cover crops on groundwater quality. *Cover crops for clean water*: 57-68.

Melling L, Hatano R, and Goh KJ, 2007. Nitrous oxide emissions from three ecosystems in tropical peatland of Sarawak, Malaysia. *Soil Science and Plant Nutrition*. 53: 792-805.

Mirsky SB, Curran WS, Mortensen DA, et al., 2009. Control of Cereal Rye with a Roller/Crimper as Influenced by Cover Crop Phenology. *Agronomy Journal*. 101: 1589-1596.

Mitchell JP, Shrestha A, Mathesius K, et al., 2017. Cover cropping and no-tillage improve soil health in an arid irrigated cropping system in California's San Joaquin Valley, USA. *Soil & Tillage Research*. 165: 325-335.

Moeller K, Stinner W, and Leithold G, 2008. Growth, composition, biological N-2 fixation and nutrient uptake of a leguminous cover crop mixture and the effect of their removal on field nitrogen balances and nitrate leaching risk. *Nutrient Cycling in Agroecosystems*. 82: 233-249.

Mondal MK, and Lenka M, 2012. Solubility of CO_2 in aqueous strontium hydroxide. *Fluid phase equilibria*. 336: 59-62.

Morales Salmeron L, Martin-Lammerding D, Tenorio Pasamon JL, et al., 2019. Effects of cover crops on soil biota, soil fertility and weeds, and Pratylenchus suppression in experimental conditions. *Nematology*. 21: 227-241.

Moreira Rovedder AP, and Foletto Eltz FL, 2008. Revegetation with cover crops for soils under arenization and wind erosion in Rio Grande do Sul state, Brazil. *Revista Brasileira De Ciencia Do Solo*. 32: 315-321.

Morimoto S, Uchida T, Matsunami H, et al., 2018. Effect of winter wheat cover cropping with no-till cultivation on the community structure of arbuscular mycorrhizal fungi colonizing the subsequent soybean. *Soil Science and Plant Nutrition*. 64: 545-553.

Mueller P, Granse D, Nolte S, et al., 2017. Top-down control of carbon sequestration: grazing affects microbial structure and function in salt marsh soils. *Ecological Applications*. 27: 1435-1450.

Muhammad I, Khan F, Khan A, et al., 2018. Soil fertility in response to urea and farmyard manure incorporation under different tillage systems in Peshawar, Pakistan. *Int. J. Agric. Biol*. 20: 1539-1547.

Muhammad I, Sainju UM, Zhao F, et al., 2019. Regulation of soil CO_2 and N_2O emissions by cover crops: A meta-analysis. *Soil & Tillage Research*. 192: 103-112.

Muhammad I, Wang J, Khan A, et al., 2021a. Impact of the mixture verses solo residue management and climatic conditions on soil microbial biomass carbon to nitrogen ratio: a systematic review. *Environmental Science and Pollution Research*. 28: 64241-64252.

Muhammad I, Wang J, Sainju UM, et al., 2021b. Cover cropping enhances soil microbial biomass and affects microbial community structure: A meta-analysis. *Geoderma*. 381.

Muhammad I, Yang L, Ahmad S, et al., 2022. Irrigation and nitrogen fertilization alter soil bacterial communities, soil enzyme activities, and nutrient availability in maize crop. *Frontiers in Microbiology*. 105.

Mukumbareza C, Muchaonyerwa P, and Chiduza C, 2016. Bicultures of oat (*Avena sativa* L.) and grazing vetch (*Vicia dasycarpa* L.) cover crops increase contents of carbon pools and activities of selected enzymes in a loam soil under warm temperate conditions. *Soil Science and Plant Nutrition*. 62: 447-455.

Mupambwa HA, 2012. *Winter Rotational Cover Crops Effects on Soil Strength, Aggregate Stability and Water Conservation of a Hardsetting Cambisol in Eastern Cape Province, South Africa/dcHupenyu Allan Mupambwa*. University of Fort Hare.

Mupambwa H, and Wakindiki I, 2012. Winter cover crops effects on soil strength, infiltration and water retention in a sandy loam Oakleaf soil in Eastern Cape, South Africa. *South African Journal of Plant and Soil*. 29: 121-126.

Nakamoto T, Komatsuzaki M, Hirata T, et al., 2012. Effects of tillage and winter cover cropping on microbial substrate-induced respiration and soil aggregation in two Japanese fields. *Soil science and plant nutrition*. 58: 70-82.

Neely CB, Rouquette FM, Jr., Morgan CL, et al., 2018. Integrating legumes as cover crops and intercrops into grain sorghum production systems. *Agronomy Journal*. 110: 1363-1378.

Negassa W, Price RF, Basir A, et al., 2015. Cover crop and tillage systems effect on soil

CO$_2$ and N$_2$O fluxes in contrasting topographic positions. *Soil & Tillage Research*. 154: 64-74.

Njeru EM, Avio L, Sbrana C, et al., 2013. First evidence for a major cover crop effect on arbuscular mycorrhizal fungi and organic maize growth. *Agronomy for Sustainable Development*. 34: 841-848.

Odhiambo JJO, and Bomke AA, 2000. Short term nitrogen availability following overwinter cereal/grass and legume cover crop monocultures and mixtures in south coastal British Columbia. *Journal of Soil and Water Conservation*. 55: 347-354.

Ozturk I, 2017. Measuring the impact of alternative and nuclear energy consumption, carbon dioxide emissions and oil rents on specific growth factors in the panel of Latin American countries. *Progress in Nuclear Energy*. 100: 71-81.

Ozturk I, and Acaravci A, 2010. CO2 emissions, energy consumption and economic growth in Turkey. *Renewable and Sustainable Energy Reviews*. 14: 3220-3225.

Pandey C, and Begum M, 2010. The effect of a perennial cover crop on net soil N mineralization and microbial biomass carbon in coconut plantations in the humid tropics. *Soil use and management*. 26: 158-166.

Pappa VA, Rees RM, Walker RL, et al., 2011. Nitrous oxide emissions and nitrate leaching in an arable rotation resulting from the presence of an intercrop. *Agriculture, ecosystems & environment*. 141: 153-161.

Parr M, Grossman JM, Reberg-Horton SC, et al., 2014. Roller-Crimper Termination for Legume Cover Crops in North Carolina: Impacts on Nutrient Availability to a Succeeding Corn Crop. *Communications in Soil Science and Plant Analysis*. 45: 1106-1119.

Peregrina F, 2016. Surface soil properties influence carbon oxide pulses after precipitation events in a semiarid vineyard under conventional tillage and cover crops. *Pedosphere*. 26: 499-509.

Peregrina F, Pilar Perez-Alvarez E, and Garcia-Escudero E, 2014. Soil microbiological properties and its stratification ratios for soil quality assessment under different cover crop management systems in a semiarid vineyard. *Journal of Plant Nutrition and Soil Science*. 177: 548-559.

Peyrard C, Mary B, Perrin P, et al., 2016. N$_2$O emissions of low input cropping systems as affected by legume and cover crops use. *Agriculture, Ecosystems & Environment*. 224: 145-156.

Pina M, 2019. Cover crops and biochar soil amendments shift microbial communities to improve soil fertility and fruit yield in a southern california organic mango orchard. *Hortscience*. 54: S18-S18.

Plaza-Bonilla D, Nolot JM, Raffaillac D, et al., 2015. Cover crops mitigate nitrate leaching in cropping systems including grain legumes: Field evidence and model simulations. *Agriculture Ecosystems & Environment*. 212: 1-12.

Ranabhat NB, Burrows ME, Miller ZJ, et al., 2015. Impact of cover crop termination methods on diseases of wheat and lentil. *Phytopathology*. 105: 116-116.

Ravenel C, Deneufbourg F, Pateau Y, et al., 2015. Tall fescue seed crops grown with legume cover crops: impact on nitrogen requirements. *Fourrages*: 287-291.

Reddy KN, 2001. Effects of cereal and legume cover crop residues on weeds, yield, and net return in soybean (Glycine max). *Weed Technology*. 15: 660-668.

Reddy KN, 2003. Impact of rye cover crop and herbicides on weeds, yield, and net return in narrow-row transgenic and conventional soybean (Glycine max). *Weed Technology*. 17: 28-35.

Reddy N, and Crohn DM, 2014. Effects of soil salinity and carbon availability from organic amendments on nitrous oxide emissions. *Geoderma*. 235: 363-371.

Rice WC, and Gowda PH, 2011. Influence of geographical location, crop type and crop residue cover on bacterial and fungal community structures. *Geoderma*. 160: 271-280.

Rodrigues Torres JL, Pereira MG, and Fabian AJ, 2008. Cover crops biomass production and its residues mineralization in a Brazilian no-till Oxisol. *Pesquisa Agropecuaria Brasileira*. 43: 421-428.

Rojas MA, Van Eerd LL, O'halloran IP, et al., 2018. Responses of spring-seeded cover crop roots by herbicide residues and short-term influence in soil aggregate stability and N cycling. *Canadian Journal of Plant Science*. 98: 990-1004.

Rose TJ, Wood RH, Gleeson DB, et al., 2016. Removal of phosphorus in residues of legume or cereal plants determines growth of subsequently planted wheat in a high phosphorus fixing soil. *Biology and Fertility of Soils*. 52: 1085-1092.

Rosolem CA, Pace L, and Crusciol CaC, 2004. Nitrogen management in maize cover crop rotations. *Plant and Soil*. 264: 261-271.

Sainju UM, and Singh B P, 2008. Nitrogen storage with cover crops and nitrogen fertilization in tilled and nontilled soils. *Agronomy Journal*. 100: 619-627.

Sainju UM, Singh BP, Whitehead WF, et al., 2007. Accumulation and crop uptake of soil mineral nitrogen as influenced by tillage, cover crops, and nitrogen fertilization. *Agronomy Journal*. 99: 682-691.

Sainju UM, Whitehead WF, Singh BP, et al., 2006. Tillage, cover crops, and nitrogen fertilization effects on soil nitrogen and cotton and sorghum yields. *European Journal of Agronomy*. 25: 372-382.

Sanchez II, Fultz LM, Lofton J, et al., 2019. Soil Biological Response to Integration of Cover Crops and Nitrogen Rates in a Conservation Tillage Corn Production System. *Soil Science Society of America Journal*. 83: 1356.

Sanchis E, Ferrer M, Torres AG, et al., 2012. Effect of water and straw management practices on methane emissions from rice fields: a review through a meta-analysis. *Environmental engineering science*. 29: 1053-1062.

Sang YK, Gutierrez J, and Kim PJ, 2012. Considering winter cover crop selection as green manure to control methane emission during rice cultivation in paddy soil. *Agriculture Ecosystems & Environment*. 161: 130-136.

Sanz-Cobena A, García-Marco S, Quemada M, et al., 2014a. Do cover crops enhance

N_2O, CO_2 or CH_4 emissions from soil in Mediterranean arable systems? *Science of the total environment.* 466: 164-174.

Sanz-Cobena A, Lassaletta L, Aguilera E, et al., 2017. Strategies for greenhouse gas emissions mitigation in Mediterranean agriculture: A review. *Agriculture, ecosystems & environment.* 238: 5-24.

Sanz-Cobena A, Sánchez-Martín L, García-Torres L, et al., 2012. Gaseous emissions of N_2O and NO and NO_3^- leaching from urea applied with urease and nitrification inhibitors to a maize (Zea mays) crop. *Agriculture, Ecosystems & Environment.* 149: 64-73.

Schmeer M, Loges R, Dittert K, et al., 2014. Legume-based forage production systems reduce nitrous oxide emissions. *Soil and Tillage Research.* 143: 17-25.

Schroth G, Salazar E, and Da Silva JP, 2001. Soil nitrogen mineralization under tree crops and a legume cover crop in multi-strata agroforestry in central Amazonia: Spatial and temporal patterns. *Experimental Agriculture.* 37: 253-267.

Setyawan D, Gilkes R, and Tongway D, 2011. Nutrient cycling index in relation to organic matter and soil respiration of rehabilitated mine sites in Kelian, East Kalimantan. *Journal of Tropical Soils.* 16 (3): 219-223. DOI: 10.5400/jts.2011.16.3.219.

Sharma V, Irmak S, and Padhi J, 2018. Effects of cover crops on soil quality: Part I. Soil chemical properties-organic carbon, total nitrogen, pH, electrical conductivity, organic matter content, nitrate-nitrogen, and phosphorus. *Journal of Soil and Water Conservation.* 73: 637-651.

Shelton RE, Jacobsen KL, and Mcculley RL, 2017. Cover Crops and Fertilization Alter Nitrogen Loss in Organic and Conventional Conservation Agriculture Systems. *Frontiers in plant science.* 8: 2260.

Sheppard L, Leith I, Leeson S, et al., 2013. Fate of N in a peatland, Whim bog: immobilisation in the vegetation and peat, leakage into pore water and losses as N_2O depend on the form of N. *Biogeosciences.* 10: 149-160.

Singh J, and Kumar S, 2021. Responses of soil microbial community structure and greenhouse gas fluxes to crop rotations that include winter cover crops. *Geoderma.* 385: 114843.

Smith S, and Read D, 2008. Colonization of roots and anatomy of arbuscular mycorrhiza. *Mycorrhizal Symbiosis. Academic Press*: London. 42-90.

Snapp SS, and Borden H, 2005. Enhanced nitrogen mineralization in mowed or glyphosate treated cover crops compared to direct incorporation. *Plant and Soil.* 270: 101-112.

Sofi JA, Dar IH, Chesti MH, et al., 2018. Effect of nitrogen fixing cover crops on fertility of apple (Malus domestica Borkh) orchard soils assessed in a chronosequence in North-West Himalaya of Kashmir valley, India. *Legume Research.* 41: 87-94.

Sogbedji JM, Van Es HM, and Agbeko KL, 2006. Cover cropping and nutrient management strategies for maize production in western Africa. *Agronomy Journal.* 98: 883-889.

Somenahally A, Dupont JI, Brady J, et al., 2018. Microbial communities in soil profile are more responsive to legacy effects of wheat-cover crop rotations than tillage systems. *Soil Biology and Biochemistry.* 123: 126-135.

Steenwerth K, and Belina KM, 2008. Cover crops and cultivation: Impacts on soil N dynamics and microbiological function in a Mediterranean vineyard agroecosystem. *Applied Soil Ecology*. 40: 370-380.

Tang H, Xiao X, Tang W, et al., 2015. Effects of winter covering crop residue incorporation on CH_4 and N_2O emission from double-cropped paddy fields in southern China. *Environmental Science and Pollution Research*. 22: 12689-12698.

Tang H, Xiao X, Tang W, et al., 2017. Returning Winter Cover Crop Residue Influences Soil Aggregation and Humic Substances under Double-cropped Rice Fields. *Revista Brasileira De Ciencia Do Solo*. 41.

Teasdale JR, 1996. Contribution of cover crops to weed management in sustainable agricultural systems. *Journal of production agriculture*. 9: 475-479.

Teasdale JR, Coffman CB, and Mangum RW, 2007. Potential long-term benefits of no-tillage and organic cropping systems for grain production and soil improvement. *Agronomy Journal*. 99: 1297-1305.

Thorup-Kristensen K, Magid J, and Jensen LS, 2003. Catch crops and green manures as biological tools in nitrogen management in temperate zones. *Advances in agronomy*. 79: 227-302.

Tian L, Dell E, and Shi W, 2010. Chemical composition of dissolved organic matter in agroecosystems: correlations with soil enzyme activity and carbon and nitrogen mineralization. *Applied Soil Ecology*. 46: 426-435.

Tian Y, Zhang X, Liu J, et al., 2011. Effects of summer cover crop and residue management on cucumber growth in intensive Chinese production systems: soil nutrients, microbial properties and nematodes. *Plant and soil*. 339: 299-315.

Tonitto C, David MB, and Drinkwater LE, 2006. Replacing bare fallows with cover crops in fertilizer-intensive cropping systems: A meta-analysis of crop yield and N dynamics. *Agriculture Ecosystems & Environment*. 112: 58-72.

Toom M, Talgre L, Maee A, et al., 2019. Selecting winter cover crop species for northern climatic conditions. *Biological Agriculture & Horticulture*. 35: 263-274.

Tosti G, Benincasa P, Farneselli M, et al., 2012. Green manuring effect of pure and mixed barley-hairy vetch winter cover crops on maize and processing tomato N nutrition. *European Journal of Agronomy*. 43: 136-146.

Tribouillois H, Constantin J, and Justes E, 2018. Cover crops mitigate direct greenhouse gases balance but reduce drainage under climate change scenarios in temperate climate with dry summers. *Global Change Biology*. 24: 2513-2529.

Tribouillois H, Cruz P, Cohan JP, et al., 2015. Modelling agroecosystem nitrogen functions provided by cover crop species in bispecific mixtures using functional traits and environmental factors. *Agriculture Ecosystems & Environment*. 207: 218-228.

Uliarte EM, Schultz HR, Frings C, et al., 2013. Seasonal dynamics of CO_2 balance and water consumption of C-3 and C-4-type cover crops compared to bare soil in a suitability study for their use in vineyards in Germany and Argentina. *Agricultural and Forest Meteorology. 181: 1-16.

Unger PW, Schomberg HH, Dao TH, et al., 1997. Tillage and crop residue management practices for sustainable dryland farming systems.

Valkama E, Rankinen K, Virkajärvi P, et al., 2016. Nitrogen fertilization of grass leys: yield production and risk of N leaching. *Agriculture, Ecosystems & Environment*. 230: 341-352.

Van Der Heijden MG, Bardgett RD, and Van Straalen NM, 2008. The unseen majority: soil microbes as drivers of plant diversity and productivity in terrestrial ecosystems. *Ecology letters*. 11: 296-310.

Venkateswarlu B, Srinivasarao C, Ramesh G, et al., 2007. Effects of long-term legume cover crop incorporation on soil organic carbon, microbial biomass, nutrient build-up and grain yields of sorghum/sunflower under rain-fed conditions. *Soil Use and Management*. 23: 100-107.

Villamil MB, Bollero GA, Darmody RG, et al., 2006. No-till corn/soybean systems including winter cover crops: Effects on soil properties. *Soil Science Society of America Journal*. 70: 1936-1944.

Wang GY, Hu YX, Liu YX, et al., 2021. Effects of Supplement Irrigation and Nitrogen Application Levels on Soil Carbon-Nitrogen Content and Yield of One-Year Double Cropping Maize in Subtropical Region. *Water*. 13: 1180.

Wang K, Hooks CR, and Marahatta SP, 2010. Use of a strip-till cover crop system to manipulate above and below ground organisms in cucurbit plantings. *Phytopathology*. 100: S133.

Wawan Dini IR, Hapsoh, and Iop, 2019. The effect of legume cover crop Mucuna bracteata on soil physical properties, runoff and erosion in three slopes of immature oil palm plantation. in *International Conference on Sustainable Agriculture for Rural Development*. 2018.

Weaver MA, Krutz LJ, Zablotowicz RM, et al., 2007. Effects of glyphosate on soil microbial communities and its mineralization in a Mississippi soil. *Pest Management Science: formerly Pesticide Science*. 63: 388-393.

Wegner BR, Chalise KS, Singh S, et al., 2018. Response of Soil Surface Greenhouse Gas Fluxes to Crop Residue Removal and Cover Crops under a Corn-Soybean Rotation. *Journal of Environmental Quality*. 47: 1146-1154.

Weinert TL, Pan WL, Moneymaker MR, et al., 2002. Nitrogen recycling by nonleguminous winter cover crops to reduce leaching in potato rotations. *Agronomy Journal*. 94: 365-372.

Werle R, Mobli A, Striegel S, et al., 2021. Large Scale Evaluation of 2,4-D Choline Off-target Movement and Injury in 2,4-D-susceptible Soybean. *Weed Technology*: 1-22.

White CM, and Weil RR, 2010. Forage radish and cereal rye cover crop effects on mycorrhizal fungus colonization of maize roots. *Plant and Soil*. 328: 507-521.

Whitehead W, and Singh BP, 2010. Impact of Inorganic Nitrogen and Legume-Non Legume Cover Crops on Aboveground Biomass Yields and Leaf Area Index of Two Sweet Corn Cultivars. *Hortscience*. 45: S172.

White KE, Brennan EB, and Cavigelli MA, 2020. Soil carbon and nitrogen data during eight years of cover crop and compost treatments in organic vegetable production. *Data in Brief.* 33: 106481.

Wortman SE, Drijber RA, Francis CA,, 2013. Arable weeds, cover crops, and tillage drive soil microbial community composition in organic cropping systems. *Applied Soil Ecology.* 72: 232-241.

Xiao X, Tang H, Nie Z, et al., 2013. Effects of winter cover crop straw recycling on soil organic carbon and soil carbon pool management index in paddy fields. *Chinese Journal of Eco-Agriculture.* 21: 1202-1208.

Xiong Z, Xing G, Tsuruta H, et al., 2002. Measurement of nitrous oxide emissions from two rice-based cropping systems in China. *Nutrient Cycling in Agroecosystems.* 64: 125-133.

Xi Z, Li H, Long Y, et al., 2010. Variation of soil microbial populations and relationships between microbial factors and soil nutrients in cover cropping system of vineyard. *Acta Horticulturae Sinica.* 37: 1395-1402.

Yan X, Akiyama H, Yagi K, et al., 2009. Global estimations of the inventory and mitigation potential of methane emissions from rice cultivation conducted using the 2006 Intergovernmental Panel on Climate Change Guidelines. *Global biogeochemical cycles.* 23.

Zhou L, Wei J, Tang X, et al., 2016. Effects of winter green manure crops with and without chicken rearing on microbial biomass and effective carbon and nitrogen pools in a double-crop rice paddy soil. *Acta Prataculturae Sinica.* 25: 103-114.

Zhou T, Jiao K, Qin S, et al., 2019. The impact of cover crop shoot decomposition on soil microorganisms in an apple orchard in northeast China. *Saudi Journal of Biological Sciences.* 26: 1936-1942.

Zhou Y, Zhu H, and Yao Q, 2017. Improving soil fertility and soil functioning in cover cropped agroecosystems with symbiotic microbes. 1: 149-171.

Zhu B, Yi L, Guo L, et al., 2012. Performance of two winter cover crops and their impacts on soil properties and two subsequent rice crops in Dongting Lake Plain, Hunan, China. *Soil & Tillage Research.* 124: 95-101.

Zibilske LM, and Makus DJ, 2009. Black oat cover crop management effects on soil temperature and biological properties on a Mollisol in Texas, USA. *Geoderma.* 149: 379-385.

Zschornack T, Da Rosa CM, Camargo ES, et al., 2016. Impact of cover crops and soil drainage in CH_4 and N_2O emissions under irrigated rice cultivation. *Pesquisa Agropecuaria Brasileira.* 51: 1163-1171.

Row Spacing Effects on Radiation Distribution, Leaf Water Status and Yield of Summer Maize

J. Q. Liu, M. D. Li and X. B. Zhou

(Agricultural College of Guangxi University, Nanning 530004, China)

Abstract: Different row spacing can affect the canopy structure, and then affect the environment of crop growth and the yield. The research aimed to investigate the effects of row spacing on light interception ratio and leaf water status of summer maize (*Zea mays* L.), and to select the reasonable planting pattern. The experiment comprised five planting population distribution patterns in the same plant population density (62,500 plant/ha) in northern China from 2011 to 2013. The following row spacing × spacing between the plant schemes were used: 40 cm × 40 cm (RS40), 50 cm × 32 cm (RS50), 60 cm × 27 cm (RS60), 70 cm × 23 cm (RS70), and 80 cm × 20 cm (RS80). A significant negative correlation was observed between row spacings and leaf relative water content (LRWC), water potential (Ψ) and yield during 3 years. RS40 and RS50 had the high total photosynthetically active radiation (PAR) capture ratio (CR) and upper CR (> 100 cm) than the others. The yield of RS50 was higher than the others during 3 years study. The narrow row spacing (RS40 and RS50) was beneficial to the CR of PAR, LRWC, and Ψ. However, compared with RS50, the RS40 could increase the evapotranspiration and decrease the lower-CR. So RS50 could be the reasonable row spacing of summer maize in Huang-huai-hai Plain in China.

Keywords: *Zea mays* L.; Evapotranspiration; Capture ratio; Leaf water potential; Leaf relative water content

1 Introduction

Different approaches have been used to increase crop yield, such as increasing the amount of fertilizer, application of high-density resistant cultivars, uniformity of row spacing (RS) distribution (Ritchie and Basso, 2008). The field plant distribution, as affected by planting density and row spacing, has drawn a great deal of attention for decades (Farnham, 2001). To attain a suitable canopy structure may be obtained through change of row spacing. For maize sown at high density, wide-narrow row planting can improve the ventilation and light environment effectively (Megowan et al., 1991). For a particular region, as the planting density is generally stable, the reasonable row spacing is crucial (Norsworthy and Shipe, 2005).

The appropriateness of row spacing has been widely studied in different crops. Determinate soy-

---

Author for correspondence: whyzxb@ gmail. com. J. Q. Liu and M. D. Li are co-first authors

bean grown in RS50 or less can produce higher yields than that in RS75 to RS100, and the narrow RS can capture light effectively (Bowers et al., 2000). Similarly, in cotton, 19 cm and 38.1 cm row spacing can capture more light than the traditional wide RS (Jost and Cothren, 2000). Inconsistent results have been produced about the studies of narrow-row maize production systems. The results vary from no yield advantage of planting maize in narrow-row to a 7% increase in yield over wider rows (Johnson et al., 1998).

To build a good canopy structure, row spacing is of great significance (Sharratt and Mcwilliams, 2005). The adjustment of crop population distribution can affect the structure and function of canopy, improve the interception rate of photosynthetically active radiation (PAR), and enhance the population productivity (Maddonni et al., 2001). The competition between individual in the light of crop canopy plays an important role in group productivity. The radiation use efficiency of crops is determined by light interception rate and efficiency of light energy conversion. Leaf water potential (Ψ) is the important factors that affect the leaf net photosynthetic rate (Peri et al., 2011).

The reasonable RS can improve the light, temperature, humidity, and water resource utilization. And then affect the photosynthetic efficiency and yield of crop. Summer maize, as one of the main crops is produced in double-crop (following winter wheat) production systems in Huang-huai-hai Plain, China. Cultural practices are generally used for non-irrigated summer maize from June to October. The precipitation of summer maize growing season accounts for 70%-80% of annual precipitation. The research aimed to investigate the effects of RS on light interception ratio and water status, to select the reasonable planting pattern. For these purposes, the PAR and moisture content of leaves were determined so that proper information could be provided for the selection and management of RS in the region.

2 Materials and Methods

The research was conducted at the Agronomy Experimental Station of Shandong Agricultural University, Tai'an, China (36°09′N, 117°09′E) in 2011, 2012 and 2013. This site represented the main summer maize growing region of Huang-huai-hai Plain in China. The type of soil was loam, which contained SOM (16.3 g/kg), N (1.3 g/kg), P (35 mg/kg), and K (95 mg/kg); the pH was 6.9. The annual average rainfall was 693.5 mm from 1971 to 2010. The rainfall (on July and August) accounted for 52.5% of the annual rainfall. The annual average rainfall was 500 mm during the growing seasons of summer maize from 1971 to 2010; the values were 572.5 mm, 337.1 mm, and 461.8 mm from 2011 to 2013, respectively (Table 1).

Table 1 Monthly rainfall (mm) from 2011 to 2013 during the summer maize growing season

Years	June*	July	August	September	Total
2011	38.7	192.0	165.8	176.0	572.5
2012	11.6	210.5	53.4	61.6	337.1
2013	8.1	399.8	42.9	11.0	461.8

Notes: * The rainfall from sowing date to June 30.

The experiments were executed during the growing seasons of June to October from 2011 to 2013. As a part of the continuous winter wheat-summer maize rotation experiment, the previous winter wheat was hand-harvested and their residues were removed. The summer maize seeds (cv. Luyu 14) were sown by hand at a seeding rate of 62,500 seeds/ha. The seeding of maize was at 3-4 cm soil depth on18 June 2011, 17 June 2012, 19 June 2013. The experiment involved five plant population distribution patterns under rainfed conditions. The following row spacing × spacing between the plant schemes were used: 40 cm × 40 cm (RS40), 50 cm × 32 cm (RS50), 60 cm × 27 cm (RS60), 70 cm × 23 cm (RS70), and 80 cm × 20 cm (RS80). Each experimental plot possessed dimensions of 4 m × 4 m; three replications were obtained in a randomized block design. The growth stage of V6, R0, R2, R3, and R4 were measured in this experiment (Ritchie et al., 1996). Fully expanded leaves were selected at V6 and ear leaves from R0 to R4. Dicot weeds in the summer maize plots were controlled chemically by applying the herbicide 0.84 kg/ha 2-methyl-4-clorophenoxyacetic acid (MCPA), other weeds were removed by hand. The air temperature during the growing seasons of 2011, 2012 and 2013 was showed in Figure 1.

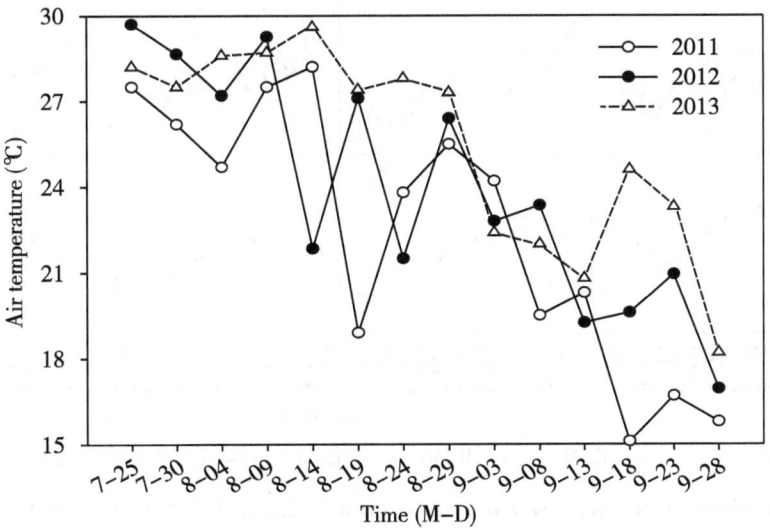

Figure 1 Air temperature during summer maize growth stage

Canopy radiation, reflection, and underlying radiation were measured. Typical sunny days were selected to measure the data by using the SunScan Canopy Analysis System (Delta T Devices Ltd., Cambridge, UK), where a 1.5 m long linear sensor was placed parallel to the row direction. The PAR capture ratio (CR) was calculated as a ratio of the difference between incident and transmitted radiations to incident radiation, the PAR penetration ratio (PR) was calculated as a ratio of transmitted radiation to incident radiation, and the PAR reflection ratio (RR) was calculated as a ratio of the PAR reflection measured above the canopy to incident radiation (Han et al., 2014).

The Ψ was measured by using a PSYPRO Water Potential System (Wescor Inc., Logan, USA) with eight L-51 sample chambers, measuring three leaves for each treatment. During the transfer of each leaf to the sample chamber, water loss was minimized by ice box in a black plastic bag immediately after excision. The leaves were cut approximately 7 mm in diameter by hole puncher and sealed in the sample chamber. Samples were equilibrated for 20 min and then the readings were recorded.

Leaf relative water content (LRWC) was measured on clear-sky days at 08:30. Three leaves per treatment were obtained from different individual plants. LRWC was calculated by the equation RWC (%) = (Fw-Dw)/(Tw-Dw) × 100 (Aydi et al., 2008).

Fw is the fresh weight, Dw is the dry weight, and Tw is the turgid weight of the leaf samples. Being excised from the plants, the leaves were weighed Fw and placed in distilled water at 4℃ in the dark to minimize respiration losses until they reached a constant weight typically after 12 h. The leaf Tw was measured, after which the leaves were dried at 80℃ for 48 h and obtained the Dw.

The soil temperatures were recorded from 0 cm, 5 cm, 10 cm, and 15 cm depth with soil thermometers buried at respective soil depths. The temperatures were read at 8:00 and 14:00 every 5 d from July 25 to September 28, the average of which was as daily temperature (Figure 2).

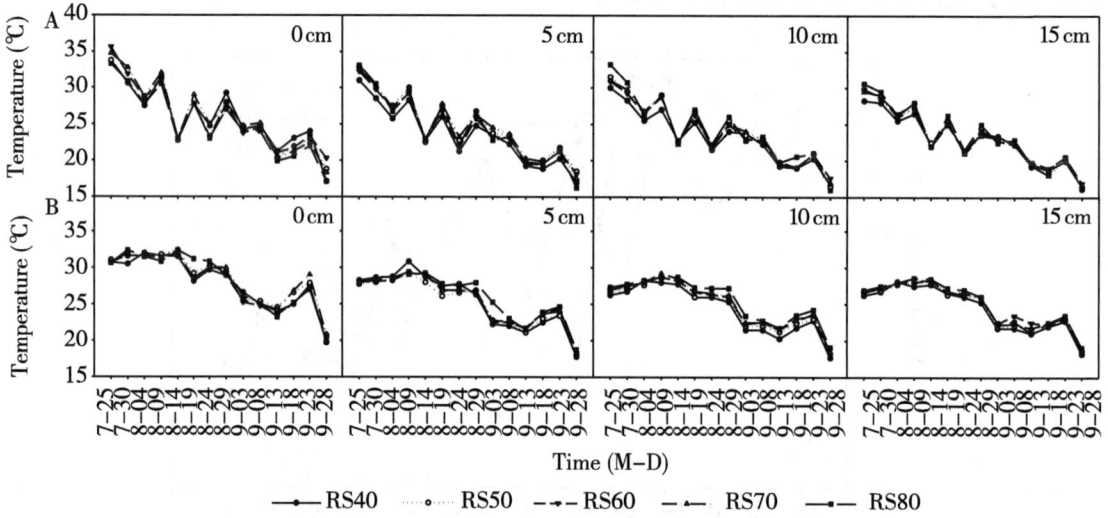

Figure 2 Soil temperature during summer maize growth season. A, B are 2012 and 2013 respectively

Grain yield was randomly recorded from an area of 2 m² in each plot on 24 September 2011, 2 October 2012, 2 October 2013.

All data were analyzed using SAS 9.2 and graphs were drawn using SigmaPlot 10.0. Experimental data were evaluated by ANOVA. The effects were considered significant in all statistical calculations if $P \leq 0.05$ based on least significant difference tests (LSDs).

3 Results

3.1 Photosynthetically available radiation distribution

During 2011-2013, CR at R3 was the highest and V6 was the lowest. CR was the highest proportion, PR was the medium, and RR was the least (Table 2).

Table 2 Effects of row spacing on the capture ratio (CR), reflection ratio (RR) and penetration ratio (PR) of PAR in 2011–2013 (%)

Row spacing (cm)	V6 CR	V6 RR	V6 PR	R0 CR	R0 RR	R0 PR	R2 CR	R2 RR	R2 PR	R3 CR	R3 RR	R3 PR	R4 CR	R4 RR	R4 PR
2011															
40	88.9a*	3.1c	8.1b	89.6a	3.0c	7.4c	90.5a	2.9c	6.5b	89.1a	3.5a	7.4b	86.1ab	3.9a	10.0bc
50	88.1ab	3.5b	8.4b	88.9a	3.0c	8.2bc	89.3a	3.0c	7.7b	86.8a	2.5b	10.7b	89.2ab	2.8b	8.0bc
60	87.8ab	4.2a	8.0b	85.2bc	4.2a	10.6ab	83.3ab	4.3a	12.4ab	87.6a	3.3a	9.1b	89.6a	3.0b	7.4c
70	82.7c	3.1c	14.2a	82.5c	3.9b	13.6a	85.2ab	3.8b	11.0ab	85.5a	3.4a	11.1b	85.1b	2.9b	12.0ab
80	85.2bc	3.5b	11.3ab	85.7b	2.8d	11.5a	81.4b	2.8c	15.8a	81.3b	2.2b	16.5a	80.4c	4.2a	15.4a
LSD (0.05)	3.16	0.26	3.40	2.98	0.21	3.16	7.39	0.38	7.70	3.72	0.36	3.99	4.31	0.32	4.48
2012															
40	65.1a	1.6ab	33.3b	89.4	2.3	8.3b	89.5b	2.7b	7.9b	89.6	4.4a	6.0ab	82.3	4.6a	13.1
50	61.6ab	1.9a	36.5ab	86.4	2.2	11.4a	88.7ab	3.7ab	7.6b	91.1	3.6c	5.3ab	84.0	2.9cd	13.0
60	60.8ab	1.4b	37.8ab	90.2	2.6	7.2b	90.2a	4.4a	5.4c	88.5	3.7bc	7.8a	83.7	3.9b	12.4
70	59.5ab	1.5ab	39.0ab	87.0	3.7	9.3ab	87.0b	3.2b	9.8b	89.1	3.2d	7.7a	83.0	3.0c	13.9
80	55.8b	1.6ab	42.6a	89.8	2.4	7.8b	88.1ab	3.3ab	8.6ab	91.9	3.9b	4.1b	85.9	2.4d	11.7
LSD (0.05)	9.06	0.52	7.88	6.00	2.08	2.68	7.88	1.18	1.92	6.60	0.30	2.70	9.31	0.61	4.75
2013															
40	81.3a	2.6	16.0c	82.2ab	1.8b	16.0cd	77.9a	3.0a	19.0b	86.2	3.2	10.6bc	74.6ab	4.2a	21.3b
50	81.4a	2.8	15.8c	84.5a	2.4a	13.1d	82.4a	3.0a	14.6c	87.2	3.6	9.2c	73.2ab	3.6abc	23.2b
60	71.8b	2.7	25.4a	79.4abc	1.4c	19.2bc	83.3a	2.8a	13.9c	84.7	3.5	11.8ab	74.9a	3.3bc	21.8b
70	74.4b	2.4	23.2ab	78.0bc	1.6bc	20.4b	81.8a	3.3a	14.9c	83.7	3.5	12.8a	71.3ab	4.1ab	24.7b
80	77.3ab	2.4	20.2b	75.0c	1.2c	23.8a	58.0b	1.6b	40.4a	83.6	3.5	13.0a	68.7b	3.0c	28.3a
LSD (0.05)	5.86	0.49	3.19	6.26	0.40	3.31	7.41	0.52	2.14	5.29	0.58	2.17	6.12	0.79	3.42

Notes: * Values followed by the same letter in a column are not significantly different according to $LSD_{0.05}$.

In 2011, CR, RR, and PR had significant difference among the treatments at the different growth stage ($P<0.05$). The CR of RS70 and RS80 was lower than RS40, RS50, and RS60, and PR was reverse. During 2012, CR (R0, R3, and R4) and PR (R0 and R4) among the treatments had no significant difference ($P>0.05$). While, at V6 in 2012, affected by less rainfall, CR was significantly lower than the other two years ($P<0.05$). During V6-R2, CR of RS70 and RS80 was lower than RS40, RS50, and RS60. In 2013, CR (R3) and RR (V6 and R3) among the treatments had no significant difference ($P>0.05$). During V6-R4, CR average of RS40 and RS50 was 11.8% higher than that of RS80; PR of RS80 was significantly higher than those of RS40 and RS50 ($P<0.05$).

From R0 to R4, CR accounted for 84.1% of the total CR in the summer maize growing season in the 3 years. The upper CR (>100 cm) was 66.6%; the lower (0-100 cm) was 17.6%. The upper of CR accounted for 79.1% of the total CR. The upper CR of RS40, RS50, RS60, RS70, and RS80 were 71.4%, 70.6%, 63.7%, 61.1%, and 66.2%; the lower were 14.2%, 15.4%, 21.4%, 22.2%, and 14.6%, respectively. The upper CR accounted for 83.4%, 82.1%, 74.9%, 73.3%, and 81.9% of the total CR (Figure 3).

3.2 Leaf water potential

The Ψ average in 2011, 2012, and 2013 was -1.54 MPa, -1.33 MPa, and -1.42 MPa, respectively (Figure 4). The precipitation of 2011 growing season was apparently higher than the other 2 years, but the low rainfall of R3 might contribute to the decrease of Ψ. The Ψ average of V6, R0, R2, R3, and R4 were -1.41 MPa, -1.36 MPa, -1.42 MPa, -1.58 MPa, and -1.40 MPa in a 3 year study; R3 was the lowest and R0 was the highest. The Ψ average of RS40, RS50, RS60, RS70, and RS80 were -1.36 MPa, -1.39 MPa, -1.42 MPa, -1.50 MPa, and -1.49 MPa from 2011 to 2013. The Ψ average of RS40 and RS50 was 8.8% higher than that of RS70 and RS80.

3.3 Leaf relative water content

Generally speaking, the LRWC of 2011 and 2012 were higher than 2013. In 2013, the LRWC decreased with the advance of the growth stage, which associated with low rainfall. The total rainfall of August and September were merely 54 mm in 2013. In the 3 years, the LRWC average of V6, R0, R2, R3, and R4 was 89.7%, 90.0%, 91.4%, 87.1%, and 83.6%; RS40, RS50, RS60, RS70, and RS80 was 88.7%, 88.8%, 88.5%, 87.4%, and 88.3%, respectively. The LRWC decreased at late growth stage, and the LRWC of RS70 and RS80 were lower than others. No significant differences were observed between RS and LRWC ($P>0.05$) (Figure 5).

3.4 Evapotranspiration

In the growing season, the total evapotranspiration of RS40, RS50, RS60, RS70, and RS80 were 433.4 mm, 479.7 mm, 463.1 mm, 473.6 mm, 475.3 mm (2011); 314.3 mm, 299.2 mm, 316.0 mm, 313.5 mm, 307.6 mm (2012); 494.6 mm, 496.3 mm, 471.1 mm, 486.3 mm, 500.3 mm (2013), respectively.

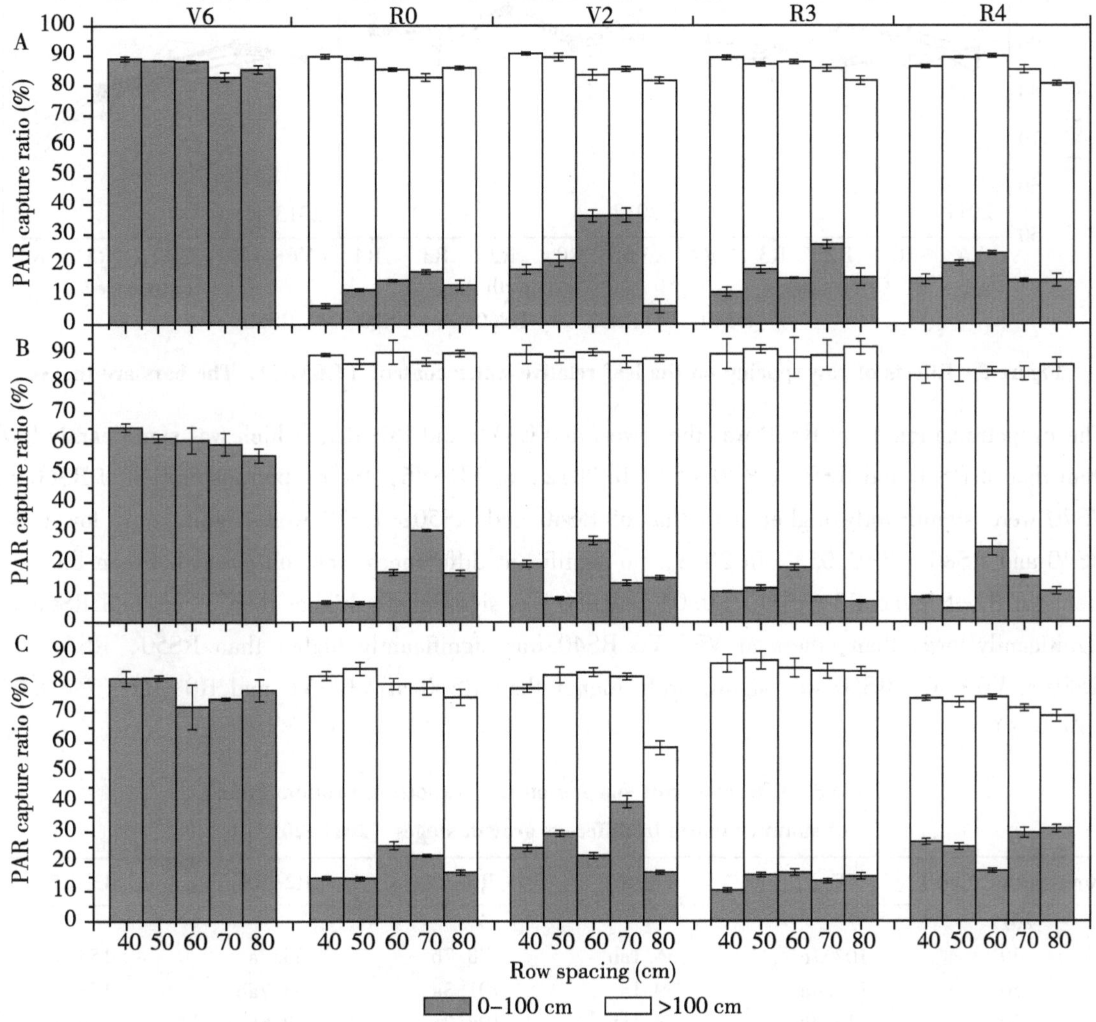

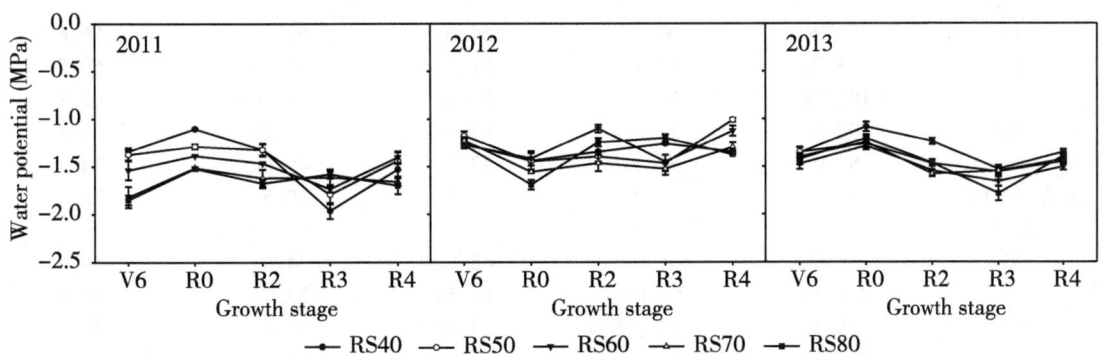

Figure 3 Effects of row spacing on PAR capture ratio at different growth stages.
A, B, C are 2011, 2012 and 2013 respectively; the bars are the SE

Figure 4 Effects of row spacing on the leaf water potential. The bars are the SE

In 2011, no significant differences were observed between different RSs at R3-R5 ($P>0.05$).

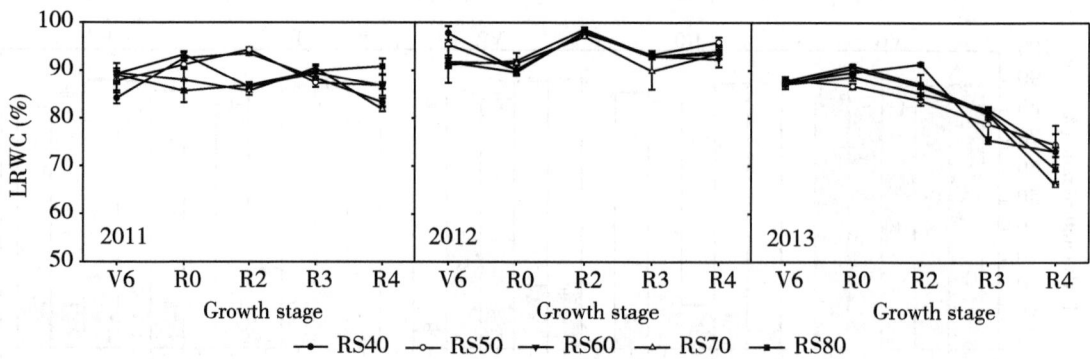

Figure 5 Effects of row spacing on the leaf relative water content (LRWC). The bars are the SE

The evapotranspiration of RS40 was the lowest at VE-V6 and R0-R2, which was significantly lower than that of RS70 and RS80 ($P<0.05$). In 2012, at VE-V6, the evapotranspiration of RS60 and RS70 were significantly higher than that of RS40 and RS50; RS70 were significantly lower than RS40 and RS80 ($P<0.05$). In 2013, no significant differences were observed between the treatments at R2-R3 and R3-R5 ($P>0.05$). RS80 was significantly higher than others and RS60 was significantly lower than others at VE-V6. RS40 was significantly higher than RS50, RS70, and RS80 at V6-R0; RS50 was significantly higher than RS60 at VE-V6 and R0-R2 ($P<0.05$) (Table 3).

Table 3 Effects of row spacing on the evapotranspiration (mm) of summer maize in different growth stages (2011-2013)

Row spacing (cm)	VE-V6	V6-R0	R0-R2	R2-R3	R3-R5
2011					
40	107.1c*	56.1ab	78.7b	35.7a	155.8
50	127.4a	59.1a	101.3a	33.7ab	158.2
60	114.4bc	58.1ab	104.2a	29.5b	156.8
70	119.2b	52.8b	108.3a	35.4a	157.9
80	117.5b	57.0ab	107.9a	36.9a	156.1
LSD (0.05)	7.5	5.6	8.1	5.8	5.8
2012					
40	76.2b	66.1	54.6	50.6a	66.8
50	83.7b	57.7	51.7	45.7ab	60.5
60	100.7a	60.8	50.8	40.9ab	62.7
70	98.5a	57.3	57.0	39.0b	61.6
80	87.0ab	57.5	57.2	49.6a	56.3
LSD (0.05)	14.2	9.7	8.9	10.5	13.8
2013					
40	286.4b	65.0a	63.8ab	34.8	44.6
50	290.2b	57.0b	66.5a	34.7	47.9
60	273.7c	63.4ab	56.3b	35.4	42.4
70	290.4b	56.9b	64.3ab	33.3	41.4
80	303.4a	56.0b	65.1ab	34.9	40.9
LSD (0.05)	7.7	7.6	9.6	6.2	11.0

Notes: * Values followed by the same letter in a column are not significantly different according to $LSD_{0.05}$.

3.5 Grain yield of summer maize

In 2011, the grain of RS40 was significantly higher than that of RS80. In 2012, RS50 was significantly higher than RS60, RS70 and RS80 ($P<0.05$), the values were 11.58%, 19.30% and 19.62%, respectively; RS50 was 8.35% higher than RS40. In 2013, RS50 was significantly higher than RS40, RS70 and RS80 ($P<0.05$); RS50 was 8.63% higher than RS60 (Figure 6). The yield average of the 3 years, RS50 was significantly higher than RS60, RS70 and RS80 ($P<0.05$), the values were 5.69%, 9.84% and 12.37%, respectively, and RS50 was 2.77% higher than RS40. The yield of RS40, RS70, and RS80 in different years followed the order of 2011>2012>2013; RS50 was 2012>2013>2011; RS60 was 2012>2011>2013. For the 3 years experiment, yield of 2012 was the highest and 2011 was the lowest.

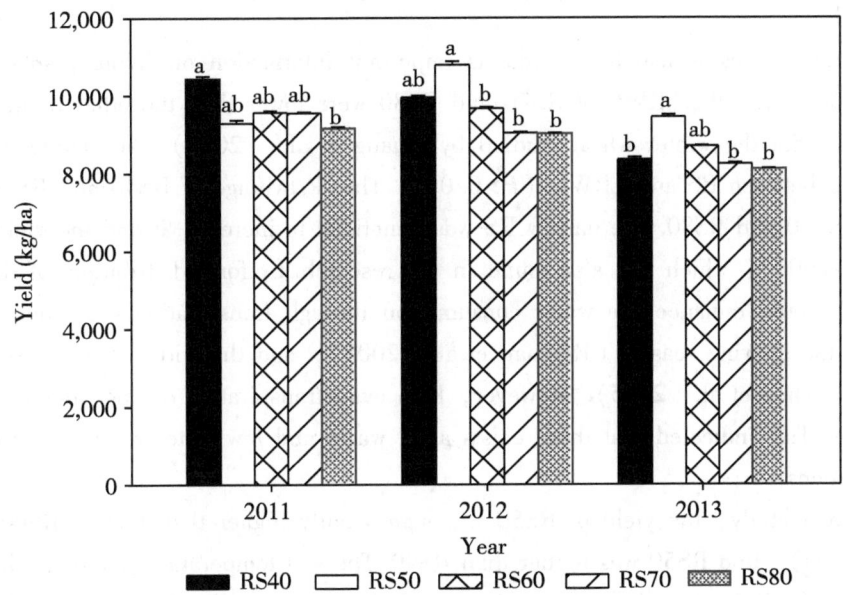

Figure 6　Effects of row spacing on the yield of summer maize during 2011–2013. The bars are the SE

The grain yield and water potential were significantly negative correlation with RS, correlation coefficients (r) were −0.9020 ($P<0.05$) and −0.9550 ($P<0.01$), respectively. The Ψ was significantly positive correlation with grain yield ($r=0.9225$, $P<0.01$). The grain yield and Ψ were positive correlation with LRWC, and r was 0.6761 and 0.8072 respectively. RS was negative correlation with LRWC ($r=-0.5894$) (Table 4).

Table 4　Correlation matrix of row spacing-grain yield, leaf relative water content (LRWC) and water potential (Ψ) of summer maize grown in 2011–2013

	Row spacing	Yield	Ψ	LRWC
Row spacing	1.0000	−0.9020*	−0.9550**	−0.5894
Yield		1.0000	0.9225**	0.6761
Ψ			1.0000	0.8072
LRWC				1.0000

Notes: * r values presented at $P<0.05$; ** r values presented at $P<0.01$.

4 Discussion

In the study, the CR of RS70 and RS80, especially the latter, were significantly lower than the other treatments. The narrow RS was beneficial to the CR of PAR, hence, observation of present study was consistent with Lehrsch et al. (1994). The upper CR (>100 cm) of different RSs was higher than that of lower. The higher total CR and upper CR (>100 cm) (like RS40 and RS50) were more favorable towards yield (Board et al., 1992). The lower CR of RS40 (14.2%) and RS80 (14.6%) were lower than the others. Accompanying with the waste of light, the increasing of RS could affect the light energy utilization rate, thus may impact yield adversely (Maddonni and Otegui, 2006). Our study was consistent with it, the wide RS could decrease the CR and yield.

Water from leaves is usually viewed as important information on living plants (Yu et al., 2000). In our study, the LRWC of RS70 and RS80 were lower than the other treatments, which was consistent with the winter wheat studied by Huang et al. (2013). No significant differences were observed between RS and LRWC ($P>0.05$). The Ψ average of RS40 and RS50 was higher than that of RS70 and RS80. The narrow RS was beneficial to increase Ψ and the yield (Sakamoto and Murata, 2000), which was also found in our research. Uniform distribution of the population (like RS40) could enhance the water consumption through transpiration thus increased the soil water during the growing season (Rahman et al., 2005), and the wide RS increased the evapotranspiration (Zhou et al., 2015). However, high evapotranspiration did not increase the yield of summer maize. This indicated that there exists a lot waste and low water resource utilization under rainfed conditions.

In a 3 year study, the yield of RS50 was significantly higher than that of RS60, RS70 and RS80 ($P<0.05$), and RS50 was higher than RS40. The soil temperature gradually decreased with the increasing of soil depth and the decreasing of RS. Row spacing can change farmland microclimate (Wang et al., 2015). The close intrarow spacing can weaken the growth of the crop (De Bruin and Pedersen, 2008), which is in row with our result of RS80. The wide RS had a higher soil temperature, which can increase the evapotranspiration and decrease the yield. Ψ is an important factor that affect the leaf net photosynthetic rate (Peri et al., 2011), which is the basis of the formation of crop yield. In our study the grain yield and Ψ were significantly negative correlation with RS. The narrow RS is beneficial to the yield of summer maize. For the 3 years, the air temperature of 2011 was the lowest and 2013 was the highest. The suitable air temperature of 2012 and the low air temperature in the late growth stage of 2011 may contribute to the yield.

5 Conclusion

The wide RS increased the soil temperature hence affected the CR of PAR, LRWC, and Ψ. However, compared with RS50, the RS40 could increase the evapotranspiration and decrease the lower-CR. In conclusion, RS50 planting pattern is a reasonable cultivation approach that could promote the yield of summer maize in Huang-huai-hai Plain in China.

Acknowledgements

This research was supported by the National High Technology Research and Development Program of China (2013AA102903), Guangxi Natural Science Foundation (2015GXNSFAA139049) and Starting Foundation for Docotors of Guangxi University (XBZ160072). The authors especially thank Huang Juan, Feng Zhibo, Wang Guoyun and Han Yuanyuan for his valuable contributions.

References

Aydi SS, Aydi S, Gonzalez E, et al., 2008. Osmotic stress affects water relations, growth, and nitrogen fixation in *Phaseolus vulgaris* plants. Acta Physiol. Plant. 30: 441-449.

Board JE, Kamal M, and Harville BG, 1992. Temporal importance of greater light interception to increase yield in narrow-row soybean. Agron. J. 84: 575-579.

Bowers GR, Rabb JL, Ashlock LO, et al., 2000. Row spacing in the early soybean production system. Agron. J. 92: 524-531.

De Bruin JL, and Pedersen P, 2008. Effect of row spacing and seeding rate on soybean yield. Agron. J. 100: 704-710.

Farnham DE, 2001. Row spacing, plant density, and hybrid effects on corn yield and moisture. Agron. J. 94: 1049-1053.

Han YY, Wang GY, Zhou XB, et al., 2014. Radiation use efficiency and yield response of winter wheat to planting patterns and irrigation in northern China. Agron. J. 106: 168-174.

Huang J, Chen YH, Zhou XB, et al., 2013. Spatial arrangement effects on soil and leaf water status of winter wheat. J. Anim. Plant Sci. 23: 1379-1384.

Johnson GA, Hoverstad TR, and Greenwald RE, 1998. Integrated weed management using narrow corn row width, herbicides, and cultivation. Agron. J. 90: 40-46.

Jost PH, and Cothren JT, 2000. Growth and yield comparisons of cotton planted in conventional and ultra-narrow row spacings. Crop Sci. 40: 430-435.

Lehrsch GA, Whisler FD, and Buehring NW, 1994. Cropping system influences on extractable water for mono-and double-cropped soybean. Agr. Water Manage. 26: 13-25.

Maddonni GA, and Otegui ME, 2006. Intra-specific competition in maize: Contribution of extreme plant hierarchies to grain yield, grain yield components and kernel composition. Field Crops Res. 97: 155-166.

Maddonni GA, Otegui ME, and Cirilo AG, 2001. Plant population density, row spacing and hybrid effects on maize canopy architecture and light attenuation. Field Crops Res. 71: 183-193.

Megowan M, Taylor HM, and Willingham J, 1991. Influence of row spacing on growth, light and water use by sorghum. J. Agr. Sci. 116: 329-339.

Norsworthy JK, and Shipe ER, 2005. Effect of row spacing and soybean genotype on main stem and branch yield. Agron. J. 97: 919-923.

Peri PL, Arena M, Pastur GM, et al., 2011. Photosynthetic response to different light intensities, water status and leaf age of two *Berberis* species (Berberidaceae) of Patagonian steppe, Argentina. J. Arid Environ. 75: 1218-1222.

Rahman MA, Chikushi J, Saifizzaman M, et al., 2005. Rice straw mulching and nitrogen response of no-till wheat following rice in Bangladesh. Field Crops Res. 91: 71-81.

Ritchie JT, and Basso B, 2008. Water use efficiency is not constant when crop water supply is adequate or fixed: The role of agronomic management. Eur. J. Agron. 28: 273-281.

Ritchie SW, Hanway JJ, and Benson GO, 1996. How a corn plant develops. Spec. Rep. 48. Rev. ed. Iowa State Univ. Coop. Ext. Serv., Ames.

Sakamoto A, and Murata N, 2000. Genetic engineering of glycinebetaine synthesis in plants: current status and implication for enhancement of stress tolerance. J. Exp. Bot. 51: 81-88.

Sharratt BS, and Mcwilliams DA, 2005. Microclimatic and rooting characteristics of narrow-row versus conventional-row corn. Agron. J. 97: 1129-1135.

Wang XY, Zhang Z, Zhou XB, et al., 2015. Planting pattern and irrigation effect on farmland microclimate and yield of winter wheat. J. Anim. Plant Sci. 25: 708-715.

Yu GR, Miwa T, Nakayama K, et al., 2000. A proposal for universal formulas for estimating leaf water status of herbaceous and woody plants based on spectral reflectance properties. Plant Soil. 227: 47-58.

Zhou XB, Chen YH, and Ouyang Z, 2015. Spacing between rows: effects on water-use efficiency of double-cropped wheat and soybean. J. Agr. Sci. Cambridge. 153: 90-101.

不同耕作方式对玉米田土壤物理性质及产量的影响

范继征[1], 闫飞燕[2], 石达金[1], 吕巨智[1], 张玉[1], 钟昌松[1], 程伟东[1], 刘永红[1]

(1. 广西农业科学院玉米研究所, 广西 南宁 530007;
2. 广西农业科学院农产品质量安全与检测技术研究所, 广西 南宁 530007)

摘要: 通过对土壤容重、土壤总孔隙度、土壤含水量、土壤田间持水量及玉米产量的测定和分析,研究一次性施肥条件下深松-旋耕、深松-免耕、常规旋耕和免耕不同耕作方式对土壤物理性状及玉米产量的影响。结果表明,在不同生育时期的不同土层中,耕作方式对于降低土壤容重、增加土壤孔隙度、增加土壤含水量和田间持水量的作用表现出不同程度的差异;不同耕作方式下玉米产量表现为深松-免耕>免耕>常规旋耕>深松-旋耕;深松-免耕处理比常规旋耕处理产量提高了 15.68%,产量最高达到了 6 829.73 kg/hm²。

关键词: 玉米;耕作方式;土壤物理性状;产量

Effect of different tillage management on soil physical properties and maize yield

Fan Jizheng[1], Yan Feiyan[2], Shi Dajin[1], Zhong Changsong[1], Lü Juzhi[1], Cheng Weidong

([1]Maize Research Institute, Guangxi Academy of Agricultural Sciences Nanning 530007, Guangxi, China;[2]Research Institute of Agro-products Quality Safetied and Testing Technology, Guangxi Academy of Agricultural Sciences Nanning 530007, Guangxi, China)

Abstract: By the mensuration and analysis of soil bulk density, soil total porosity, soil moisture content, soil moisture capacity and maize yield, the effects of subsoiling and rotary tilling, subsoiling and no-tillage, conventional rotary tillage and no-tillage on soil physical characters and maize yield through field experiments in a single fertilization were researched. The results showed that the effects of tillage methods on the soil bulk density, soil total porosity, soil moisture content, soil moisture capacity showed were differences, in the different soil layer of different growth period. The tillage managements have a great effect on maize yield, with order of subsoiling and no-tillage, no-tillage, conventional rotary tillage, subsoiling and rotary tilling. After subsoiling and no-tillage, production increased 15.68% than conventional rotary tillage, achieved 6,829.73

基金项目: 广西农业科学院基本科研业务专项 (桂农科 2014YZ20);国家现代玉米产业技术体系南宁综合试验站 (CARS-02-73);粮食安全关键技术研究与应用示范项目 (桂科攻 1123001-1J);广西农业科学院科技发展基金 (2015JZ12);

作者简介: 范继征 (1982—), 女, 河南长葛人, 硕士, 从事玉米栽培与生理研究。E-mail: fiona-fiona-happy@163.com。

kilograms per hectare.

Keywords：Maize；Tillage management；Soil physical property；Yield

近些年来，随着农村劳动力转移和城镇化的发展，一方面，提高了农村劳动者的收入，缓解了人地矛盾关系，促进了城市的繁荣发展；另一方面，农作劳动力过度流失也影响到农业生产[1]。由于劳动力缺乏及生产成本增加，广西许多地区群众自发采用了玉米免耕技术和一次性施肥的生产方式。此外，受机械动力及传统观念因素的影响，土壤耕层深度一般在15~25 cm，特别是近几年采用免耕的方式，土壤耕层深度更浅，年复一年耕层变浅、犁底层变坚硬，从而限制了作物的生长发育，不但土壤蓄水保水能力下降，而且作物易倒伏、早衰。目前，国内已有大量关于施肥方式和耕作方式的研究报道，主要内容有耕作方式对土壤特性的影响[2]、一次性施肥技术[3]、深耕对土壤和产量的影响[4]、深耕技术的优势[5]、平作与垄作的产量表现[6]、耕作方式对光合特性的影响[7]、常规耕作、免耕、深翻、深松对土壤和玉米产量的影响[8]等方面。针对广西玉米施肥与耕作技术应用研究相对滞后的现状，开展一次性施肥条件下不同耕作方式对秋玉米田间土壤理化性质及产量的影响研究，以期为广西玉米生产提供科学依据。

1 材料与方法

1.1 试验区概况

试验于2013年秋季在广西壮族自治区农业科学院明阳基地（22°36′34″N，108°14′33″E）进行，年平均年降水量为1 100~1 500 mm，2013年玉米全生育期（8月19日播种，12月20日收获）降水量为481.2 mm，图1所示为2013年试验期间日降水量。供试土壤为黏土，肥力中等，试验前作为玉米。试验地0~20 cm土层含全氮0.079%、全磷0.085%、全钾0.751%、碱解氮57.5 mg/kg、速效磷34 mg/kg、速效钾181.8 mg/kg、有机质18.8 mg/kg、pH值6.66。

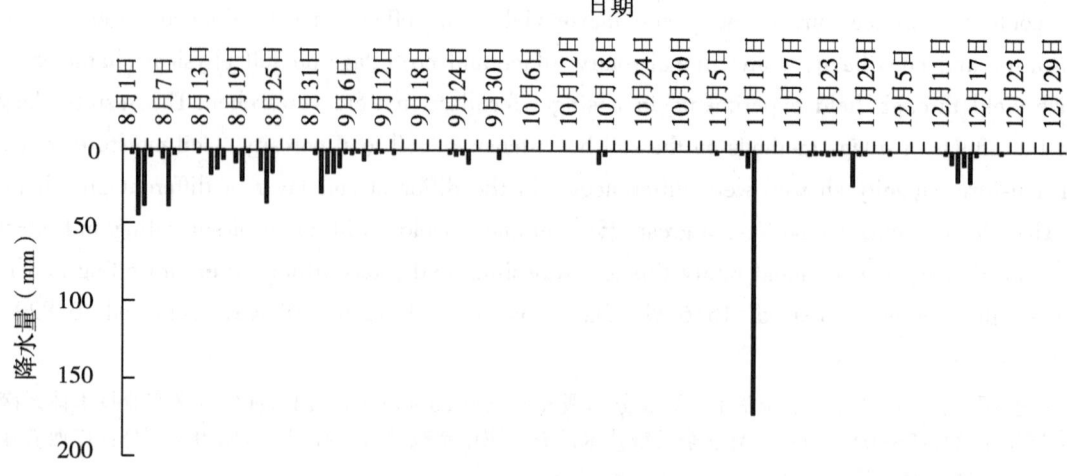

图1 2013年试验期间日降水量

1.2 试验设计

试验设 4 个处理：（1）深松 35 cm-旋耕 25 cm（SS-RT）；（2）深松 35 cm-免耕（SS-NT）；（3）常规旋耕 25 cm（RT）；（4）免耕（NT）。试验玉米品种为正大 619。采用随机区组试验设计，三次重复。小区面积为 7 m×33 m=231 m^2，种植密度 57 000 株/hm^2。深松处理于 2013 年 8 月 1 日秋玉米播种前。施肥方式统一在播种时一次性施入缓效肥恩泰克（N：P$_2$O$_5$：K$_2$O=21：7：11）750 kg/hm^2。试验田四周设保护行，田间管理同当地大田生产。

1.3 测定项目与方法

在玉米播种前、拔节期、抽雄期、灌浆期，采用 5 点取样法，采集深度 0~15 cm、16~25 cm、26~35 cm、36~45 cm 土样测定土壤含水量、容重、持水量和总孔隙度。土壤容重采用环刀法测定；土壤总孔隙度由公式计算得出[9]；土壤含水量采用烘干法；土壤田间持水量采用环刀法测定。玉米完熟期采取实收中间 4 行计产，并折算成标准含水量（14%）的产量。

1.4 数据处理

试验中所获得的数据采用 Microsoft Excel 2013 和 DPS 7.05 统计分析软件进行数据处理与统计分析，显著性检验采用 LSD 法，作图采用 Origin8.0 软件。

2 结果与分析

2.1 不同耕作方式对土壤容重的影响

容重是土壤的一个重要物理性质，一般作为衡量土壤紧实程度的一个指标。表 1 结果表明，在播种前期和拔节期不同耕作方式对各个土层土壤容重的影响表现为差异不显著；在抽雄期仅对 16~25 cm 土层的影响达到了差异显著水平，主要体现在深松-旋耕处理土壤容重显著低于免耕处理；在灌浆期，不同耕作方式对各个土层土壤容重的影响表现为差异不显著。综合整个玉米生育期来看，深松-旋耕和深松-免耕处理在各个土层的土壤容重低于常规旋耕处理和免耕处理。

表 1 不同耕作方式对土壤容重的影响

处理	土壤容重（g/cm^3）							
	播种前期				拔节期			
	0~15 cm	16~25 cm	26~35 cm	36~45 cm	0~15 cm	16~25 cm	26~35 cm	36~45 cm
深松-旋耕 SS-RT	1.69±0.13a	1.72±0.17a	1.72±0.08a	1.67±0.07a	1.75±0.07a	1.77±0.11a	1.73±0.08a	1.69±0.10a
深松-免耕 SS-NT	1.69±0.06a	1.74±0.09a	1.72±0.08a	1.71±0.05a	1.75±0.09a	1.77±0.11a	1.74±0.04a	1.73±0.08a
常规旋耕 RT	1.72±0.06a	1.74±0.10a	1.71±0.10a	1.72±0.02a	1.75±0.07a	1.80±0.13a	1.76±0.20a	1.75±0.19a
免耕 NT	1.71±0.11a	1.80±0.05a	1.77±0.11a	1.74±0.08a	1.80±0.01a	1.80±0.14a	1.82±0.12a	1.79±0.09a

(续表)

处理	抽雄期				灌浆期			
	0~15 cm	16~25 cm	26~35 cm	36~45 cm	0~15 cm	16~25 cm	26~35 cm	36~45 cm
深松-旋耕 SS-RT	1.77±0.10a	1.76±0.04a	1.73±0.03a	1.73±0.05a	1.75±0.04a	1.72±0.08a	1.72±0.10a	1.68±0.02a
深松-免耕 SS-NT	1.74±0.07a	1.79±0.04ab	1.73±0.04a	1.72±0.04a	1.76±0.16a	1.73±0.15a	1.69±0.15a	1.70±0.09a
常规旋耕 RT	1.77±0.11a	1.81±0.11ab	1.76±0.08a	1.73±0.08a	1.81±0.14a	1.81±0.13a	1.75±0.18a	1.71±0.08a
免耕 NT	1.80±0.05a	1.87±0.01b	1.82±0.03a	1.75±0.06a	1.75±0.12a	1.82±0.06a	1.78±0.08a	1.73±0.10a

注：不同字母表示处理间差异显著性，大写字母（$P<0.01$），小写字母（$P<0.05$）。

2.2 不同耕作方式对土壤总孔隙度的影响

土壤总孔隙度即土壤孔隙容积占土体容积的百分比，反映了土壤孔隙状况和松紧程度。从表2可以看出，在播种前期不同耕作方式对土壤各个土层的总孔隙度影响表现为差异显著，主要体现在免耕处理显著高于常规旋耕处理，此外，其他处理的总孔隙度均高于常规旋耕处理；在拔节期不同耕作方式对各个土层土壤总孔隙度的影响表现为差异不显著；在抽雄期仅对0~15 cm土层总孔隙度的影响达到了差异显著水平，主要体现在深松-免耕处理的土壤总孔隙度显著高于常规旋耕处理；在灌浆期仅26~35 cm土层土壤总孔隙度表现为差异显著，主要体现在深松-旋耕处理显著高于免耕处理。综合整个玉米生育期来看，不同耕作方式对土壤总孔隙度的影响差异不明显。

表2 不同耕作方式对土壤总孔隙度的影响

处理	土壤总孔隙度（%）							
	播种前期				拔节期			
	0~15 cm	16~25 cm	26~35 cm	36~45 cm	0~15 cm	16~25 cm	26~35 cm	36~45 cm
深松-旋耕 SS-RT	37.18±2.25ab	37.23±2.45ab	38.48±3.03a	38.69±4.23ab	32.76±0.66a	33.63±0.71a	34.39±1.13a	35.70±0.35a
深松-免耕 SS-NT	38.65±5.64ab	37.23±4.70ab	37.34±2.27ab	38.12±2.92ab	32.98±1.46a	33.45±1.92a	34.36±0.62a	34.86±1.32a
常规旋耕 RT	36.10±2.47b	35.44±2.48b	36.14±3.93b	37.09±3.16b	32.95±0.63a	33.72±1.26a	33.78±1.67a	34.16±1.08a
免耕 NT	40.46±1.45a	38.83±0.92a	39.02±1.13a	39.54±1.26a	33.82±2.36a	33.59±2.80a	34.50±2.55a	35.90±1.75a
处理	抽雄期				灌浆期			
	0~15 cm	16~25 cm	26~35 cm	36~45 cm	0~15 cm	16~25 cm	26~35 cm	36~45 cm
深松-旋耕 SS-RT	38.63±0.70b	38.70±1.11a	40.61±1.53a	41.18±0.94a	35.36±0.92a	35.52±1.59a	37.96±2.54a	36.72±2.13a
深松-免耕 SS-NT	42.13±1.26a	41.08±0.11a	42.42±1.79a	43.09±1.25a	35.55±1.30a	36.86±1.20a	36.22±0.64ab	36.91±0.47a
常规旋耕 RT	38.51±2.40b	38.73±1.12a	40.04±1.69a	40.45±0.65a	36.26±0.33a	37.24±3.35a	36.16±0.46ab	36.64±1.39a
免耕 NT	39.59±0.69ab	38.74±1.37a	39.36±1.42a	41.28±1.51a	35.67±1.47a	36.61±0.40a	35.90±0.56b	36.28±1.03a

注：不同字母表示处理间差异显著性，大写字母（$P<0.01$），小写字母（$P<0.05$）。

2.3 不同耕作方式对土壤含水量的影响

土壤含水量也称土壤含水率,即 100 g 烘干土中含有若干克水分,是农业生产中的一个重要参数。由图 2 可知,在播种前期,深松-旋耕不同耕作方式对增加土壤含水量的影响程度整体表现为深松-旋耕>深松-免耕>旋耕>免耕,深松-免耕与常规旋耕处理差异不明显,与常规旋耕处理相比,深松-旋耕和深松-免耕处理土壤含水量在 0~25 cm 土层增加明显,表明深松和旋耕提高了降水入渗量;拔节期深松-旋耕处理在各个土层的土壤含水量明显高于其他处理,深松-免耕处理在 26~35 cm 深层土壤高于常规旋耕处理和免耕处理,表明深松可增加深层土壤的含水量;抽雄期深松-免耕处理在 16~45 cm 土层的土壤含水量仍高于其他处理,深松-旋耕处理在 36~45 cm 土层的土壤含水量高于常规旋耕处理和免耕处理;灌浆期深松-免耕处理的土壤含水量在 0~25 cm 土层与免耕处理相近,在 26~35 土层要高于其他处理,在 36~45 cm 土层与深松-旋耕处理相近。综合整个玉米生育期来看,在播种前期和拔节期,深松-旋耕处理在各个土层含水量高于其他处理;在抽雄期和灌浆期,深松-免耕处理在 16~45 cm 土层的土壤含水量要高于其他处理。

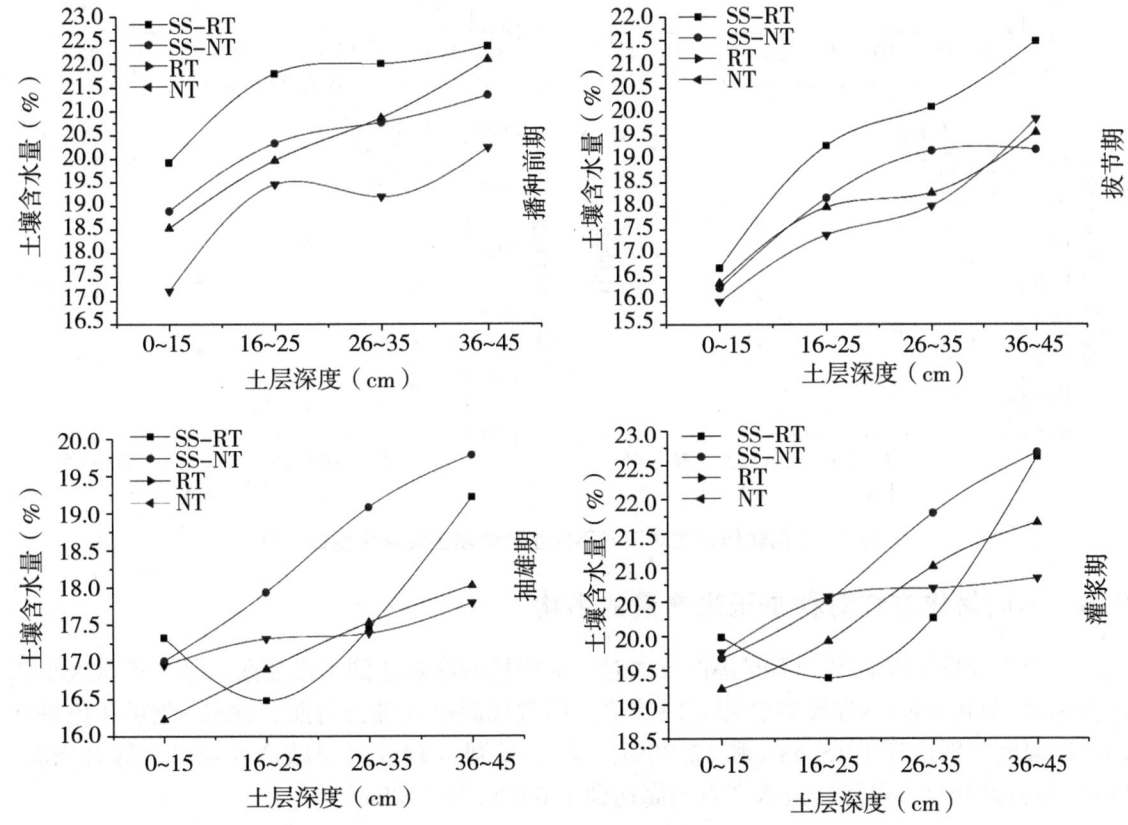

图 2 不同耕作方式对玉米不同生长时期土壤含水量的影响

2.4 不同耕作方式对土壤田间持水量的影响

土壤田间持水量在一定程度上反映了土壤储存水资源的能力。由图 3 可以看出:在播种前期,免耕处理的土壤持水量在各个土层明显低于其他处理,表明不同耕作处理可以提高降水的入渗量，从而提高土壤的持水量；在拔节期，常规旋耕处理和免耕处理土持水量随着土层深度增加表现为先降低后升高的趋势，而深松-旋耕处理和深松免耕处理则表现出逐渐升高的趋势；在抽雄期，深松-免耕处理的土壤持水量在0~25 cm土层高于其他处理，在26~45 cm土层，深松-旋耕和深松-免耕处理的土壤持水量相近，但高于常规旋耕处理和免耕处理；在灌浆期，深松-旋耕处理和常规旋耕处理的土壤持水量随着土层深度的增加表现为降低-升高-降低的趋势，而深松-免耕处理和免耕处理表现为升高-降低-升高的趋势。综合整个生育期来看，深松-旋耕和深松-免耕处理增加了26~35 cm的土壤持水量。

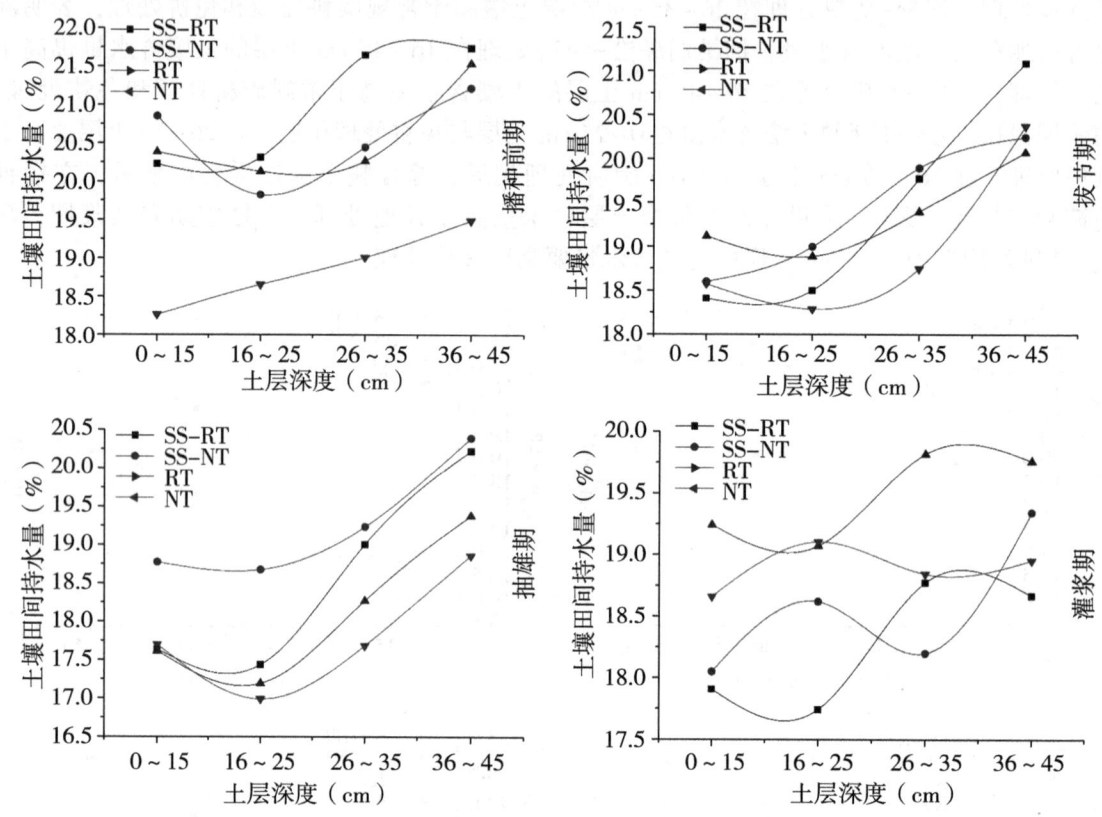

图3 不同耕作方式对玉米不同生长时期土壤持水量的影响

2.5 不同耕作方式对秋季玉米产量的影响

从表3分析可以看出，不同耕作方式对玉米产量的影响达到了极显著水平，产量大小依次为深松-免耕>免耕>常规旋耕>深松-旋耕。以常规旋耕处理为对照，深松-免耕和免耕处理增产幅度分别达到了15.68%和12.20%，深松-旋耕处理较对照减产3.45%。综合来看，深松-免耕处理效果最优，玉米产量最高达到了6 829.73 kg/hm²。

表3 不同施肥与耕作方式对玉米产量的影响

处理	平均产量 (kg/hm²)	差异显著性		比对照 (±%)
		5%水平	1%水平	
深松-旋耕 SS-RT	5 699.81	b	B	-3.45
深松-免耕 SS-NT	6 829.73	a	A	15.68

（续表）

处理	平均产量（kg/hm²）	差异显著性		比对照（±%）
		5%水平	1%水平	
常规旋耕 RT	5 903.77	b	B	—
免耕 NT	6 624.01	a	A	12.20

注：不同字母表示处理间差异显著性，大写字母（$P<0.01$），小写字母（$P<0.05$）。

3　讨论与结论

综合分析可以看出，在不同生育时期的不同土层中，耕作方式对于降低土壤容重、增加土壤孔隙度、增加土壤含水量和田间持水量的作用表现出不同程度的差异；不同耕作方式下玉米产量表现为深松-免耕>免耕>常规旋耕>深松-旋耕；深松-免耕处理比常规旋耕处理产量提高了 15.68%，产量最高达到了 6 829.73 kg/hm²。

深松能够打破犁底层，显著降低土层容重，增加土壤通透性。在本研究中，4 种不同耕作方式对不同时期、不同土层的土壤容重和土壤总孔隙度影响不同，深松-旋耕和深松-免耕处理在各个土层的土壤容重值低于常规旋耕处理和免耕处理，但差异不是很明显，尤其是土壤总孔隙度，主要原因在于深松处理后播种前，降雨日 8 d，降水量累计达到了 171 mm，一定程度上影响了深松的效应。前人研究也证明了这一点[10-12]。

深松可以提高土壤蓄水保墒能力及作物水分利用效率，从而促进根系生长发育。在本研究中，从整个玉米生育期来看，与常规旋耕处理相比，深松-旋耕处理增加了播种前期和拔节期 0~45 cm 土层的土壤含水量；深松-免耕处理增加了抽雄期和灌浆期 16~45 cm 土层的土壤含水量；深松-旋耕和深松-免耕处理增加了播种前期、拔节期和抽雄期 26~35 cm 土层的土壤持水量。这与孙晓明等的研究结果相一致[13]。

深松可以提高作物产量。本研究中，4 种不同耕作方式的玉米产量大小依次为深松-免耕>免耕>常规旋耕>深松-旋耕。以常规旋耕处理为对照，深松-免耕和免耕处理增产幅度分别达到了 15.68% 和 12.20%，深松-旋耕处理较对照减产 3.45%。综合来看，深松-免耕处理增产效果最优。这与宫秀杰等的研究结果相一致[14]。

参考文献

[1] 鲁奇，杨春悦，张超阳．少数民族地区农村劳动力转移的调查研究——以广西壮族自治区为例［J］．山西大学学报：哲学社会经济版，2007，30（4）：1-6.

[2] 许迪，Schmid R，Mermoud A．夏玉米耕作方式对耕层土壤特性时间变异性的影响［J］．水土保持学报，2000，14（1）：64-70.

[3] 高强，李德忠，汪娟娟，等．春玉米一次性施肥效果研究［J］．玉米科学，2007，15（4）：125-128.

[4] 闫惊涛，康永亮，田志浩．土壤耕作深度对旱地冬小麦生长和水分利用的影响［J］．河南农业科学，2011，40（10）：81-83.

[5] 王景琴，朱秀章，刘通．耕地深松深耕技术的优势及完善措施［J］．现代农业科技，2011（19）：137-139.

[6] 刘玉涛, 王宇先, 张树权, 等. 耕作方式对半干旱地区玉米生长和产量的影响 [J]. 黑龙江农业科学, 2012 (7): 19-21.

[7] 刘武仁, 郑金玉, 罗洋, 等. 不同耕作方式对玉米叶片冠层光合特性的影响 [J]. 玉米科学, 2012, 20 (6): 103-106, 111.

[8] 李永平, 王孟本, 史向远, 等. 不同耕作方式对土壤理化性状及玉米产量的影响 [J]. 山西农业科学, 2012, 40 (7): 723-727.

[9] 柏炜霞, 李军, 王玉玲, 等. 渭北旱塬小麦玉米轮作区不同耕作方式对土壤水分和作物产量的影响 [J]. 中国农业科学, 2014, 47 (5): 880-894.

[10] Raper R L, Bergtold J S. In-row subsoiling: a review and suggestions for reducing cost of this conservation tillage operation [J]. Applied Engineering in Agriculture, 2007, 23 (4): 463-471.

[11] Camp C R, Sadler E J. Irrigation, deep tillage, and nitrogen management for a corn-soybean rotation [J]. Transactions of the ASAE, 2002, 45 (3): 601-608.

[12] Coates W. Minimum tillage systems for irrigated cotton: is subsoiling necessary [J]. Applied Engineering in Agriculture, 1997, 13 (2): 175-179.

[13] 孔晓民, 韩成卫, 曾苏明, 等. 不同耕作方式对土壤物理性状及玉米产量的影响 [J]. 玉米科学, 2014, 22 (1): 108-113.

[14] 宫秀杰, 钱春荣, 于洋, 等. 深松免耕技术对土壤物理性状及玉米产量的影响 [J]. 玉米科学, 2009, 17 (5): 134-137.

不同耕作方式对土壤水分及玉米生长发育的影响

吕巨智[1]，钟昌松[1]，范继征[1]，石达金[1]，程伟东[1]，刘永红[2]，闫飞燕[1]

(1. 广西农业科学院玉米研究所，广西 南宁 530227；
2. 四川省农业科学院作物研究所，四川 成都 610066)

摘要：针对广西玉米主产区土壤耕层变浅、犁底层坚硬的问题，以当地习惯的耕作方式旋耕为对照，设置深松35 cm+免耕、深松35 cm+旋耕、深松25 cm+免耕、深松25 cm+旋耕和免耕5种耕作方式，分析各处理对土壤水分及玉米生长发育的影响。研究结果表明：深松处理均提高土壤含水量，其中以深35 cm+旋耕和深松35 cm+免耕的保墒效果最佳；深松处理均增加了叶面积、根冠比；而深松处理的玉米穗长、穗粗、秃尖长、穗行数、行粒数、百粒重等产量构成因素均较对照有不同程度的改善。在研究设定的5种耕作方式中，以深松35 cm+旋耕为最优组合。

关键词：玉米；耕作方式；土壤水分；产量

Effects of different tillage managements on soil moisture and growth and development of maize

Lü Juzhi[1], Zhong Changsong[1], Fan Jizheng[1], Shi Dajin[1],
Cheng Weidong[1], Liu Yonghong[2], Yan Feiyan[1]

([1]Maize Research Institute, Guangxi Academy of Agricultural Sciences, Nanning 530007, China;
[2]Crop Research Institute, Sichuan Academy of Agricultural Sciences, Chengdu 610066, China)

Abstract: The characters of Guangxi province the main maize production region are that soil layer shoal and plow pan is hard. According to the characters, 5 tillage managements were designed to study effects of different tillage managements on soil moisture and development of maize. CK is the rotary tillage, the others are 35 cm subsoiling depth and no-tillage, 35 cm subsoiling depth and rotary tillage, 25 cm subsoiling depth and no-tillage, 25 cm subsoiling depth and rotary tillage. The results showed that soil water contents of different tillage managements were all higher than CK. Among of which the 35 cm subsoiling depth and no-tillage and 35 cm subsoiling depth and rotary tillage in the fall were more significant. All the tillage managements increased the leaf area, the root/shoot ratio, otherwise, after the subsoiling treatment, maize yield components, such as corn

---

基金项目：国家现代农业产业技术体系南宁玉米综合试验站项目（CARS-02-73）；广西科学研究与技术开发项目（桂科攻1123001-1J）。

作者简介：吕巨智（1984—），男，湖北武穴人，硕士，研究方向为玉米栽培生理与育种研究工作。E-mail: lvjuzhi520@sina.com。

ear length, ear diameter, barren-tip length, number of rows per ear, kernels per row, 100-grain weight, are improved in different degrees than CK. Among them, the combination of 35 cm subsoiling depth and rotary tillage is the optimal solution.

Keywords：Maize；Cultivation methods；Soil water content；Yield

近年来，广西玉米主产区采用耕作方式不合理，导致耕层变浅，犁底层紧实，容重增加。导致土壤蓄水和透水能力降低，根系下扎阻力加大，从而增加了玉米旱灾、内涝、倒伏、中后期营养不足等风险，直接影响了玉米产量[1-3]。增加玉米产量，首先要打破犁底层，增加耕层厚度，提高土壤保水性能。许多研究表明，合理的耕作措施可以改善土壤结构，提高土壤的持水性能，增加作物对水分及养分的吸收，有利于作物的生长发育[4,5]。本文针对广西玉米主产区的实际情况，设置不同耕作方式，分析其对土壤水分、玉米生长发育及产量等指标的影响，为探索最佳的耕作方式以提高玉米产量和建立合理的耕作制度提供理论依据和技术支撑。

1 材料与方法

1.1 试验地概况

试验在广西壮族自治区农业科学院玉米研究所试验田进行。供试土壤为黏壤土，土壤基本肥力见表1。

表1 试验地土壤化学性状

土层 （cm）	全氮 （g/kg）	碱解氮 （mg/kg）	全磷 （g/kg）	速效磷 （mg/kg）	全钾 （g/kg）	速效钾 （mg/kg）	有机质 （%）	pH 值
0~15	0.09	0.07	0.13	56.00	30.47	175.00	15.33	6.37
15~25	0.08	0.07	0.13	55.67	34.57	147.67	16.53	6.34
25~35	0.10	0.07	0.15	65.33	37.67	150.00	17.71	6.56
35~45	0.09	0.07	0.12	62.67	34.50	164.67	17.90	6.57

1.2 试验材料

供试材料：选择广西当前主推玉米品种正大619为试验用品种，该品种发芽率达95%以上。

1.3 试验设计

本试验于2012年3月28日至2012年7月27日进行，设6个处理。处理1（CK）：传统旋耕，玉米人工收获，秸秆移走，旋耕2遍灭茬后用牛开行人工播种；处理2（T1）：深松35 cm+免耕，玉米人工收获，秸秆移走，深松35 cm后免耕，开播种沟人工播种；处理3（T2）：深松35 cm+旋耕，玉米人工收获，秸秆移走，深松35 cm后旋耕灭茬，用拖拉机开行，人工播种；处理4（T3）：深松25 cm+免耕，玉米人工收获，秸秆移走，深松25 cm后免耕播种，开播种沟人工播种；处理5（T4）：深松25 cm+旋耕，玉米人工收获，秸秆移走，深松 25 cm 后旋耕灭茬，拖拉机开行，人工播种；处理 6（T5）：免耕，玉米人工收获，秸秆移走，开播种沟人工播种。

本试验采用随机区组排列，共 6 个处理，3 次重复，10 行区，行长 33 m，行距 0.7 m，重复间留走道 1.0 m，小区行间不留走道，四周设保护行，密度为 3 800 株/亩。

肥力设置：（1）基肥。供试肥料为陶氏益农复合肥 150 kg/hm²（N∶P∶K=16∶16∶16，芬兰产，湖北华丰公司分装）。（2）追肥。定苗后结合中耕除草，施陶氏益农复合肥 10 kg/亩，尿素 150 kg/hm²（总 N≥46.4%，贵州宜化化工有限责任公司）；大喇叭口期结合大培土施用攻苞肥，施陶氏益农复合肥 375 kg/hm²。

1.4 测定项目及方法

在玉米播种前期、苗期、开花吐丝期、成熟期等四个生育时期，采用土钻取土，分别取 0~15 cm、15~25 cm、25~35 cm、35~45 cm 四个层次，每个小区取 3 个点，土壤养分均按常规分析法测定，土壤含水量用环刀法测定[6]。田间指标测定及取样时间分别为 4 月 13 日、4 月 18 日、5 月 2 日、5 月 16 日、6 月 9 日，株高、叶长、叶宽、地上部、根系测定为每处理 3 次重复，每个重复测定 3 株。叶面积采用长宽系数法计算；干物重测定采用烘干法，灌浆期每个小区选取有代表性的植株 10 株，测定株高和穗位高。在玉米达到完全成熟时进行测产，每点测产面积 17.5 m²，收获时每小区选取代表性果穗 10 穗进行室内考种，测定穗长、穗粗、秃尖长、穗行数、行粒数和百粒重等农艺性状。

2 结果与分析

2.1 不同耕作方式对土壤含水量的影响

土壤水分是土壤的重要组成部分之一，它不仅是作物生长需水的主要给源，而且还深刻地影响着土壤内养分转化和生物活动过程。从图 1 至图 4 可以看出，不同处理生育期土壤含水量的总体变化趋势存在显著差异，成熟期 35~45 cm 土层土壤含水量最大，其中，深松处理为 17.50%~19.33%，免耕处理为 17.03%，传统耕作处理为 15.28%。各个处理不同土层的不同生育时期土壤含水量总的变化趋势为：播种前变化不大，苗期普遍呈现降低趋势，开花期和成熟期不同耕作方式的土壤含水量均随着土壤深度的增大呈增加趋势。总之，各个处理土壤含水量由高到低的顺序为：深松>免耕>传统耕作。

从图 1 可以看出 0~10 cm 土层在播种前期，6 个处理的含水量经方差分析显示，三个处理之间未达到显著差异。随着生育期的推进，在开花期和成熟期气温逐渐升高，蒸发量加大，土壤水分散失速率加快，导致土壤含水量普遍降低。而深松处理的土壤含水量高于传统耕作。深松处理为间隔扰动土壤，同样具有较强接纳和保蓄水分的能力，也具有较高的土壤含水量；传统耕作处理为表层土壤破碎程度严重，不容易保水，导致雨水散失较多，土壤含水量较低。

从图 2 可以看出 15~25 cm 土层在播种前期，传统耕作的含水量高于其他处理，但其余各个测定时期均表现为深松的土壤含水量高于传统耕作，6 个处理达到显著差异。

从图 3 可以看出 25~35 cm 土层在播种前期变化不大，苗期呈现下降趋势，其余两个测定时期均表现为深松处理的土壤含水量高于传统耕作，6 个处理达到显著差异。

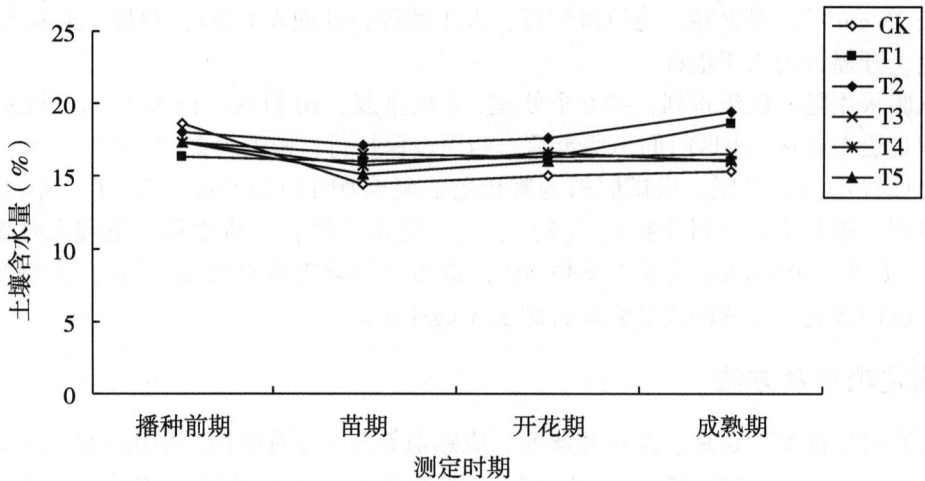

图1 不同耕作方式对各生育时期 0~15 cm 土壤水分含量影响

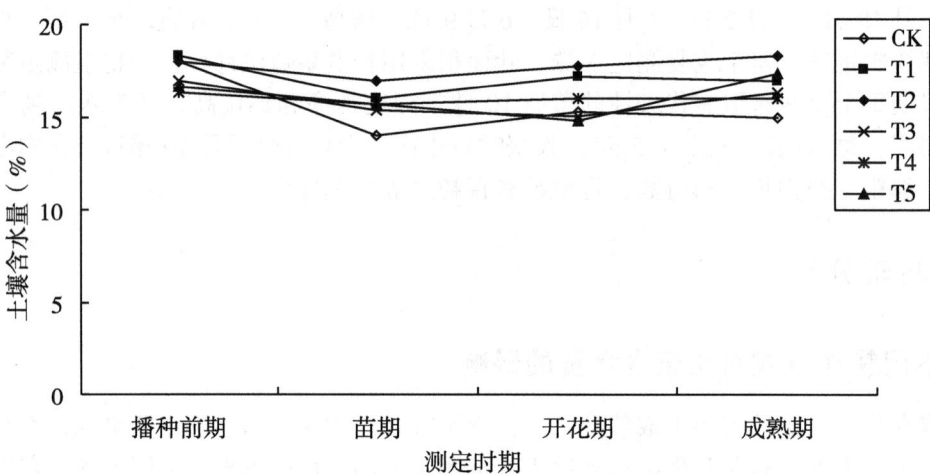

图2 不同耕作方式对各生育时期 15~25 cm 土壤水分含量影响

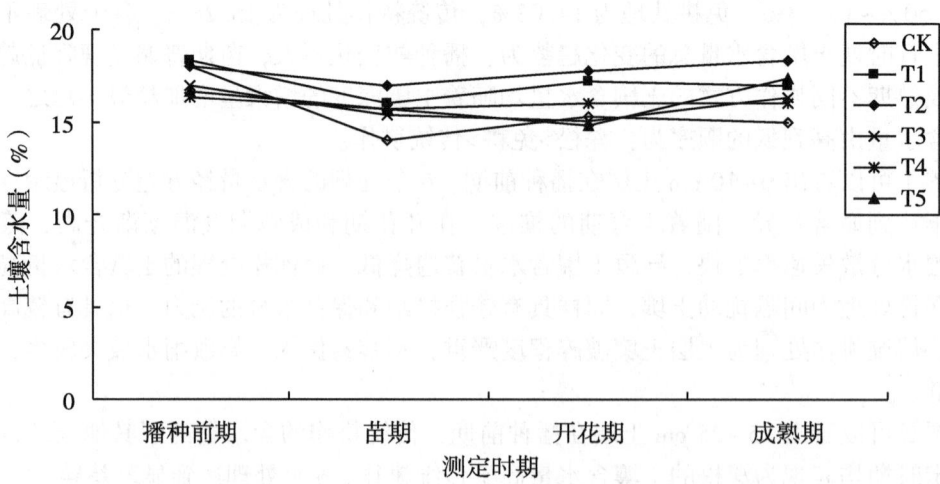

图3 不同耕作方式对各生育时期 25~35 cm 土壤水分含量影响

从图 4 可以看出 35~45 cm 土层在播种前期变化不大,苗期呈现下降趋势,其余两个测定时期均表现为逐步升高的趋势,在成熟期达到最大。深松的土壤含水量高于传统耕作,6 个处理达到显著差异。

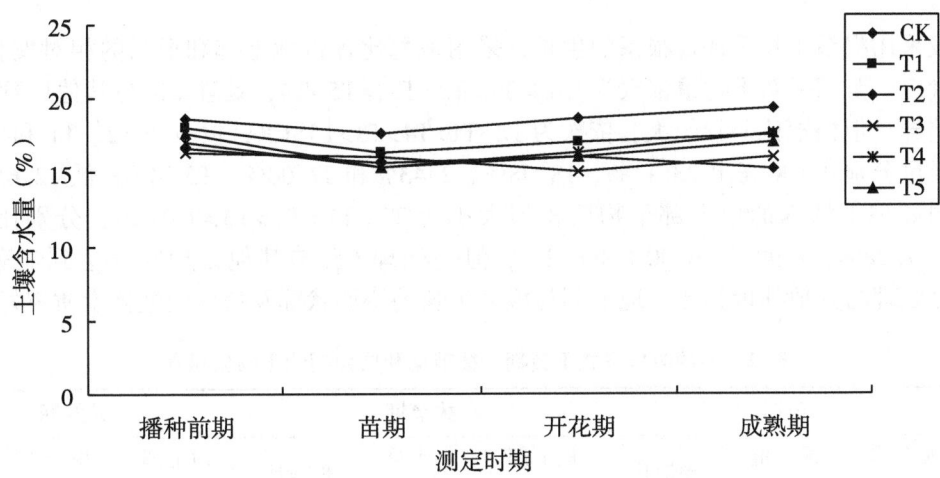

图 4　不同耕作方式对各生育时期 35~45 cm 土壤水分含量影响

2.2　不同耕作方式对单株叶面积的影响

由图 5 可以看出,在玉米的整个生育期内,各处理叶面积在拔节期后上升速度最快,在开花吐丝期达到最大值,灌浆期开始下降。这主要由于拔节后迅速长出新叶,开花吐丝期后叶片全部伸出,而后随着子粒灌浆加快和营养供应不足,叶片逐渐衰老、死亡脱落,从而导致叶面积下降。

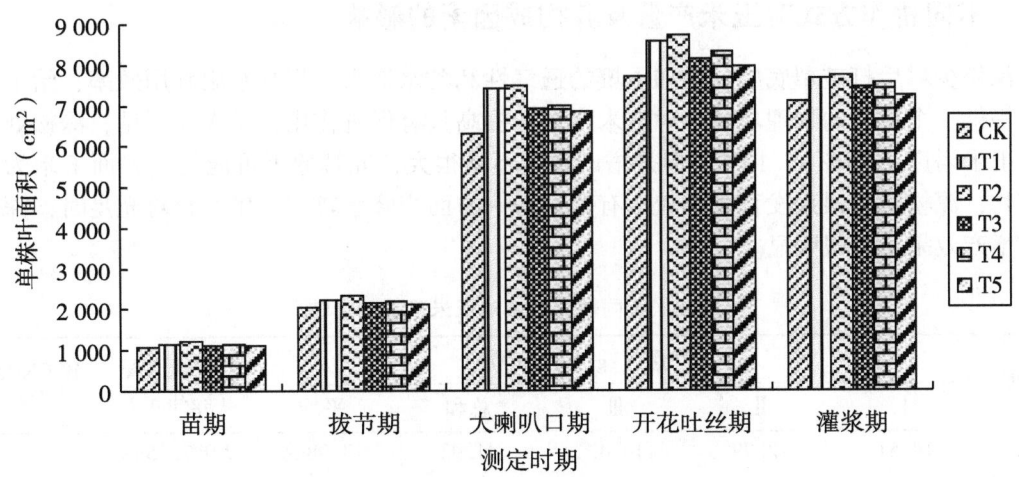

图 5　不同耕作方式对单株叶面积的影响

5 个处理各生育期的单株叶面积大小都是 T2>T1>T4>T3>T5>CK,开花吐丝期时达到各测定时期的最大,分别比 CK 增加 13.76%、11.24%、7.85%、5.62%、3.25%,说明深松处理在苗期、拔节期、大喇叭口期、开花吐丝期和灌浆期的单株叶面积都有一定幅度的增长,其中前期、后期增长幅度较小,中期增长幅度较大,说明深松能够促进玉米的茎叶生长，且具有"前缓-中快-后稳"的特点，可以有效增加叶面积，延缓叶片衰老，延长叶片功能期，有利于光合产物的形成和向库的运输，从而促进籽粒的充实。

2.3 不同耕作方式对植株干物质积累及根冠比的影响

一般采用根系干物重评价根系的生长，采用根冠比评价地上部和根系的相对发育状况。从表2看出，苗期根系干物重依次为CK≥T1>T2>T3≥T5>T4，处理CK与其他处理间未达到显著差异，但根冠比差异较大，依次为T2>T1>T4>T3>T5>CK；拔节期T2、T1和T3、T4处理根干重分别比CK高出23.11%、18.98%、2.43%和27.00%，T5比CK低12.65%；成熟期不同处理下植株的地上部干物质积累大小为T2>T1>T4>T3>T5>CK，分别比CK高11.10%、7.98%、4.90%、6.00%和8.2%，但是处理CK与其他处理间未达到显著差异。表明深松处理与其他处理相比，地上部与根系生长的协调状况及植株的整体素质较好。

表2 不同耕作方式下苗期、拔节期和成熟期的干物质积累

处理	苗期			拔节期			成熟期		
	地上部 (g)	根干重 (g)	根冠比	地上部 (g)	根干重 (g)	根冠比	地上部 (g)	根干重 (g)	根冠比
CK	0.35 aA	0.08aA	0.219	12.49 aA	1.37 aA	0.110	366.89aA	17.42dD	0.047
T1	0.28abAB	0.08aA	0.274	10.51 bcAB	1.63 aA	0.155	396.19aA	33.51bB	0.085
T2	0.25bAB	0.07aA	0.284	10.39 cAB	1.69 aA	0.162	407.63aA	42.05aA	0.103
T3	0.26bAB	0.06aA	0.234	10.81bcAB	1.40 aA	0.130	384.85aA	26.81cBC	0.070
T4	0.22bAB	0.05aA	0.239	12.32 abA	1.74 aA	0.141	388.93aA	29.50bcBC	0.076
T5	0.27bB	0.06aA	0.225	9.36 cB	1.20 aA	0.128	371.46aA	23.74cCD	0.064

2.4 不同耕作方式对玉米产量及其构成因素的影响

深松少耕后打破犁底层，增加土壤的透气性和贮水能力，提高水肥利用效率，给玉米的生长提供一个良好的物理环境，对玉米产量的提高具有促进作用，从表3可见，深松处理比对照玉米增产1.47%~3.13%。但是增产幅度不是很大，究其原因可能是一方面玉米为须根系作物，深松对其根系发育影响较小有关，另一方面当降水较多、供水相对充足时，深松的蓄水增产效果不是很明显。

表3 不同耕作方式对玉米产量的影响

处理	小区产量（kg/17.5 m²）					折公顷产量 (kg/hm²)	较CK增产 (%)
	I	II	III	总和	平均		
CK	14.51	12.99	14.42	41.93	13.98aA	7 987.35aA	
T1	12.84	15.73	14.58	43.15	14.38aA	8 219.25aA	2.90
T2	14.62	15.03	13.60	43.25	14.42aA	8 238.00aA	3.13
T3	14.11	14.06	13.79	41.96	13.99aA	8 157.45aA	2.12
T4	13.33	13.87	15.76	42.96	14.32aA	8 182.5aA	2.45
T5	13.26	14.82	13.41	41.50	13.83aA	8 150.51aA	1.47

由表4可知，深松处理的玉米穗长、穗粗、秃尖长、穗行数、行粒数、百粒重等产量构成因素均较对照有不同程度的改善。深松处理的株高、穗位高最高，穗长最长，行粒数最多；各处理在穗粗、穗行数和百粒重等方面均未出现显著差异。

表4　不同耕作方式对玉米主要性状的影响

处理	株高 (cm)	穗位高 (cm)	穗长 (cm)	穗粗 (cm)	秃尖长 (cm)	穗行数	行粒数	百粒重 (g)
CK	263.53 aA	115.00 aA	20.27 aA	4.57 aA	0.79 aA	14.20 aA	35.63 aA	34.45 aA
T1	272.43 aA	122.80 aA	20.57 aA	4.65 aA	1.15 aA	14.13 aA	36.73 aA	33.69 aA
T2	266.90 aA	151.47 aA	20.67 aA	4.60 aA	1.00 aA	14.33 aA	36.80 aA	34.44 aA
T3	276.83 aA	121.37 aA	20.33 aA	4.67 aA	1.43 aA	14.33 aA	36.17 aA	34.46 aA
T4	273.33 aA	121.30 aA	20.53 aA	4.65 aA	1.67 aA	14.47 aA	35.83 aA	34.10 aA
T5	268.70 aA	116.50 aA	19.43 aA	4.53 aA	1.60 aA	14.27 aA	34.10 aA	34.07 aA

3　结论与讨论

土壤水分是土壤的重要组成部分之一，它不仅是作物生长需水的主要给源，而且还深刻地影响着土壤内养分转化和生物活动过程。前人对不同作物的研究表明，深松对保持土壤水分具有显著的作用[7-9]。本试验条件下，深松35 cm+旋耕、深松35 cm+免耕、深松25 cm+免耕和深松25 cm+旋耕比当地习惯性耕作方式提高土壤含水量。不同层次的土壤在不同生育期呈现"前低—中高—后低"的趋势，在根系分布密集的15~45 cm耕层中，由于深松疏松了土壤，利于增强降雨入渗率，增加土壤水分含量，可以将伏雨贮存，减少地面径流，能够扩大土壤水库容，这与前人研究结果基本一致[10-14]。

根和冠是对作物有机体最基本的划分。根、冠关系可视为环境因素对其作用后，经过作物体内许多基本变化过程及自适应、自调节后所表现出的综合效应，是一个整体功能的问题[15-16]。较高的根冠比为作物创造了良好的营养生长条件，较多的根系有利于植株对水分和矿质营养的吸收。特别是深松35 cm+旋耕的耕作方式具有较大根干重，延缓了叶片的衰老，从而延长叶片的功能期。试验结果表明，深松促进了根系向下生长，增加了叶面积和植株干物质。

深松对产量及产量构成因素的影响不尽相同[17-21]。本研究中，因为深松处理的玉米穗长、穗粗、秃尖长、穗行数、行粒数、百粒重和粒率等产量构成因素均较对照有不同程度的改善，产量为深松>免耕>传统耕作，但是增产幅度不是很大。究其原因，可能是一方面首先是玉米所试验地年年翻犁，从而导致耕层土壤已经很疏松，其次玉米为须根系作物，深松对其根系发育影响较小，另一方面当降水较多、供水相对充足时，深松的蓄水增产效果不是很明显。2013年我们研究室在靖西、都安、天三个粮食重点示范县做了深松试验，结果表明深松区平均每公顷产量比传统耕作区增产691.5~829.5 kg，比常规区增产14.3%~15.6%，增产的原因可能是由于三个粮食重点示范县的玉米地多年没有犁耙翻动，土壤很紧实，犁底层很厚，深松后打破了犁底层，增加土壤通透性和贮水能力，提高水肥利用效率，给玉米的生长提供一个良好的物理环境，从而达到增产的效果。

试验分析结果表明，从提高土壤保水能力、促进植株生长、增加玉米产量的角度，结合广西玉米主产区地区耕层浅、犁底层厚且坚硬的现状，保水、健株、增产的耕作措施以"深松35 cm+旋耕"为比较适宜的耕作方式。2012年玉米生育期间降水较多，增产幅度不大，在干旱年份增产幅度如何，深松对须根系作物根生物量的影响较小，那么对于直根系作物根生物量的影响如何，由于是一年试验，这两个问题有待进一步研究，其结果有待进一步验证。

参考文献

[1] 边少锋，马虹，薛飞，等．吉林省西部半干旱区深松蓄水耕作技术研究［J］．玉米科学，2008（1）：67-68.

[2] 赵红岩，李钦，王洪利．东北黑土区的土壤深松与玉米增产［J］．东北农机，2008（9）：64-65.

[3] 朱凤武，王景利，潘世强，等．土壤深松技术研究进展［J］．吉林农业大学学报，2003，25（4）：457-461.

[4] 丁昆仑．深松耕作对土壤水分物理特性作物生长的影响［J］．中国农村水利水电，1997（7）：13-16.

[5] 隋华，贾兰英，徐建坡，等．土壤深松对玉米效应的试验研究［J］．天津农林科技，2002（4）：1-4.

[6] 鲍士旦．土壤农化分析［M］．北京：中国农业科学技术出版社，2000：100-130.

[7] 金复鑫，彭文英，张科利，等．北京保护性耕作条件下土壤水分动态变化研究．土壤通报，2009，40（1）：28-33.

[8] 刘洋，孙占祥，白伟，等．不同耕法对土壤含水量、玉米生长发育及产量的影响［J］．辽宁农业科学，2011（2）：10-14.

[9] 张晓平，方华军，杨学明，等．免耕对黑土春夏季节温度和水分的影响［J］．土壤通报，2005，36（3）：313-316.

[10] 黄毅，邹洪涛，虞娜，等．东北半干旱区秋后玉米地不同处理方式对土壤水分状况的影响［J］．水土保持研究，2006，13（2）：34-37.

[11] 张西群，齐新，董文旭，等．玉米深松免耕播种对土壤性状及玉米生长发育的影响［J］．河北农业科学，2010，14（3）：26-28.

[12] 刘绪军，荣建东．深松耕法对土壤结构性能的影响［J］．水土保持应用技术，2009（1）：9-11.

[13] 肖孔操，黄道友，刘守龙，等．不同轮作制度下红壤旱地水分时空变化对稻草覆盖的响应．水土保持学报，2009，23（2）：219-222.

[14] Alvarez R, Steinbach H S. A review of the effects of tillage systems on some soil physical properties, water content, nitrate avail ability and crops yield in the Argentine Pampas. Soil & Tillage Research, 2009 (104): 1-15.

[15] 邵国庆，李增嘉，宁堂原，等．不同水分条件下常规尿素和控释尿素对玉米根冠生长及产量的影响［J］．作物学报，2009，35（1）：118-123.

[16] 葛体达，隋方功，李金政，等．干旱对夏玉米根冠生长的影响［J］．中国农学通报，2005，21（1）：103-109.

[17] 赵伟．不同深松处理对玉米产量及其产量构成因素的影响［J］．黑龙江农业科

学，2011 (12)：35-37.
[18] 李洪文，陈君达，李问盈，等．保护性耕作条件下深松技术研究 [J]．农业机械学报，2000，31 (6)：42-43.
[19] 智建奇，贾志森，郑联寿，等．不同保护性耕作方式对旱地玉米的增产效应 [J]．玉米科学，2006，14 (2)：112-114.
[20] Cox W J, Cherney J H, Hanchar J H. Zone tillage depth affects yield and economics of corn silage production. Agronomy Journal, 2009 (101): 1093-1098.
[21] Berenguer M J, Faci J M. Sorhhum (*Sorghum bicolor* L. Moench) yield compensation processes under different plant densities and variable water supply. European Journal of Agronomy, 2001, 15: 43-55.

水氮条件对南亚热带玉米产量及农田土壤有机碳氮组分的影响

刘涌鑫，毛祥敏，周勋波

(广西大学农学院，广西 南宁 530004)

摘要：为探究水氮条件对玉米(Zea mays L.)产量及农田土壤碳氮的影响，试验在广西一年两季玉米种植区，于2018年春玉米和秋玉米生长季进行。品种为万川1306，种植密度为52 500 株/hm², 行株距为60 cm × 28 cm, 小区面积为16.8 m²。试验采用裂区试验设计，主处理水分条件分别为雨养和灌溉，副处理施氮量分别为0 kg/hm² (N0)、150 kg/hm² (N1)、200 kg/hm² (N2)、250 kg/hm² (N3)和300 kg/hm² (N4)。在玉米成熟期采集土壤耕层(0~20 cm)样品，测定土壤有机碳、土壤微生物量碳、土壤有机氮组分。试验结果表明：同一处理条件下，土壤有机氮各组分含量表现为氨基酸态氮≈氨态氮>酸解未知氮>氨基糖态氮。春玉米，灌溉的产量较雨养提高了15.7% ($P<0.05$)，秋玉米，灌溉与雨养产量基本持平。施氮量与作物产量成极显著正相关关系($P<0.01$)，但过量施氮不能显著提高作物产量。土壤微生物量碳与作物产量成极显著正相关关系($P<0.01$)。广西春玉米在阶段性干旱条件下，适时合理补水结合250 kg/hm²施氮量，具有较好的土壤供碳氮潜力，并获得较高产量。

关键词：水分；氮素；有机碳；有机氮；产量

Effects of Water and Nitrogen Conditions on Yield of Subtropical Maize and Soil Organic Carbon and Nitrogen Components

Liu Yongxin, Mao Xiangmin, Zhou Xunbo

(Guangxi University Agricultural College, Nanning 530004, China)

Abstract: To study the effect of water and nitrogen on maize (Zea mays L.) yield and soil nitrogen composition, The experiment was carried out in the maize planting region of two crops a year in Guangxi, and the spring and autumn maize growing season in 2018. The variety is Wanchuan 1306, the planting density is 52,500 plants/ha, row spacings was 60 cm, and the plot area is 16.8 m². Two-factor split plot design, the main plot is water condition: rainfed and irrigation, the split-plot is nitrogen levels: 0 kg/ha (N0)、150 kg/ha (N1)、200 kg/ha (N3)、250 kg/ha

---

项目基金：国家自然科学基金(31760354)和广西自然科学基金(2017GXNSFAA198036, 2019GXNSFAA185028)；

作者简介：刘涌鑫(1997—)，男，硕士在读，主要从事玉米栽培研究 E-mail: yosinl@foxmail.com。

(N4)、300 kg/ha (N5). Soil samples in 0~20 cm were collected at maize maturity stage, and soil organic carbon, soil microbial biomass carbon and soil organic nitrogen components were determined. The result showed that the order of soil organic nitrogen was amino acid nitrogen ≈ ammonia nitrogen>acid hydrolysis unknown nitrogen>amino sugar nitrogen. In spring maize, yield of irrigation increased by 15.7% compared with rainfed ($P<0.05$), and autumn maize, yield of irrigation and rainfed were not obvious difference. There was significant positive correlation between nitrogen application and crop yield ($P<0.01$), too high nitrogen application could not significantly increase crop yield. There was very significant positive correlation between soil microbial biomass carbon and crop yield ($P<0.01$). Under the drought condition, timely and reasonable water supplement combined with 250 kg/ha nitrogen application had improved soil carbon and nitrogen content and high yield.

Keywords: Water; Nitrogen; Organic carbon; Organic nitrogen component; Yield

玉米（Zea mays L.）是我国重要的粮食作物，是目前种植面积最广、产量最高的谷类作物，位居三大粮食作物之首[1]。当前，世界粮食安全问题面临土壤退化的巨大挑战，土壤退化与土壤侵蚀、土壤肥力下降、土壤管理和耕作系统等密切相关[2]。田间管理措施能改变土壤的理化性状，直接或间接地影响土壤有机碳，最终影响土壤肥力[3]，因此，农艺措施对作物生产有重要影响。

施肥能促进根系生长，是作物生长发育过程中必不可少的农艺措施[4]。氮素是植物生长发育过程所必需的元素之一，是土壤肥力中最活跃的因素，适量施氮可以促进作物生长[5]，施氮量与作物干物质积累量相关，增施氮肥对玉米实现较高产量有重要作用[6]。土壤有机碳含量是土壤肥力的一部分，施氮能丰富土壤有机碳含量[7]，加速产生作物根系分泌物[8]，同时土壤微生物活性数量和酶活性提高，加速土壤有机碳分解[9]；外源氮素投入也影响土壤-植物系统中碳氮的积累与分配[10]。长期施肥能够提高土壤有机氮组分含量，单施化肥处理中酸解铵态氮含量最高，酸解氨基糖氮含量最低[11]。

水分是维持作物生长的重要因子，土壤水可以影响土壤理化性质，进而影响作物生长[12]。适当供水可明显提高地上部干物质重，最终增加玉米的产量[13]；在适宜的水分条件下土壤氮和碳的矿化速率可以得到加快，从而固定更多的碳素，增加土壤有机碳含量，但在缺水或过量灌溉的条件下土壤氮和碳的矿化速率将会下降，氮和碳损失增加[14]。降水量高能够降低酸解铵态氮含量，增加酸解氨基糖态氮含量[15]。广西玉米种植区光热资源丰富，日照时间长，降水充沛，但由于多年连种，导致土壤养分不平衡[16]。

本试验在南亚热带广西一年两作玉米种植区进行，研究在阶段性干旱条件下适当灌溉和不同氮处理对土壤有机碳氮组分及产量的影响，通过改善水分和氮素条件确定土壤供碳氮潜力最大的农艺栽培措施，在达到高产的同时又减少对环境的压力，为作物生产活动提供一定的理论依据。

1 材料与方法

1.1 试验地概况

本试验于2018年在广西壮族自治区南宁市广西大学农学实验田（22°50′N，108°17′

E）进行。该区域属于玉米一年两季种植区，为湿润的亚热带季风气候，年平均气温 22.0℃，年均降水量为 1 300~2 000 mm，平均相对湿度为 79%。2018 年月降水量、月均最高温度和最低温度如图 1 所示。玉米试验地的土壤类型为黏土，土壤理化性质如表 1 所示。

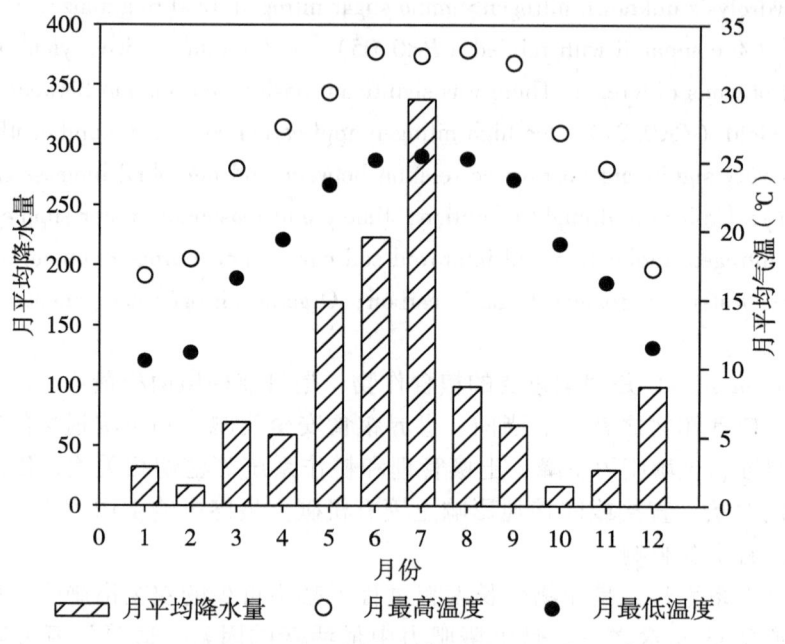

图 1　2018 年月降水量、月均最高温度及最低温度

表 1　试验前 0~20 cm 耕作层土壤基本理化性质

pH	速效氮 (mg/kg)	速效磷 (mg/kg)	速效钾 (mg/kg)	土壤容重 (g/cm^3)	土壤有机质 (g/kg)	田间持水量 (V%)
5.4	126.2	40.0	124.5	1.50	17.5	37.2

1.2　试验设计

玉米试验品种为万川 1306，行株距为 60 cm×28 cm，小区面积为 16.8 m^2。试验分为春玉米和秋玉米两季进行，春秋玉米分别于 2018 年 3 月 22 日和 8 月 11 日使用播种机按 2~3 株/穴播种，深度为 2~3 cm，分别于 2018 年 4 月 15 日和 8 月 26 日进行间苗，保证试验地基本苗数为 52 500 株/hm^2。试验设置两因素，分别为主因素水分（雨养和灌溉），根据降水和土壤含水量情况采用软管补水滴灌，无论任何生育时期土壤含水量保持不低于田间最大持水量的 60%，水分速测仪 TDR 100 测定土壤水分，生长季灌溉量如图 2 所示；副因素施氮量，分别为纯氮 0 kg/hm^2（N0）、150 kg/hm^2（N1）、200 kg/hm^2（N2）、250 kg/hm^2（N3）和 300 kg/hm^2（N4），其中播前基肥占总施氮量的 2/3，大喇叭口期追加剩余的 1/3 氮肥。磷（P$_2$O$_5$）、钾（K$_2$O）肥均作为基肥按 100 kg/hm^2 一次施入。分别于 2018 年 7 月 12 日和 12 月 16 日收获测产，每个小区选取 2 m^2（约 11 株）用于测产。

1.3　测定项目与方法

于 2018 年 7 月 10 日和 11 月 30 日玉米成熟期采用 5 点取样法取土样，每个处理 3 次重

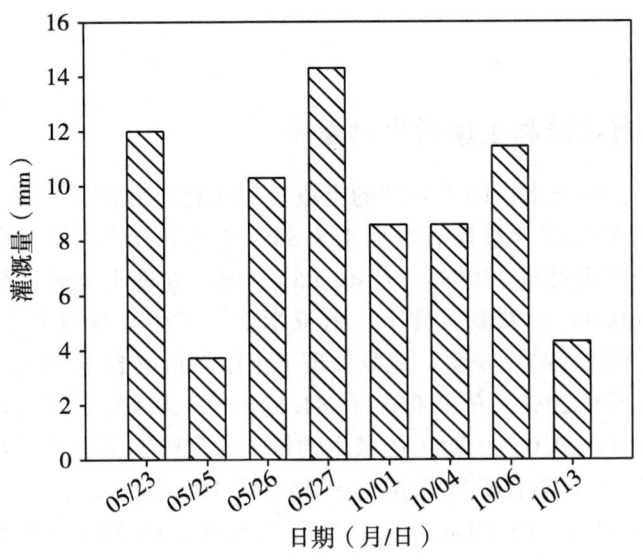

图 2 2018 年月灌溉量

复,取完将土壤分成 2 份,一份鲜土用于土壤微生物量碳的测定;另一份风干土用于土壤有机碳和有机氮组分的测定。风干的土样采用四分法磨土,过 100 目筛,塑封袋保存用于后期土壤相应指标的测定。

1.3.1 土壤微生物量碳

采用氯仿熏蒸-硫酸钾浸提的方法[17]。鲜土过 2 mm 筛,称取 11.0 g 过筛土于培养皿,再放入干燥罐封闭黑暗预培养 7 d。预培养的土壤取 3 份加入氯仿抽提熏蒸 24 h 再取出浸提滴定,3 份不熏蒸的可直接取出浸提滴定。

$$有机碳量(\mu g\ C/g) = 12 \times 10^3 \times (V0-V) \times M \times f/W$$

M 为 $FeSO_4$ 浓度(mol/L),V、V0 分别为空白和样品消耗的 $FeSO_4$ 体积(mL),f 为稀释倍数,W 为烘干土壤重量(g)。

土壤微生物量碳:$Bc = Ec/k_{EC}$,$Ec = 熏蒸-未熏蒸$,$k_{EC} = 0.38$。

1.3.2 土壤有机碳

用风干土测定。称取 0.20 g 100 目过筛土,参考鲍士旦[18]土壤农化分析重铬酸钾容量法-外加热法。

1.3.3 土壤有机氮组分

称取 2.50 g(风干土)100 目过筛土壤样品于水解瓶中,参考 Bremner[19]酸水解法、鲁如坤[20]酸水解-蒸馏法以及刘延美和刘小虎[21]的方法进行测定。

1.3.4 数据统计分析

试验数据采用 SPSS Statistics 21.0 进行统计分析(Duncan 法)。用 Origin 8.0 和 SigmaPlot 10.0 软件作图。

2 结果与分析

2.1 成熟期土壤有机碳和土壤微生物量碳

从表 2 可以看出，春玉米，雨养处理的土壤有机碳含量较灌溉处理显著提高了 18.36%（$P<0.05$），土壤微生物量碳含量显著增加 158.34%（$P<0.05$）。秋玉米，雨养条件下土壤有机碳含量显著高于灌溉处理 1.09%（$P<0.05$）。雨养处理土壤微生物量碳较灌溉处理显著增加 13.33%（$P<0.05$），灌溉条件下，秋玉米土壤微生物量碳较春玉米显著增加 20%（$P<0.05$）。两季玉米试验结果表明，灌溉处理下土壤有机碳含量和土壤微生物量碳含量均值均显著低于雨养处理 9.24%（$P<0.05$）和 84.62%（$P<0.05$），水分对土壤有机碳含量的影响达到了显著水平（$P<0.05$），对土壤微生物量碳含量影响不显著（$P>0.05$）。

春玉米，N3 条件下土壤有机碳含量达到最大值 16.61 g/kg，分别高于 N0、N1、N2、N4 处理 9.85%（$P<0.05$）、16.40%（$P<0.05$）、2.85%、16.32%（$P<0.05$），均高于 N0、N1、N2、N4 处理 9.85%（$P<0.05$）、16.40%（$P<0.05$）、2.85%、16.32%（$P<0.05$），均高于其他处理，随着施氮量的增加，土壤微生物量碳呈现先升高后下降的趋势，在 N3 处理时达到最大值 0.25 mg/kg，并且均显著高于其他施氮水平（$P<0.05$），分别高于 N0、N1、N2、N4 水平 31.58%、19.05%、13.64%、19.05%。秋玉米，随着施氮量的增加土壤有机碳呈先上升后下降的趋势，而土壤有机碳含量则呈上升趋势。两季玉米土壤有机碳含量平均值表现为 N2>N3>N1>N0>N4，土壤微生物量碳含量表现为 N4>N3>N2>N1>N0；与 N0 相比，N3 处理的土壤有机碳含量和土壤微生物量碳含量分别提高了 9.1% 和 50.0%。施氮量对土壤有机碳和土壤微生物量碳的影响达到了极显著水平（$P<0.01$）。

水氮互作对土壤有机碳和土壤微生物量碳含量的影响均达到极显著水平（$P<0.01$），水分和施氮均能影响土壤有机碳组分，雨养结合 N3 处理有利于土壤有机碳和土壤微生物量碳的积累。

表 2 玉米成熟期水分条件和施氮量对土壤有机碳及微生物量碳的影响

处理	土壤有机碳（g/kg）			微生物量碳（mg/kg）		
	春玉米	秋玉米	平均值	春玉米	秋玉米	平均值
水分条件						
雨养	16.57±0.16a	15.83±0.01a	16.20±0.08a	0.31±0.01a	0.17±0.01a	0.24±0.01a
灌溉	14.00±0.36b	15.66±0.01b	14.83±0.18b	0.12±0.02b	0.15±0.01a	0.13±0.05b
施氮水平						
N0	15.12±0.11b	14.68±0.07e	14.90±0.02bc	0.19±0.03c	0.07±0.02d	0.14±0.01d
N1	14.27±0.13c	16.36±0.03b	15.32±0.08b	0.21±0.02bc	0.12±0.01c	0.16±0.03c
N2	16.15±0.47a	16.66±0.09a	16.41±0.28a	0.22±0.01b	0.13±0.03c	0.18±0.02c
N3	16.61±0.15a	15.89±0.04c	16.25±0.05a	0.25±0.01a	0.18±0.02b	0.21±0.02b
N4	14.28±0.03c	15.13±0.11d	14.70±0.02c	0.21±0.04bc	0.31±0.01a	0.25±0.01a
水分条件	0.0293	0.0092	0.0265	0.0001	0.2053	0.0062
施氮水平	0.0001	0.0001	0.0001	0.0001	0.0001	0.0001
水分和氮素	0.0001	0.0001	0.0001	0.0001	0.0001	0.0003

注：同一列不同小写字母表示差异显著（$P<0.05$）。

2.2 有机氮组分含量

试验表明，雨养和灌溉处理的酸解总氮含量均值分别为 874.62 mg/kg 和 905.01 mg/kg，灌溉的酸解总氮含量较雨养提高 3.47%（$P<0.05$）。春玉米，雨养处理较灌溉处理氨基糖态氮显著增加 16.42%，氨基酸态氮灌溉处理高于雨养 12.60%，灌溉处理氨态氮显著高于雨养处理 9.31%（$P<0.05$），酸解未知氮雨养处理高于灌溉 13.61%，表明灌溉利于酸解有机氮各组分含量提高。秋玉米，灌溉处理较雨养处理氨基糖态氮、酸解未知氮显著增加，增幅分别为 13.37%（$P<0.05$）、14.89%（$P<0.05$），这表明灌溉结合 N3 利于氨基糖态氮和酸解未知氮提高，最终酸解总氮较高。雨养处理较灌溉处理氨基酸态氮增加 0.93%，氨态氮显著增加 1.56%（$P<0.05$）。

春玉米，氨基糖态氮呈现升高后下降再升高的趋势，在 N4 时达到最大值 83.13 mg/kg，并显著高于 N0、N1、N3 施氮水平（$P<0.05$），氨基酸态氮 N4 处理显著高于 N0、N1、N2、N3 处理（$P<0.05$），分别高于 17.74%、9.88%、19.11%、15.56%，氨态氮在 N2 处理时达到最大值 348.6 mg/kg，酸解未知氮呈现先升高后下降的趋势，在 N3 处理时达到最大值为 179.6 mg/kg，并显著高于其他处理（$P<0.05$）。秋玉米，氨基酸态氮随着施氮量的升高而升高，在 N4 时取得最大值 318.98 mg/kg，且显著高于其他施氮水平（$P<0.05$），呈现 N4>N3>N2>N1>N0 的趋势。两季玉米来看，水分和氮素共同作用对氨基糖态氮、氨基酸态氮、氨态氮、酸碱未知氮均有极显著影响（$P<0.01$）。

两季玉米表明，灌溉和施氮均提高了酸解有机氮组分总含量。各处理的土壤酸解有机氮各组分含量为：氨基酸态氮（AAN）≈氨态氮（NHN）>酸解未知氮（HUN）>氨基糖态氮（ASN）；氨基酸态氮和氨态氮约占酸解总氮的 80%，氨基酸态氮和氨态氮是主要的酸解氮形态（图3）。酸解总氮含量变化范围 800~1 000 mg/kg，占土壤全氮的 50% 左右。

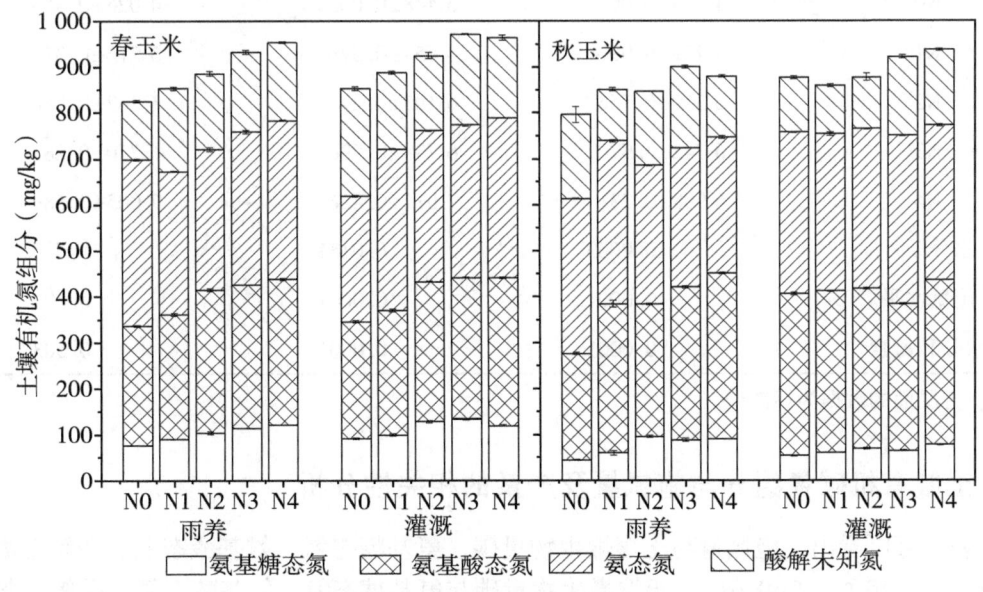

图 3　玉米成熟期水分条件和施氮量对有机氮组分的影响
误差线为标准误。

2.3 水分条件和施氮量对产量的影响

从表 3 可以看出，春玉米，雨养或灌溉处理后产量均值分别为 5 165 kg/hm² 和 7 015 kg/hm²，灌溉的产量较雨养提高了 35.82%（$P<0.05$），秋玉米，雨养条件下作物产量均值较灌溉增加 132 kg/hm²，增幅为 2.38%，增幅显著（$P<0.05$）。两季试验可以看出灌溉处理较雨养处理作物产量增加 15.83%。

产量随施氮量的增加而增加。春玉米，N4 产量均值为 6 998 kg/hm²，显著高于 N0 和 N1 施氮水平（$P<0.05$），增幅分别为 52.03% 和 21.09%。秋玉米，N4 产量均值最大为 7 011 kg/hm²，显著高于 N0 水平（$P<0.05$），产量约为 N0 水平的两倍。两季玉米来看，N0、N1、N2、N3、N4 施氮水平产量均值分别为 4 025 kg/hm²、5 216 kg/hm²、6 262 kg/hm²、6 770 kg/hm²、7 005 kg/hm²。可以看出，施氮量与作物产量呈显著正相关关系（$P<0.05$）。N0、N1、N2 处理间差异显著（$P<0.05$），N2、N3 和 N4 处理间差异不显著（$P>0.05$），进一步表明过量施氮不能显著增加作物产量。

表 3　玉米成熟期不同水分条件和施氮量对产量的影响

处理	产量（kg/hm²）		
	春玉米	秋玉米	平均值
水分条件			
雨养	5 165±0.08a	5 687±0.48a	5 426±0.80a
灌溉	7 015±0.58b	5 555±0.22b	6 285±0.18a
施氮水平			
N0	4 603±0.00a	3 448±0.06a	4 025±0.45a
N1	5 779±0.99b	4 654±0.57b	5 216±0.32b
N2	6 329±0.75c	6 195±0.27c	6 262±0.95c
N3	6 742±0.97d	6 798±0.59d	6 770±0.28c
N4	6 998±0.23e	7 011±0.82e	7 005±0.58c
水分条件	0.0003	0.0151	0.5455
施氮水平	0.0001	0.0001	0.0002
水分和氮素	0.0001	0.0001	0.5037

注：土层面的数据为标准误。

2.4 土壤有机碳氮组分与施氮量及产量的相关性分析

从表 4 可以看出，施氮量与土壤微生物量碳、氨基糖态氮、氨基酸态氮、酸解总氮、产量呈极显著正相关（$P<0.01$）。土壤微生物量碳与氨基糖态氮、氨基酸态氮、酸解总氮、产量呈极显著正相关（$P<0.01$）。氨基糖态氮与氨基酸态氮、酸解总氮、产量呈极显著正相关（$P<0.01$）。氨基酸态氮与酸解总氮、产量呈极显著正相关（$P<0.01$）。酸解未知氮与酸解总氮呈极显著正相关（$P<0.01$）。酸解总氮与产量呈极显著正相关（$P<0.01$）。

表 4 灌浆参数与施氮量及产量之间的相关分析

变量	N	SOC	MBC	ASN	AAN	NHN	HUN	THUN	Yield
N	1.0000	0.1592	0.8618**	0.9191**	0.9758**	-0.0779	0.2995	0.8769**	0.9887**
SOC		1.0000	-0.1641	0.3888	-0.0067	-0.4344	0.1979	0.1501	0.2799
MBC			1.0000	0.7707**	0.8953**	0.0012	0.3950	0.8687**	0.8334**
ASN				1.0000	0.8693**	-0.4138	0.3957	0.8432**	0.9542**
AAN					1.0000	-0.0739	0.1925	0.8152**	0.9434**
NHN						1.0000	-0.0626	-0.0195	-0.1576
HUN							1.0000	0.7070**	0.3514
THUN								1.0000	0.8874**
Yield									1.0000

注：N：氮素；SOC：土壤有机碳；MBC：土壤微生物量碳；ASN：氨基糖态氮；AAN：氨基酸态氮；NHN：硝态氮；HUN：水解未知有机氮；THUN：水解有机氮；Yield：产量。

** 相关系数显著水平为 $P<0.01$。

3 讨论

已有研究表明，施肥能增加凋落物的产生，从而增加土壤碳的积累[22,23]；合理施用无机氮肥土壤的 C/N 比下降，使得土壤中原有有机碳的分解速度加快，导致土壤有机碳含量下降、土壤微生物减少、降低土壤微生物量碳含量[24-25]；当氮素施用量超过 140 kg/hm² 时，土壤有机碳和施氮量成负相关[14]。本试验结果表明，当施氮量超过 250 kg/hm²（N3）时，土壤有机碳含量降低，可能是由于施氮量的增加改变了土壤理化性质，使土壤微生物活性上升，促进土壤中微生物活动，加快有机质分解，土壤有机碳含量降低[14]；南亚热带高温高湿条件下灌溉会导致土壤碳含量下降[14]，两季玉米试验结果表明，灌溉处理的土壤有机碳和土壤微生物量碳显著低于雨养（$P<0.05$）。

土壤中酸解氨基酸态氮的含量及其分配比例受水氮的调控影响较为复杂[26]。研究表明，单施化肥处理耕作层酸解铵态氮含量增加显著[11]。姬景红等研究表明[27]，合适的水分条件能改善土壤中有机氮比例。相同土层有机氮含量均以酸解氮为主，灌溉和施肥处理的有机氮各比例表现为酸解未知态氮>酸解氨态氮>氨基酸态氮>氨基糖态氮[28]。本试验结果进一步证实，水分和施氮均能影响有机氮组分总含量，并且灌溉结合 N3 利于氨基糖态氮、氨基酸态氮和酸解未知氮提高，最终形成酸解总氮。

在适宜的水肥范围内，灌溉与施肥对产量存在显著正效应，但水氮过量时产量下降，出现报酬递减的规律[29]。秋玉米数据表明，由于土壤有机碳和作物产量呈现正相关关系，灌溉处理较雨养处理土壤有机碳下降，因此秋玉米灌溉处理和雨养处理玉米产量相差不大，可能的原因为广西秋玉米生长季降水量较高，导致水分条件的改变对产量的影响不显著；水氮互作对玉米优质高产影响显著，产量随水氮增加而增加，接近某一限制值时，产量增加逐渐放缓[30]。水、肥配合显著提高了作物产量和效益[29]报道施氮量 243.27 kg/hm² 结合灌溉下限为田间持水量的 65.6%时，产量最高；黄金生等研究表明[31]，在广西的赤红壤区施氮量超过 240 kg/hm² 时，秋玉米的产量不再明显增加，反而呈下降趋势。本试验结果表明，水分和施氮均能影响作物产量，灌溉和 N3 处理下作物产量达到最大，表明在南亚热带广西发生阶段性干旱时，春玉米适量灌溉结合合理施氮利于玉米干物质运输分配，促进产量提高。

4 结论

灌溉条件下土壤有机碳和土壤微生物量碳显著低于雨养；250 kg/hm² (N3) 施氮处理下土壤有机碳和土壤微生物量碳均值能达到最大，因此雨养条件下 N3 处理有利于土壤有机碳和土壤微生物量碳的积累。土壤有机氮各组分含量表现为氨基酸态氮≈氨态氮>酸解未知氮>氨基糖态氮，酸解总氮中的绝大部分（80%）为氨基酸态氮和氨态氮。酸解总氮含量变化范围 800~1 000 mg/kg，占土壤全氮的 50% 左右。灌溉处理配合 N3 处理有利于提升土壤有机氮各组分含量。因此广西春玉米适量灌溉和合理施氮结合能显著提高玉米产量，而秋玉米灌溉效果还有待于进一步研究。

参考文献

[1] 张宇飞，刘立志，马昱萱，等. 耕作和秸秆还田方式对玉米产量及钾素积累转运的影响 [J]. 作物杂志，2019，189 (2)：122-127.

[2] Lal R. Enhancing ecosystem services with no-till [J]. Renewable Agriculture and Food Systems, 2013, 28 (2): 102-104.

[3] Al-Kaisi M M, Yin X, Licht M A. Soil carbon and nitrogen changes as influenced by tillage and cropping systems in some Iowa soils [J]. Agriculture Ecosystems and Environment, 2005, 105 (4): 635-647.

[4] 褚天铎. 简明施肥技术手册 [M]. 北京：金盾出版社，2014.

[5] 张俊华，张家宝，贾科利. 氮素和盐碱胁迫下作物与土壤光谱特征研究 [M]. 银川：宁夏人民出版社，2016.

[6] 宁芳. 施氮量对渭北旱地春玉米田土壤水肥利用、玉米生长和产量的影响 [D]. 杨凌：西北农林科技大学，2019.

[7] Zhang X B, Sun Z G, Liu J, et al. Simulating greenhouse gas emissions and stocks of carbon and nitrogen in soil from a long-term no-till system in the North China Plain [J]. Soil and Tillage Research, 2018, 178: 32-40.

[8] Gärdenäs A I, Ågren G I, Bird J A, et al. Knowledge gaps in soil carbon and nitrogen interactions-from molecular to global scale [J]. Soil Biology and Biochemistry, 2010, 43 (4): 702-717.

[9] Russell A E, Cambardella C A, Laird D A, et al. Nitrogen fertilizer effects on soil carbon balances in midwestern U.S. agricultural systems. [J]. Ecological Applications, 2009, 19 (5): 1102-1113.

[10] 刘畅，唐国勇，童成立，等. 不同施肥措施下亚热带稻田土壤碳、氮演变特征及其耦合关系 [J]. 应用生态学报，2008，19 (7)：1489-1493.

[11] 徐阳春，沈其荣，茆泽圣，等. 长期施用有机肥对土壤及不同粒级中酸解有机氮含量与分配的影响 [J]. 中国农业科学，2002，35 (4)：403-409.

[12] 陈亚，代先强，袁玲，等. 水氮耦合对土壤理化性状及作物生长的影响研究进

展［J］. 河南农业科学，2009，38（5）：11-15.

［13］张富仓，严富来，范兴科，等. 滴灌施肥水平对宁夏春玉米产量和水肥利用效率的影响［J］. 农业工程学报，2018，34（22）：111-120.

［14］俞华林，张恩和，王琦，等. 灌溉和施氮对免耕留茬春小麦农田土壤有机碳、全氮和籽粒产量的影响［J］. 草业学报，2013，22（3）：227-233.

［15］Tian J H, Wei K, Condron L M, et al. Effects of elevated nitrogen and precipitation on soil organic nitrogen fractions and nitrogen-mineralizing enzymes in semi-arid steppe and abandoned cropland［J］. Plant and Soil, 2017（417）：217-229.

［16］廖东声，万艳. 广西玉米产业生产成本控制问题分析［J］. 经济研究参考，2016（29）：75-79.

［17］李振高，骆永明，滕应. 土壤与环境微生物研究法［M］. 北京：科学出版社，2008.

［18］鲍士旦. 土壤农化分析［M］. 3版. 北京：中国农业出版社，2000.

［19］Bremner J M. Methods of soil analysis：Part II-Chemical and microbiological［M］. Black C A. ASA：Madison W I 1965：1238-1255.

［20］鲁如坤. 土壤农业化学分析方法［M］. 北京：中国农业科技出版社，2000.

［21］刘延美，刘小虎. 土壤酸水解氨基酸和氨基糖态氮测定方法的比较研究［J］. 吉林农业科学，2010，35（2）：19-23.

［22］Schuma G E, Janzen H H, Herrick J E. Soil carbon dynamics and potential carbon sequestration by rangelands［J］. Environmental Pollution, 2002, 116（3）：391-396.

［23］Li C X, Ma S C, Shao Y, et al. Effects of long-term organic fertilization on soil microbiologic characteristics, yield and sustainable production of winter wheat［J］. Journal of Integrative Agriculture, 2018, 17（1）：210-219.

［24］金兰淑，郑佳，徐慧，等. 施氮及灌溉方式对玉米地土壤硝化潜势及微生物量碳的影响［J］. 水土保持学报，2009，23（4）：218-220，226.

［25］薛仁风，丰明，赵阳，等. 不同生物有机肥对绿豆生长与生理特性的影响［J］. 东北农业科学，2019，44（4）：9-12，71.

［26］张玉树，丁洪，王飞，等. 长期施用不同肥料的土壤有机氮组分变化特征［J］. 农业环境科学学报，2014，33（10）：1981-1986.

［27］姬景红，张玉龙，黄毅，等. 灌溉方法对保护地土壤有机氮组分及剖面分布的影响［J］. 水土保持学报，2007，21（6）：99-104.

［28］孙文涛，孙占祥，王聪翔，等. 滴灌施肥条件下玉米水肥耦合效应的研究［J］. 中国农业科学，2006，39（3）：563-568.

［29］詹其厚，陈杰. 水肥配合对玉米产量及其利用效率的影响［J］. 土壤肥料，2005（4）：14-18.

［30］尚文彬，张忠学，郑恩楠，等. 水氮耦合对膜下滴灌玉米产量和水氮利用的影响［J］. 灌溉排水学报，2019，38（1）：49-55.

［31］黄金生，周柳强，曾艳，等. 广西赤红壤区玉米氮肥效应及适宜施氮量研究［J］. 西南农业学报，2019，32（3）：551-558.